PTC Creo™ Parametric 5.0

5.0

Part 2 (Lessons 13-22)

Louis Gary Lamit

ISBN-13: 978-1720784500

ISBN-10: 1720784507

Table of Contents (Part 1- Lessons 1-12)

Part Two (Lessons 13-22)

About the Author

Louis Gary Lamit teaches at De Anza College (since 1984) in Cupertino, CA. He is the founder of Scholarships for Veterans at www.scholarshipsforveterans.org. Mr. Lamit has worked as a drafter (1966), designer, numerical control (NC) programmer, technical illustrator, and engineer in the automotive, aircraft, and piping industries. A majority of his work experience is in mechanical and piping design. He started as a drafter in Detroit (as a job shopper) in the automobile industry, doing tooling, dies, jigs and fixture layout, and detailing at Koltanbar Engineering, Tool Engineering, Time Engineering, and Premier Engineering for Chrysler, Ford, AMC, and Fisher Body. Mr. Lamit has worked at Remington Arms and Pratt & Whitney Aircraft as a designer, and at Boeing Aircraft and Kollmorgan Optics as an NC programmer and aircraft engineer. He also owns and operates his own consulting firm (CAD-Resources.com- Lamit and Associates), and has been involved with advertising, and patent illustration. He is the author of over 40 books, journals, textbooks, workbooks, tutorials, and handbooks, including children's journals and books (www.walkingfishbooks.com). Mr. Lamit received a BS degree from Western Michigan University in 1970 and did Masters' work at Wayne State University and Michigan State University. He has also done graduate work at the University of California at Berkeley and holds an NC programming certificate from Boeing Aircraft. Since leaving industry, Mr. Lamit has taught at all levels (Melby Junior High School, Warren, Mi.; Carroll County Vocational Technical School, Carrollton, Ga.; Heald Engineering College, San Francisco, Ca.; Cogswell Polytechnical College, San Francisco and Cupertino, Ca.; Mission College, Santa Clara, Ca.; Santa Rosa Junior College, Santa Rosa, Ca.; Northern Kentucky University, Highland Heights, Ky.; and De Anza College, Cupertino, Ca.).

His textbooks include:

- *Industrial Model Building*, with Engineering Model Associates, Inc. (1981),
- *Piping Drafting and Design* (1981),
- *Piping Drafting and Design Workbook* (1981),
- *Descriptive Geometry* (1983),
- *Descriptive Geometry Workbook* (1983), and
- *Pipe Fitting and Piping Handbook* (1984), Prentice-Hall.
- *Drafting for Electronics* (3rd edition, 1998),
- *Drafting for Electronics Workbook* (2nd edition 1992), and
- *CADD* (1987), Charles Merrill (Macmillan-Prentice-Hall Publishing).
- *Technical Drawing and Design* (1994),
- *Technical Drawing and Design Worksheets and Problem Sheets* (1994),
- *Principles of Engineering Drawing* (1994),
- *Fundamentals of Engineering Graphics and Design* (1997),
- *Engineering Graphics and Design with Graphical Analysis* (1997), and
- *Engineering Graphics and Design Worksheets and Problem Sheets* (1997), West Publishing (ITP/Delmar).
- *Basic Pro/ENGINEER in 20 Lessons* (1998) (Revision 18) and
- *Basic Pro/ENGINEER (with references to PT/Modeler)* (1999), PWS.
- *Pro/ENGINEER 2000i* (1999), and
- *Pro/ENGINEER 2000i² (Pro/NC and Pro/SHEETMETAL)* (2000), Brooks/Cole Publishing (ITP).
- *Pro/ENGINEER Wildfire* (2003), Brooks/Cole Publishing (ITP).
- *Introduction to Pro/ENGINEER Wildfire 5.0* (2004), SDC.
- *Moving from 2D to 3D CAD for Engineering Design* (2007), BookSurge, eBook by MobiPocket.
- *Pro/ENGINEER Wildfire 3.0 Tutorial* (2007), BookSurge, eBook by MobiPocket.
- *Pro/ENGINEER Wildfire 3.0* (2007), Cengage.
- *Pro/ENGINEER Wildfire 4.0 Tutorial* (2008), BookSurge eBook by MobiPocket.
- *Pro/ENGINEER Wildfire 4.0* (2008), Cengage.
- *Pro/ENGINEER Wildfire 5.0* (2010), Cengage.
- *Creo Parametric (2012), Cengage.*
- *PTC Creo Parametric 2.0 (2015), Cengage.*
- *PTC Creo Parametric 3.0 (2015), Cengage*
- *PTC Creo Parametric 4.0 (Part 1- Lessons 1-12) (2017), CAD-Resources.*
- *PTC Creo Parametric 4.0 (Part 2- Lessons 13-22) (2017), CAD-Resources.*
- *PTC Creo Parametric 5.0 (Part 1- Lessons 1-12) (2018), CAD-Resources.*
- *PTC Creo Parametric 5.0 (Part 2- Lessons 13-22) (2018), CAD-Resources.*

Mr. Lamit also owns and writes for books and journals though WalkingFish Books (Amazon publications).

- *Fishing Journals, Golfing Journals, Bird Watching Journals, Garden Journals.*
- *Wally the WalkingFish children's books.*

Dedication

This book is dedicated to my granddaughter: **Thuy Dao Lamit**

gate gate pāragate pārasagate bodhi svāhā

Preface

PTC Creo Parametric 5.0 is one of the most widely used CAD/CAM software programs in the world today. Any aspiring engineer will greatly benefit from the knowledge contained herein, while in school or upon graduation as a newly employed engineer.

Significant changes, upgrades, and new capabilities including have made PTC Creo Parametric 5.0 a unique product. *This is not a revised textbook* but a new book covering all the necessary subjects needed to master this high-level CAD software. There are few if any comprehensive texts on this subject so we hope this text will fill the needs of both schools and professionals alike.

The text involves creating a new part, an assembly, or a drawing, using a set of commands that walk you through the process systematically. Lessons and Projects all come from industry and have been tested for accuracy and correctness as per engineering standards. Projects are downloadable as a PDF with live links and 3D embedded models.

Resources

Lesson Projects (downloadable PDF) are not included in the printed version to keep the length and cost to the user down- lessons and projects not in the printed portion can be downloaded at: **www.cad-resources.com**

A complete set of **Recorded Lectures** (approximately 30 hours) for this book is available for *free* on the authors **YouTube site Louis Gary Lamit- lgl@cad-resources.com**. The Lectures are in WMV format. The Lectures are the exact content presented in the classroom (or online class) by the author.

PTC Creo Parametric 5.0 Free Schools Edition software is available at www.ptc.com.

Contact

If you wish to contact the author concerning orders, questions, changes, additions, suggestions, comments, or to get on our email list, please send an email to one of the following:

Web Site: **www.cad-resources.com**
Email: **lgl@cad-resources.com**

Lesson 13 Patterns and Weldments

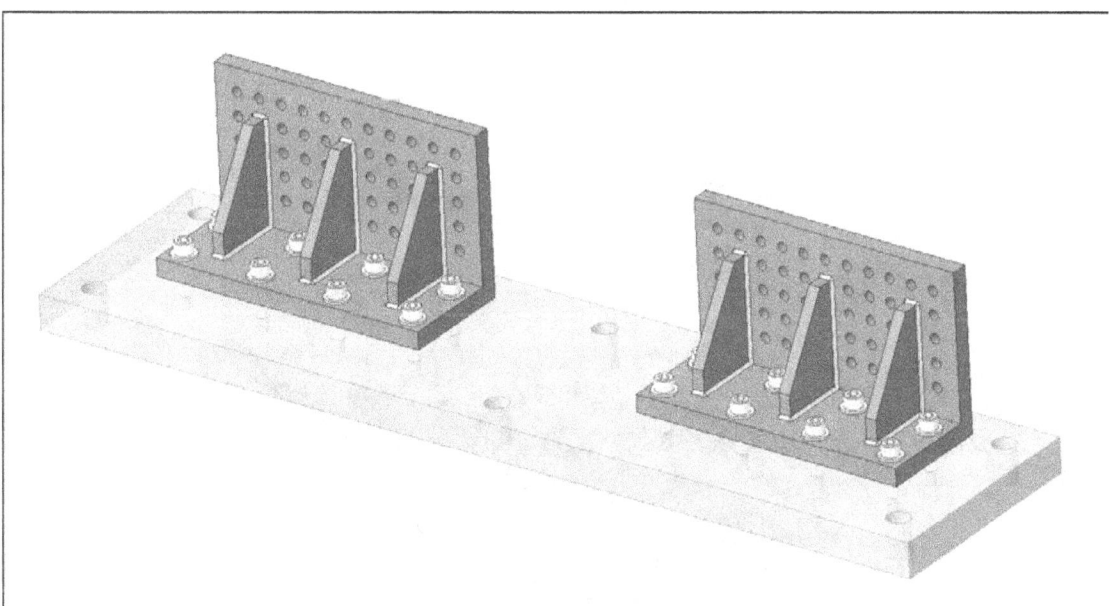

Figure 13.1 Mounting System Weldment

OBJECTIVES

- Create **directional** and **dimensional patterns**
- **Pattern components** on an assembly
- Insert multiple standard parts using a **reference pattern**
- Use **fill** to pattern a feature
- Insert **welds** on a model
- Utilize the **Screw** and **Dowel Tools**

REFERENCES AND RESOURCES

For **Resources** go to **www.cad-resources.com** > click on the PTC Creo Parametric 5.0 Book cover

- Lesson 13 Lecture at **YouTube Creo Parametric Lecture Videos**

- Lesson 13 3D PDF models embedded in a PDF

Patterns and Weldments

Creating a pattern is a quick way to reproduce a feature, or a component in an assembly (Fig. 13.1). A pattern is parametrically controlled. Therefore, you can modify a pattern by changing pattern parameters, such as the number of instances, spacing between instances, and original feature dimensions. Modifying patterns is more efficient than modifying individual features. In a pattern, when you change dimensions of the original feature, Creo Parametric automatically updates the whole pattern. This lesson will introduce you to variations of the **Pattern Tool** and introduce the **Welding** application.

Mounting Bracket

For Lessons 13-18, step-by-step commands are sometimes limited to new software commands introduced or enhanced in that lesson. You are expected to do many of the modeling using commands and practices mastered from Lessons 1-12 without repeated detailed explanations.

The Mounting Bracket will be the first component (Fig. 13.2).

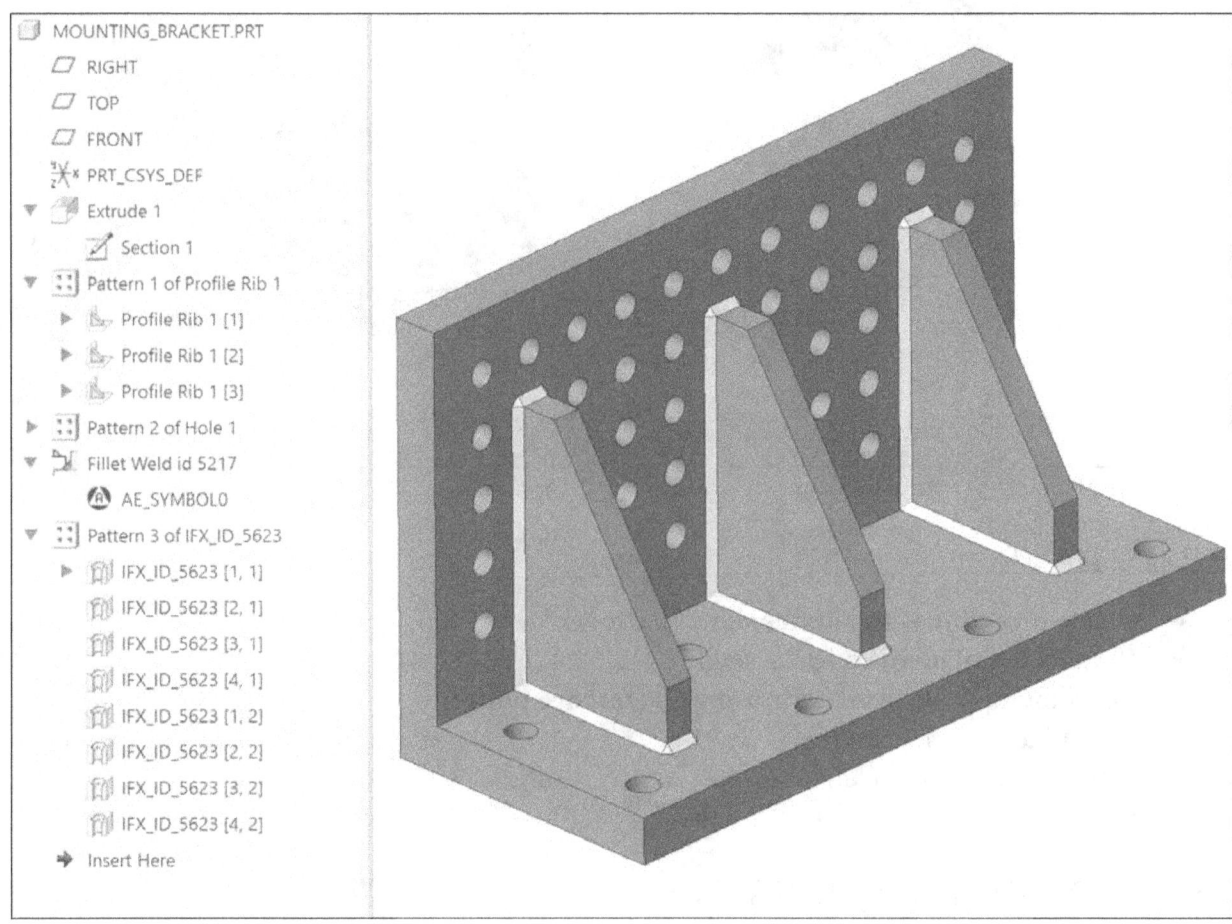

Figure 13.2 Mounting Bracket

Lesson 13 Steps

Press: **Ctrl+N > mounting_bracket > OK > File > Prepare > Model Properties >** Units **change** > [Inch lbm Second (Creo Parametric Default)] **> Close > Close >** [Extrude] **> RMB > Define Internal Sketch >** select datum **FRONT > Sketch >** [icon] > sketch and dimension the L-shaped section [Fig. 13.3(a)] > [✓] > modify the depth to **12** [Fig. 13.3(b)] > [✓] > [💾] **> OK** > change the color

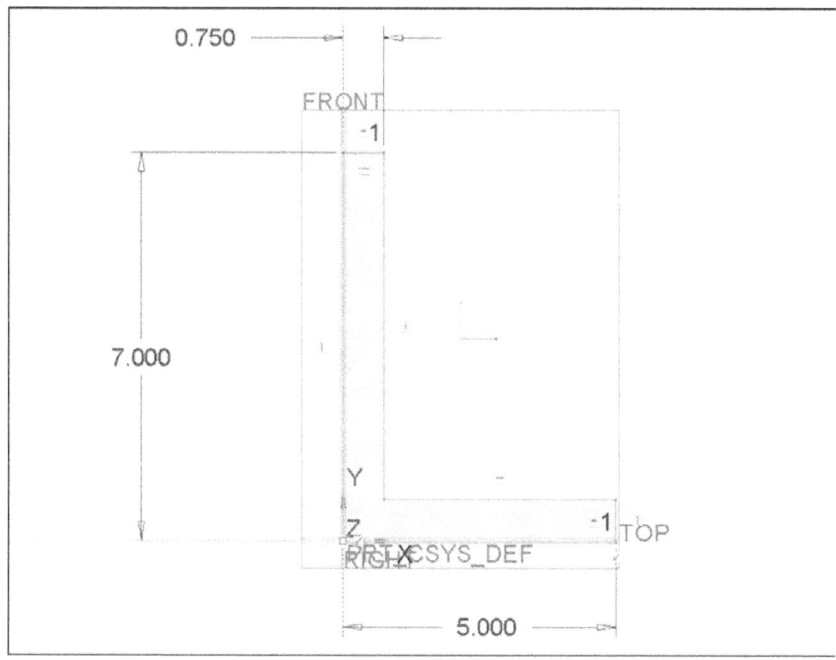

Figure 13.3(a) Sketch the L-shaped Section

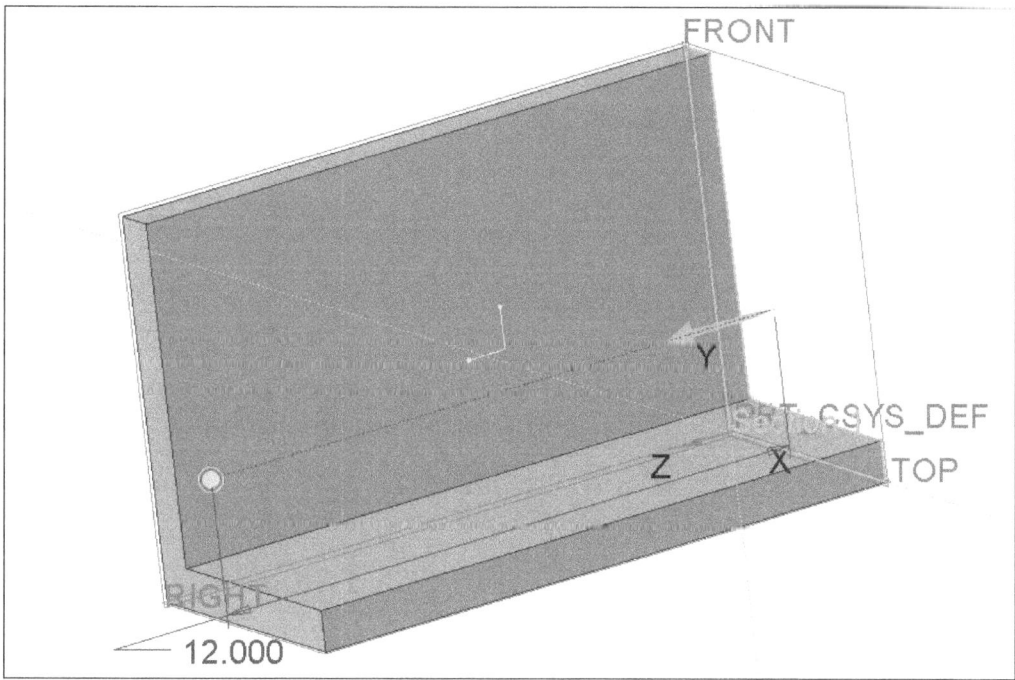

Figure 13.3(b) Mounting Bracket will be **12.00** inches in Depth

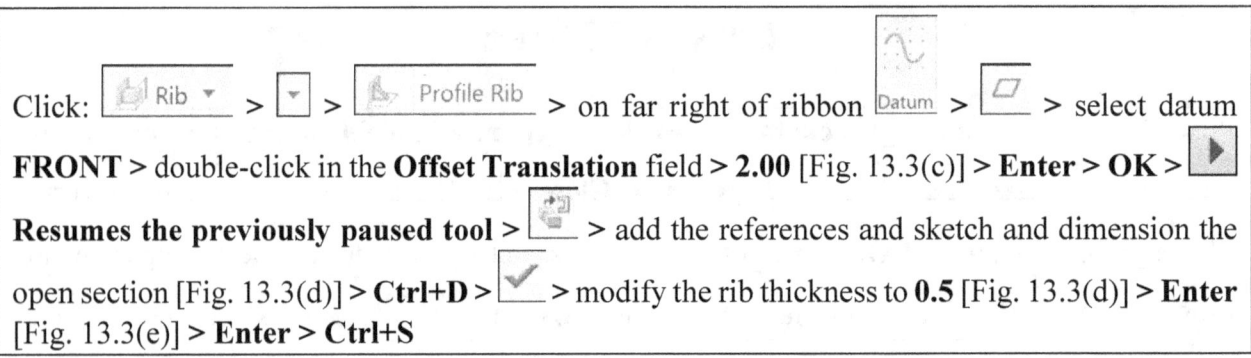

Click: [Rib ▾] > [▾] > [Profile Rib] > on far right of ribbon [Datum] > [▱] > select datum **FRONT** > double-click in the **Offset Translation** field > **2.00** [Fig. 13.3(c)] > **Enter** > **OK** > [▶]

Resumes the previously paused tool > [icon] > add the references and sketch and dimension the open section [Fig. 13.3(d)] > **Ctrl+D** > [✓] > modify the rib thickness to **0.5** [Fig. 13.3(d)] > **Enter** [Fig. 13.3(e)] > **Enter** > **Ctrl+S**

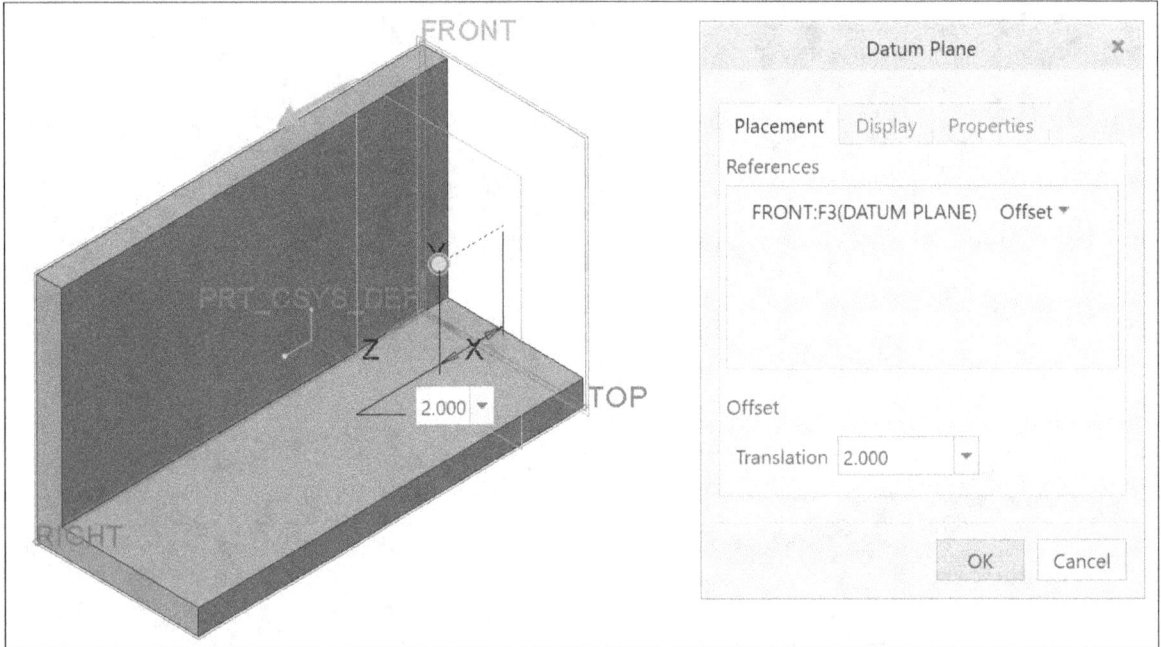

Figure 13.3(c) Offset Datum Plane (Note that you can also double-click on the value in the Graphics Window)

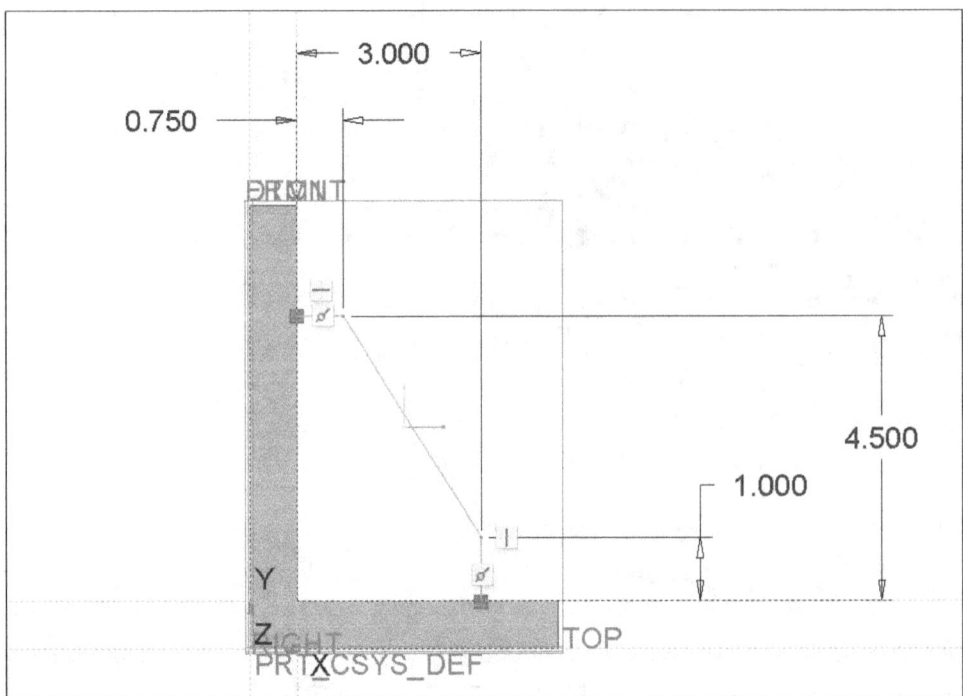

Figure 13.3(d) Open Section for the Rib (Note the Reference Lines)

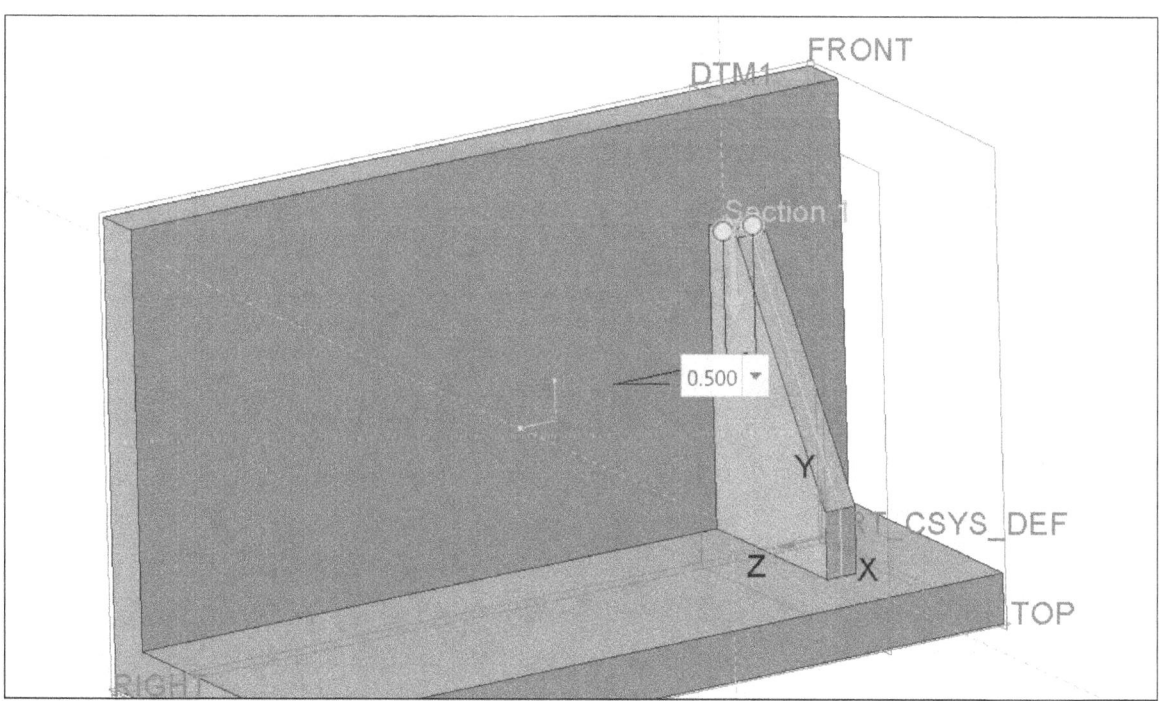

Figure 13.3(e) Rib Preview

With the rib still selected, press: **RMB** > ⊞ **Pattern** > ▾ > **Dimension** > **Dimensions** tab > in the Graphics Window, select **2.00** > *type* **4.00** > **Enter** > *left of* 1 item(s), *type* **3** > **Enter** [Fig. 13.3(f)] > ✓ > **Ctrl+S**

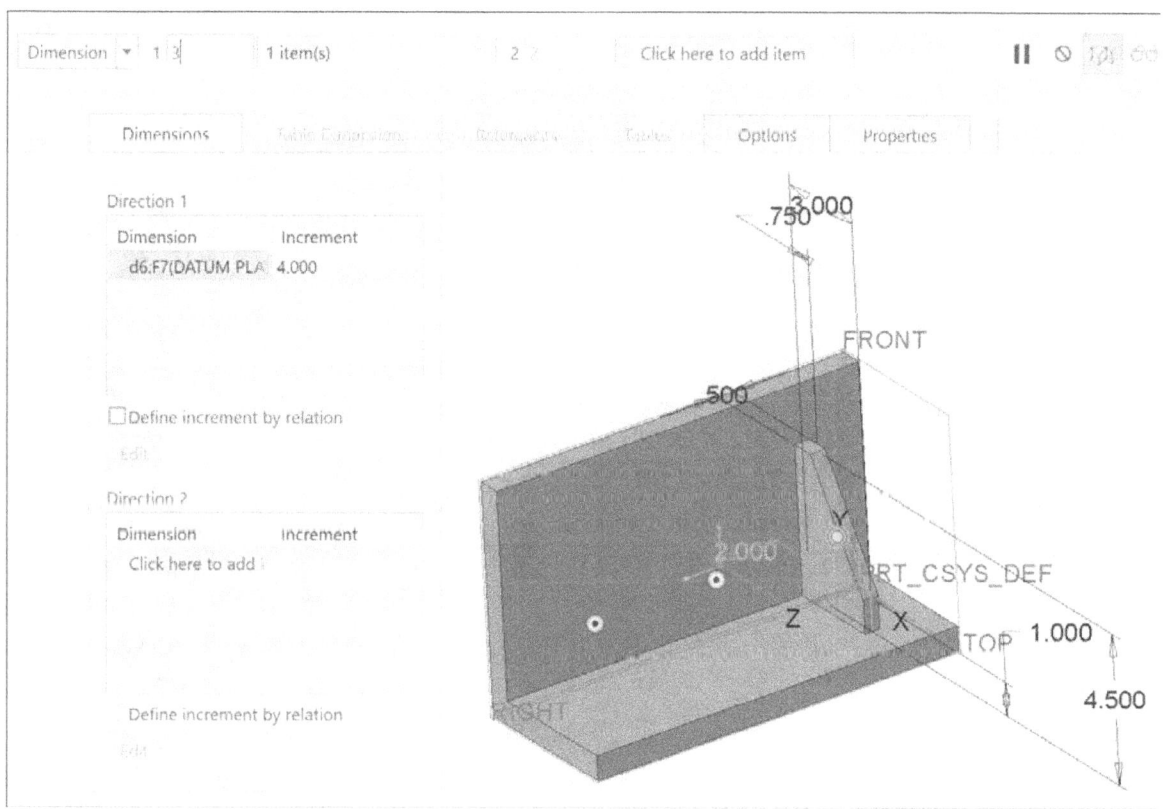

Figure 13.3(f) Dimensional Pattern

Click: [Hole] > [1/2-13 ▼] add the hole as per the Placement [Figs. 13.3(g)] and Shape requirements [Fig. 13.3(h)] > [✓] > **Ctrl+D > Ctrl+S**

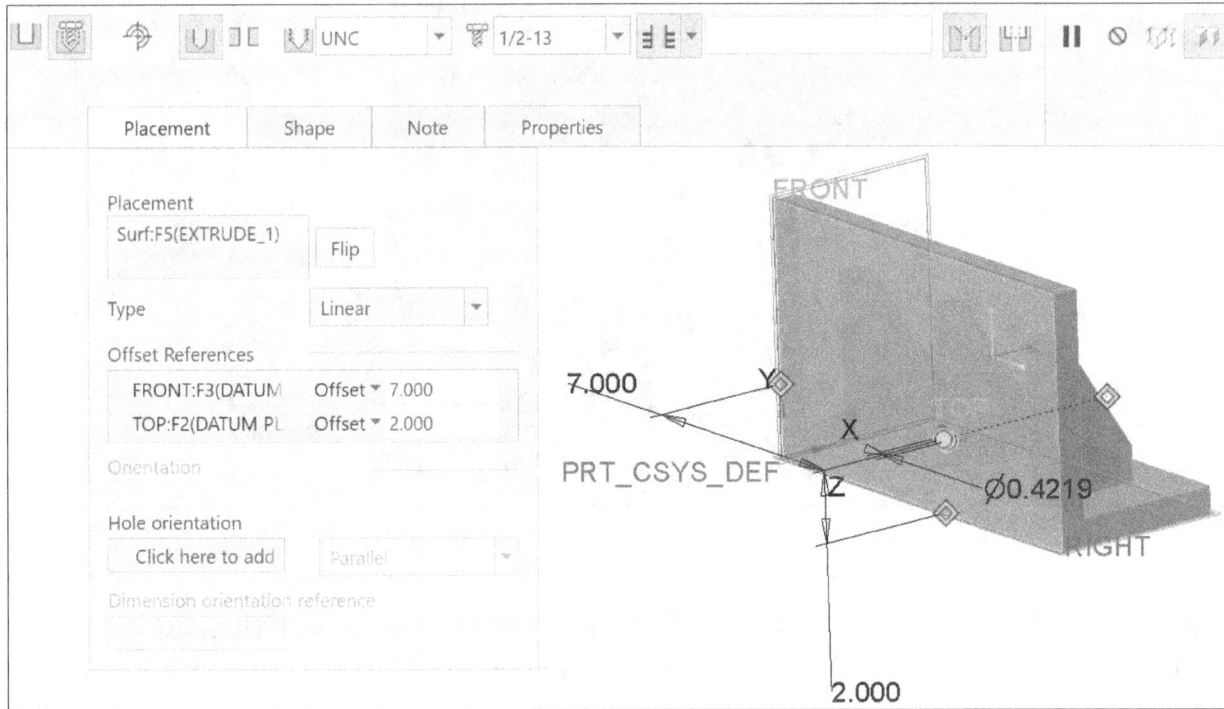

Figure 13.3(g) 7.00 from Datum FRONT and **2.00** from Datum TOP

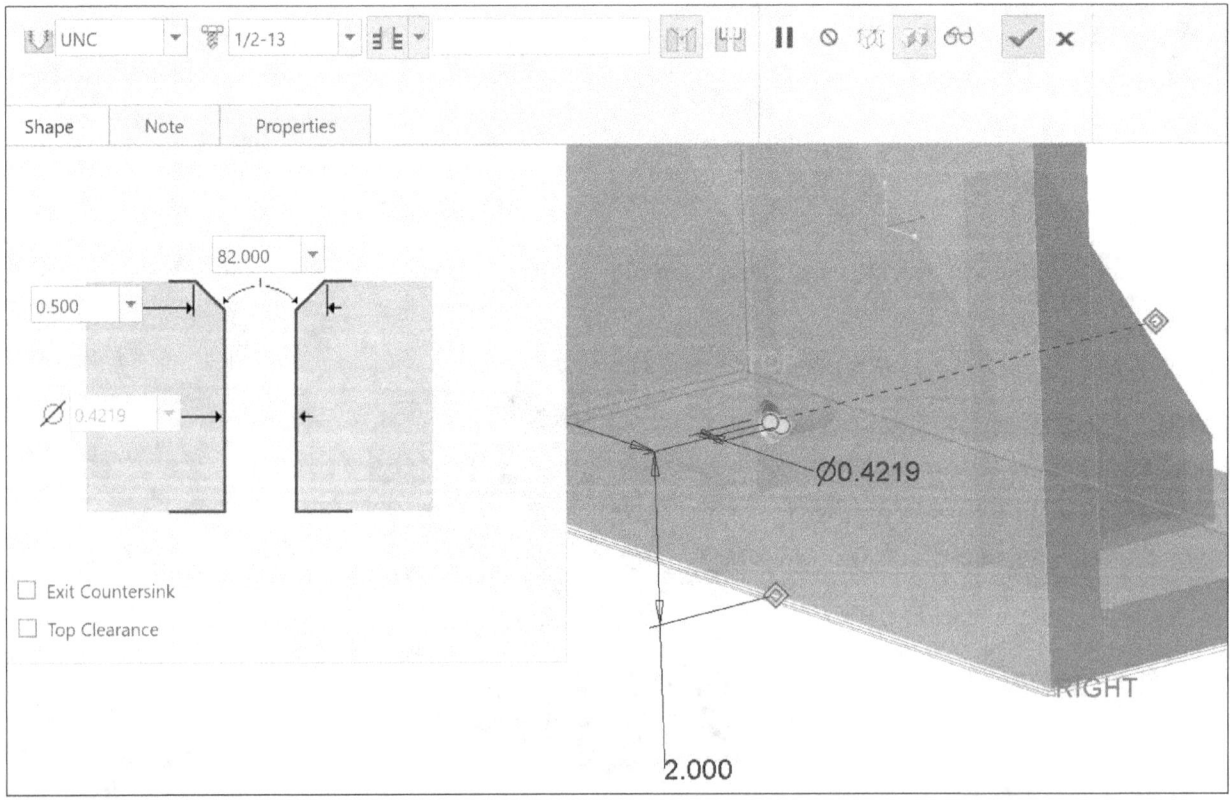

Figure 13.3(h) Shape of **.500-13** Hole

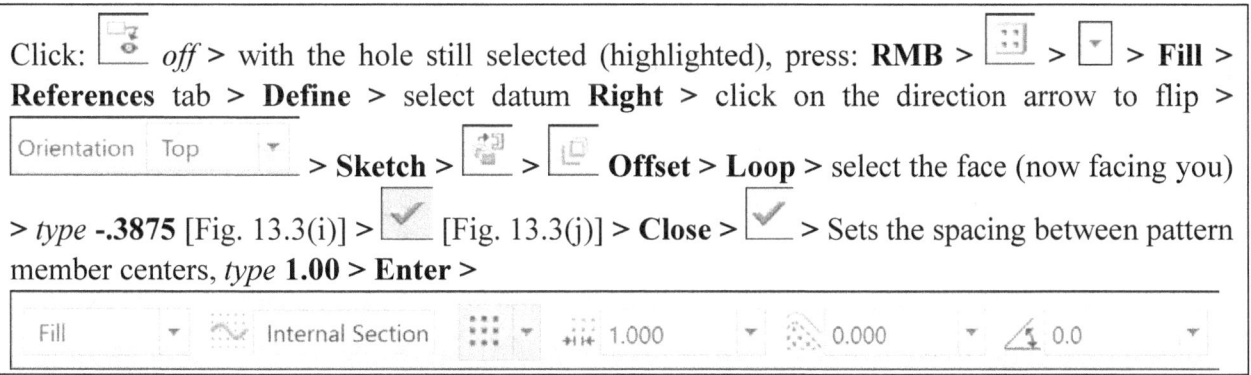

Click: *off* > with the hole still selected (highlighted), press: **RMB** > ⊞ > ▾ > **Fill** > **References** tab > **Define** > select datum **Right** > click on the direction arrow to flip > Orientation | Top ▾ > **Sketch** > ⊞ > ⊞ **Offset** > **Loop** > select the face (now facing you) > *type* **-.3875** [Fig. 13.3(i)] > ✓ [Fig. 13.3(j)] > **Close** > ✓ > Sets the spacing between pattern member centers, *type* **1.00** > **Enter** >

Fill	▾	⁓ Internal Section	⠿ ▾	⠿ 1.000	▾	⠿ 0.000	▾	◹ 0.0	▾

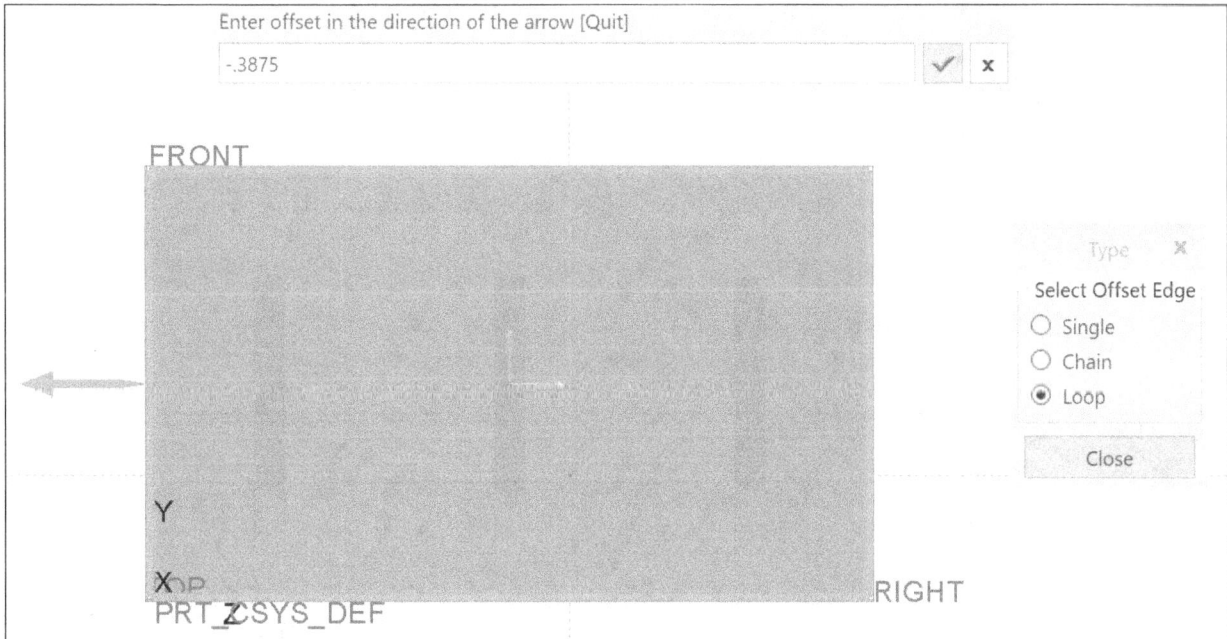

Figure 13.3(i) Offset **-.3875**

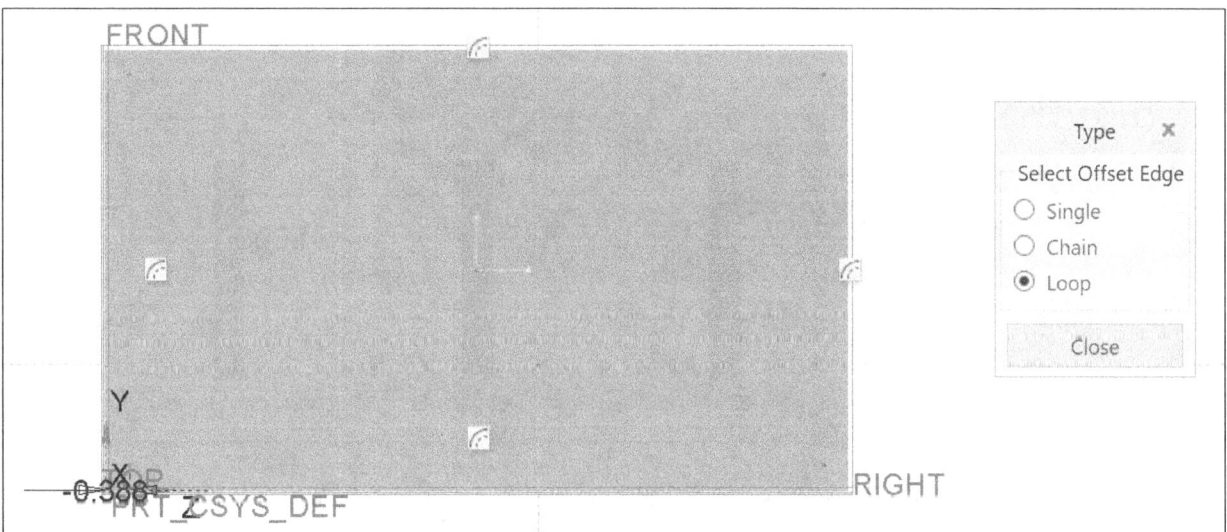

Figure 13.3(j) Offset Loop

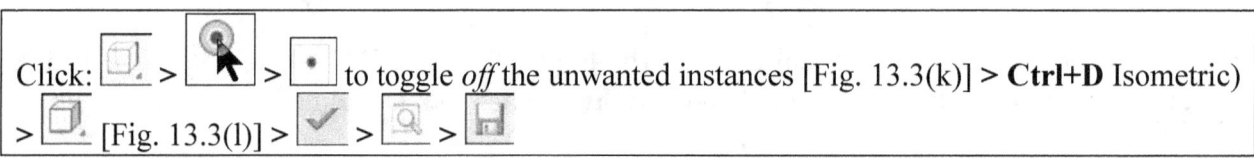

Click: [icon] > [icon] > [icon] to toggle *off* the unwanted instances [Fig. 13.3(k)] > **Ctrl+D** Isometric) > [icon] [Fig. 13.3(l)] > [icon] > [icon] > [icon]

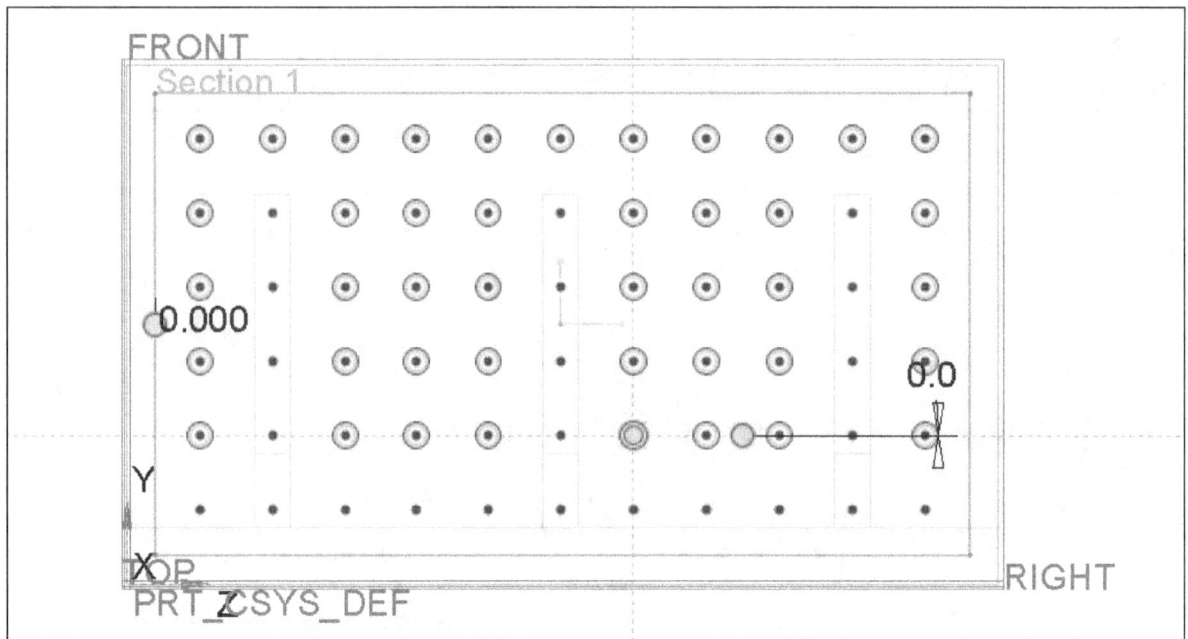

Figure 13.3(k) Toggle Off Unwanted Instances of the Pattern

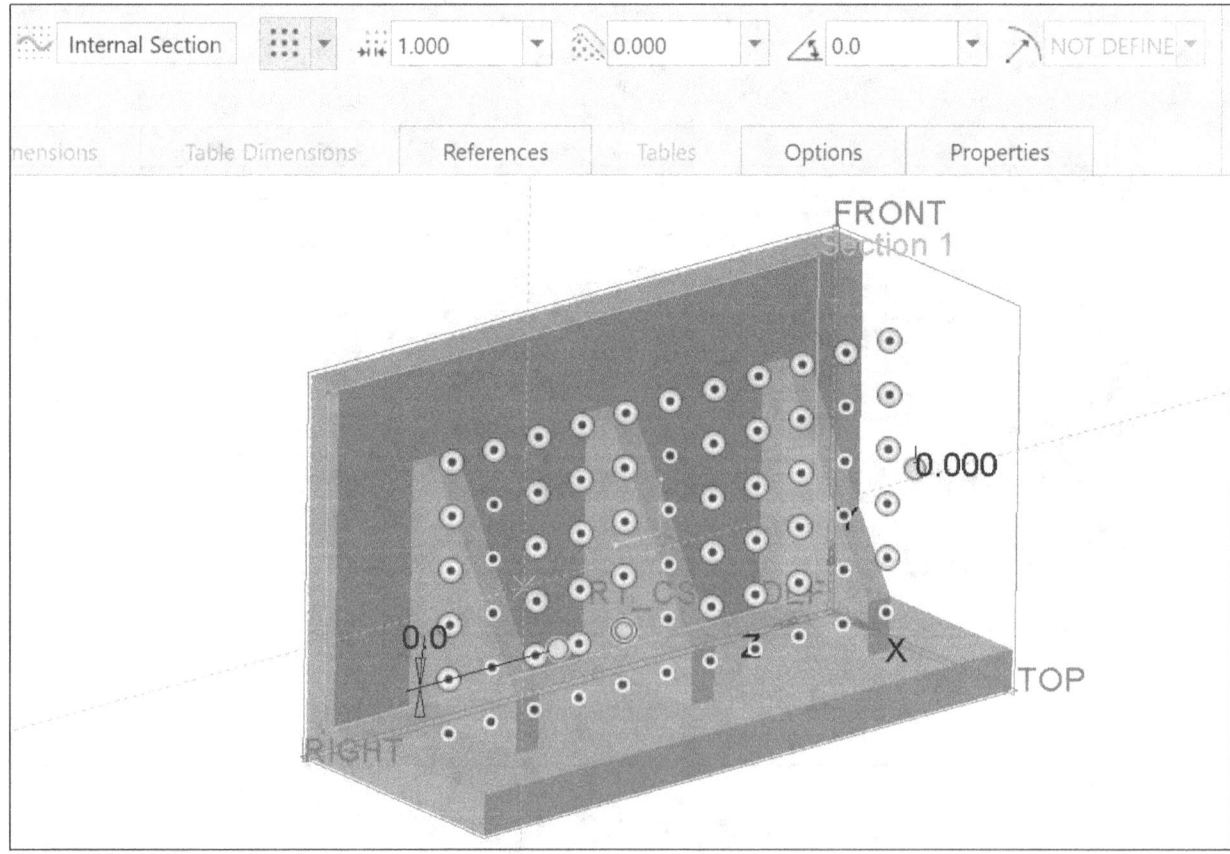

Figure 13.3(l) Pattern Preview

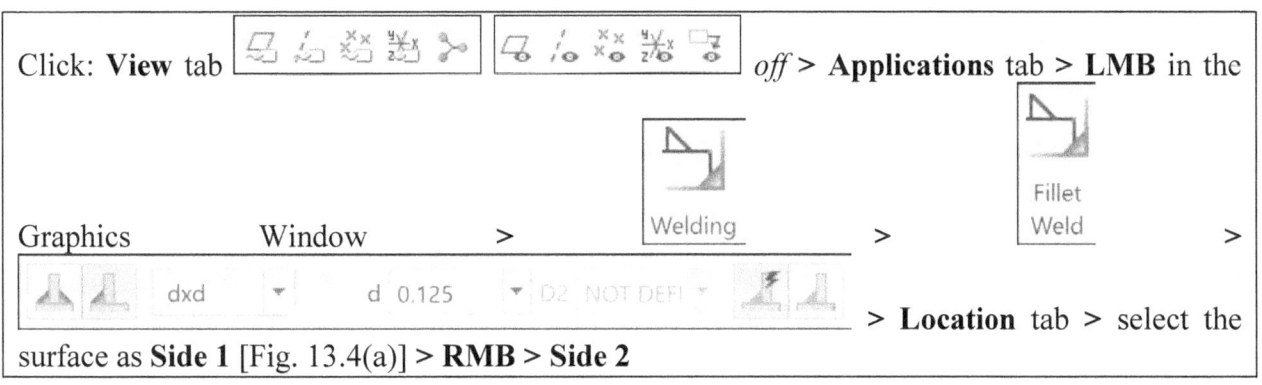

Click: **View** tab *off* > **Applications** tab > **LMB** in the Graphics Window > Welding > Fillet Weld > dxd d 0.125 D2 NOT DEFI > **Location** tab > select the surface as **Side 1** [Fig. 13.4(a)] > **RMB** > **Side 2**

| DxD | d 0.125 | D2 NOT DEFINE | NOT DEFINE | # | 0 |

Location Options Symbol Properties

Weld sets   **Side 1**   **Side 2**

Set 1   Individual Surfaces   ⊙ Click here to add item
*New set

                   Details...   Details...
                   ☑ Allow tangent propagation
                   ☐ Treat side surfaces as contact surfaces
                   Weld joints

                   Joint   Status   Attachment

                   Details... Restore Default

Surf:F5(EXTRUDE_1)

Individual Surfaces

Figure 13.4(a) Select Side 1 of the Weld Set 1

591

Press and hold the **Ctrl** key and select the 9 vertical surfaces of the ribs > release the **Ctrl** key > **RMB** > **New Set** [Fig. 13.4(b)]

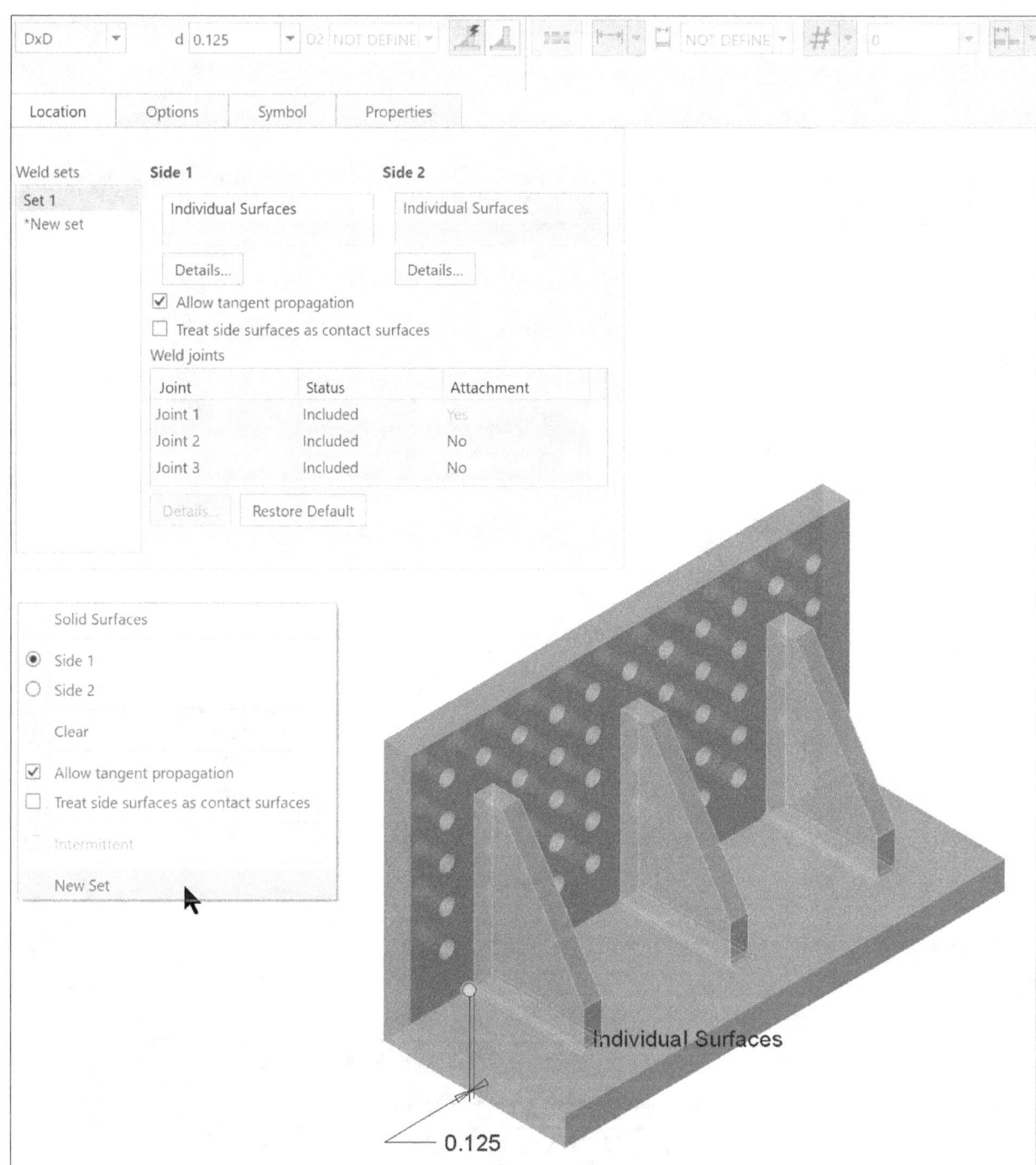

Figure 13.4(b) Select the (nine) Vertical Surfaces for Side 2 of the Weld Set 1

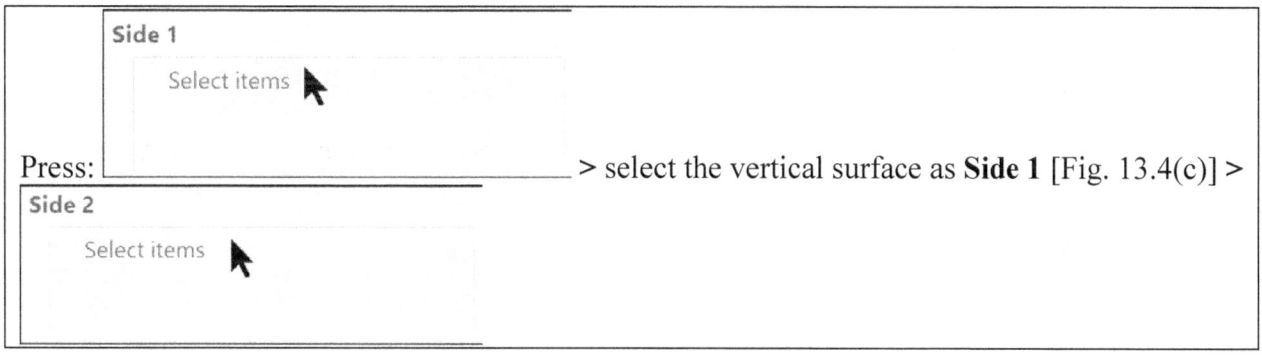

Press: _____ > select the vertical surface as **Side 1** [Fig. 13.4(c)] >

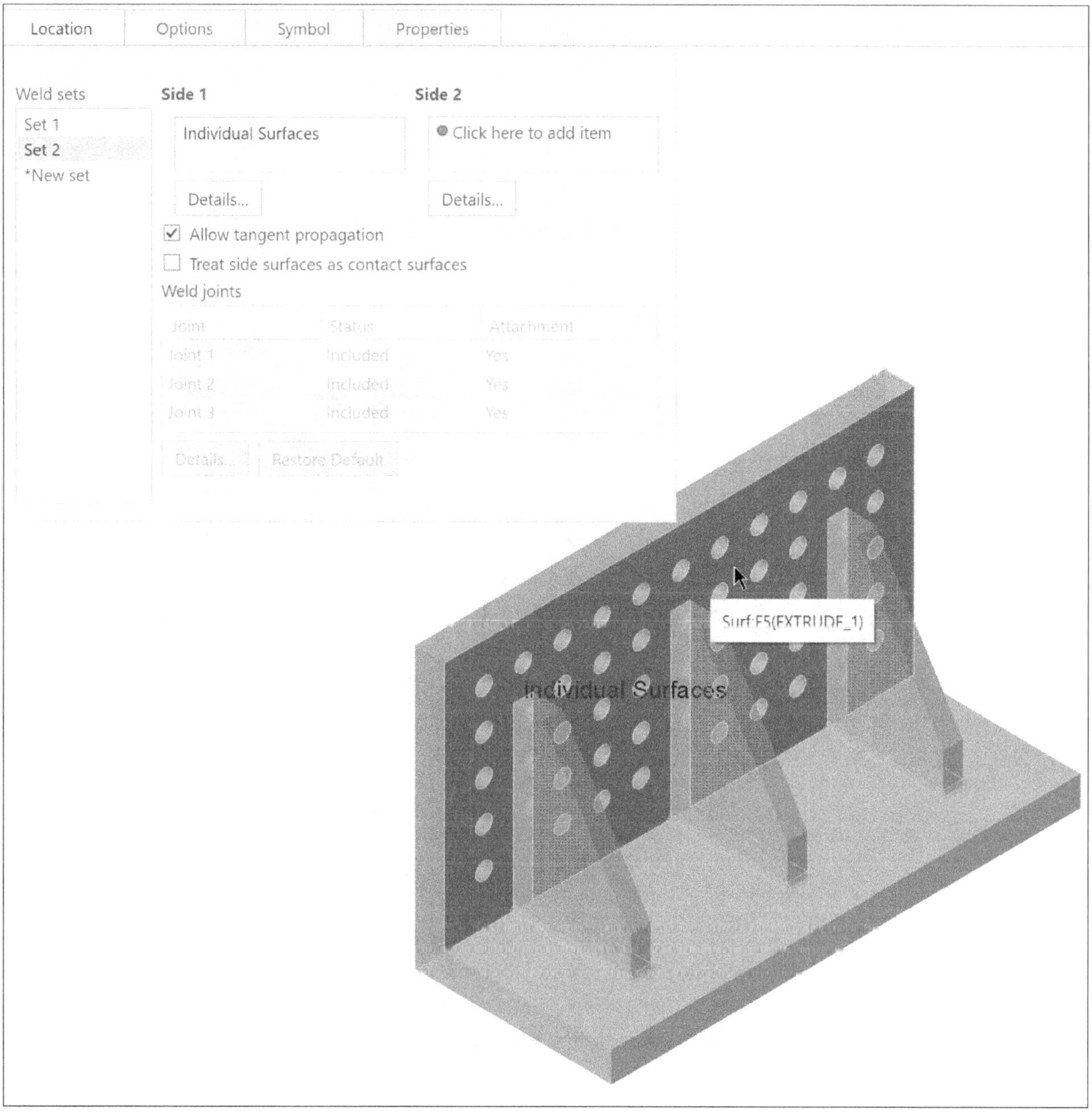

Location	Options	Symbol	Properties

Weld sets | **Side 1** | **Side 2**

Set 1
Set 2
*New set

Individual Surfaces | ● Click here to add item

Details... | Details...

☑ Allow tangent propagation
☐ Treat side surfaces as contact surfaces
Weld joints

Joint	Status	Attachment
Joint 1	Included	Yes
Joint 2	Included	Yes
Joint 3	Included	Yes

Details... | Restore Default

Figure 13.4(c) Select Side 1 of the Weld Set 2

Press and hold the **Ctrl** key and select the surfaces of the ribs that touch the vertical holed wall > release the **Ctrl** key > 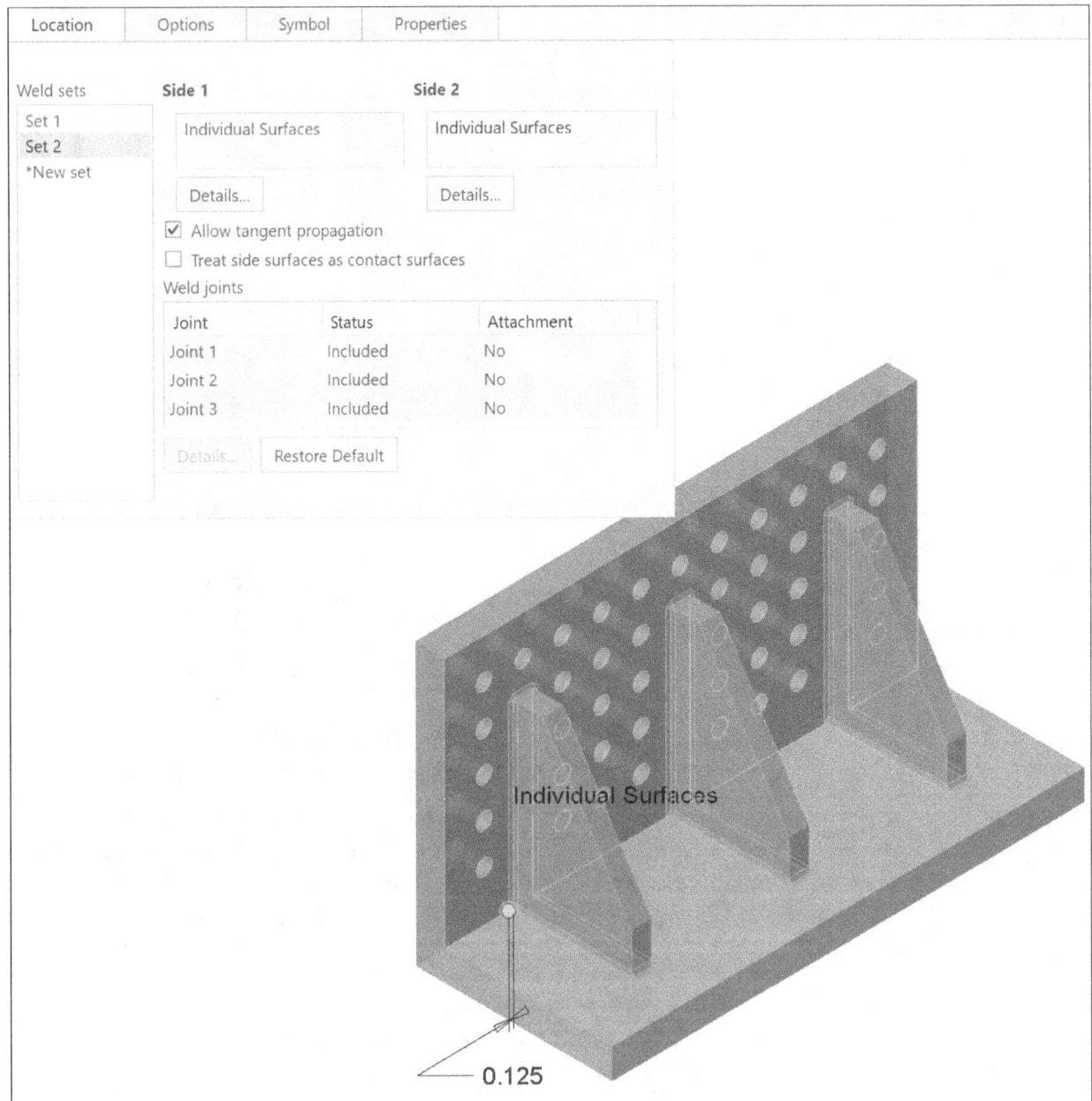 *on* [Fig. 13.4(d)] >

If you select the wrong item, select the appropriate Set # > select the appropriate Details button > place the pointer over the item in the appropriate dialog box > RMB > Remove > Ctrl key > select another surface.

Figure 13.4(d) Select the (nine) Surfaces for Side 2 of the Weld Set 2

Click: 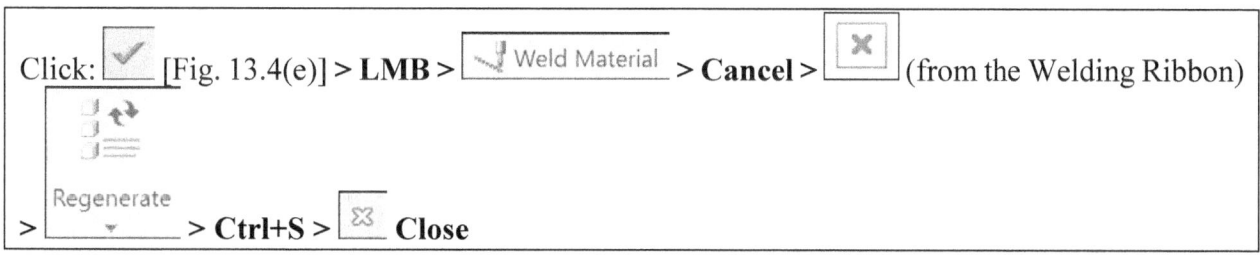 [Fig. 13.4(e)] > **LMB** > ☐ Weld Material > **Cancel** > ☒ (from the Welding Ribbon)

> Regenerate > **Ctrl+S** > ⊠ **Close**

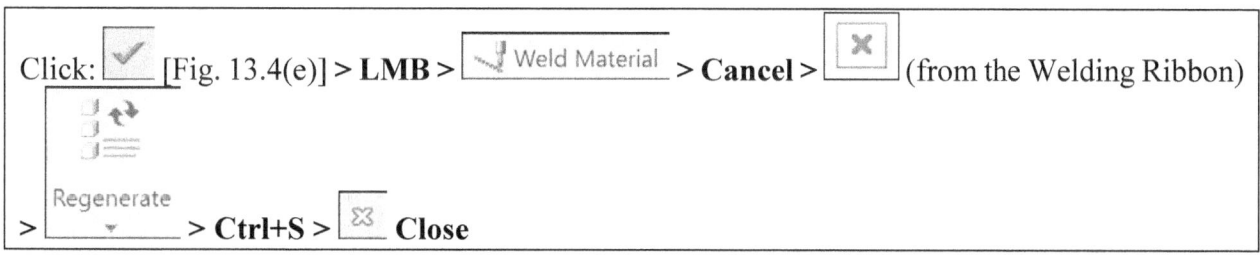

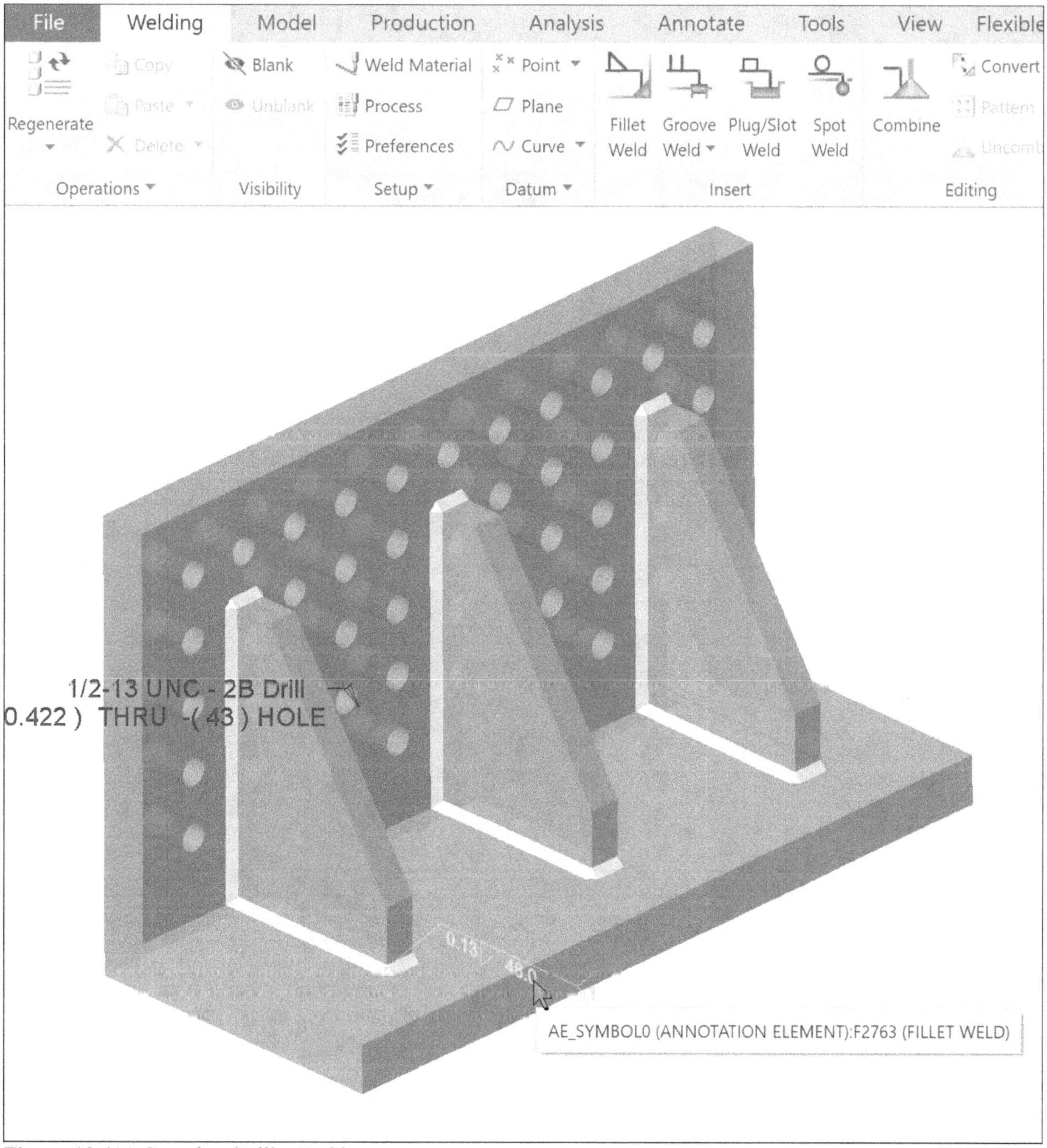

Figure 13.4(e) Completed Fillet Weld

Base Plate

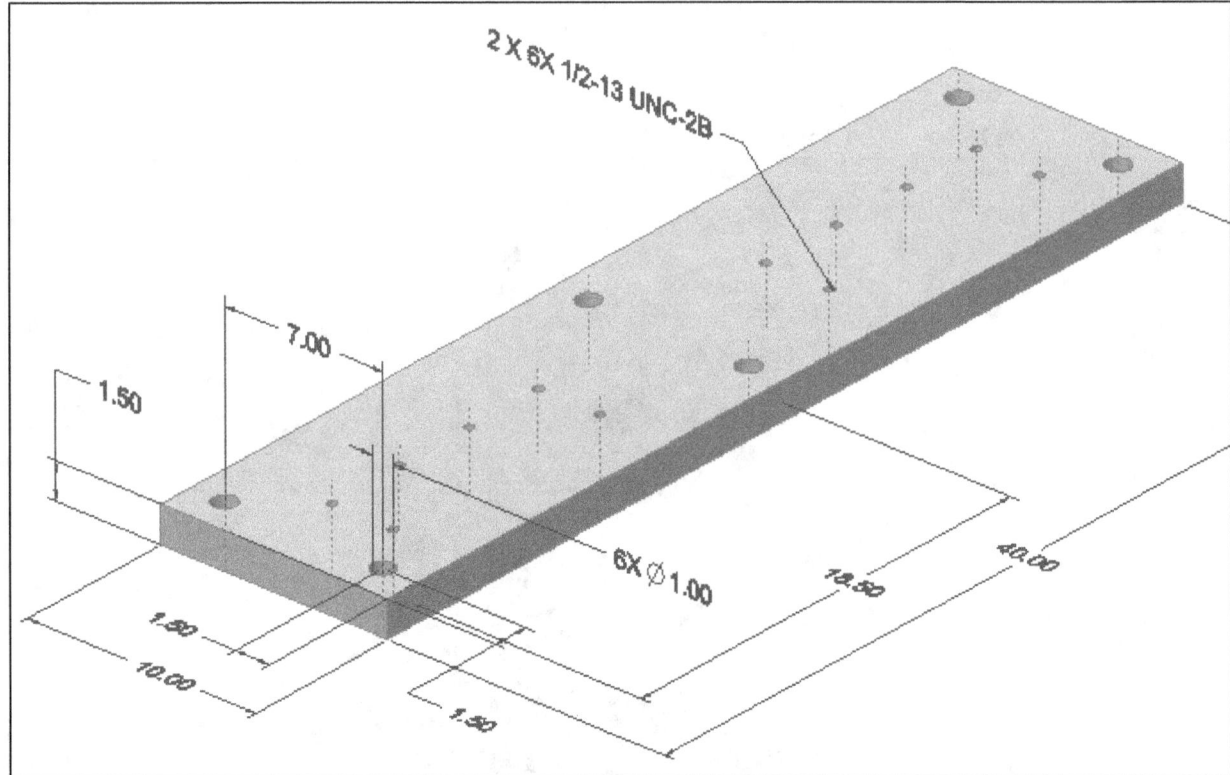

Figure 13.5(a) Base Plate

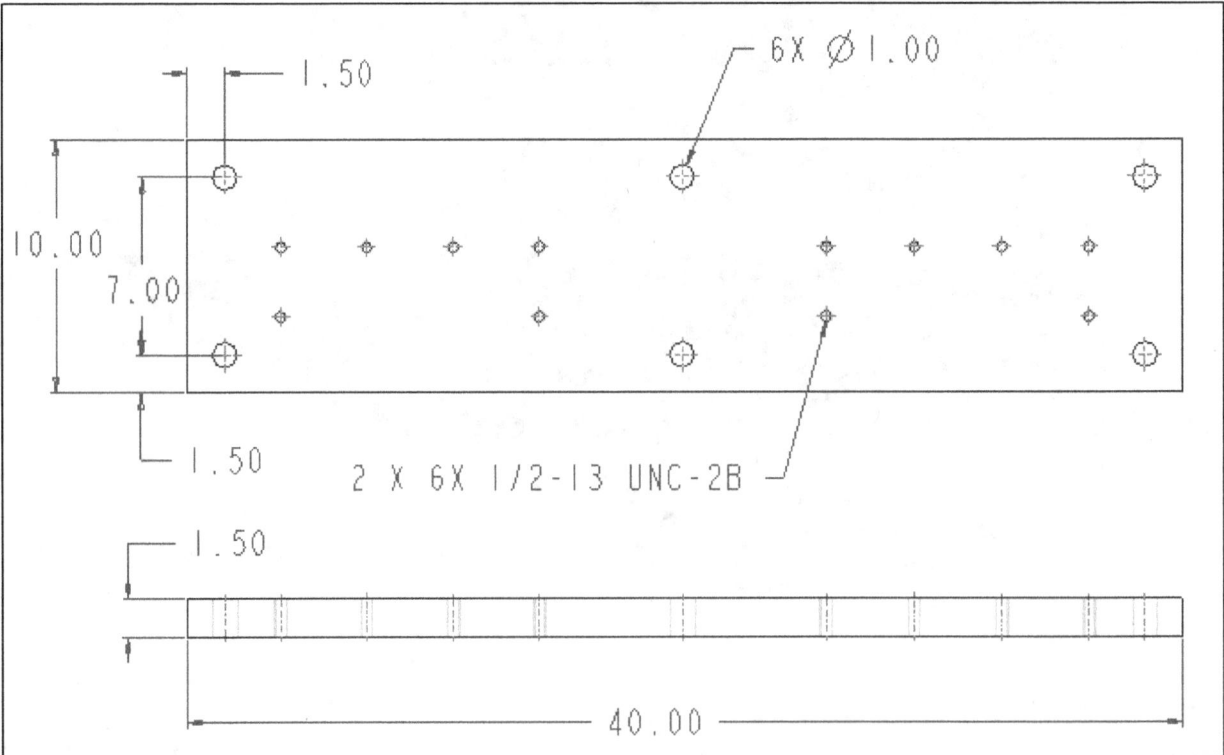

Figure 13.5(b) Base Plate Detail

Base Plate dimensions [Figs. 13.5(a-b)]. Press: **Ctrl+N** > *type* **base_plate** > **OK** > **File** > **Options** > **Customize** > **Ribbon** > **Import** > **Import customization file** > select your previously saved **.ui** file from Lesson 2 (creo_parametric_customization.ui) > **Open** > **Import** > **Configuration Editor** > **Import/Export** > **Import configuration file** > select your previously saved file from Lesson 2 (**creo_textbook.pro**) > Import for all modes > **Import** > **Open** > **OK** > **View** tab > *on* > **Model** tab > select datum **TOP** > > > > Center Rectangle > sketch and dimension the rectangular section [Fig. 13.5(c)] > ✓ > modify the depth to **1.50** inches [Fig. 13.5(d)] > **RMB** > **Show Section Dimensions** > **Enter** > **MMB** > **LMB** > **Ctrl+S** > **OK**

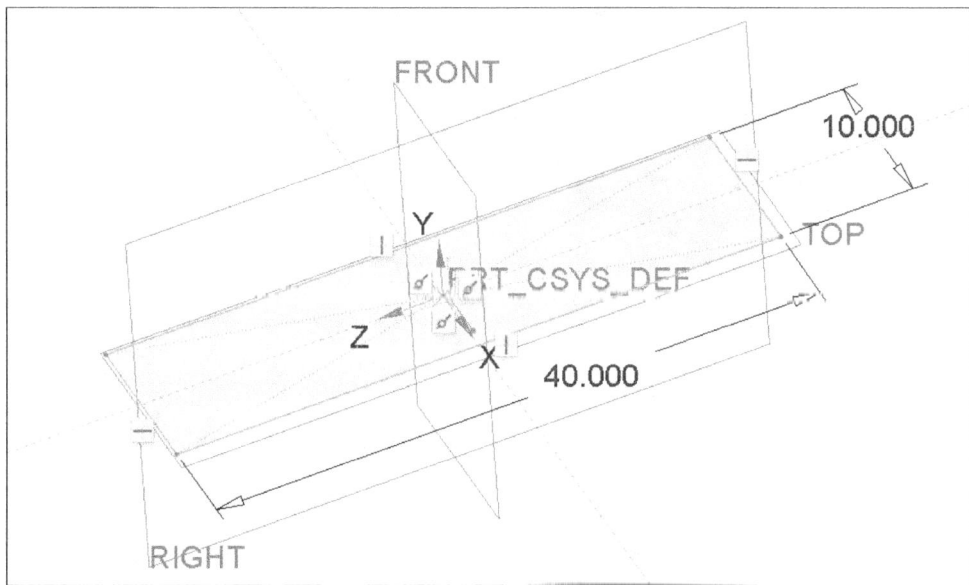

Figure 13.5(c) Sketch the Rectangular-shaped Section

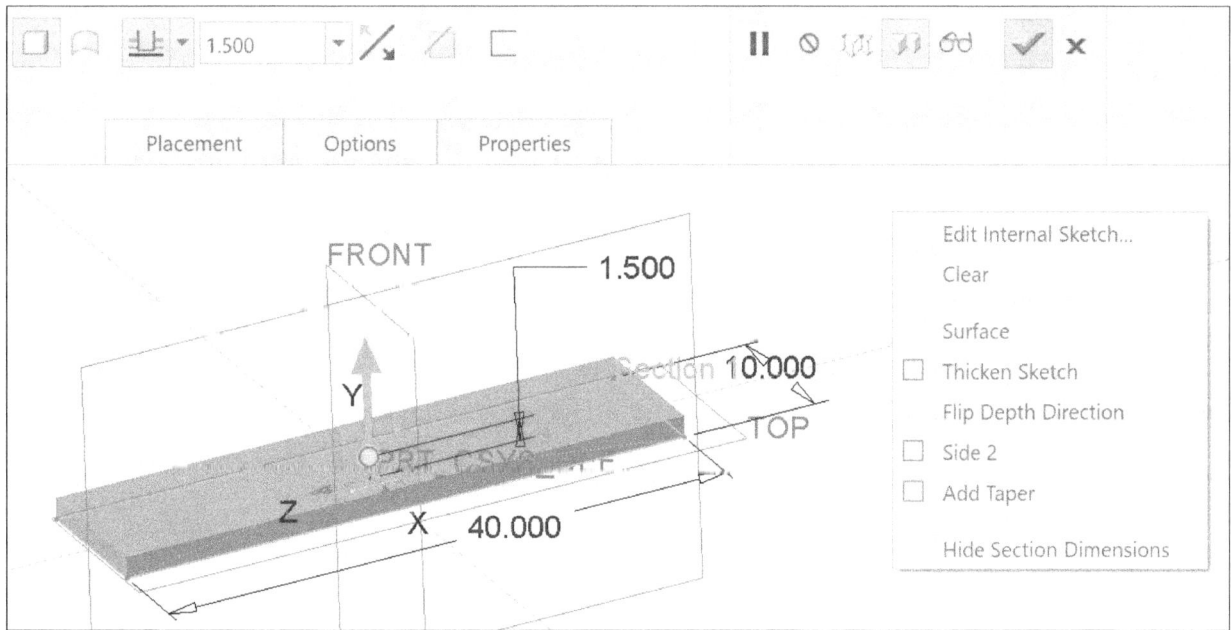

Figure 13.5(d) Depth of **1.50**

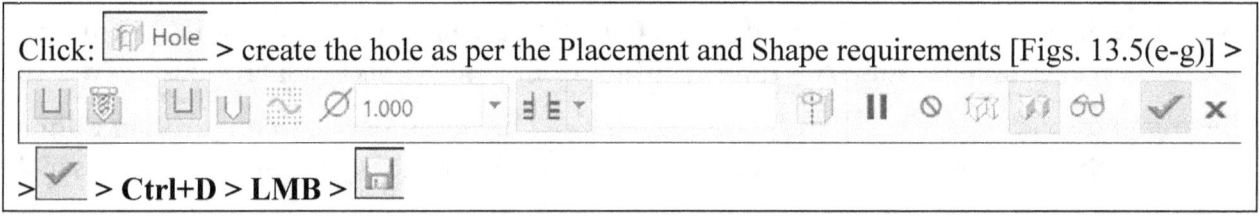

Click: 🔲 Hole > create the hole as per the Placement and Shape requirements [Figs. 13.5(e-g)] >

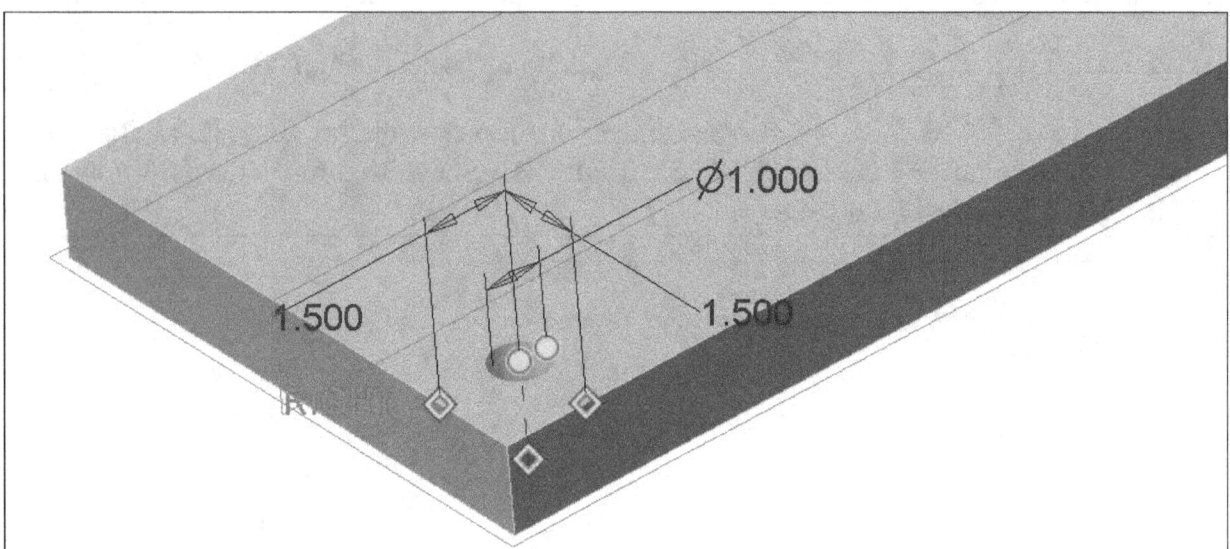

> ✓ > **Ctrl+D > LMB >** 🔲

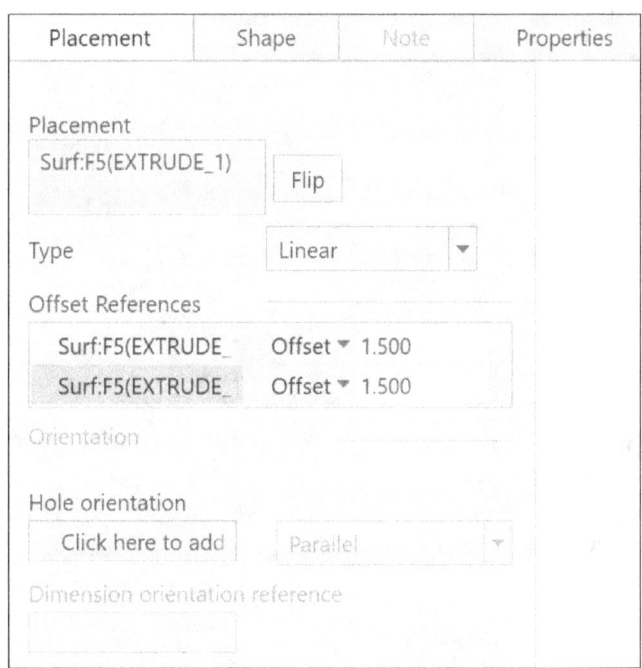

Figure 13.5(e) 1.00 Diameter Hole

Figure 13.5(f) Hole Placement

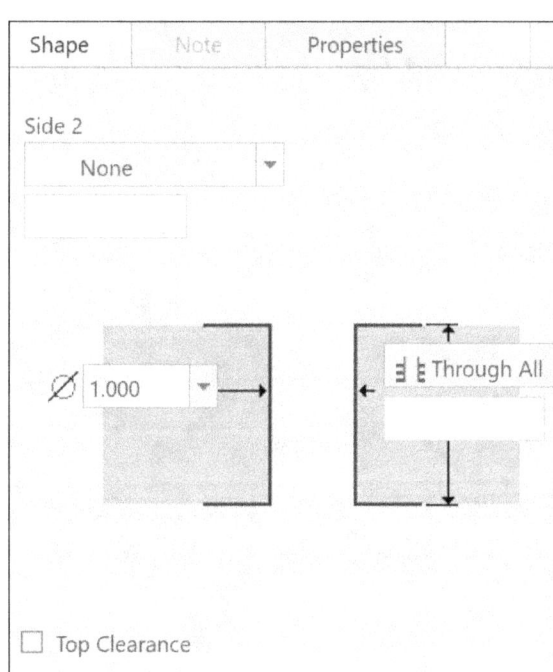

Figure 13.5(g) Hole Shape

Select **Hole** in Model Tree > [⊞] [Fig. 13.5(h)] > [▾] > **Direction** > select two reference direction surfaces > [✓] > **Annotate** tab > in the Model Tree, expand the Pattern > select the Extrude, Pattern, and Hole features > **RMB** > [⊞] **Show Annotations** > [▢] tab > [☑] > **OK** > move the dimensions as needed [Fig. 13.5(i)]

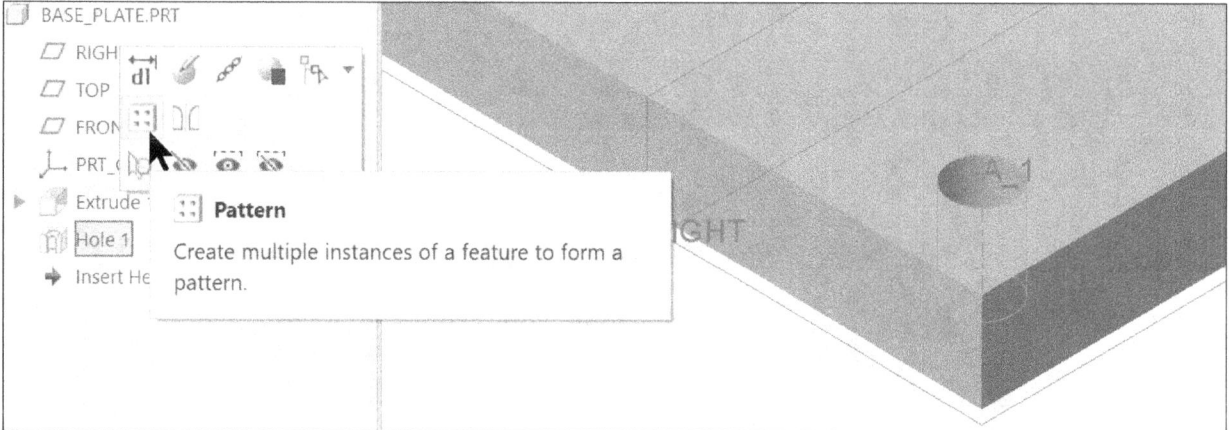

Figure 13.5(h) Pattern

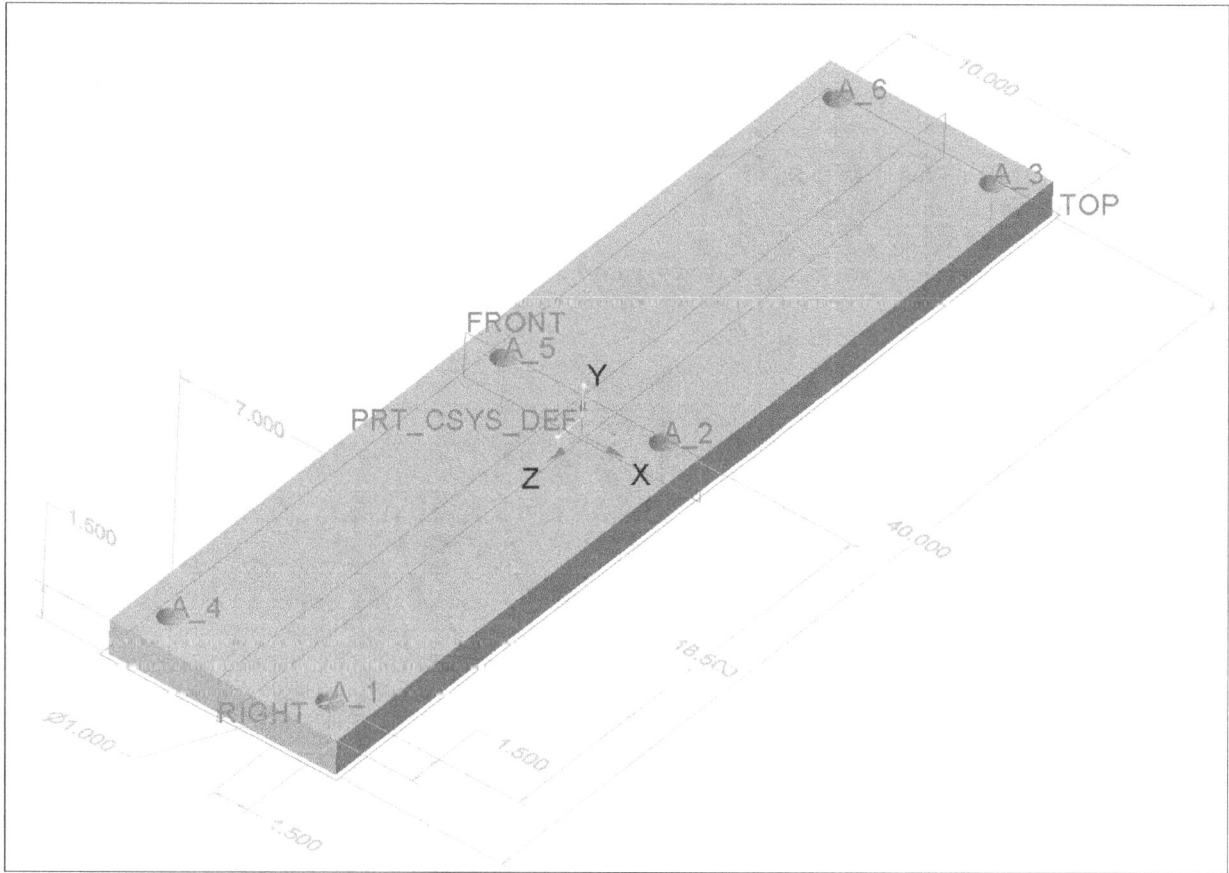

Figure 13.5(i) Pattern Preview

Click: **LMB** to deselect > select the **1.00** diameter dimension > Dimension Ribbon > **Dimension Text** >

> Dimension Text:
>
> 6 X @D

[Fig. 13.5(j)] > **LMB** > **View** tab > **Appearances** > change color as desired > [image icon] [Fig. 13.5(k)] > [toolbar icons] > select dimensions > [arrow icon] **Flip Arrows** as needed > **Ctrl+D** > **Ctrl+S** > **File** > **Close**

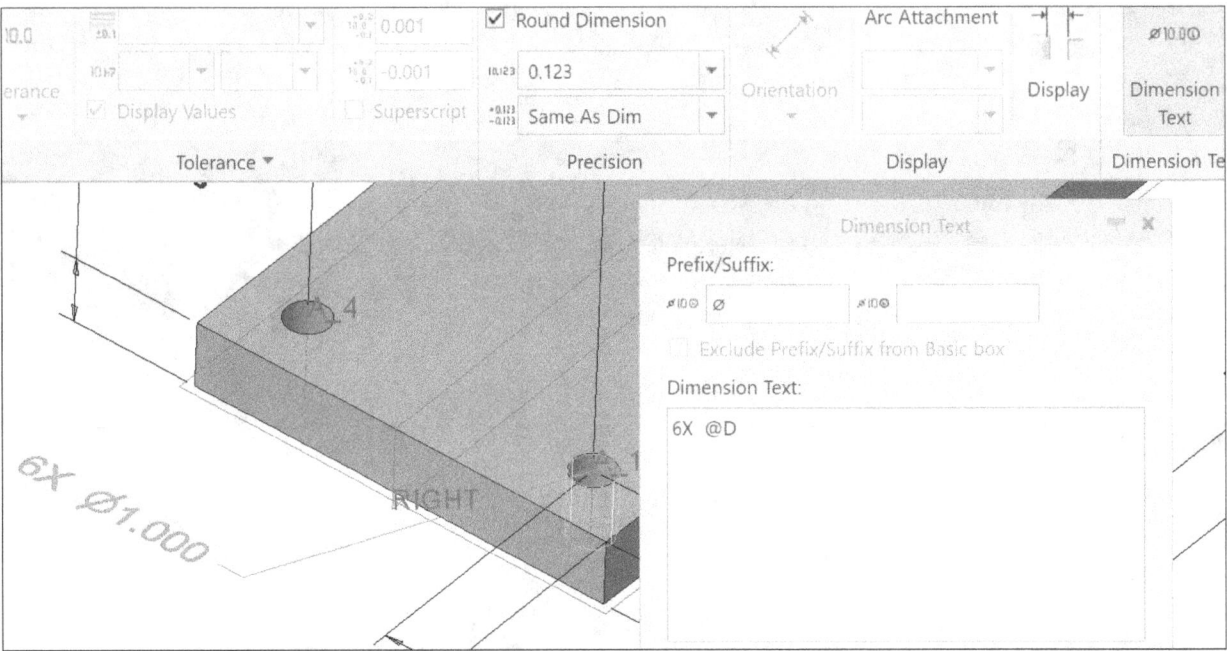

Figure 13.5(j) Dimension Properties Tab

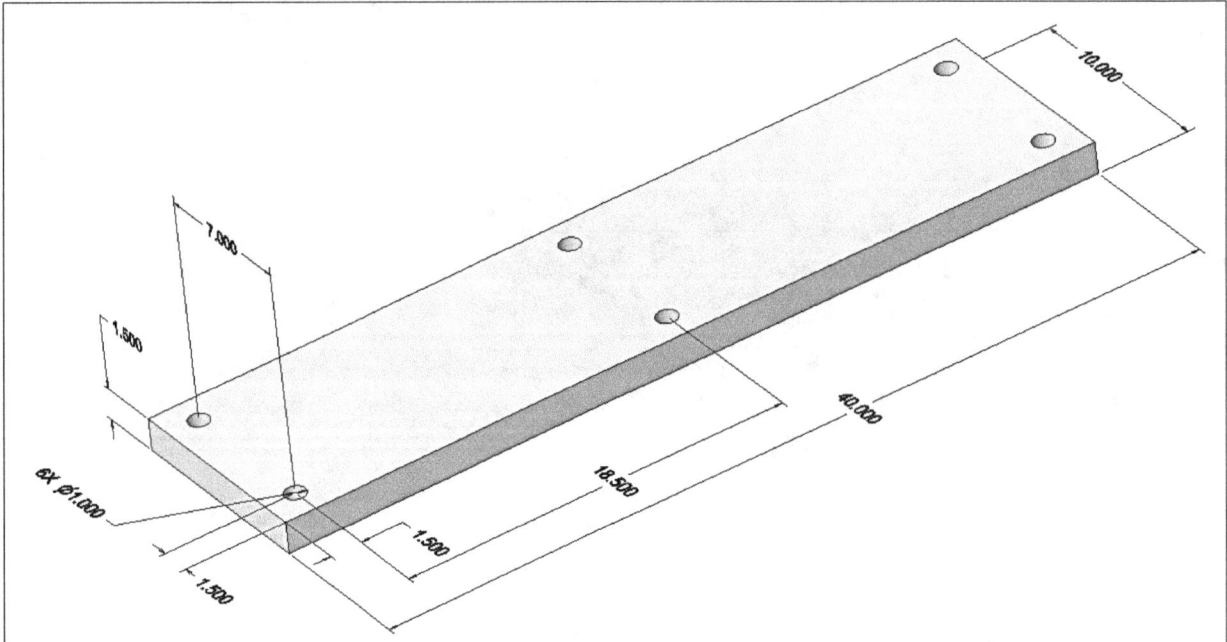

Figure 13.5(k) Annotations

Mounting System Assembly

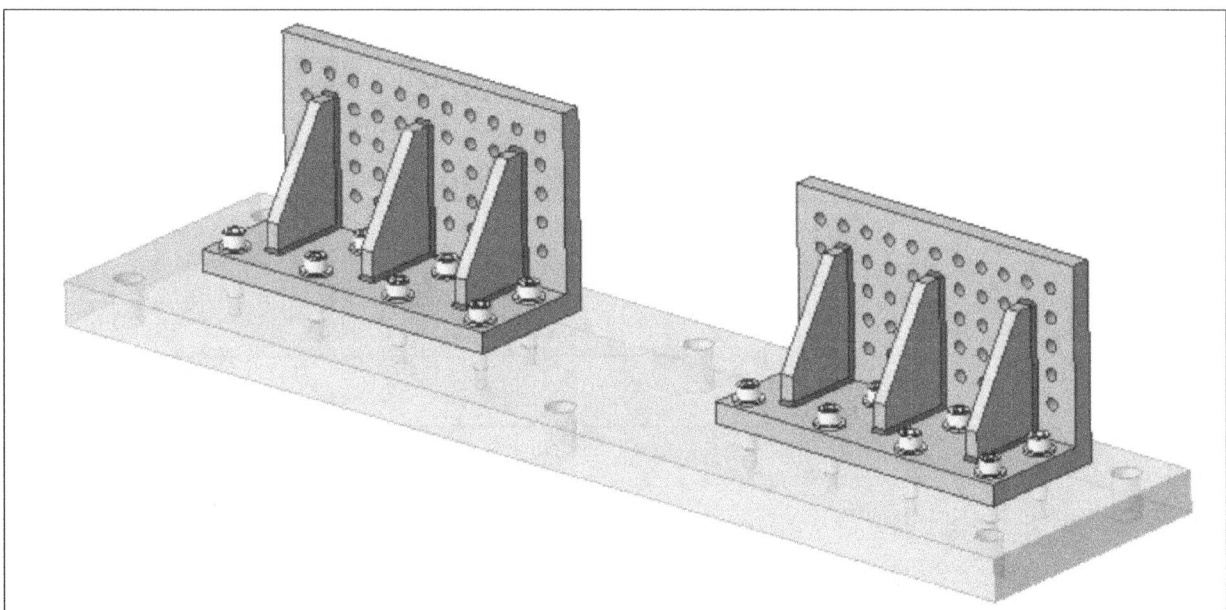

Figure 13.6(a) Mounting System Assembly

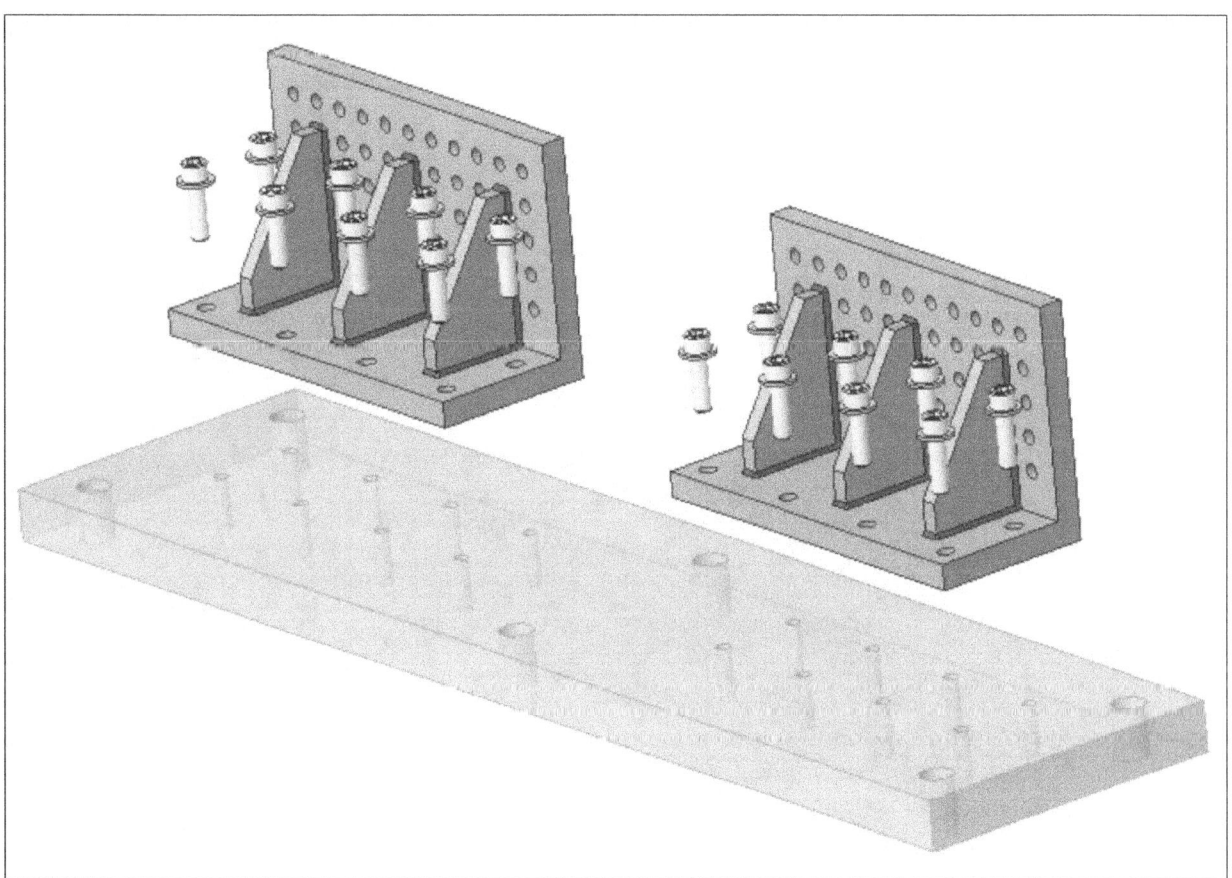

Figure 13.6(b) Mounting System Assembly Exploded

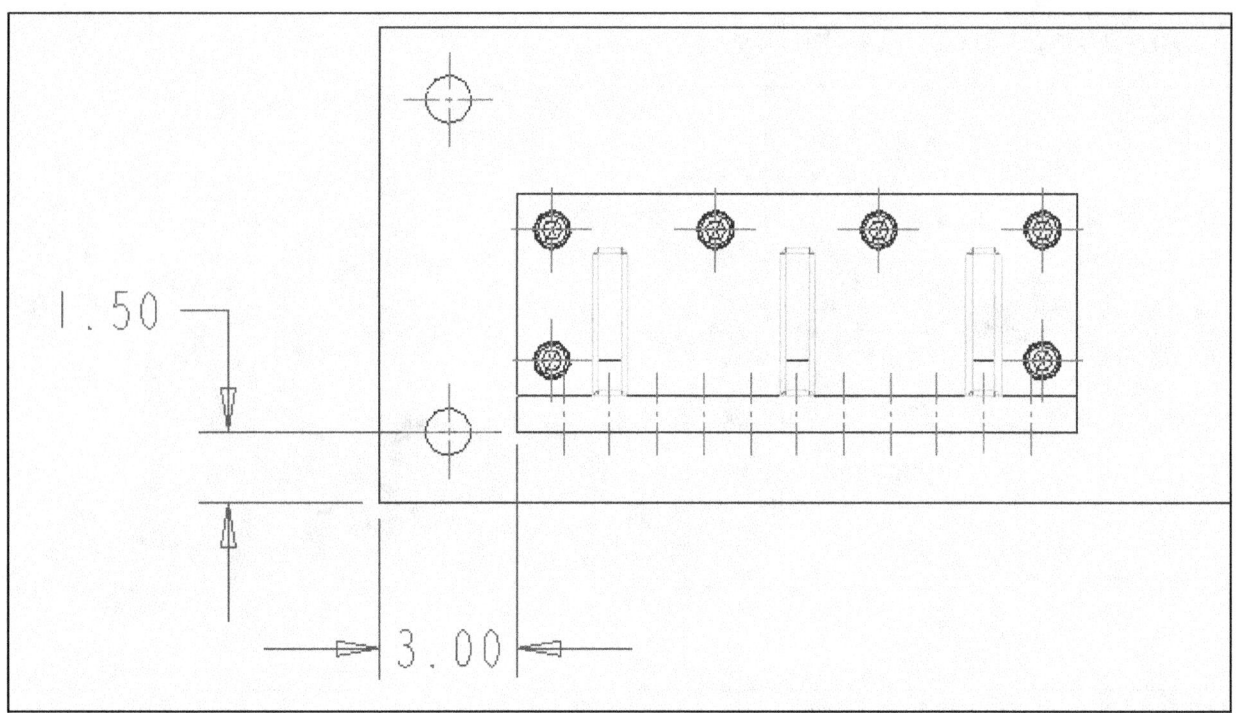

Figure 13.6(c) Mounting System Assembly Component Placement Location Dimensions

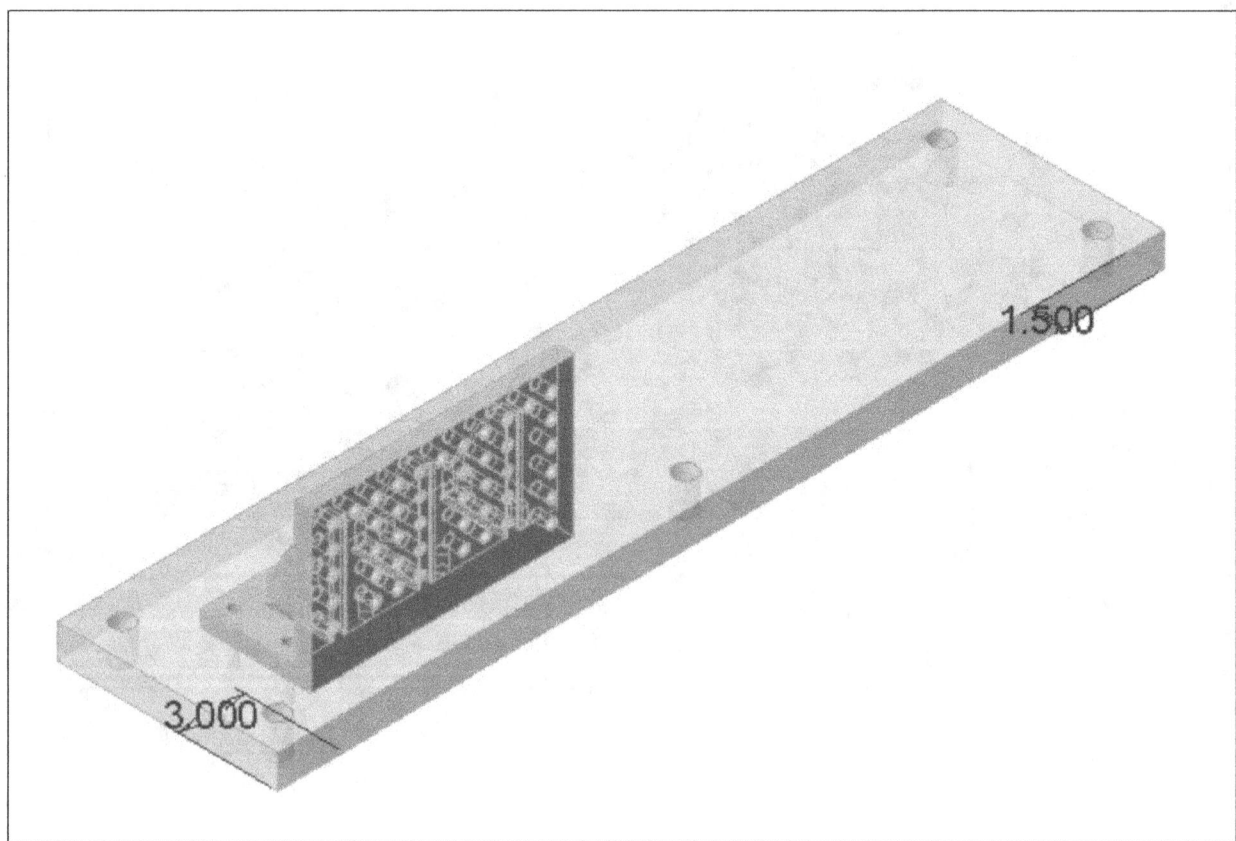

Figure 13.6(d) Mounting System Assembly Component Placement Location in the Assembly Model

602

Create the Mounting System Assembly shown in Figures 13.6(a-d).

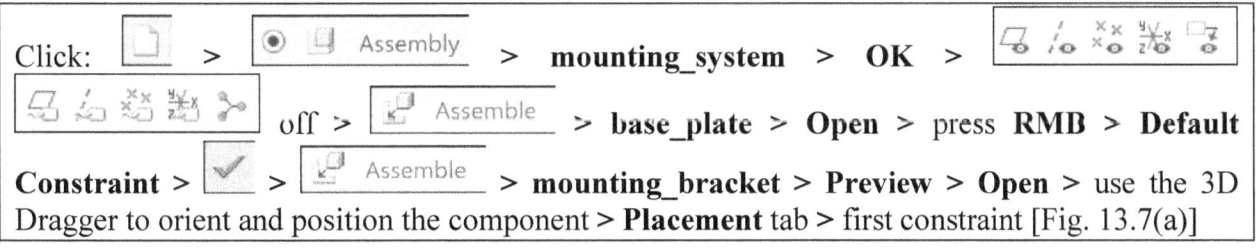

Click: [icon] > [◉ Assembly] > **mounting_system** > **OK** > [icons] [icons] off > [Assemble] > **base_plate** > **Open** > press **RMB** > **Default Constraint** > [✓] > [Assemble] > **mounting_bracket** > **Preview** > **Open** > use the 3D Dragger to orient and position the component > **Placement** tab > first constraint [Fig. 13.7(a)]

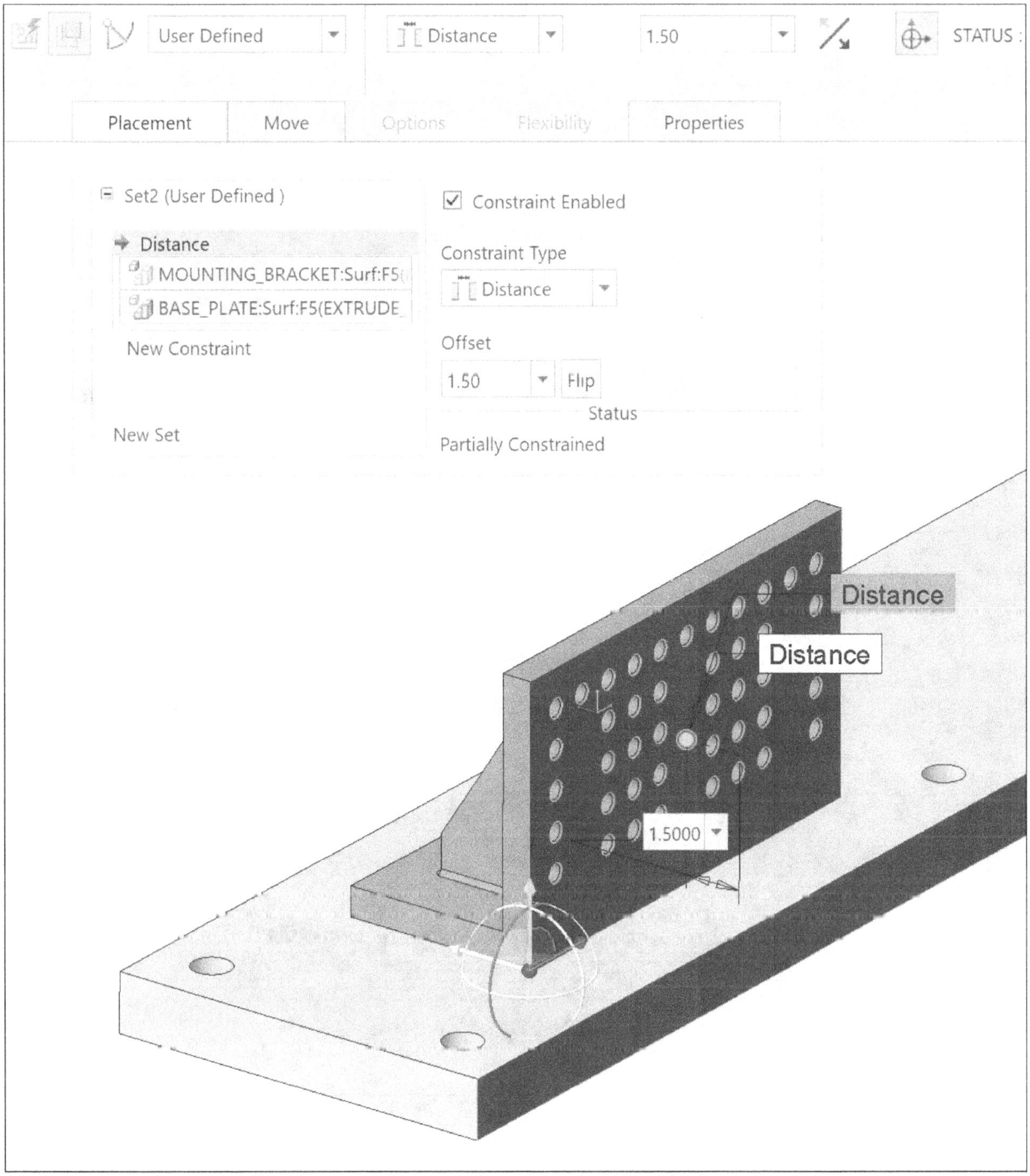

User Defined
Distance
1.50
STATUS :
Placement
Move
Options
Flexibility
Properties
Set2 (User Defined)
Distance
MOUNTING_BRACKET:Surf:F5(
BASE_PLATE:Surf:F5(EXTRUDE_
New Constraint
New Set
Constraint Enabled
Constraint Type
Distance
Offset
1.50
Flip
Status
Partially Constrained
Distance
Distance
1.5000

Figure 13.7(a) First Constraint Distance Offset **1.50** (negative value if required)

Click: New Constraint in the Placement window > second constraint [Fig. 13.7(b)]

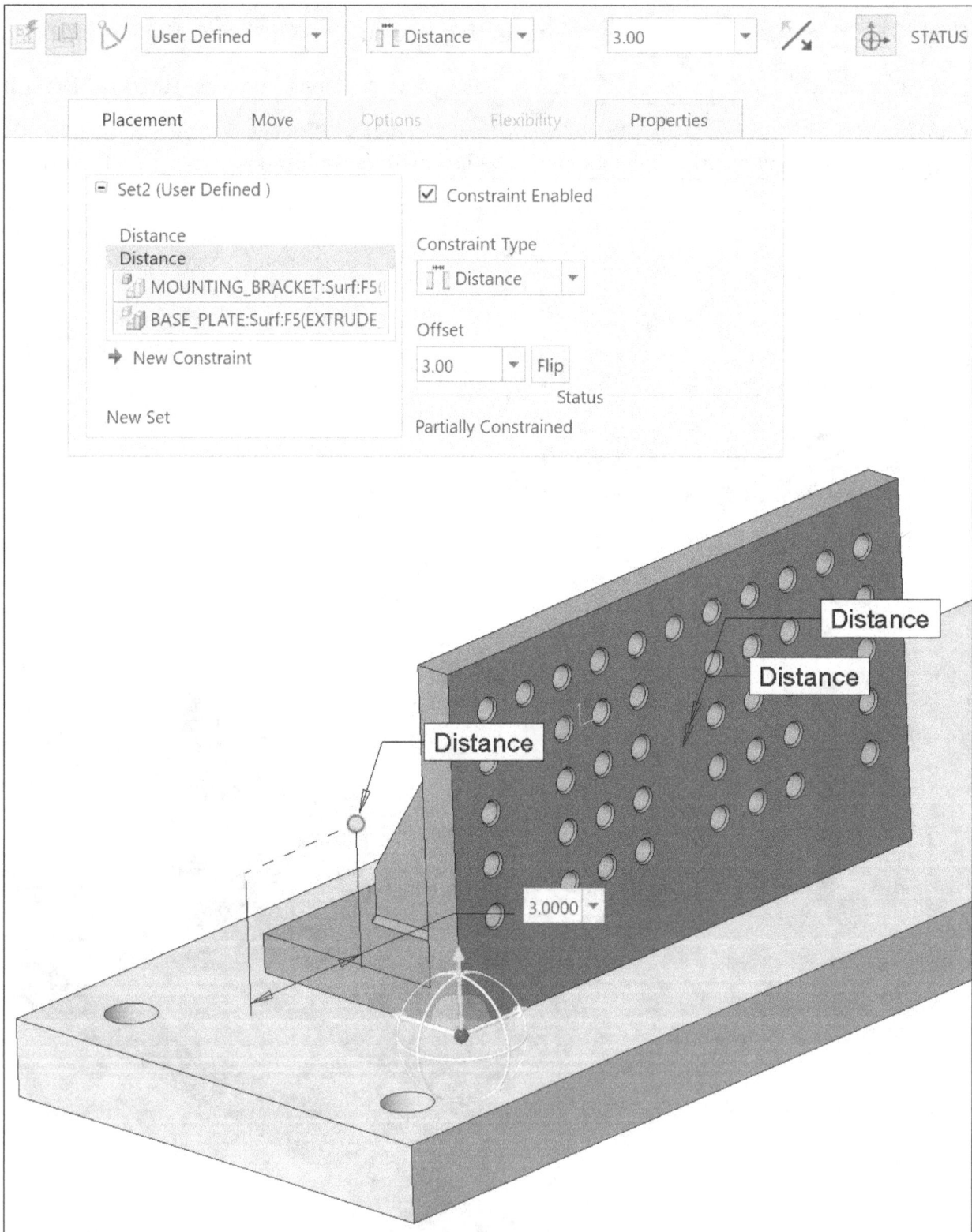

Figure 13.7(b) Distance Offset **3.00** (Enter a Negative Value if Required)

Select: 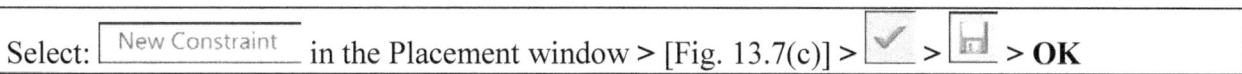 in the Placement window > [Fig. 13.7(c)] > ✓ > 🖫 > **OK**

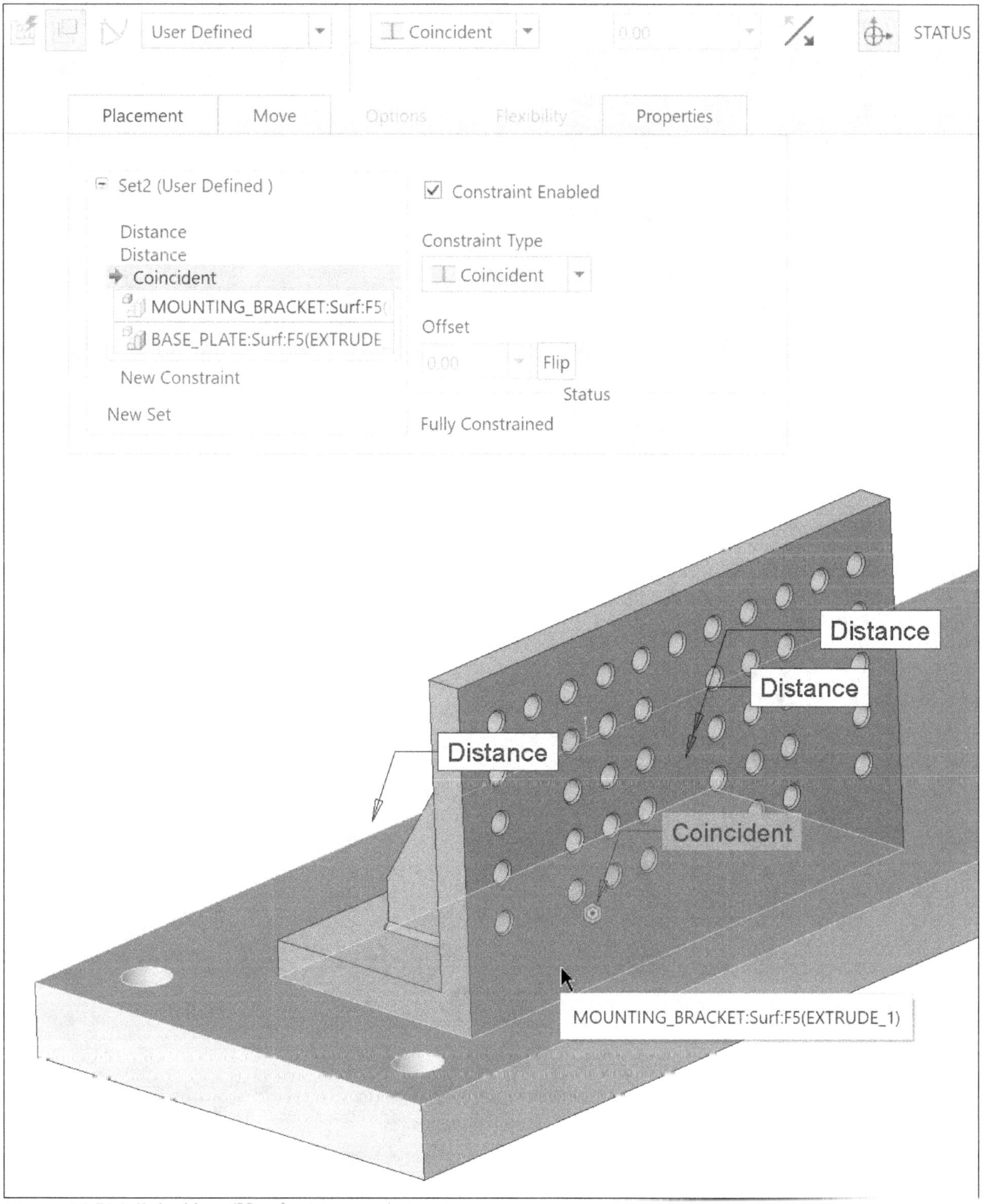

| | | | User Defined | ▼ | | I Coincident | ▼ | 0.00 | ▼ | ⬚ | ⊕ | STATUS |

| Placement | Move | Options | Flexibility | Properties |

⊟ Set2 (User Defined)

☑ Constraint Enabled

Distance
Distance
➡ Coincident
　　🗋 MOUNTING_BRACKET:Surf:F5(
　　🗋 BASE_PLATE:Surf:F5(EXTRUDE
New Constraint

New Set

Constraint Type
I Coincident ▼

Offset
0.00 ▼ Flip

Status
Fully Constrained

Distance

Distance

Distance

Coincident

MOUNTING_BRACKET:Surf:F5(EXTRUDE_1)

Figure 13.7(c) Coincident (Use the **RMB** and click through the features and select the bottom face of the Mounting Bracket)

Set: 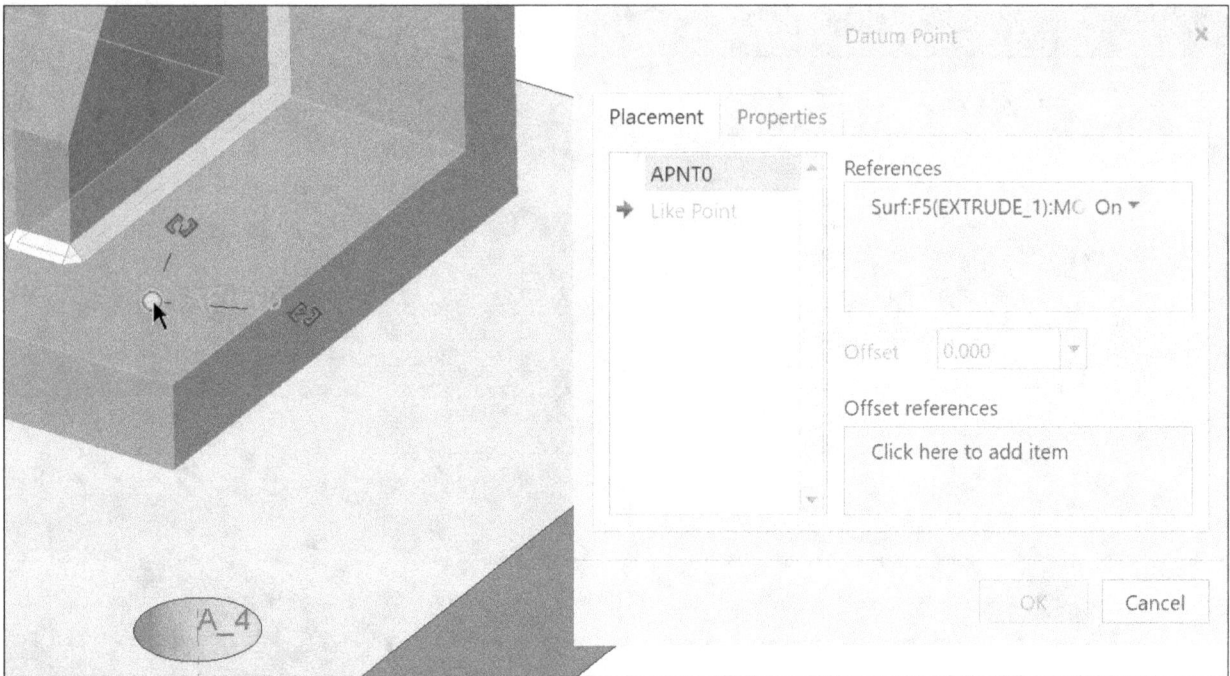 > $\overset{\times\times}{\times}$ Point ___ Create a datum point > pick a position on front corner of the **Mounting_bracket** [Fig. 13.8(a)] > locate the point at **.750** from each vertical face [Fig. 13.8(a)] > **OK** > **Ctrl+S**

Figure 13.8(a) Point

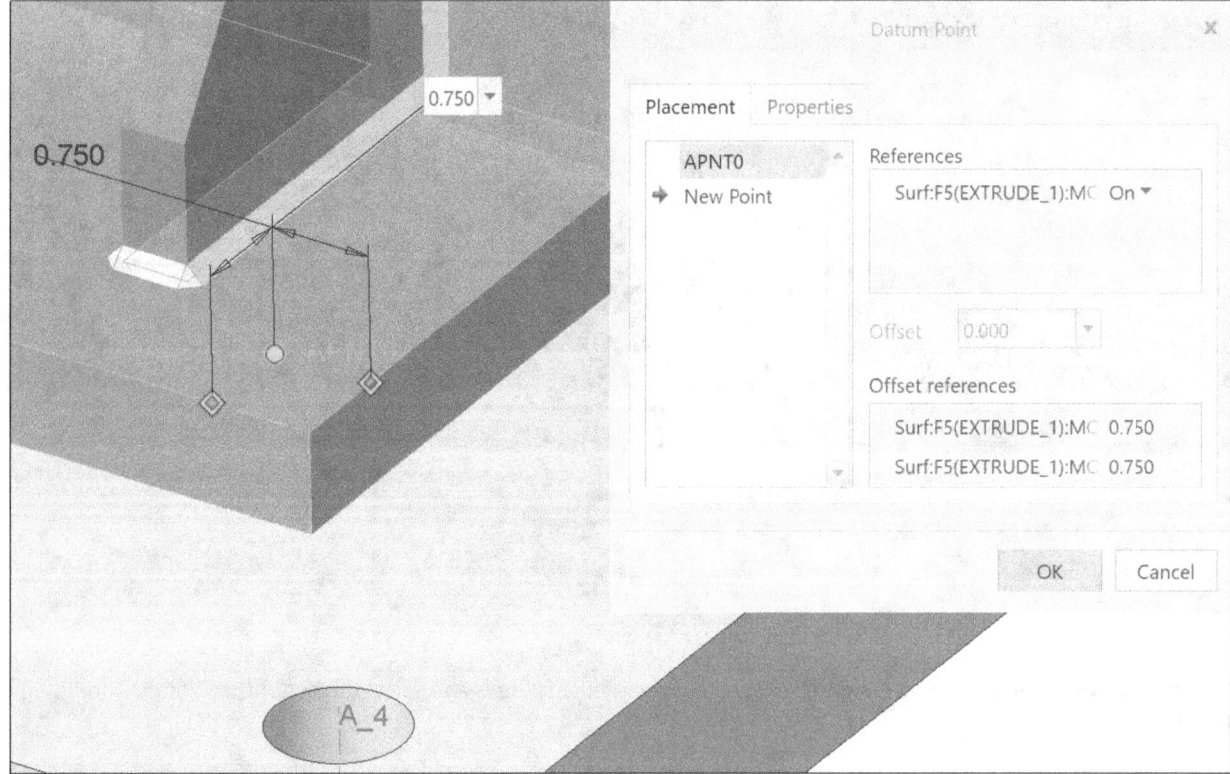

Figure 13.8(b) Point Position

With the point still selected: | ⁄ Axis | Create a datum axes [Fig. 13.8(c)] > press **Ctrl** key > select the surface [Fig. 13.8(d)] > **OK > LMB > Ctrl+S**

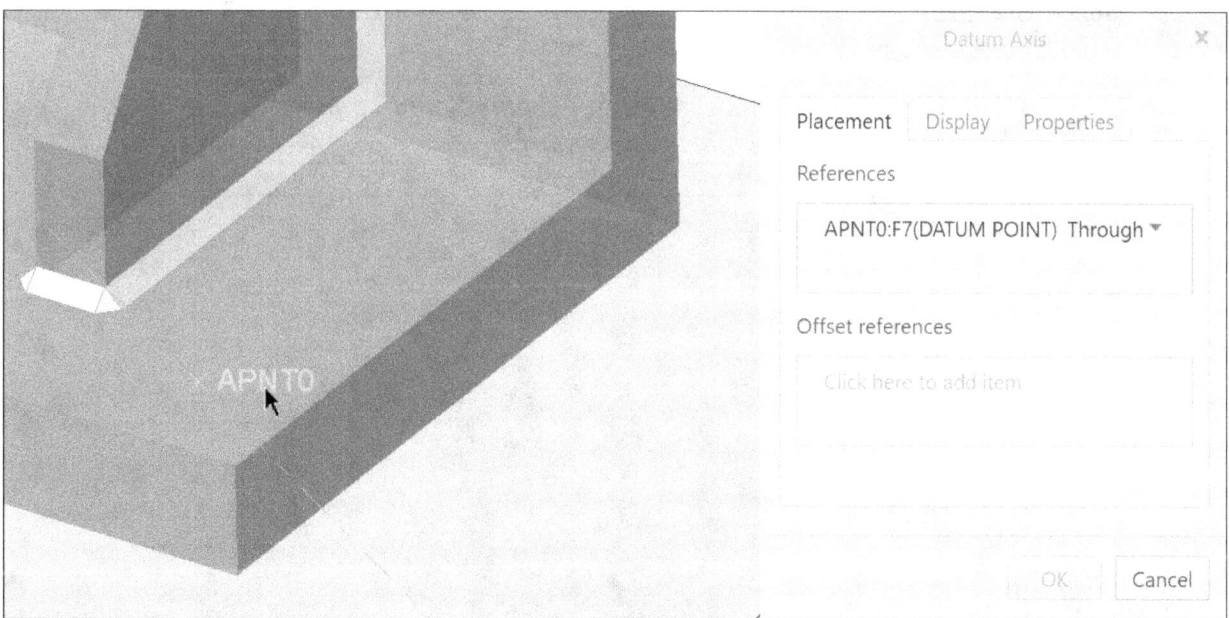

Figure 13.8(c) Select the Point

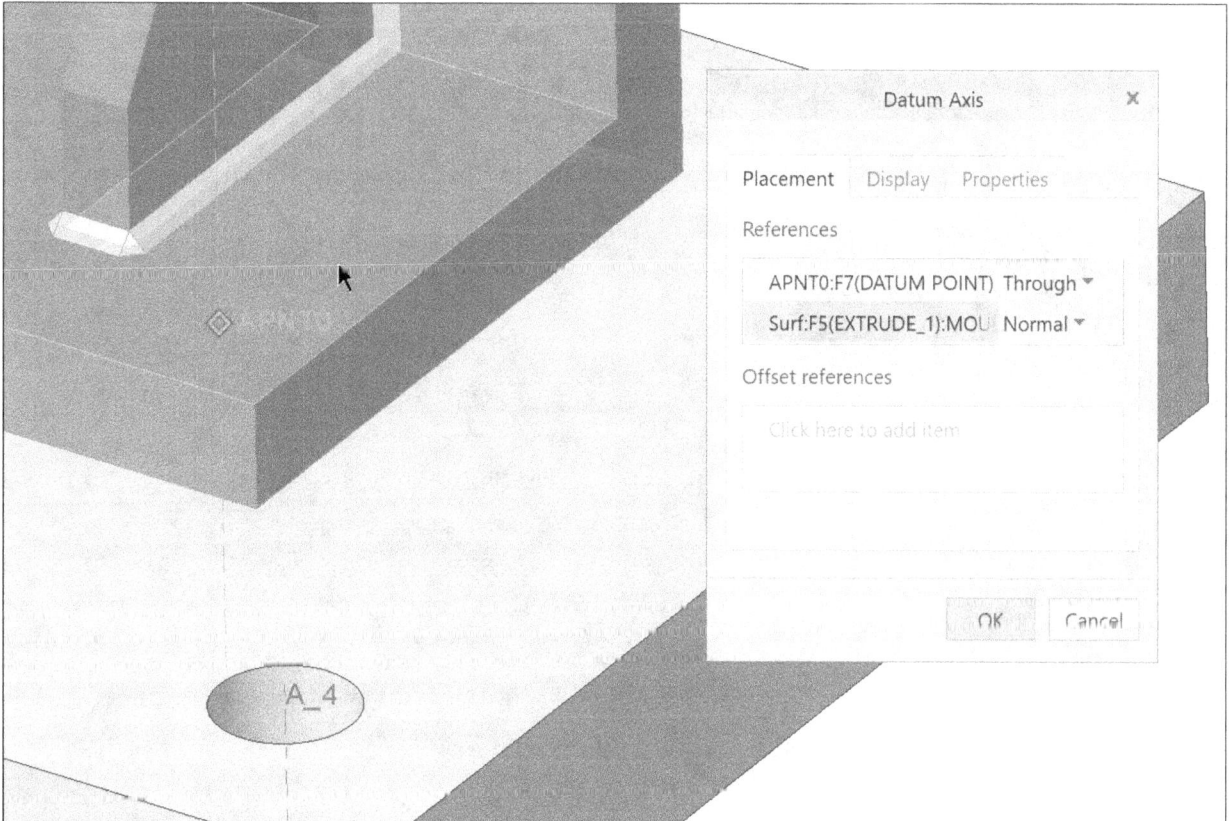

Figure 13.8(d) Select the Surface

Select **Tools tab** > 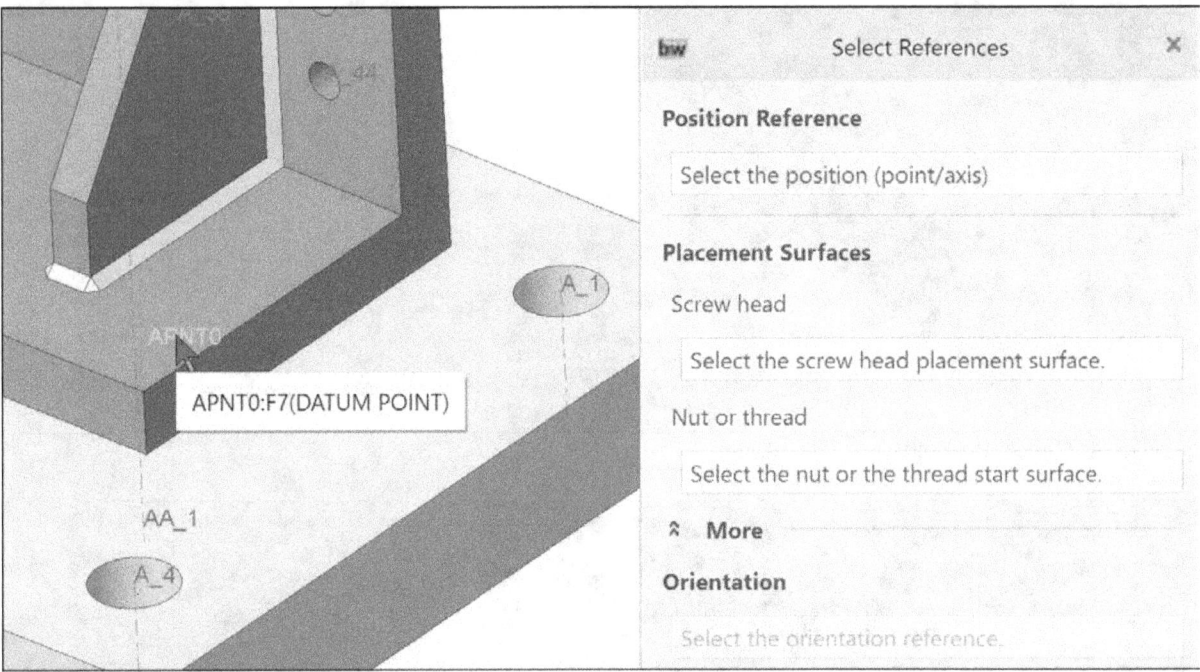 Screw Assemble on reference > select **APTN0** [Fig. 13.9(a)] > Select the screw head placement surface [Fig. 13.9(b)]

If your Creo version does not have this capability, skip the Screw commands shown here. You may add screws from the Library or model them separately.

Figure 13.9(a) Select the Point

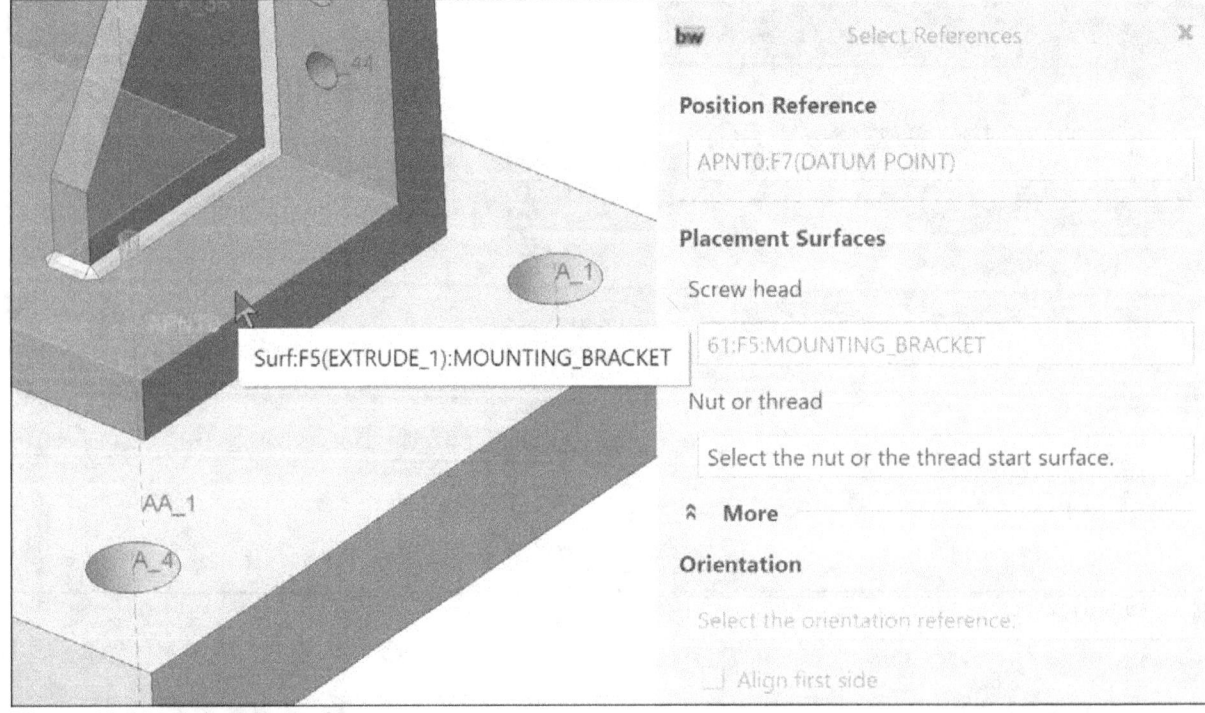

Figure 13.9(b) Select the Surface

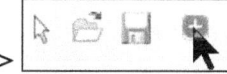
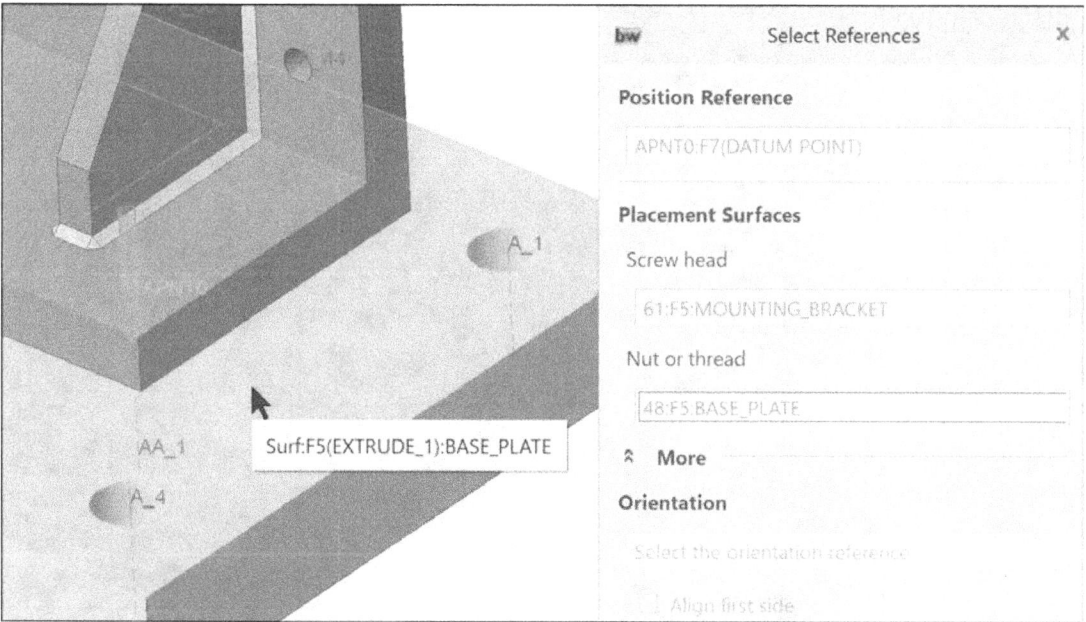

Figure 13.9(c) Select the Thread Start Surface

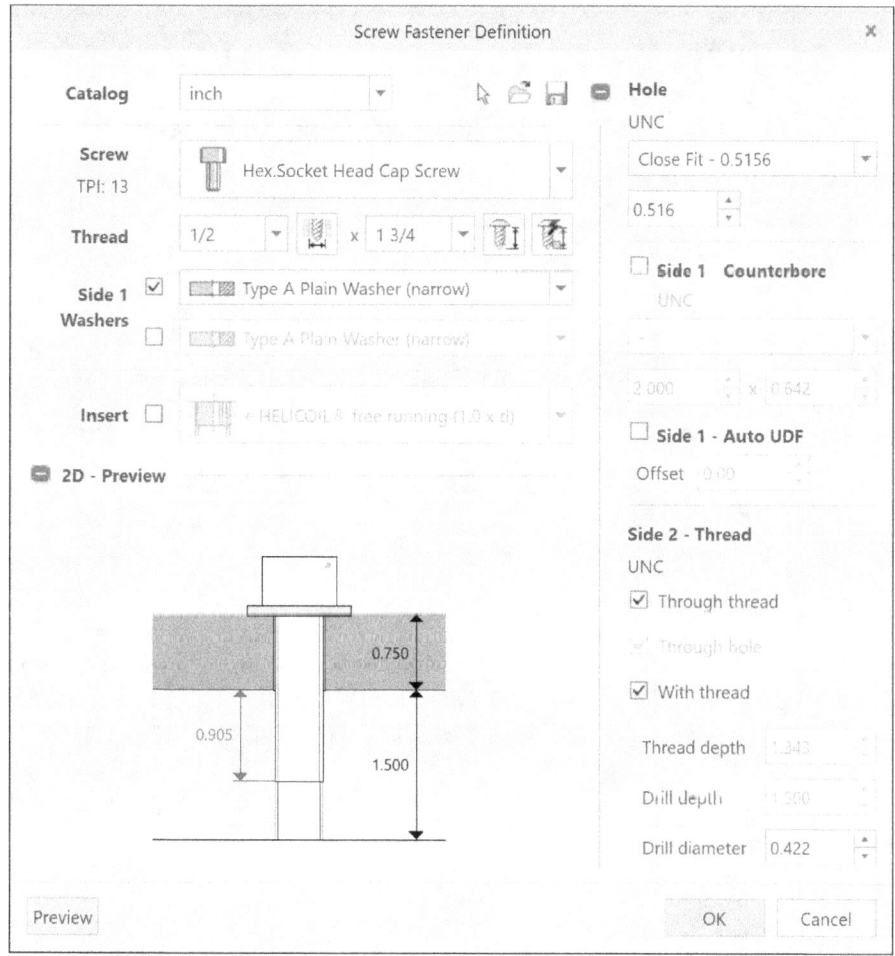

Figure 13.9(d) Screw Fastener Definition

In the Model Tree and graphics window note the features automatically generated on the parts and in the assembly.

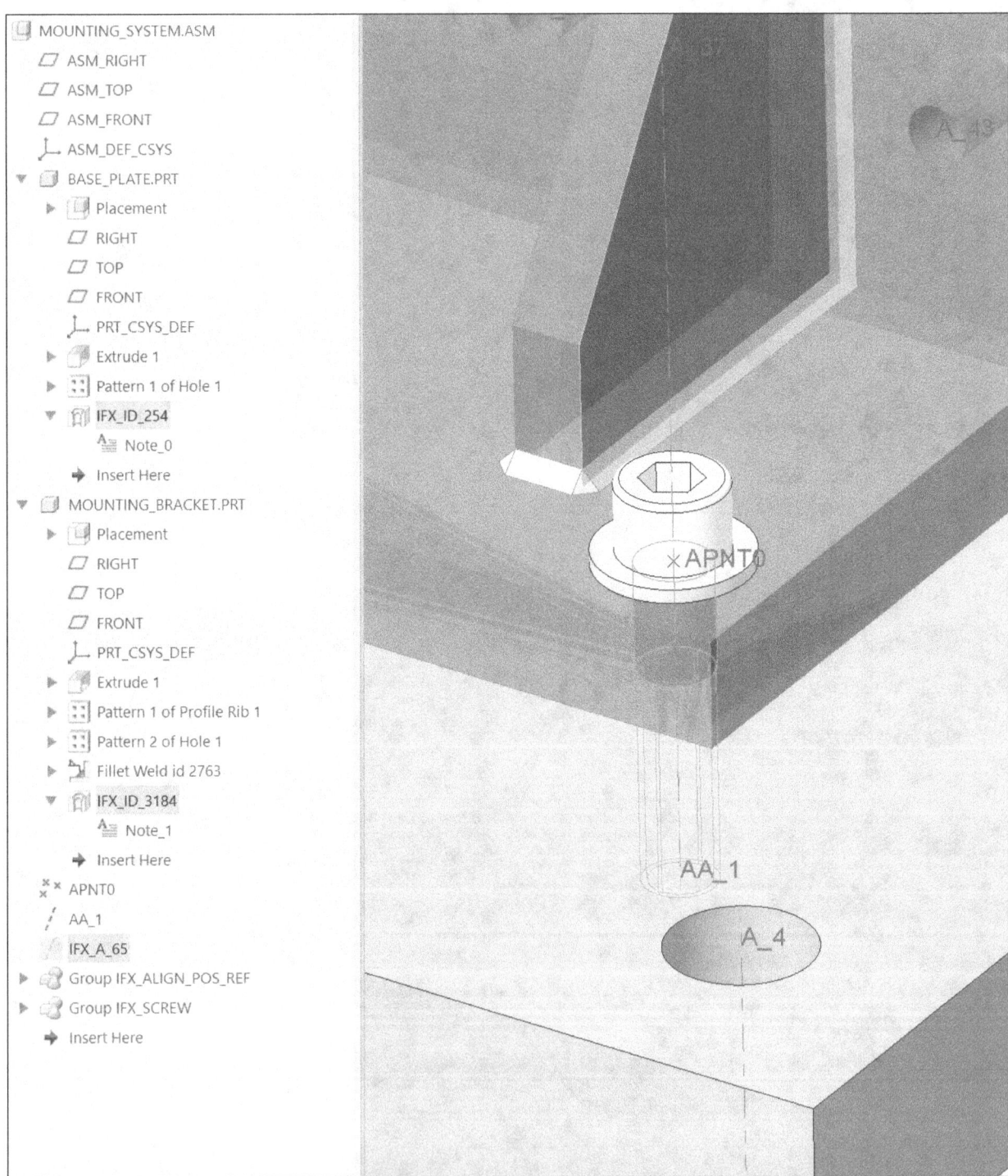

Figure 13.9(e) Holes, Threads, Screw and Washer

Select the point, axes, and groups from the Model Tree > **RMB** >  **Group** [Fig. 13.10(a)] > **RMB** > **Pattern** > select the **.750** dimension in the long direction > **4** items > select the other **.750** dimension [Fig. 13.10(b)] > ✓ > **LMB**

Figure 13.10(a) Group

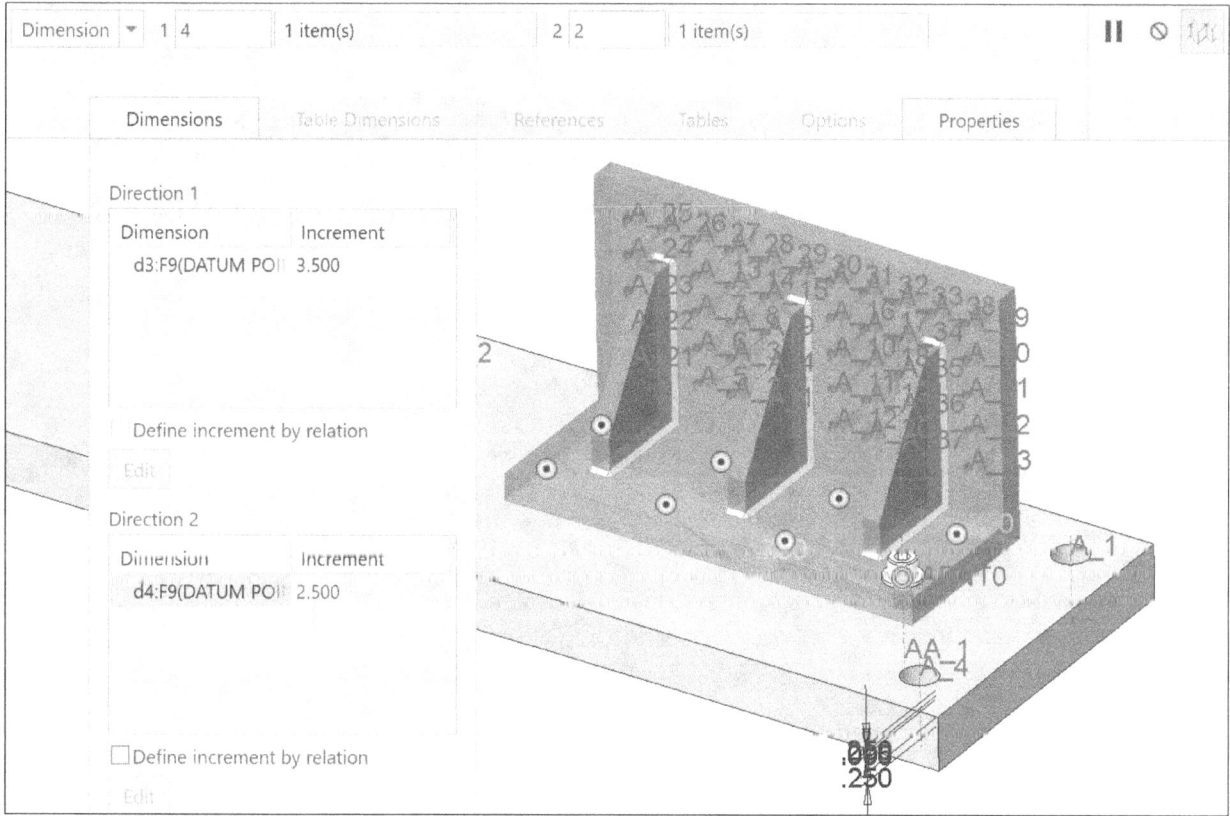

Figure 13.10(b) Pattern

Select: [IFX_ID_5623] of the Mounting Bracket in the Model Tree > **RMB** > [⊞] > **Reference** > [✓] > select [IFX_ID_253] of the Base Plate in the Model Tree > **RMB** > [⊞] [Fig. 13.10(c)] > [✓] (the axis and holes in the base plate have been added) > **LMB** > **Ctrl+S**

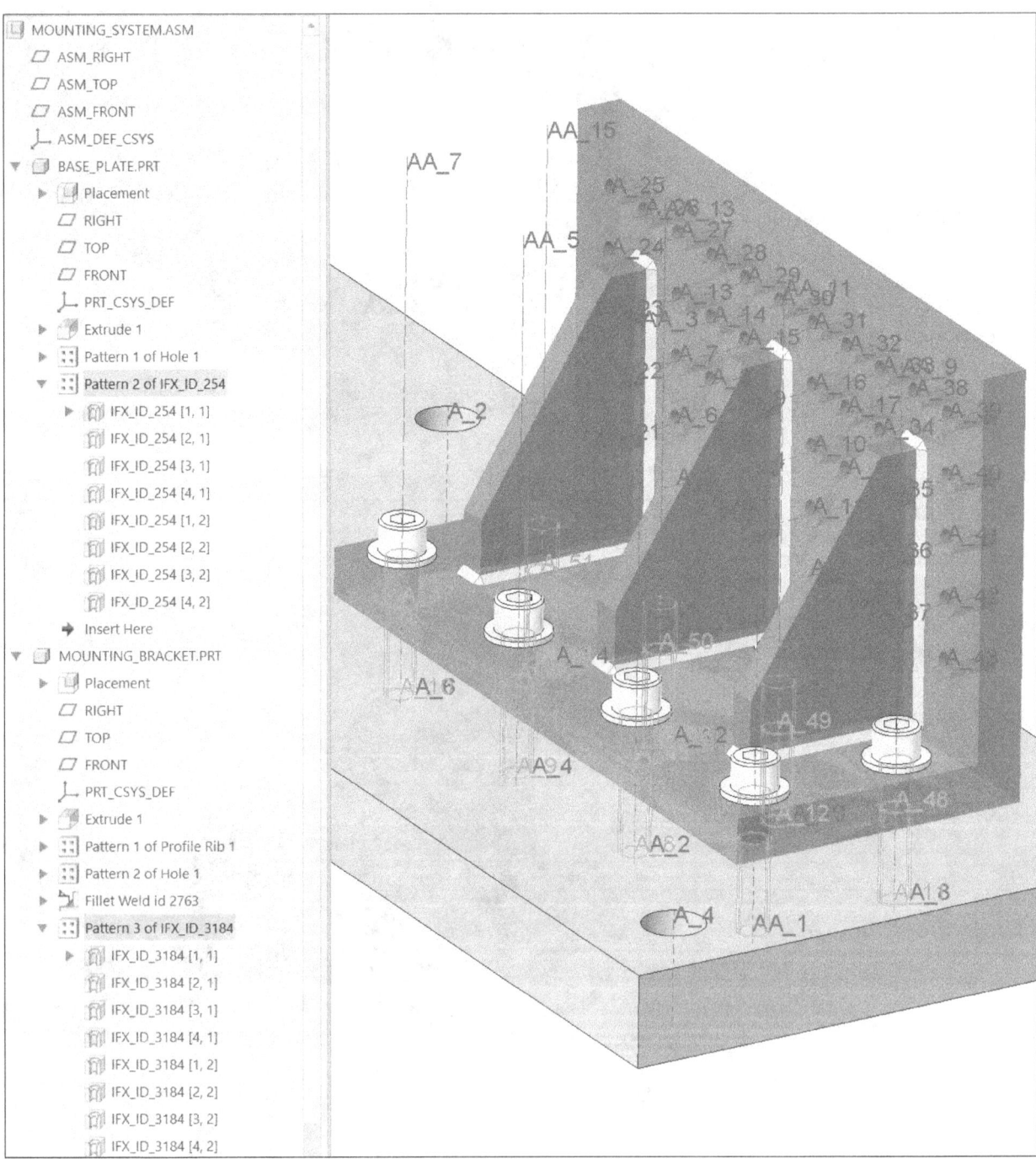

Figure 13.10(c) Pattern

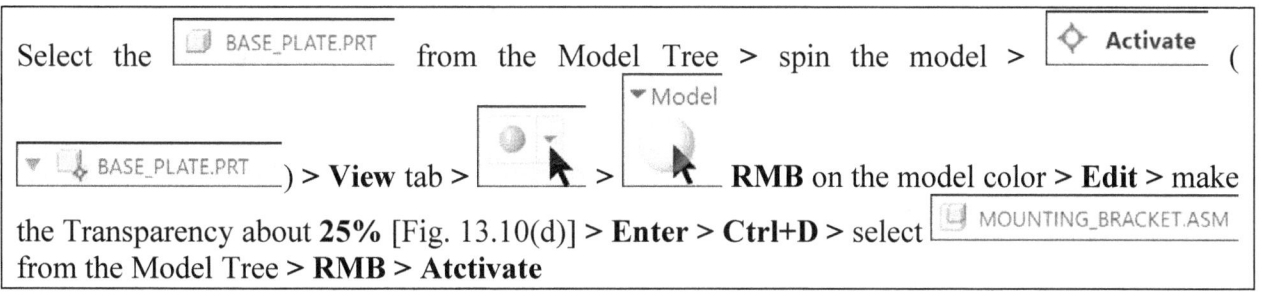

Select the `BASE_PLATE.PRT` from the Model Tree > spin the model > **Activate** (`BASE_PLATE.PRT`) > **View** tab > `Model` > `Model` **RMB** on the model color > **Edit** > make the Transparency about **25%** [Fig. 13.10(d)] > **Enter** > **Ctrl+D** > select `MOUNTING_BRACKET.ASM` from the Model Tree > **RMB** > **Atctivate**

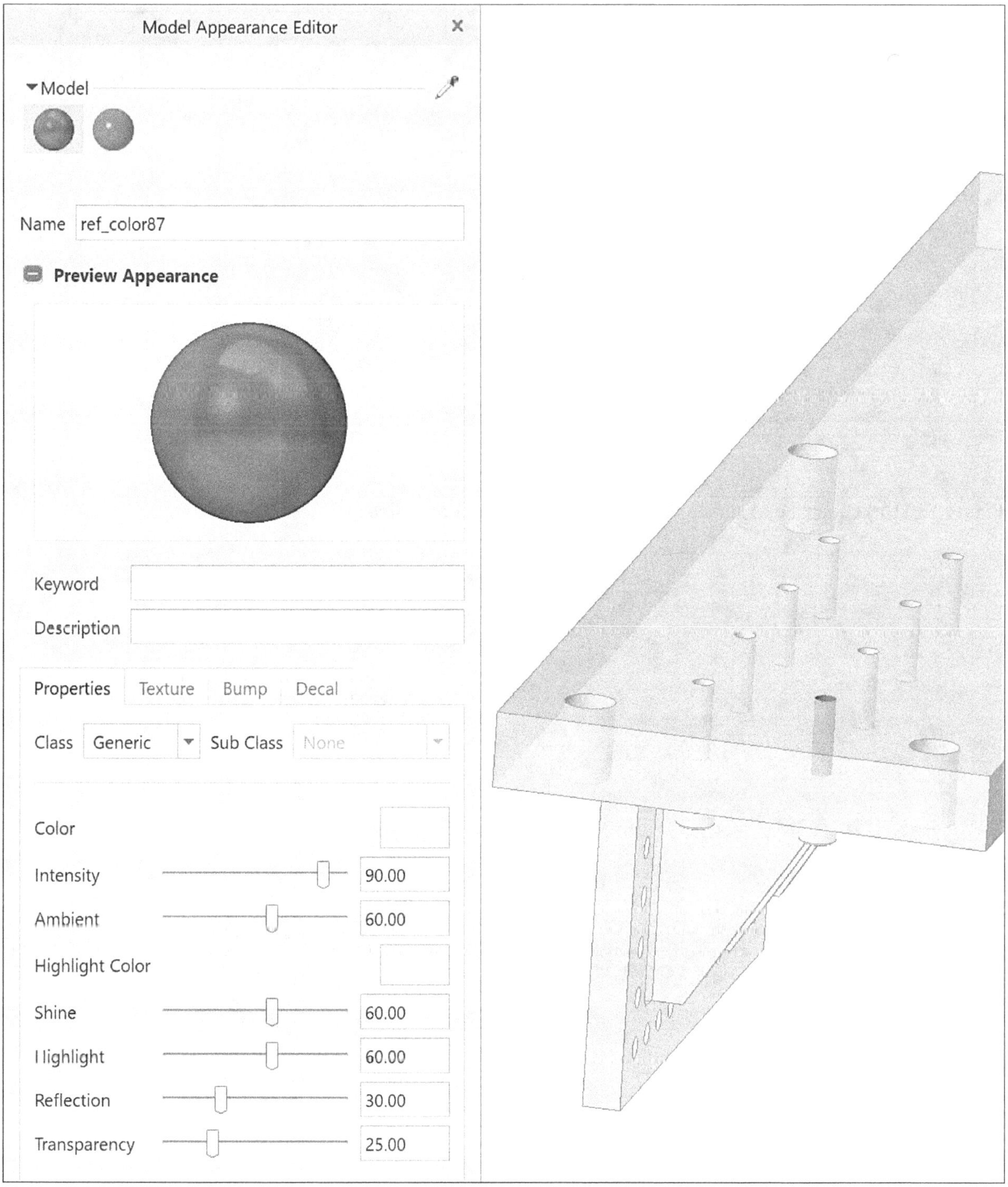

Model Appearance Editor	✕

▼ Model

Name ref_color87

Preview Appearance

Keyword

Description

Properties	Texture	Bump	Decal

Class Generic ▼ Sub Class None ▼

Color

Intensity ———————▯——— 90.00

Ambient ————▯———— 60.00

Highlight Color

Shine ————▯———— 60.00

Highlight ————▯———— 60.00

Reflection ——▯—————— 30.00

Transparency ——▯—————— 25.00

Figure 13.10(d) Transparency

613

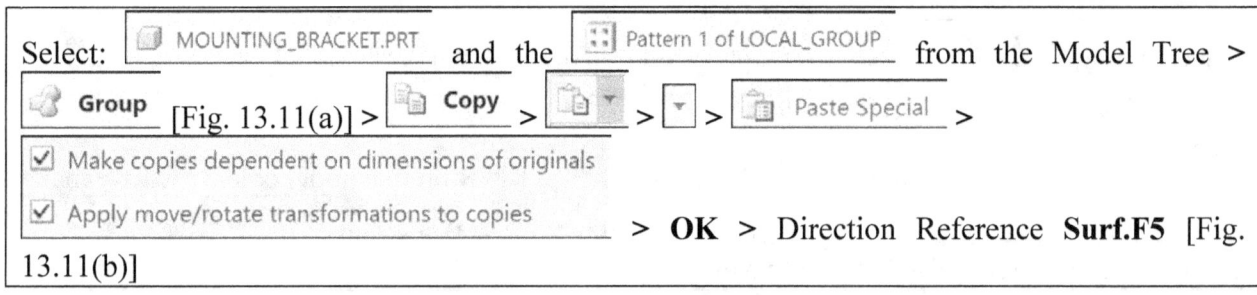

Select: ☐ MOUNTING_BRACKET.PRT and the ⟦ ⟧ Pattern 1 of LOCAL_GROUP from the Model Tree >

🔵 **Group** [Fig. 13.11(a)] > ▤ **Copy** > ▤ ▾ > ▾ > ▤ Paste Special >

☑ Make copies dependent on dimensions of originals

☑ Apply move/rotate transformations to copies > **OK** > Direction Reference **Surf.F5** [Fig. 13.11(b)]

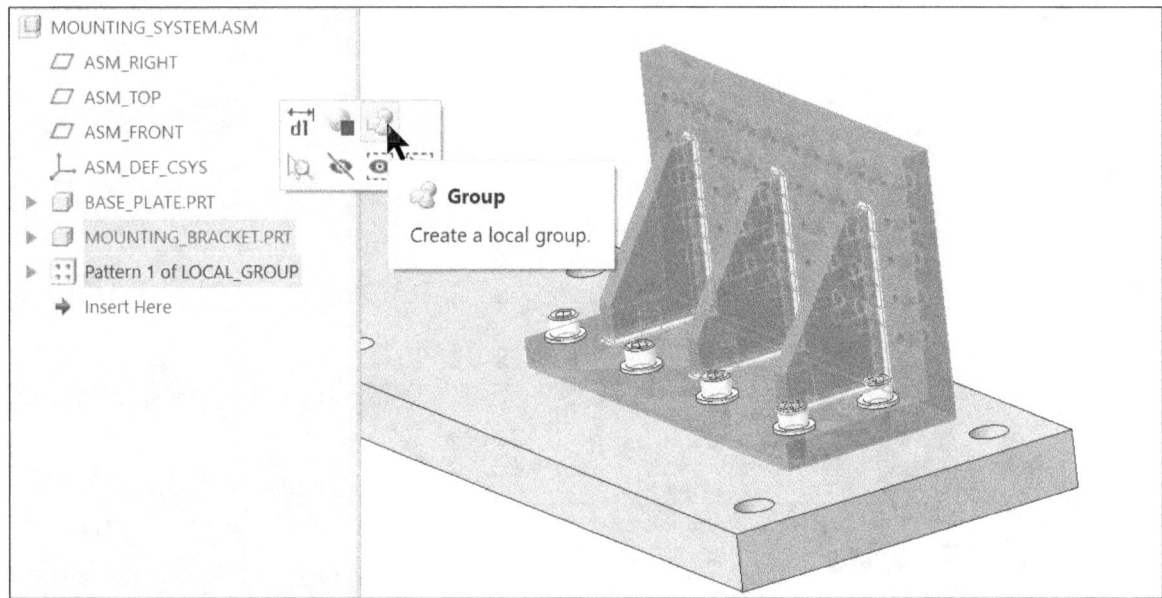

Figure 13.11(a) Group the Mounting_Bracket and Pattern 1 of Local_Group

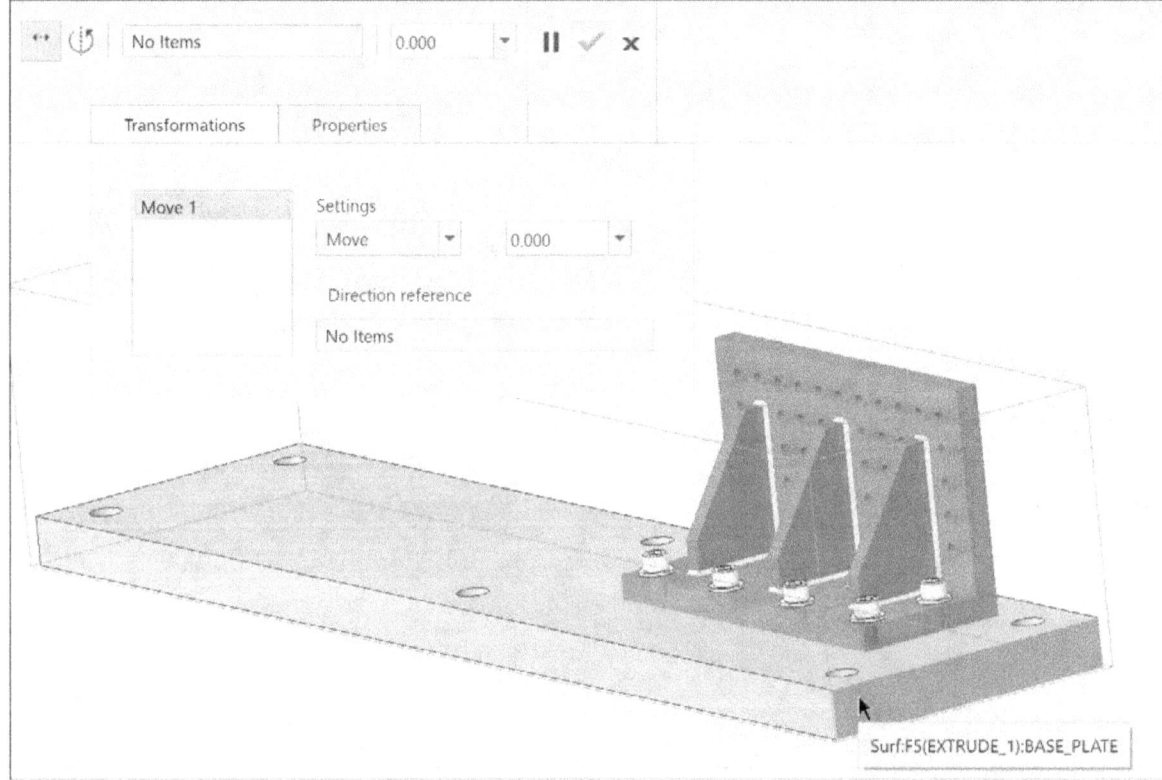

Figure 13.11(b) Copy > Paste Special > Select the Surface

Drag the Group **22** inches [Fig. 13.11(c)] > ✓ > **LMB** [Fig. 13.11(d)] > **Ctrl+S**

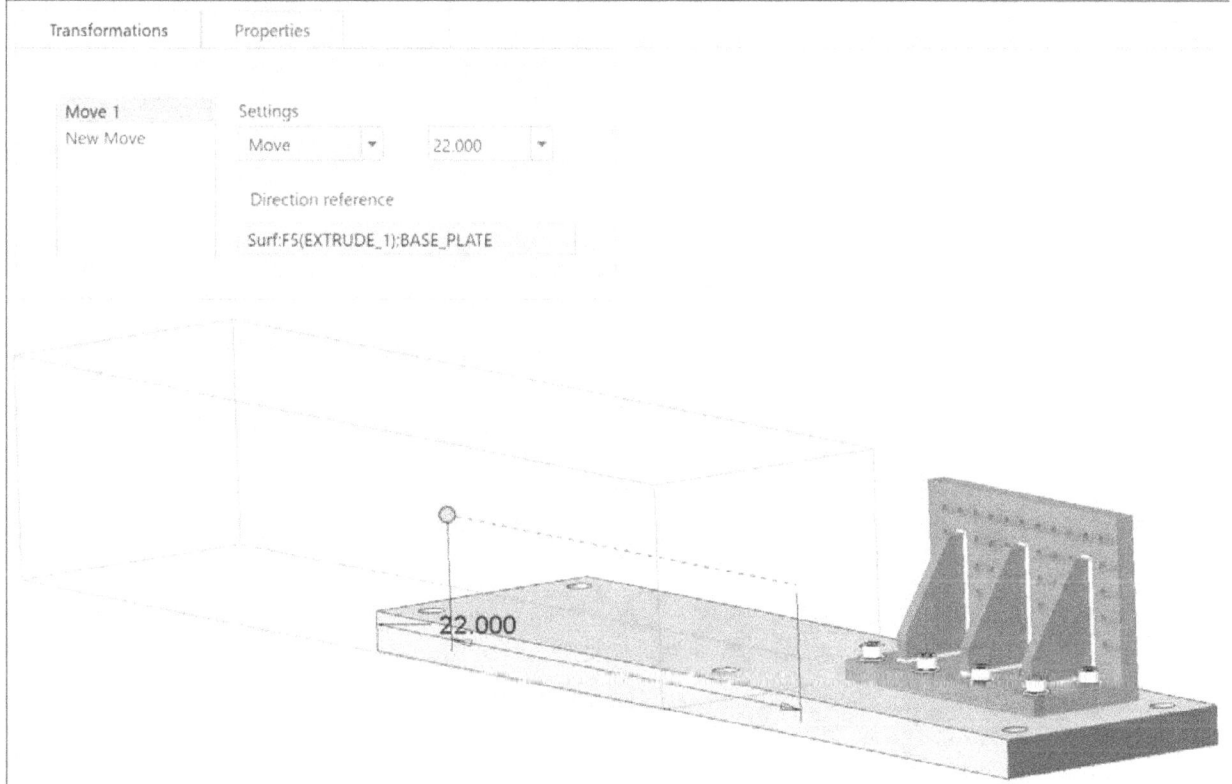

Figure 13.11(c) Drag the Group

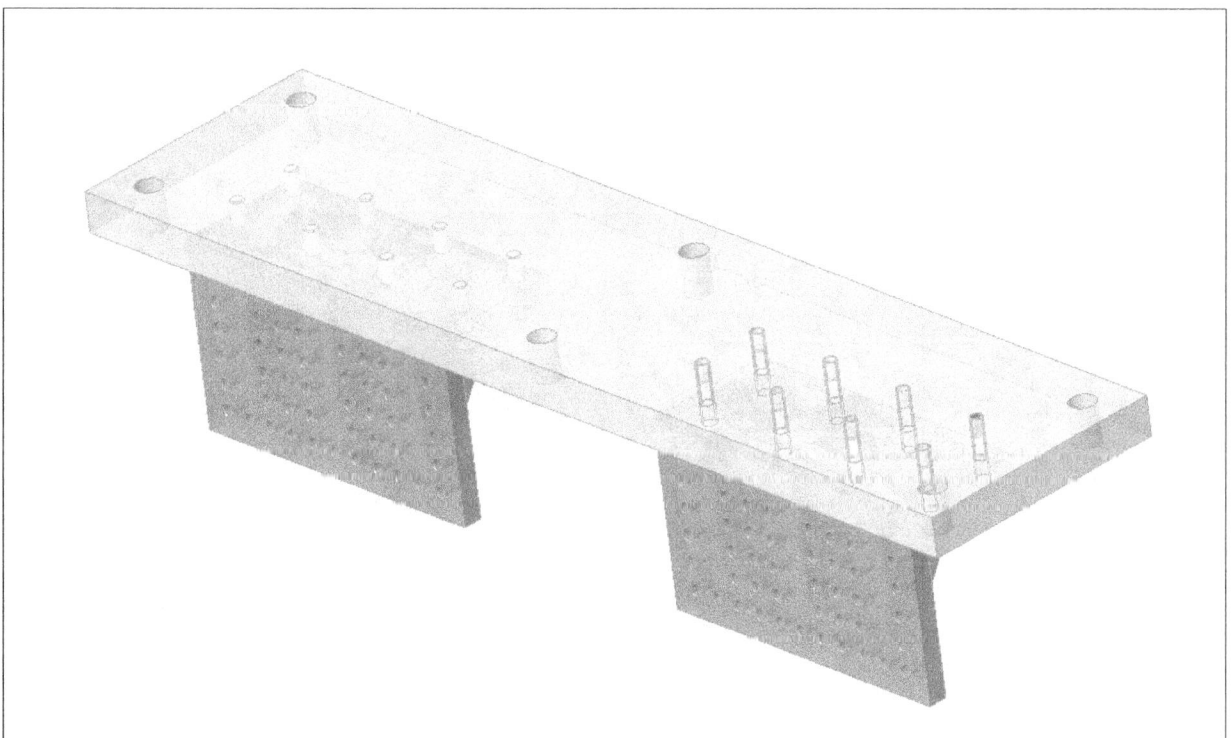

Figure 13.11(d) Assembly (holes are missing for the new copied features)

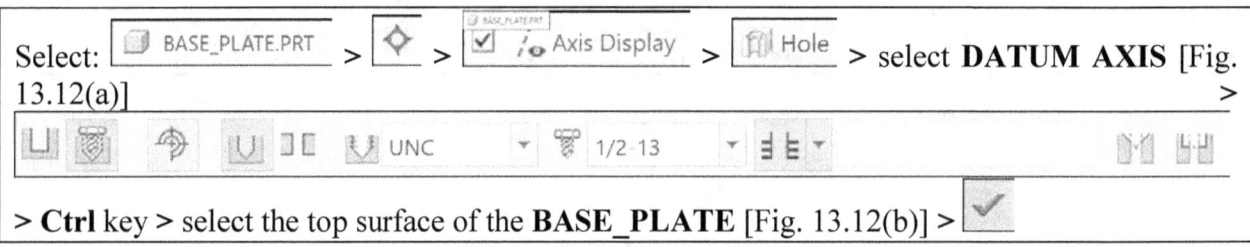

Select: [BASE_PLATE.PRT] > [◇] > [✓ /o Axis Display] > [Hole] > select **DATUM AXIS** [Fig. 13.12(a)] >

> **Ctrl** key > select the top surface of the **BASE_PLATE** [Fig. 13.12(b)] > [✓]

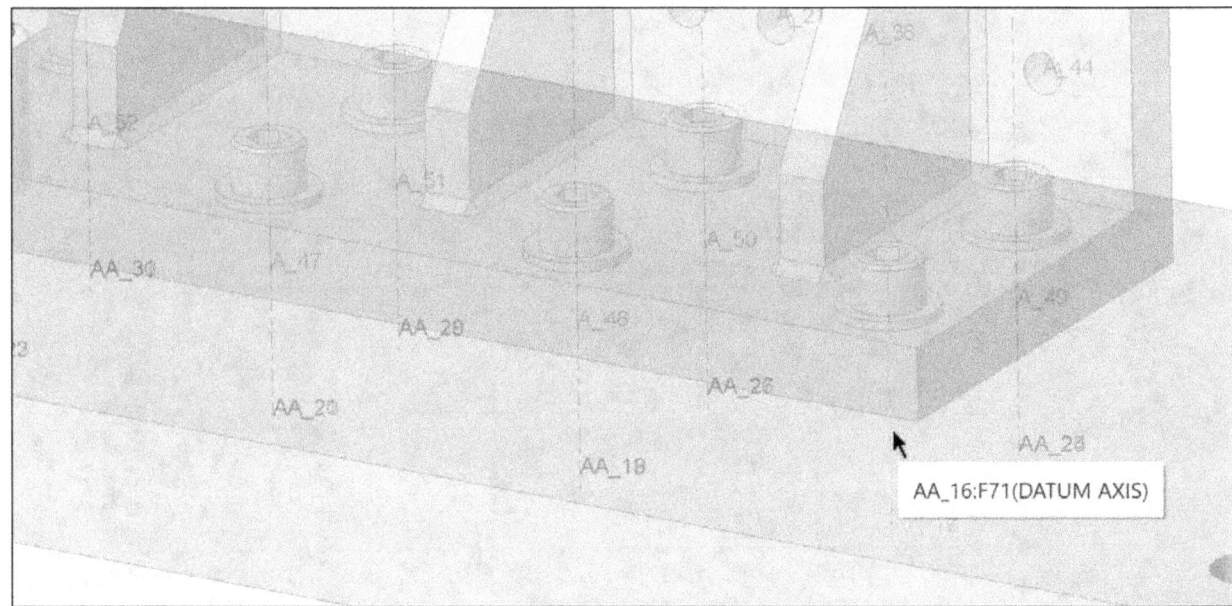

Figure 13.12(a) Select the Axis

Figure 13.12(b) Threaded Hole

616

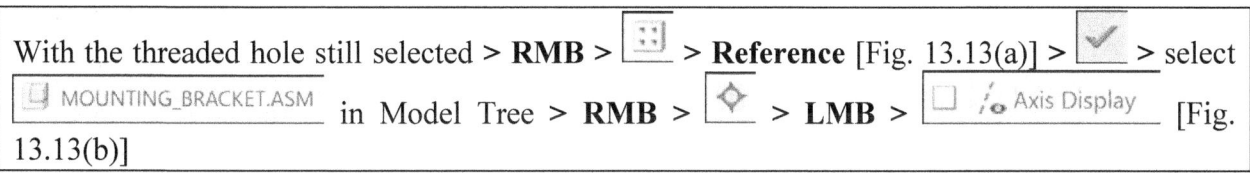

With the threaded hole still selected > **RMB** > ⊞ > **Reference** [Fig. 13.13(a)] > ✓ > select ▱ MOUNTING_BRACKET.ASM in Model Tree > **RMB** > ◇ > **LMB** > ☐ Axis Display [Fig. 13.13(b)]

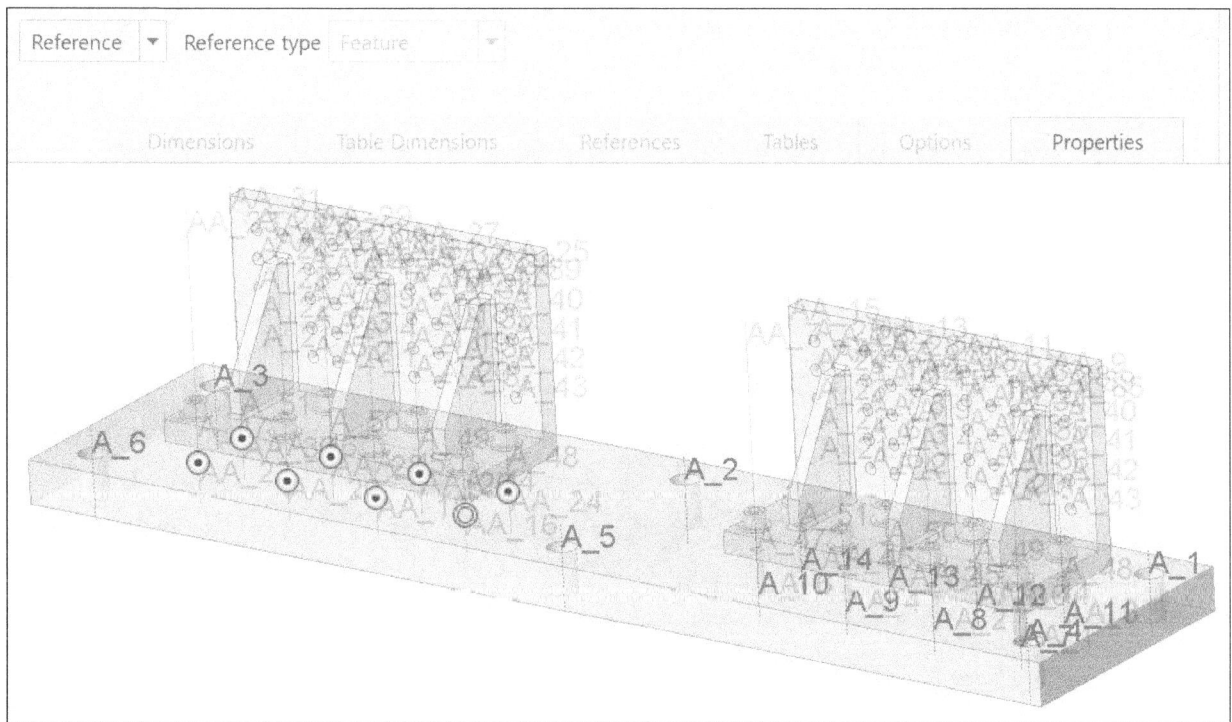

Figure 13.13(a) Pattern

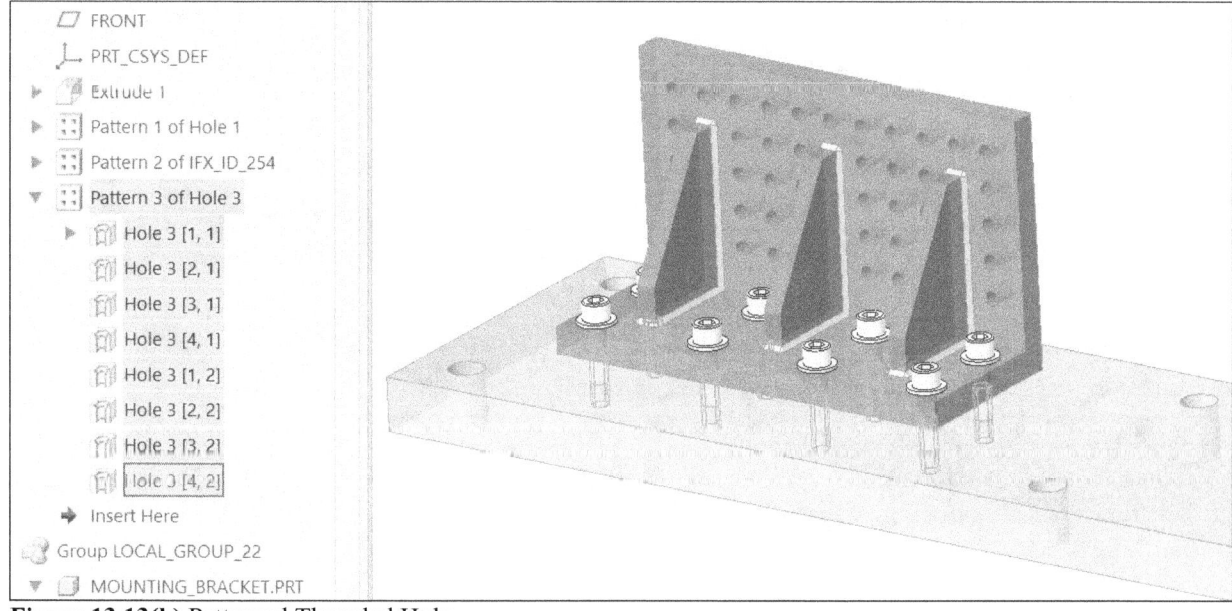

Figure 13.13(b) Patterned Threaded Hole

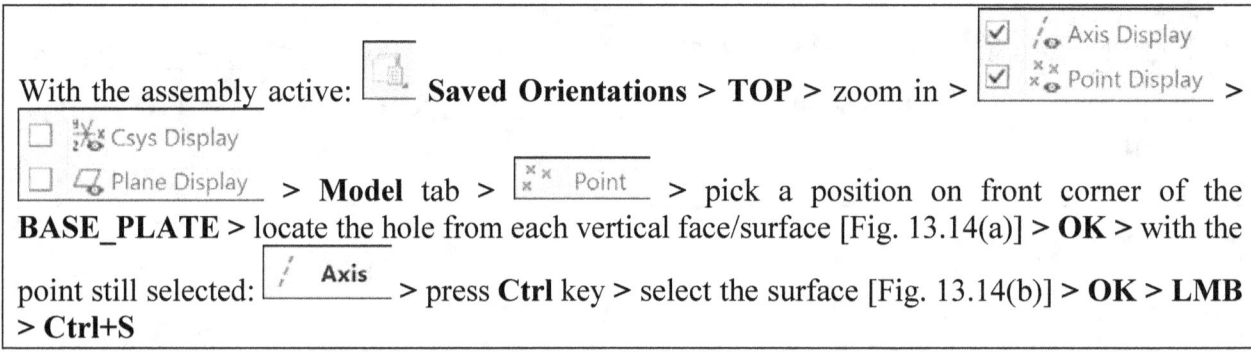

With the assembly active: **Saved Orientations > TOP** > zoom in > ☑ Axis Display ☑ Point Display > ☐ Csys Display ☐ Plane Display > **Model** tab > Point > pick a position on front corner of the **BASE_PLATE** > locate the hole from each vertical face/surface [Fig. 13.14(a)] > **OK** > with the point still selected: Axis > press **Ctrl** key > select the surface [Fig. 13.14(b)] > **OK** > **LMB** > **Ctrl+S**

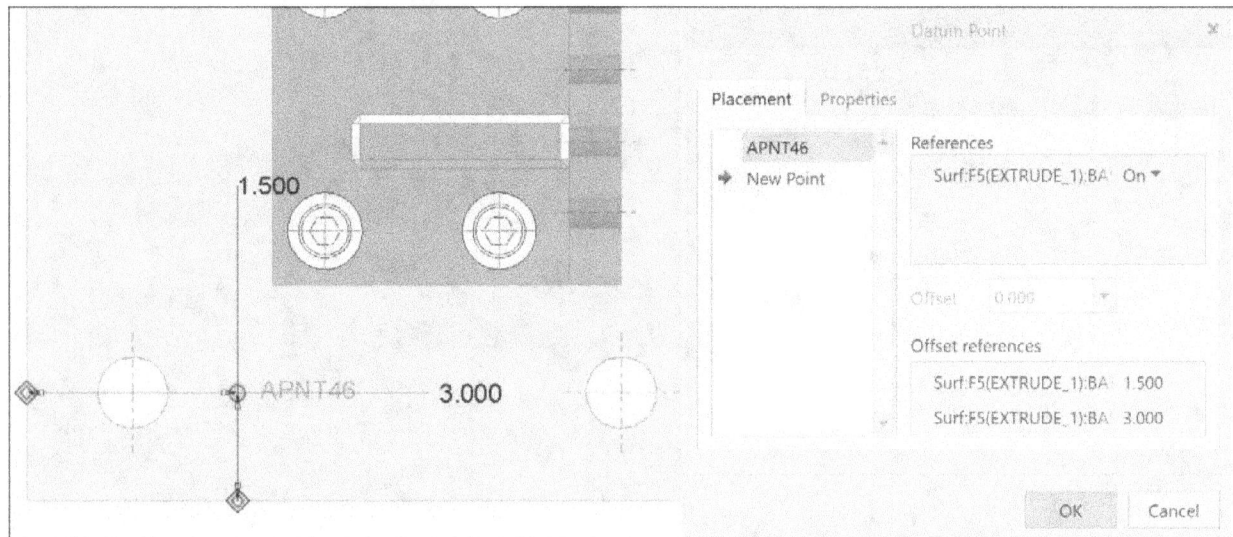

Figure 13.14(a) Datum Point

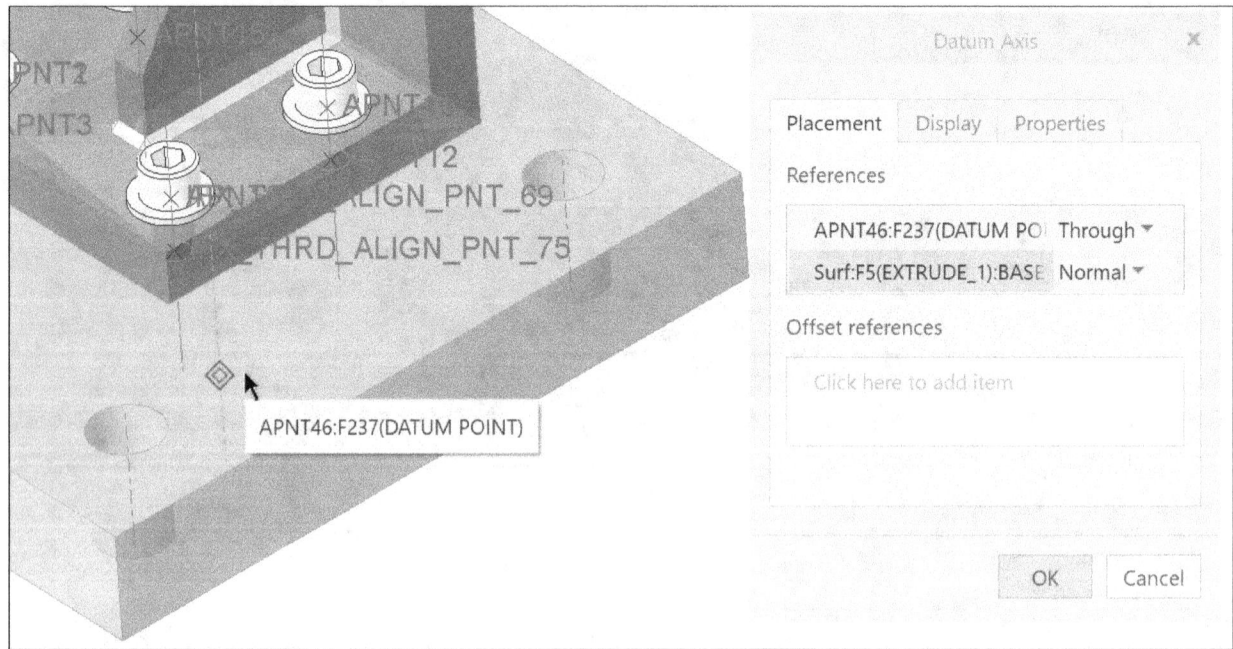

Figure 13.14(b) Datum Axis

Select **Tools** tab > Dowel Pin ▾ **Dowel** > Assemble on reference > Select the position (point/axis) **APTN16 >** > Select the dowel pin placement surface [Fig. 13.15(a)] > **OK** > Set the Dowel Pin Fastener Definition > ⊞ **Show/Hide Hole Layout** [Fig.13.15(b)] > **OK** > **Ctrl+S**

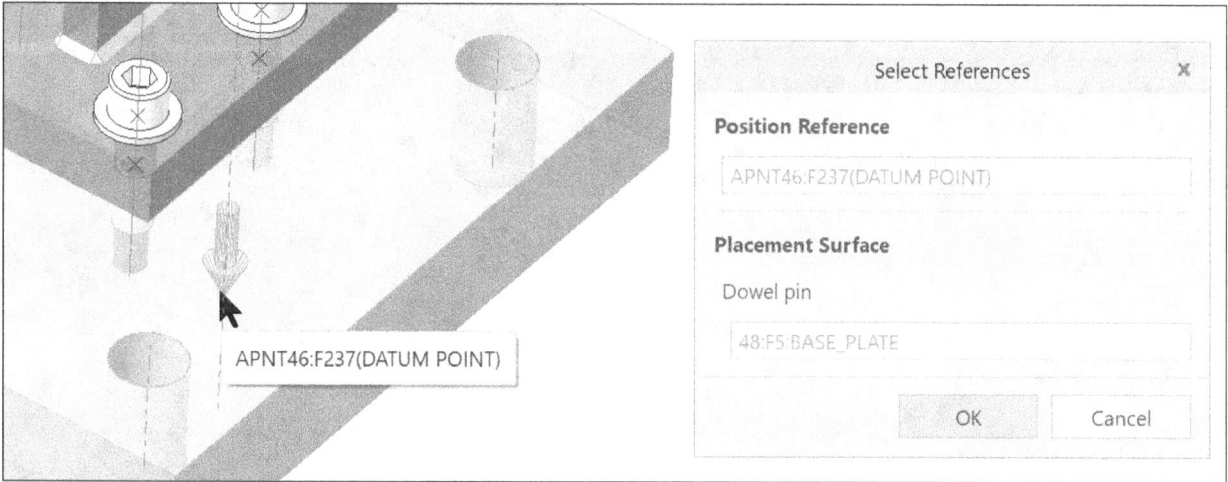

Figure 13.15(a) Point and Surface Selection for Dowel Placement

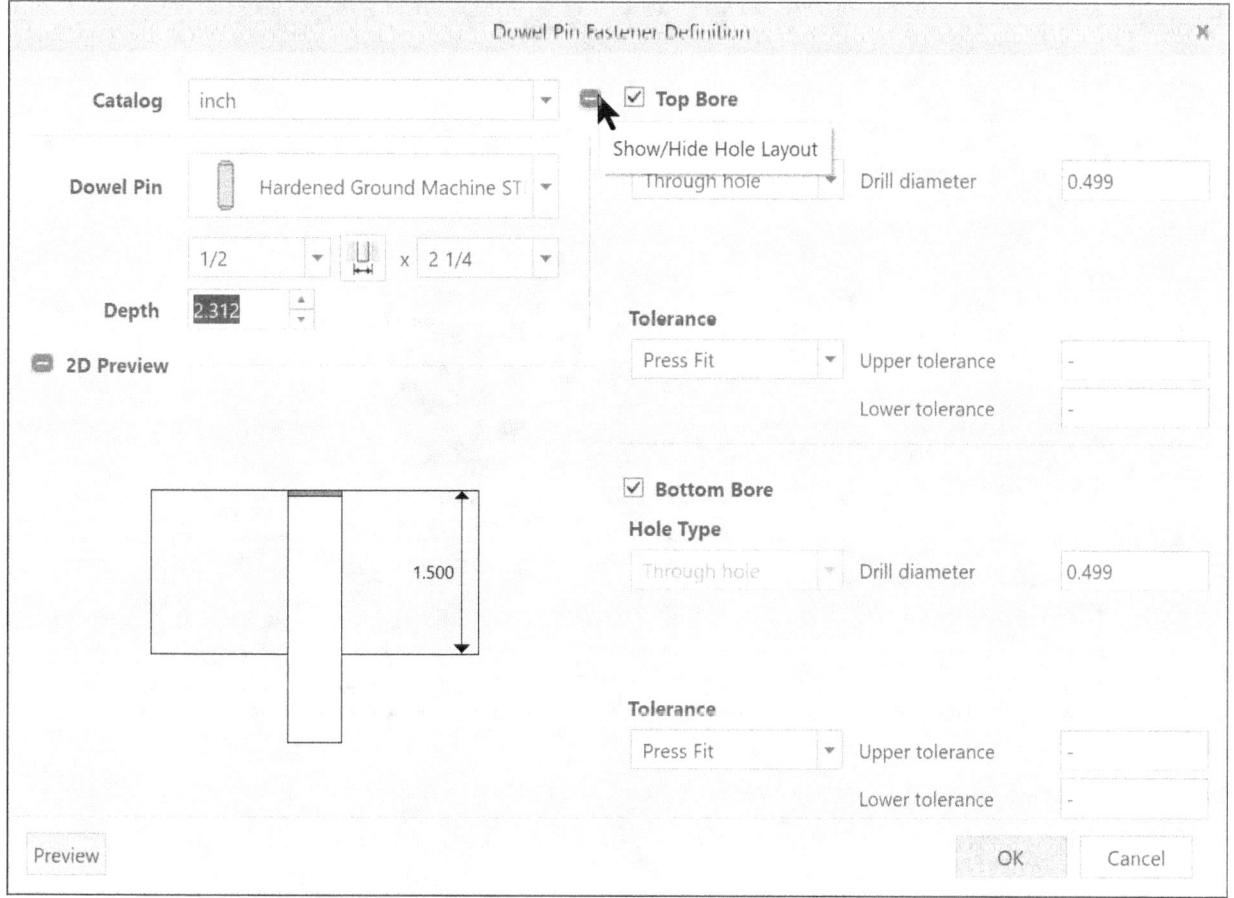

Figure 13.15(b) Dowel Pin Fastener Definition

Select the new features and create a group > **Model** tab > **Pattern** Fig.13.16(b)] > ☑ > **Ctrl+S**

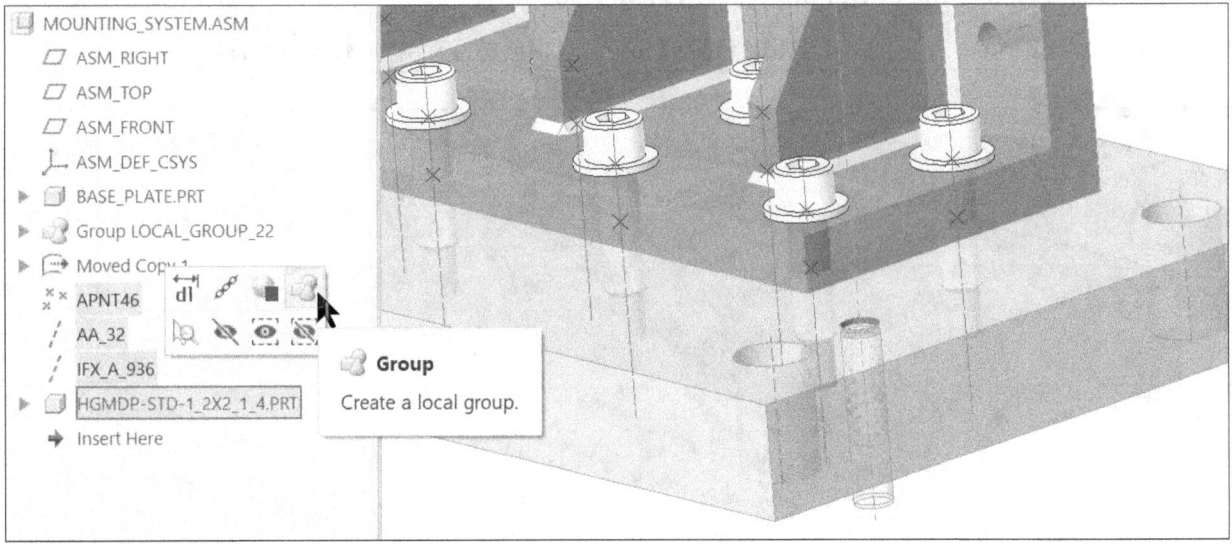

Figure 13.16(a) Group

Figure 13.16(b) Group Pattern (2 of 4 positions)

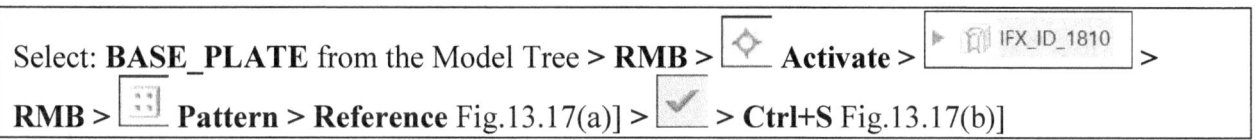

Select: **BASE_PLATE** from the Model Tree > **RMB** > ◇ **Activate** > ▸ 🗇 IFX_ID_1810 > **RMB** > ▦ **Pattern** > **Reference** Fig.13.17(a)] > ✓ > **Ctrl+S** Fig.13.17(b)]

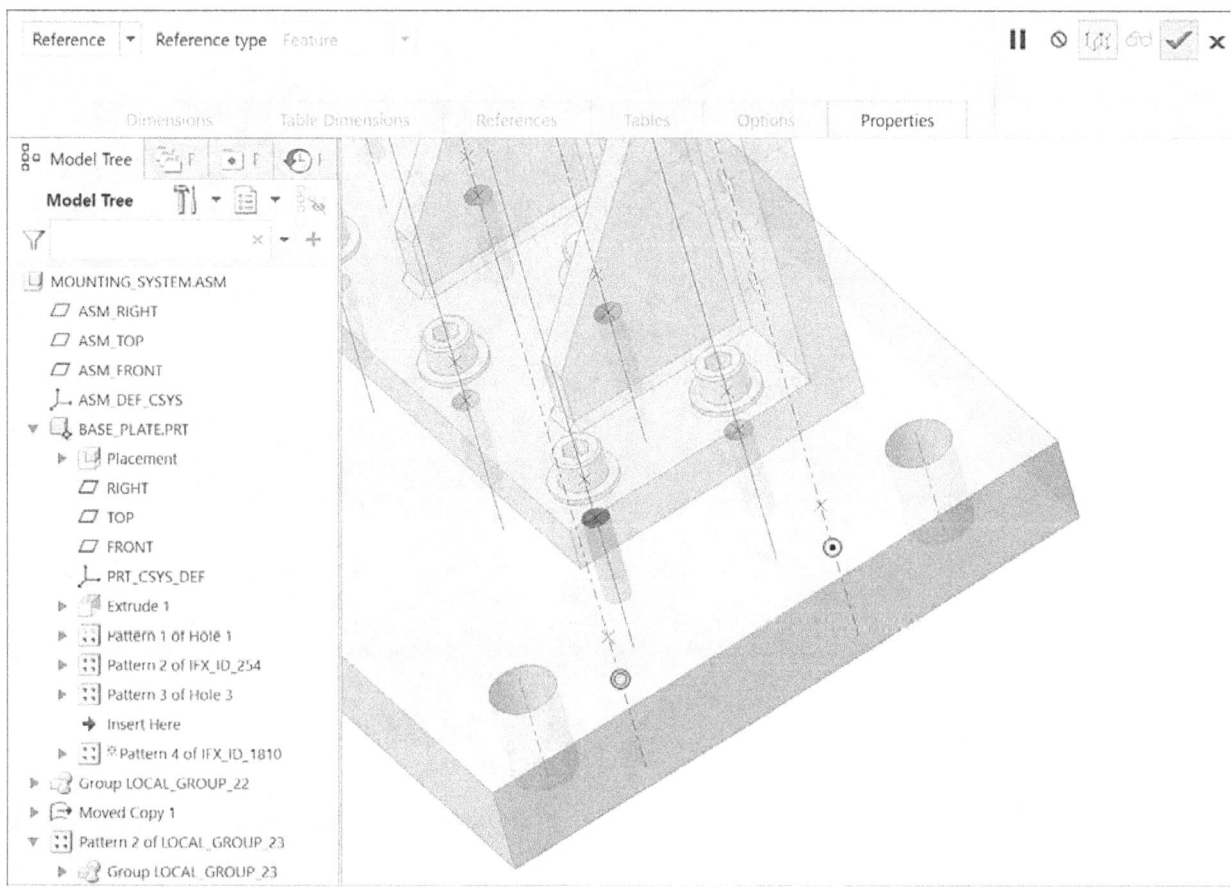

Figure 13.17(a) Pattern the Hole

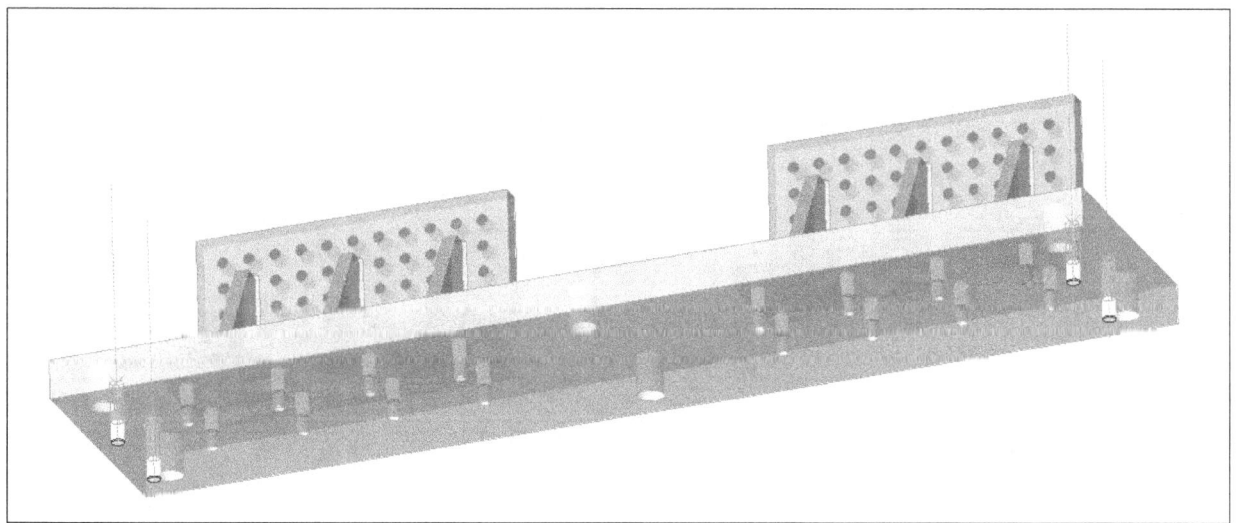

Figure 13.17(a) Dowels

621

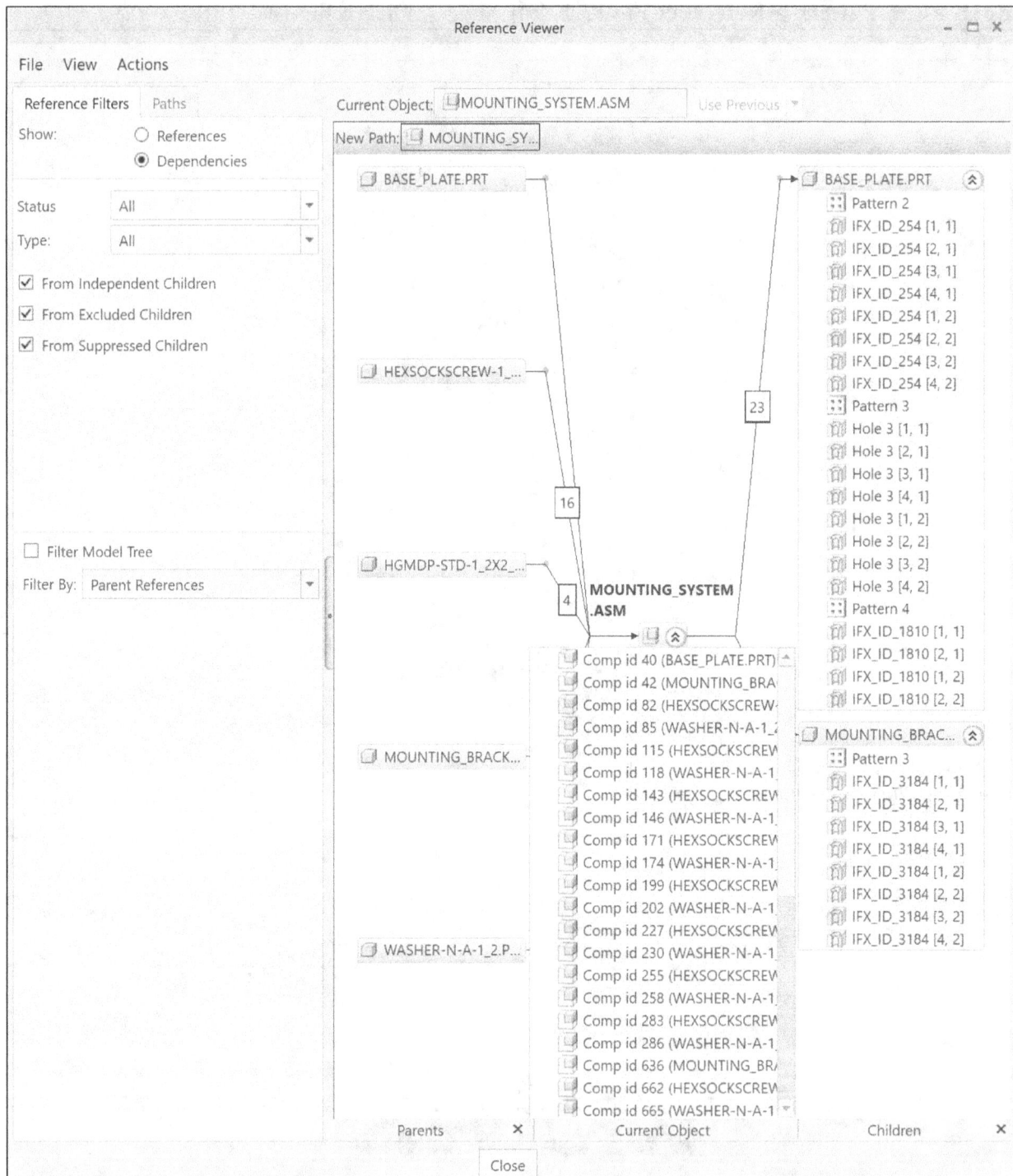

Figure 13.18 Reference Viewer

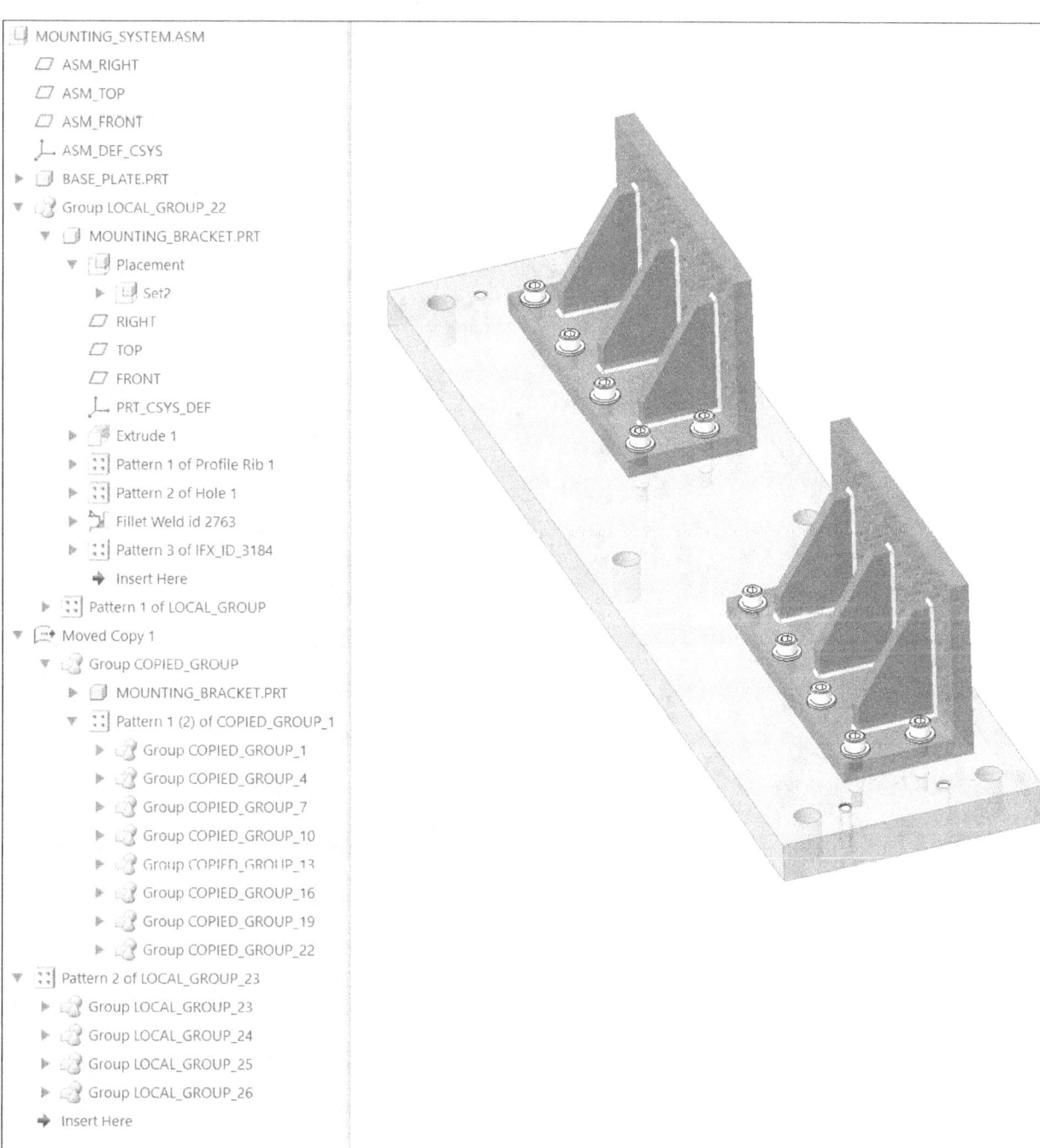

- MOUNTING_SYSTEM.ASM
  - ASM_RIGHT
  - ASM_TOP
  - ASM_FRONT
  - ASM_DEF_CSYS
  - ▶ BASE_PLATE.PRT
  - ▼ Group LOCAL_GROUP_22
    - ▼ MOUNTING_BRACKET.PRT
      - ▼ Placement
        - ▶ Set2
      - RIGHT
      - TOP
      - FRONT
      - PRT_CSYS_DEF
      - ▶ Extrude 1
      - ▶ Pattern 1 of Profile Rib 1
      - ▶ Pattern 2 of Hole 1
      - ▶ Fillet Weld id 2763
      - ▶ Pattern 3 of IFX_ID_3184
      - ➜ Insert Here
    - ▶ Pattern 1 of LOCAL_GROUP
  - ▼ Moved Copy 1
    - ▼ Group COPIED_GROUP
      - ▶ MOUNTING_BRACKET.PRT
      - ▼ Pattern 1 (2) of COPIED_GROUP_1
        - ▶ Group COPIED_GROUP_1
        - ▶ Group COPIED_GROUP_4
        - ▶ Group COPIED_GROUP_7
        - ▶ Group COPIED_GROUP_10
        - ▶ Group COPIED_GROUP_13
        - ▶ Group COPIED_GROUP_16
        - ▶ Group COPIED_GROUP_19
        - ▶ Group COPIED_GROUP_22
  - ▼ Pattern 2 of LOCAL_GROUP_23
    - ▶ Group LOCAL_GROUP_23
    - ▶ Group LOCAL_GROUP_24
    - ▶ Group LOCAL_GROUP_25
    - ▶ Group LOCAL_GROUP_26
    - ➜ Insert Here

Figure 13.19 Temporary Shading

Ctrl+N > **Drawing** > **mounting_system** > Template **Browse** > from Lesson 12 select **ASM_FORMAT_E** > **OK** > delete the top view > **SCALE .5** > add a **Projection View** from the **Front** view > **RMB** > **Sheet Setup** > **Browse** > **E Size** > **OK** [Fig. 13.20(a)] > add a **New Sheet** and **General View** > **SCALE .75** [Fig. 13.20(a)]

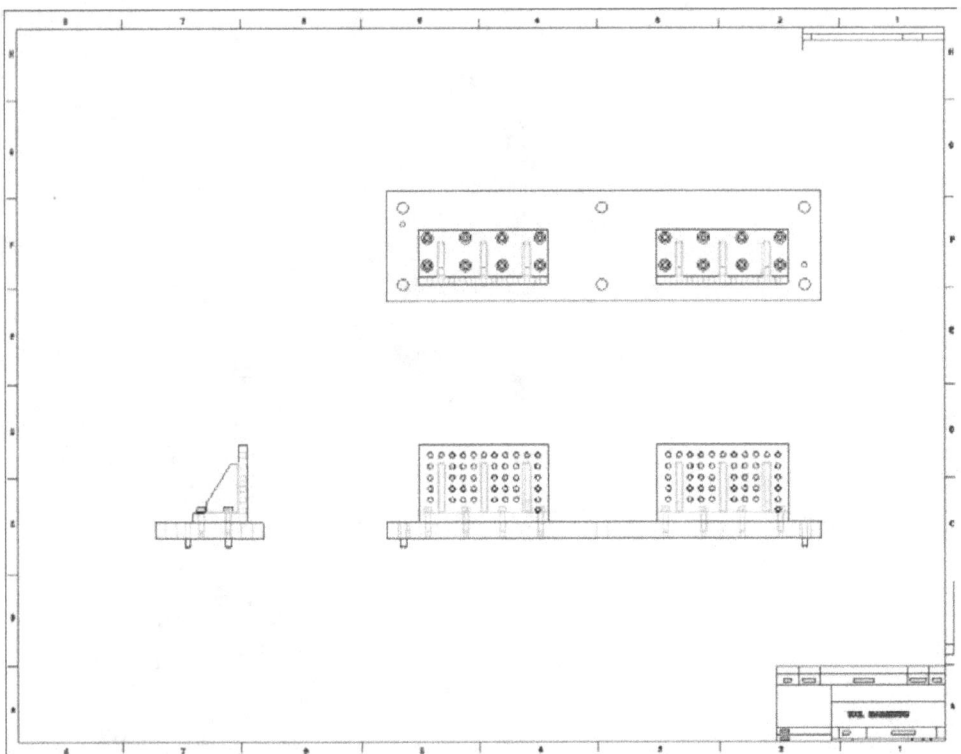

Figure 13.20(a) Sheet 1

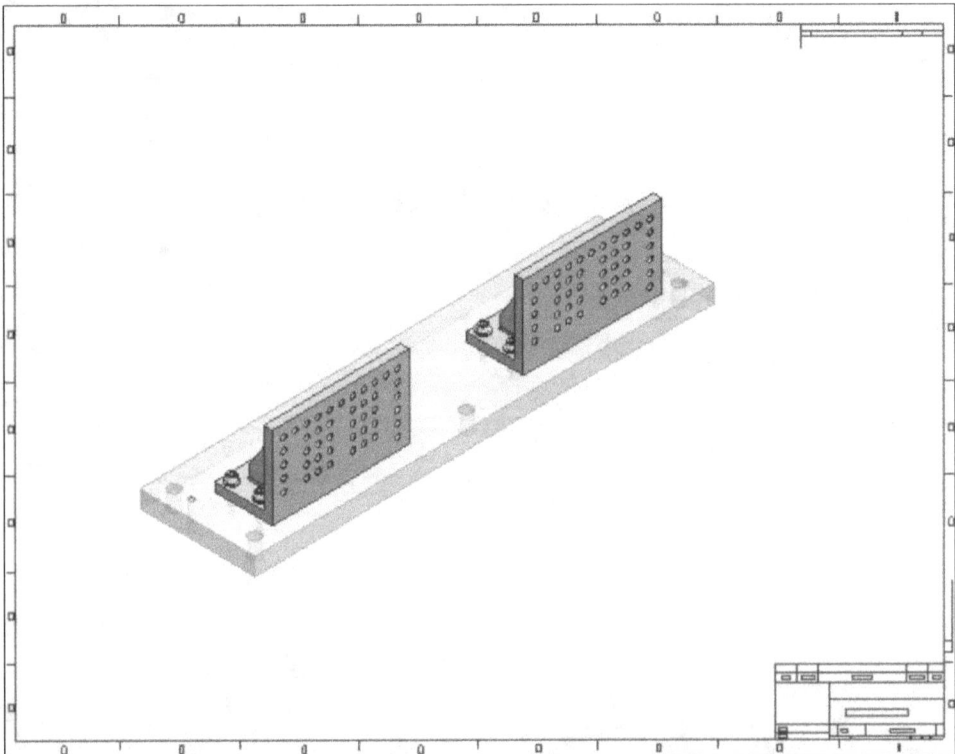

Figure 13.20(b) Sheet 1

624

Double click on the view: **Angles > Edge/Axis** > select a vertical edge reference > **180** [Fig. 13.21(a)]

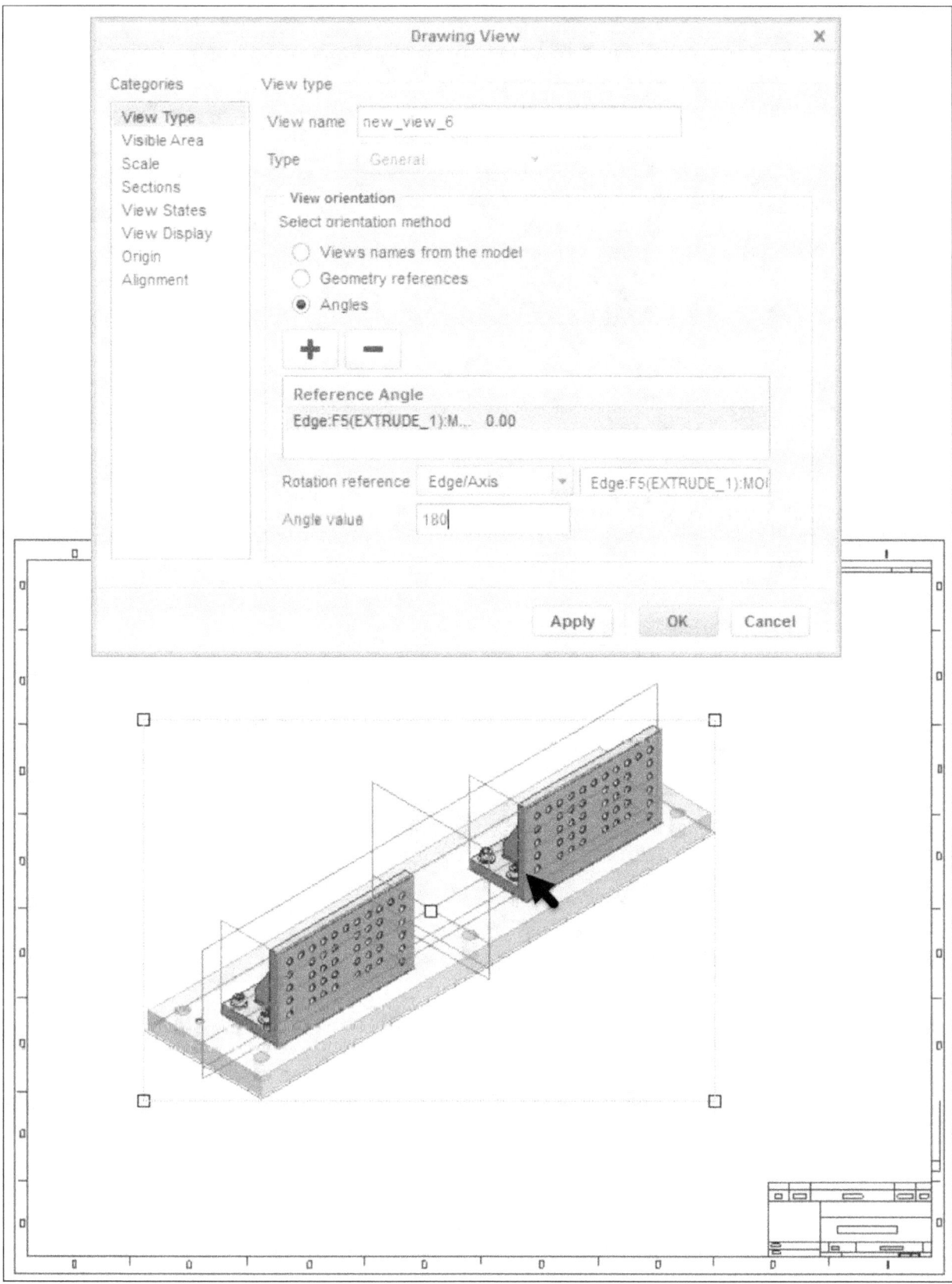

Figure 13.21(a) Reorient the Model

Click: **Apply** [Fig. 13.21(b)] **> OK > Ctrl+S**

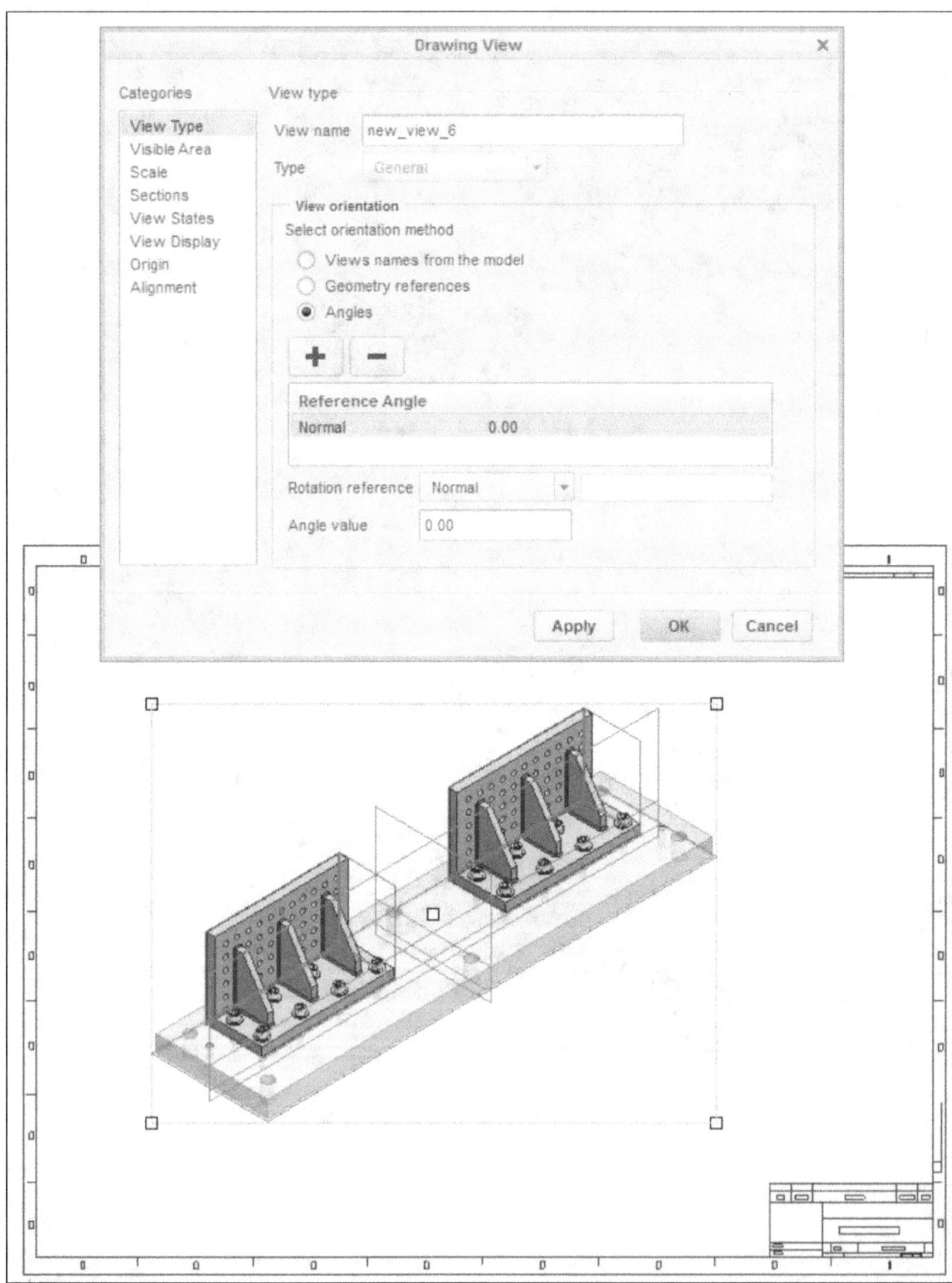

Figure 13.21(a) Reoriented View

Ctrl+S > File > Manage File > Delete Old Versions > Enter > File > Save As > Type ▢ **> Zip File (*.zip)** [Fig. 13.22(a)] *(The Zipped file includes all components, the assembly, and the drawing.)* **> OK > upload** the zip file to your course interface or attach to an email and send to your instructor and/or yourself [Fig. 13.22(b)] **> File > Close > File > Exit >Yes**

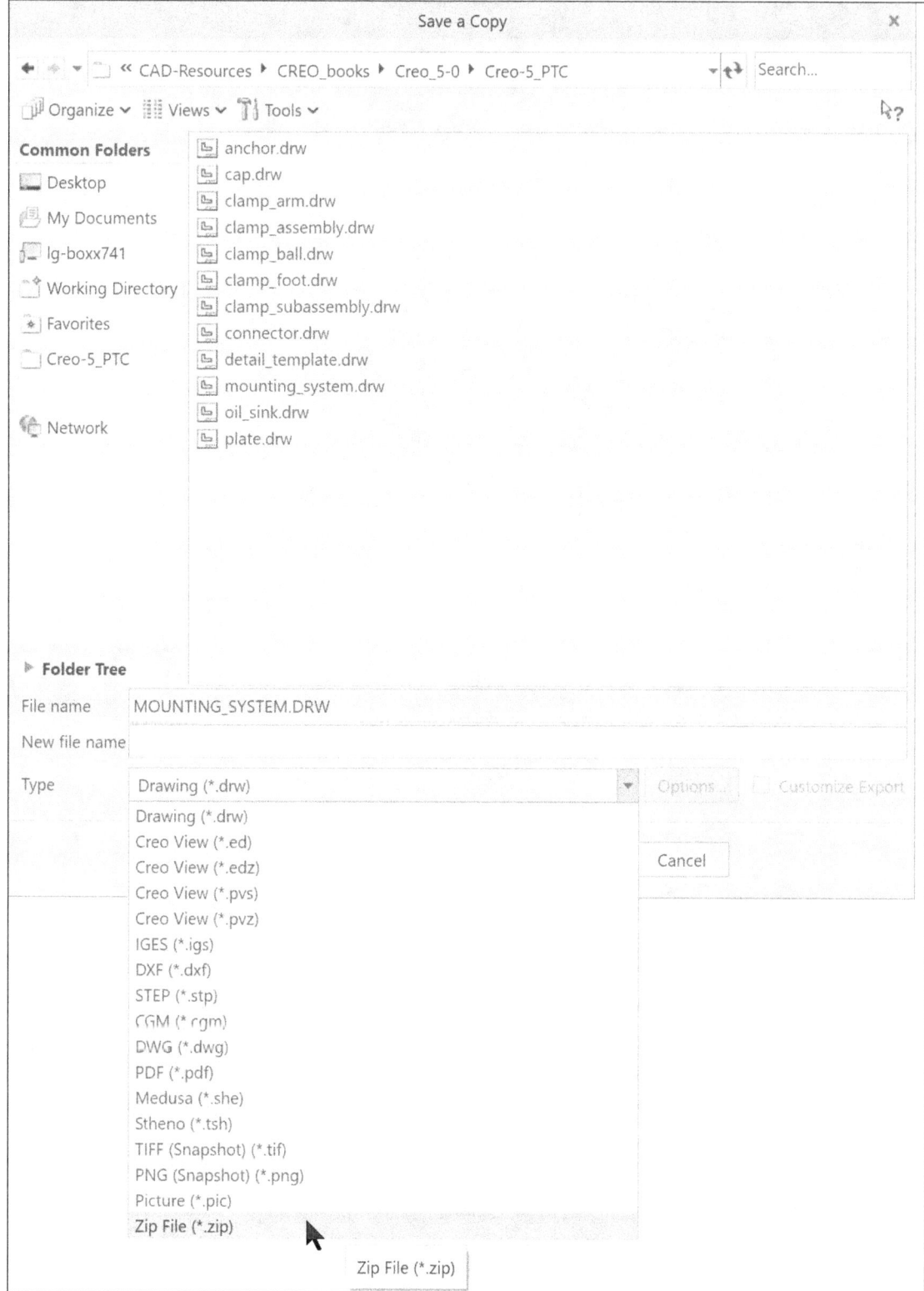

Figure 13.22(a) Zip File

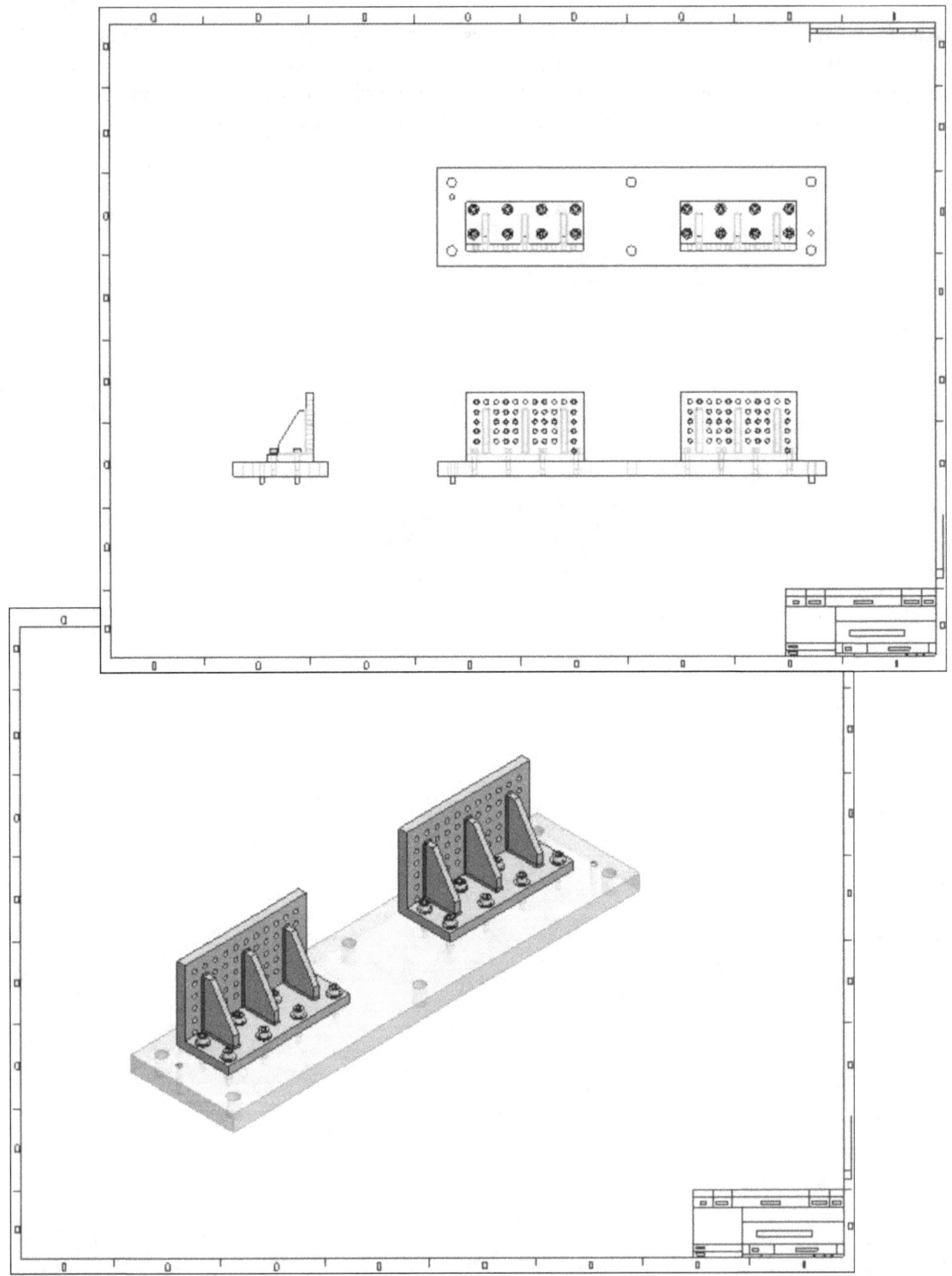

Figure 13.22(a) Zipped file includes all components, the assembly, and the drawing

For additional projects, see *www.cad-resources.com*.

Lesson 14 Blends

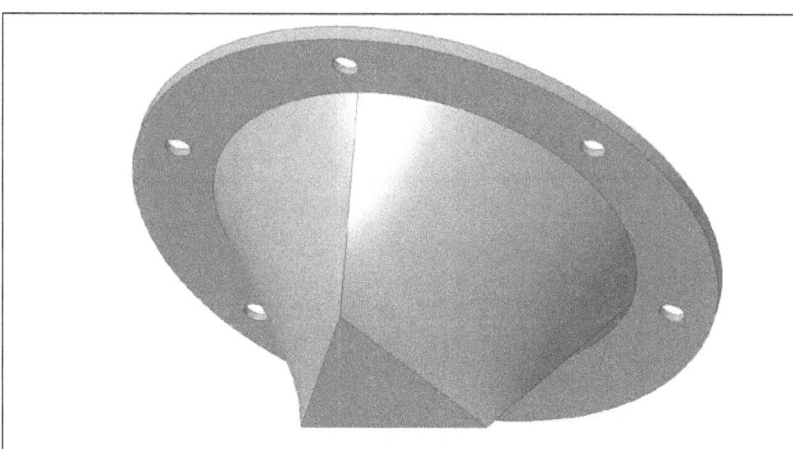

Figure 14.1 Cap

OBJECTIVES

- Create a **Blend** feature
- Use the **Shell Tool**
- Create a **Hole Pattern**
- Complete many of the command sequences without the step-by-step picks provided. At this point you should be able to do many of the functions without guidance.

REFERENCES AND RESOURCES

For **Resources** go to www.cad-resources.com > click on the PTC Creo Parametric 5.0 Book cover

- Lesson 14 Lecture at **YouTube Creo Parametric Lecture Videos**

- Lesson 14 3D PDF models embedded in a PDF

BLENDS

A blended feature consists of a series of at least two planar sections that are joined together at their edges with transitional surfaces to form a continuous feature. The Cap in Figure 14.1 uses a simple blend feature in its design. A Blend can be created as a **Parallel Blend** as used here, or you can construct a **Swept Blend**.

Blend Sections

Blends are created between the corresponding sections. Figure 14.2 shows a parallel blend for which the *section* consists of several *subsections*. Each segment in the subsection is matched with a segment in the following subsection; to create the transitional surfaces; Creo Parametric connects the *starting points* of the subsections and continues to connect the vertices of the subsections in a clockwise manner. By changing the starting point of a blend subsection, you can create blended surfaces that twist between the subsections. The default starting point is the first point sketched in the subsection.

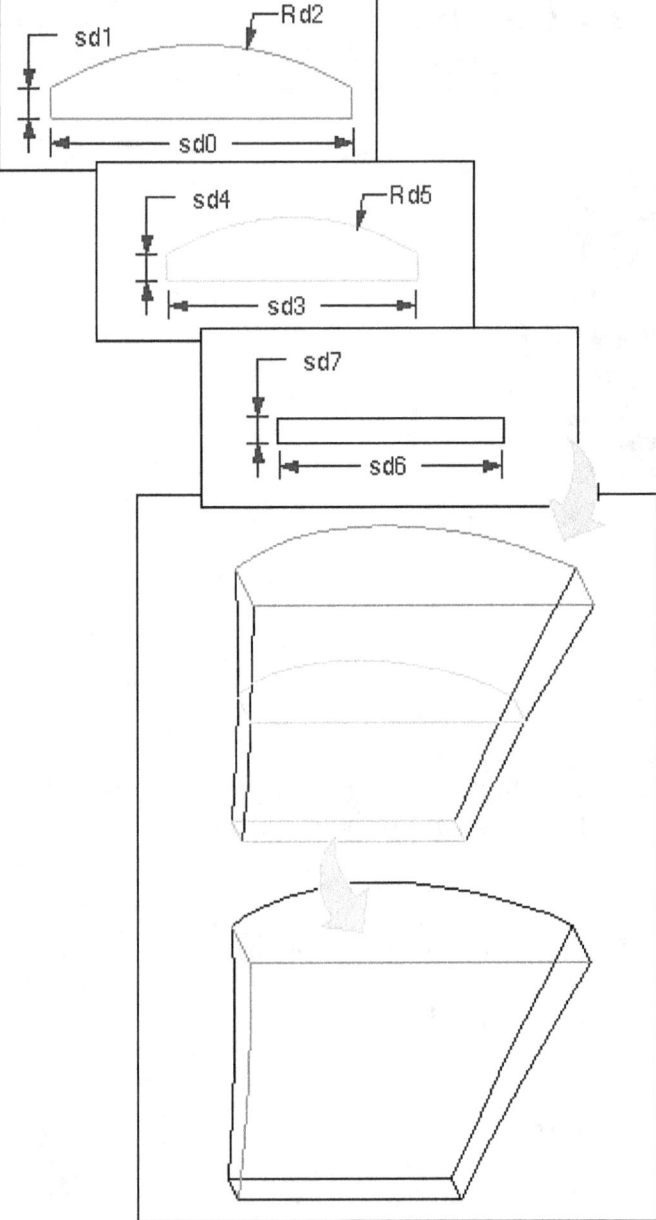

Figure 14.2 Blend Sections

Blend Options

Blends (Fig. 14.3) use one of the following transitional surface options:

- **Straight** Create a straight blend by connecting vertices of different subsections with straight lines. Edges of the sections are connected with ruled surfaces.
- **Smooth** Create a smooth blend by connecting vertices of different subsections with smooth curves. Edges of the sections are connected with ruled (spline) surfaces.
- **Parallel** All blend sections lie on parallel planes in one section sketch.
- **Rotational** The blend sections are rotated.
- **Select Section** Select section entities.
- **Sketch Section** Sketch section entities.

630

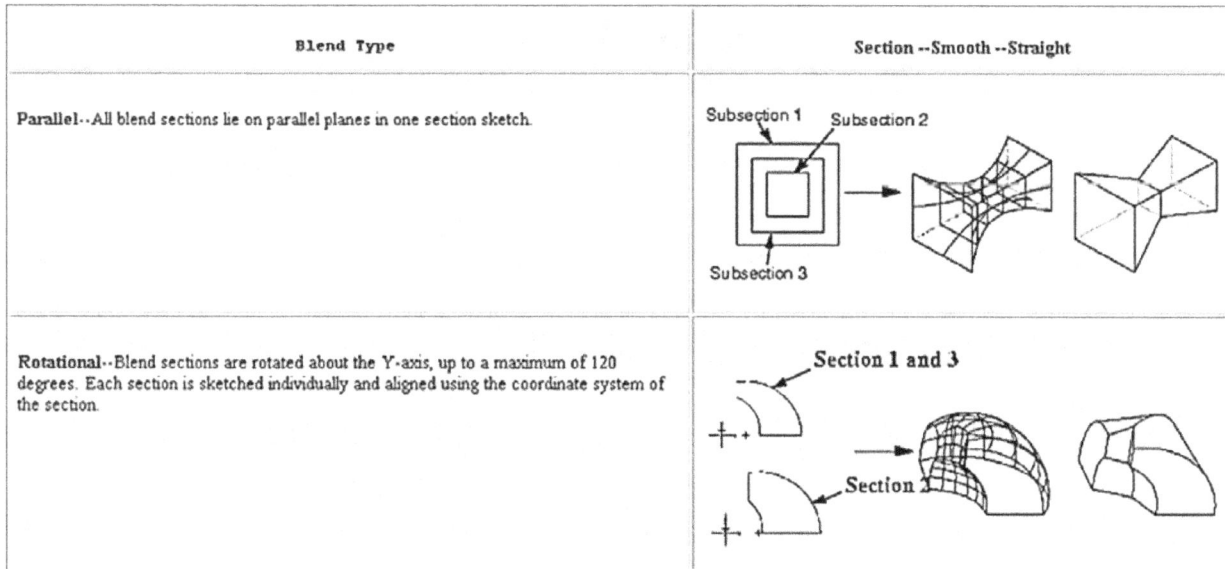

Blend Type	Section --Smooth --Straight
Parallel--All blend sections lie on parallel planes in one section sketch.	Subsection 1 Subsection 2 Subsection 3
Rotational--Blend sections are rotated about the Y-axis, up to a maximum of 120 degrees. Each section is sketched individually and aligned using the coordinate system of the section.	Section 1 and 3 Section 2

Figure 14.3 Blend Sections

Parallel Blends

A parallel blend is created from a single section that contains multiple sketches called *subsections* (Fig. 14.4). A first or last subsection can be defined as a point resulting in a blend vertex. The starting point for each subsection must be selected as per the design requirements including the starting points (Fig. 14.5).

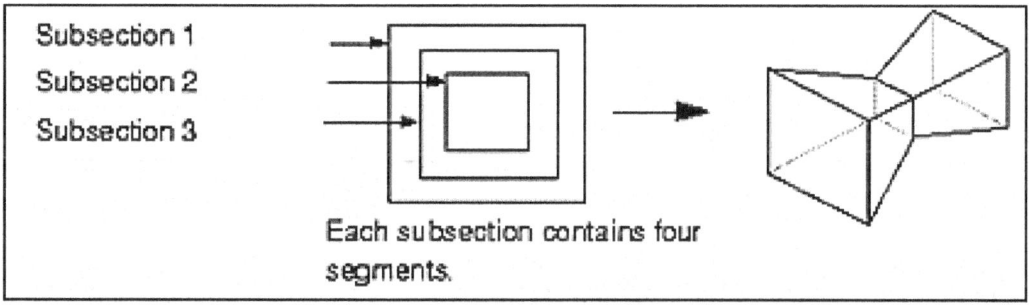

Figure 14.4 Starting Points

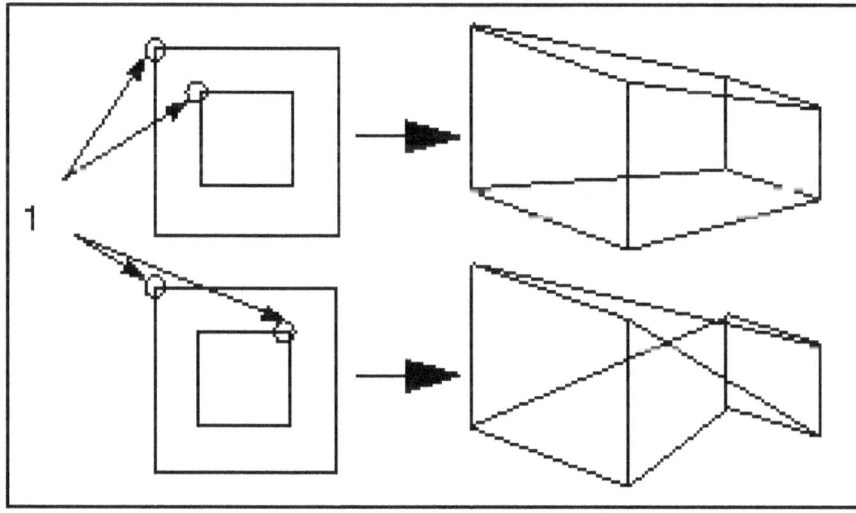

Figure 14.5 Starting Points

631

Lesson 14 STEPS

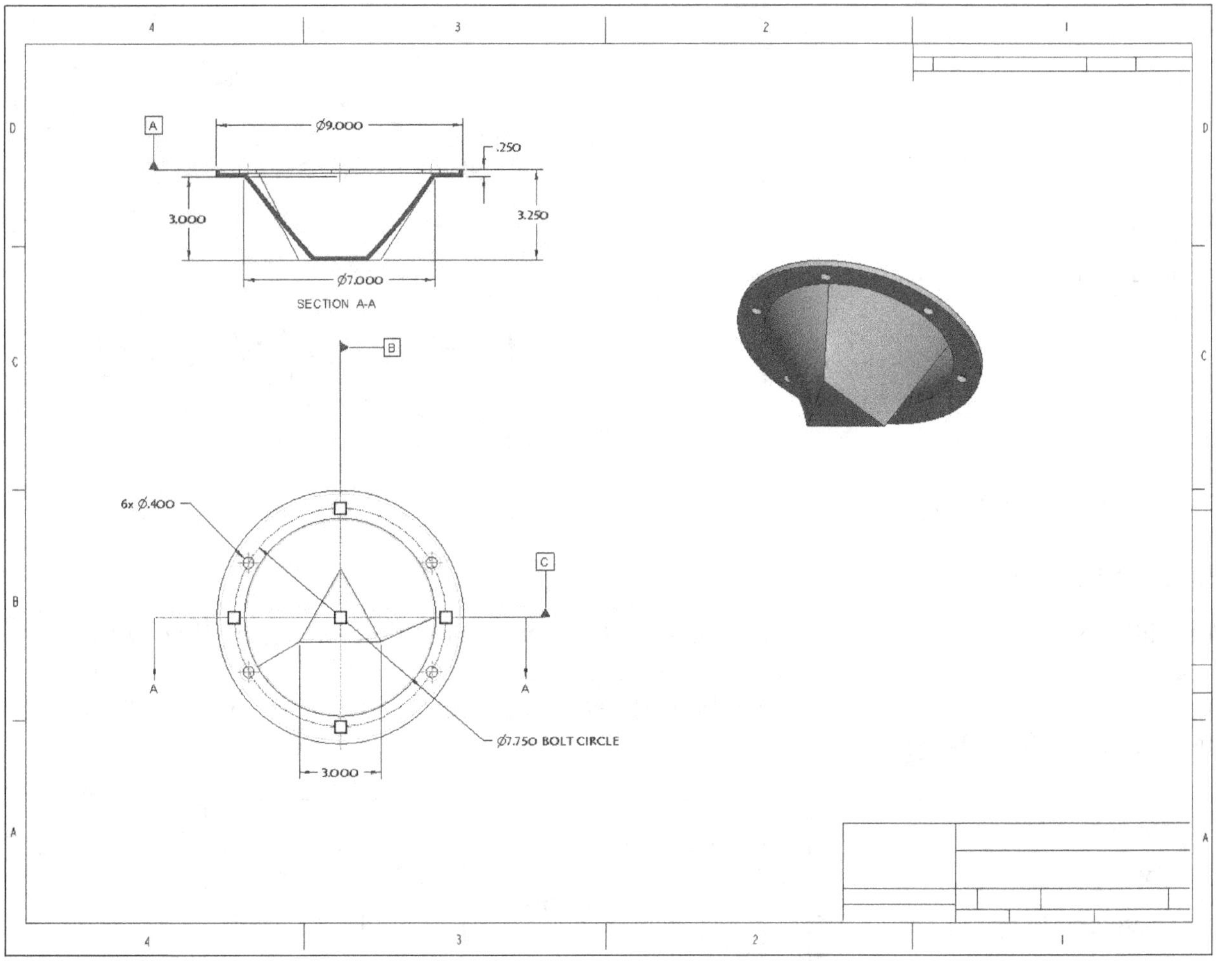

Figure 14.6 Cap Drawing

Cap

The Cap (Fig. 14.6) is a part created with a Parallel Blend (Fig. 14.7). The blend sections are a circle and a triangle. Because *the sections of a blend must have equal segments*, the "circle" is actually three equal arcs [Fig. 14.8(a)]. The part is shelled as the last feature in its creation [Fig. 14.12(b)].

Figure 14.7 Cap

Start a new part. Click: ☐ > Name **cap** > **OK**

Model Properties:
- **Material** = AL2014
- **Units** = Inch lbm Second

Set Datum ⌐-A-⌐ and **Rename** the default datum planes and coordinate system:
- Datum FRONT = **A**
- Datum RIGHT = **B**
- Datum TOP = **C**
- Coordinate System = **CSYS_CAP**

File > Options > Configuration Editor > | Show: Current Session | **> Find >**
1. Type keyword, *type: default_dec_places* > **Find Now** > 3. Set value, *type:* **3 > Enter > Close > OK**

For Lessons 13-18, step-by-step commands are limited to new software commands introduced or enhanced in that lesson. You will be expected to do most of the modeling using commands and practices mastered from Lessons 1-12 without repeated detailed explanations. Refer to Figures 14.8(a-b) for the Cap dimensions.

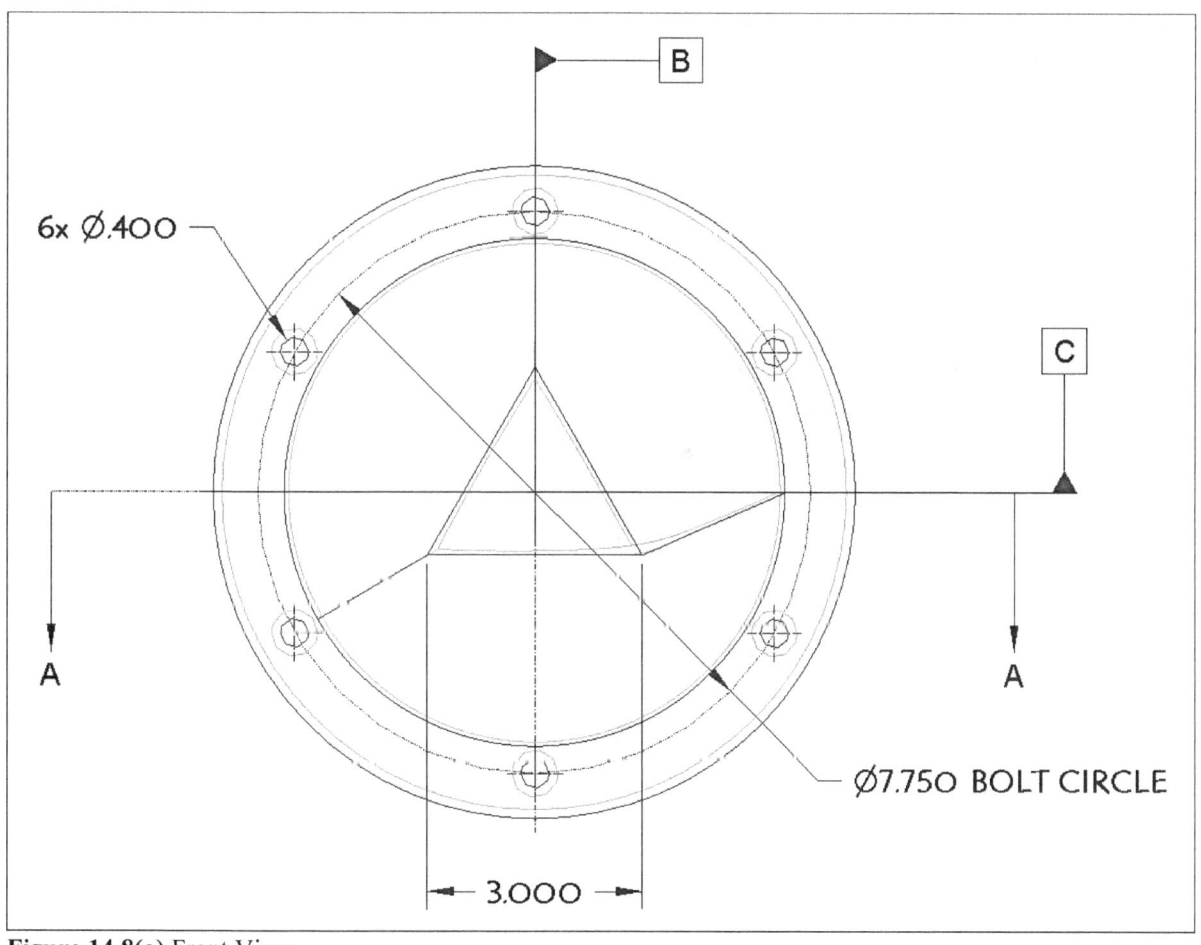

Figure 14.8(a) Front View

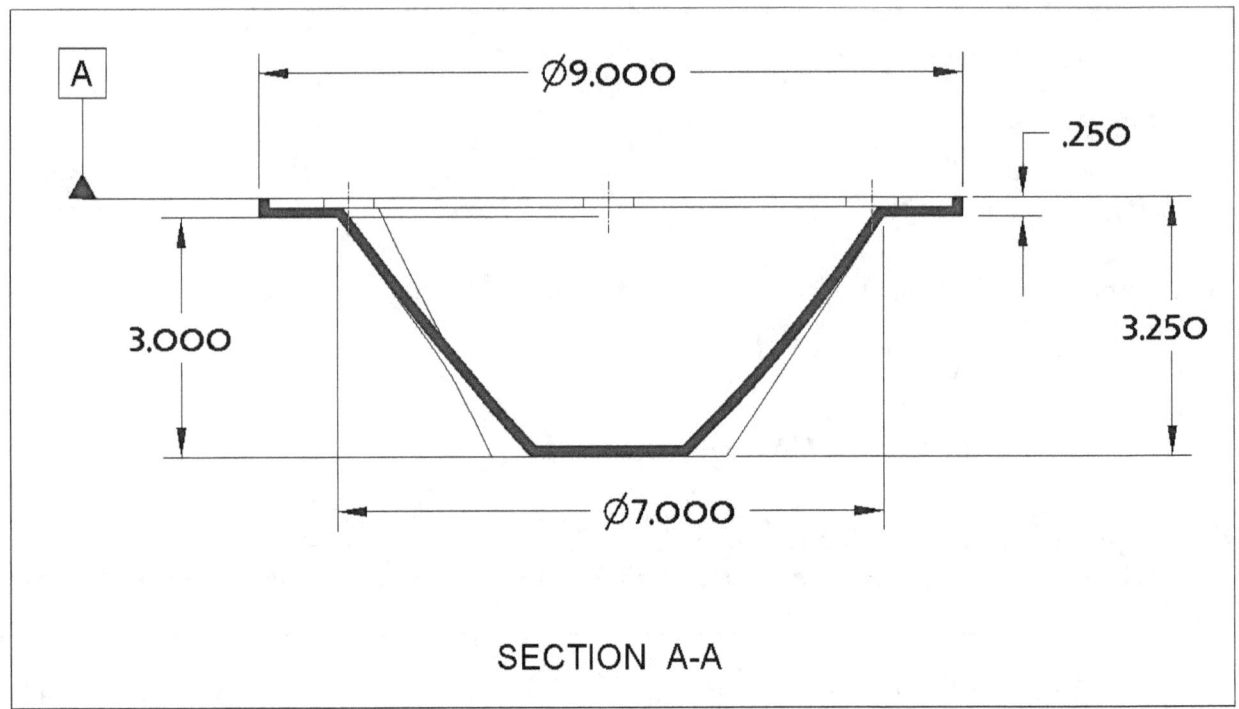

Figure 14.8(b) Right Side View

Model the circular protrusion that is ∅9.00 by .25 thick [Fig. 14.9(a)]. Sketch the first extrusion on datum **A** (**FRONT**) and centered on **B** (**RIGHT**) and **C** (**TOP**). > **Save**

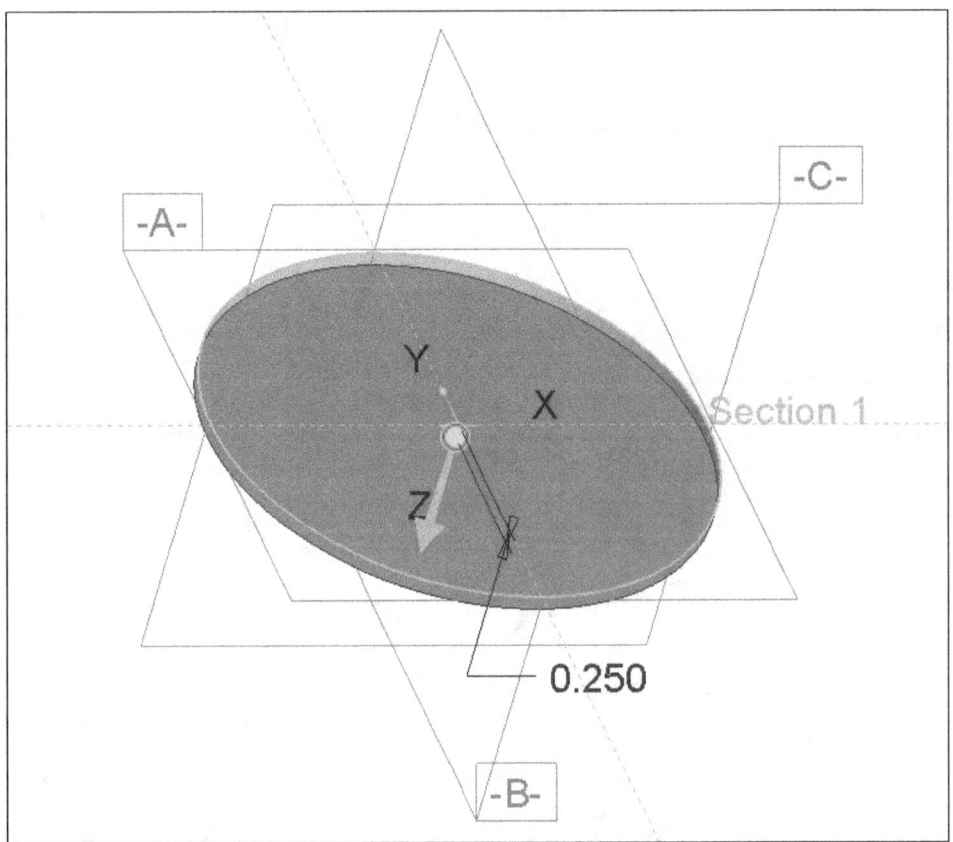

Figure 14.9(a) First Extrusion

View tab > **Appearances** > **ptc_std_aluminum_polished** [Fig. 14.9(b)] > select the **CAP.PRT** in the Model Tree > **Model** tab > Shapes ▾ Group > ⬡ Blend > **Options** tab > **Smooth** > **Sections** tab > **Define** > select the forward face [Fig. 14.10(a)] > **Sketch** >

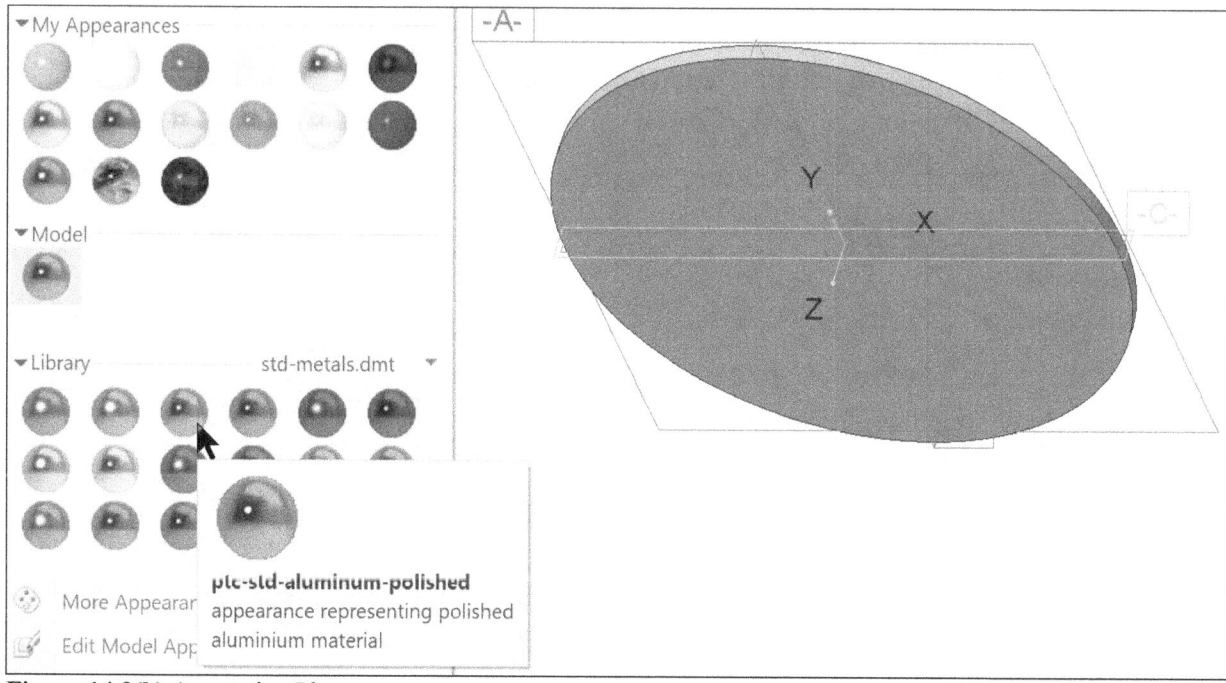

Figure 14.9(b) Annotation Plane

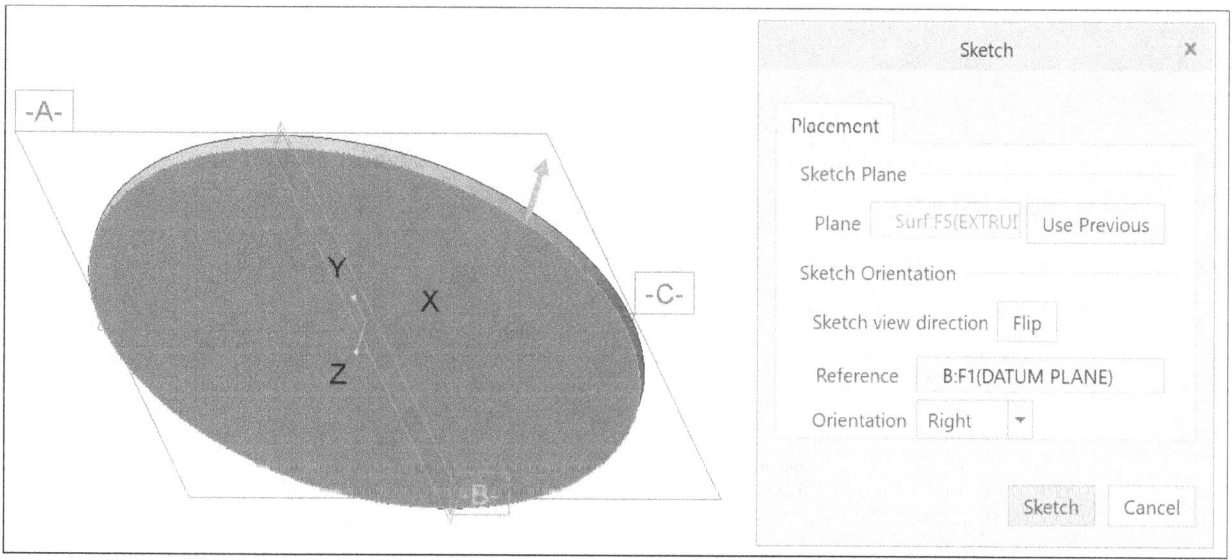

Figure 14.10(a) Blend Feature Sketch Surface (select the forward face of the first extrusion)

Click: **File** > **Options** > **Sketcher** > ☑ Show the grid > ☑ Snap to grid > **OK** > **No** > **Sketch** tab > **Setup** Group > **Display** > **Grid Display** > **Setup** Group > **Grid Settings** [Fig. 14.10(b)] > Type: ⊙ Polar > **Spacing** ⊙ Static > Radial: 0.500000 > **Enter** > Angular: 30.000000 > **Enter** > Lines: 12 > **Enter** [Fig. 14.10(c)] > **OK** > Sketch View > Hidden Line [Fig. 14.10(d)] > **RMB** > **Construction Centerline** > add vertical and horizontal construction centerlines, starting at the origin [Fig. 14.10(e)]

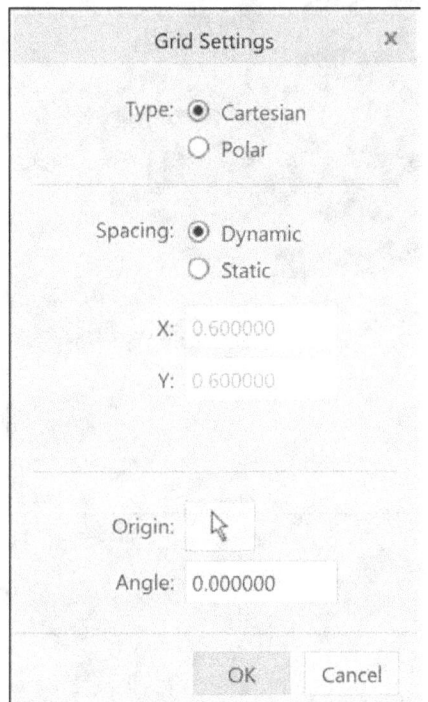

Figure 14.10(b) Grid Settings Dialog Box

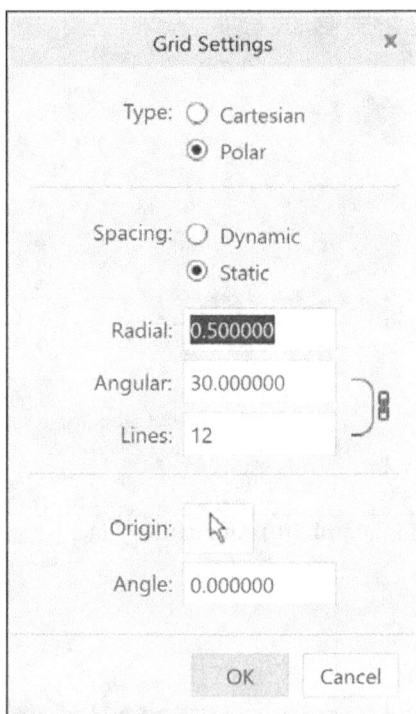

Figure 14.10(c) Polar Grid Settings

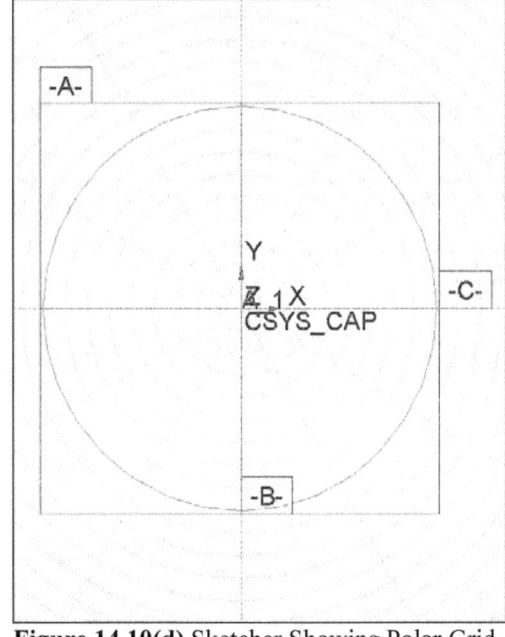

Figure 14.10(d) Sketcher Showing Polar Grid

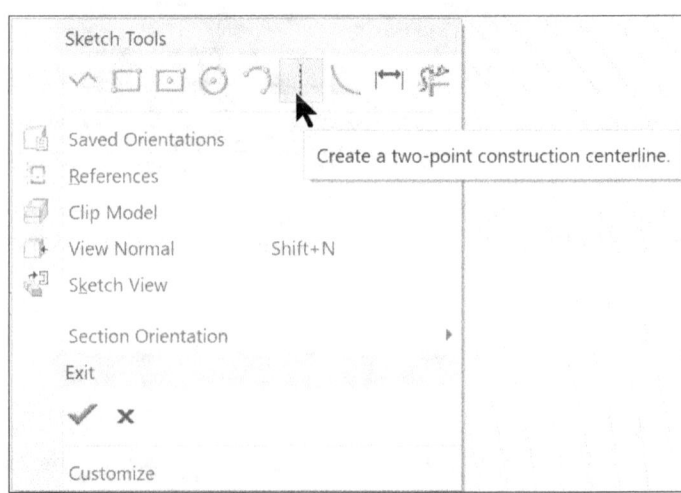

Figure 14.10(e) Add Vertical and Horizontal Centerlines

636

Add two angled construction centerlines, start the construction centerlines at the origin [Figs. 14.10(f-g)] (your angles may display as 30 or 60 degrees)

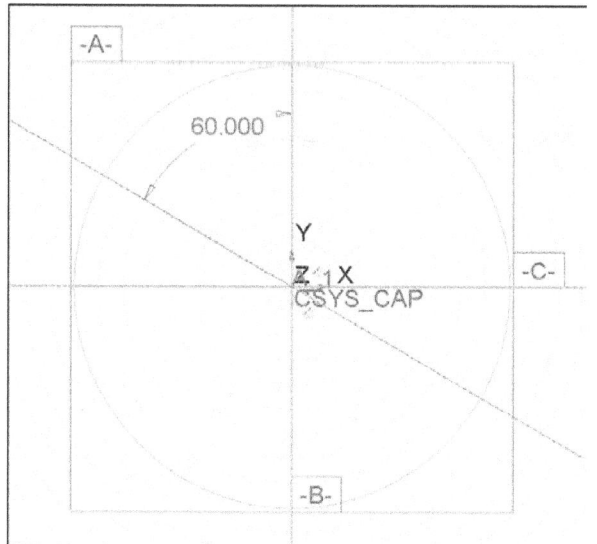

Figure 14.10(f) Sketch the first Angled Centerline

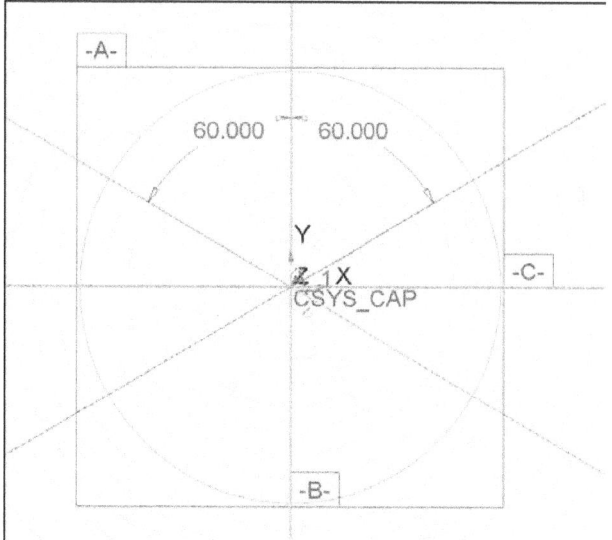

Figure 14.10(g) Sketch the Second Angled Centerline

Sketch the *first section* of the blend by creating *three* equal **120°** arcs. Click: Arc ▾ > ▾ > Center and Ends **Create an arc by picking its center and endpoints** > sketch the first arc by picking its center, it is the first end point along datum B, moving the pointer in a counter clockwise direction, and its last end point along one of the angled centerlines [Fig. 14.10(h)] > sketch the second arc by picking its center and end points following a counter clockwise direction [Fig. 14.10(i)]

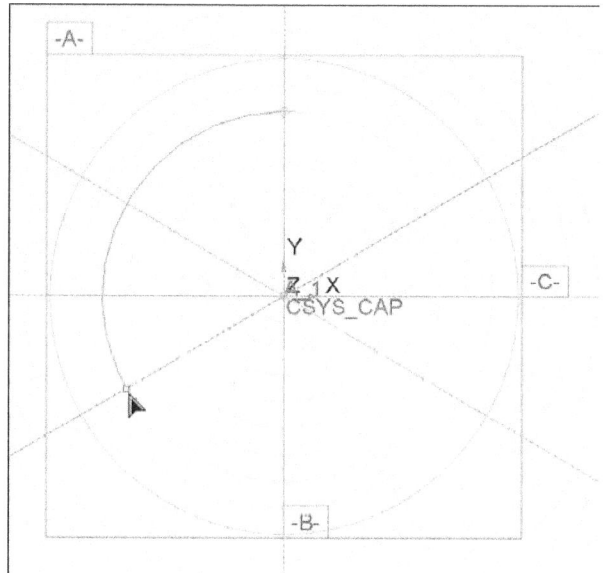

Figure 14.10(h) First **120**-degree Arc

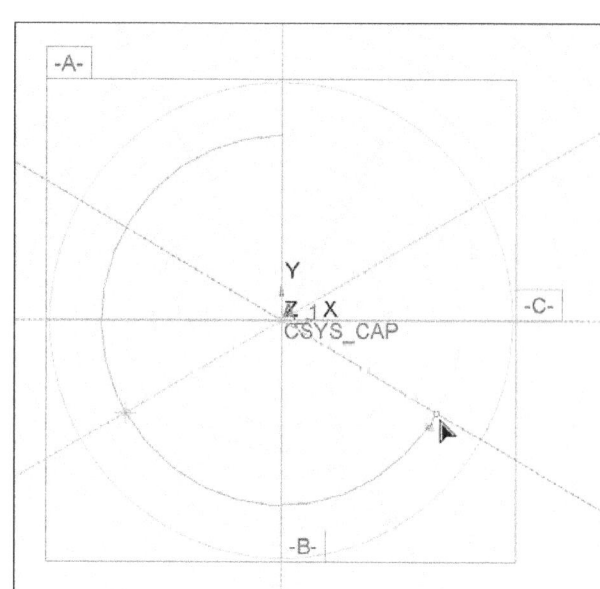

Figure 14.10(i) Second **120**-degree Arc

Sketch the third arc by picking its center and end points following a counter clockwise direction > **MMB** to end the current tool [Fig. 14.10(j)] (the arc dimension is a radius value) > click on the

3.50 radius dimension > [⟳] >

> **Diameter**
> Toggle the radius dimension to diameter.

> **LMB** > **MMB** > add in the

two 30-degree dimensions (Note the location of the *start point* for this section ⬅) [Fig. 14.10(k)] > ✓

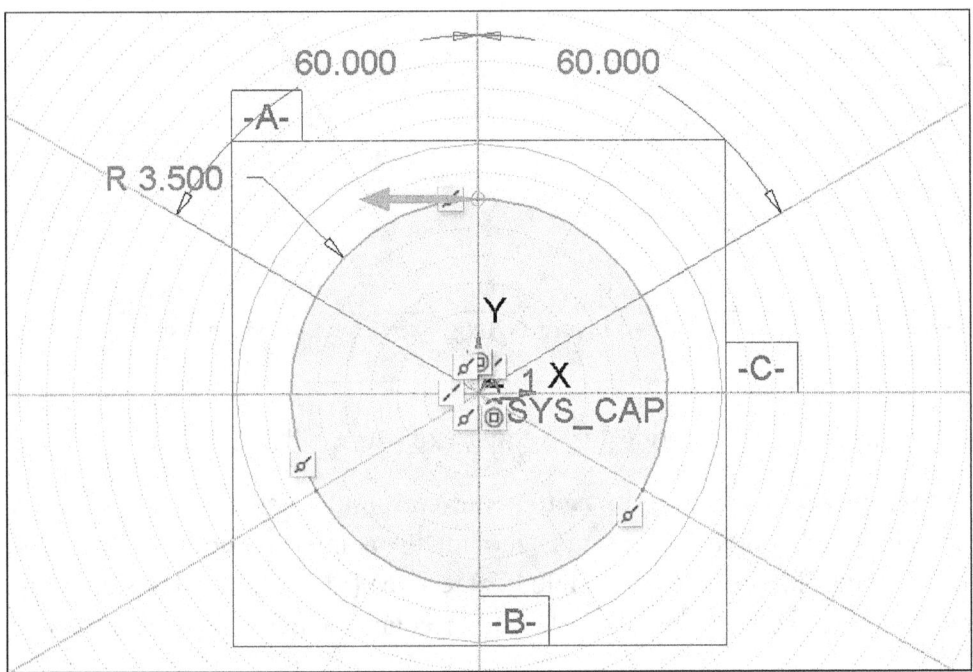

Figure 14.10(j) Third **120**-degree Arc

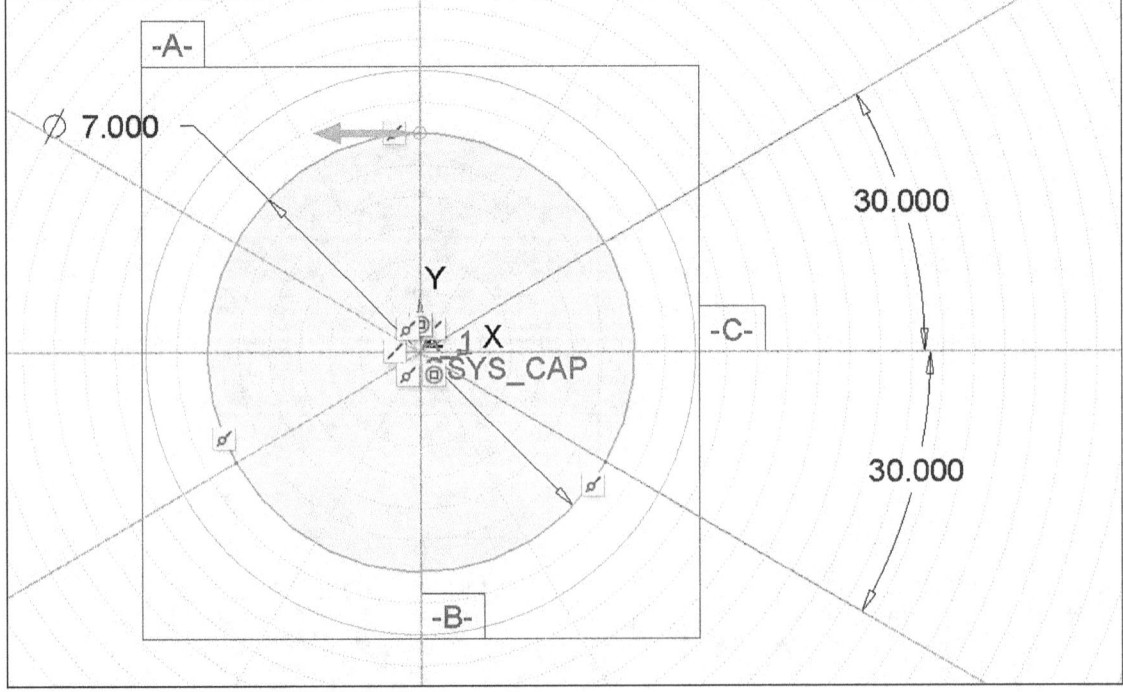

Figure 14.10(k) Diameter Dimension

Click: **File** > **Options** > **Sketcher** > set values and options as shown in Figure 14.10(l) > **OK** > **No** > **RMB** > ✓ > **Sections** tab > Section 2 > ⦿ Offset dimension > Offset from **3.00** > **Enter** > **Sketch** [Fig. 14.10(m)]

Accuracy and Sensitivity

Number of decimal places for dimensions: 4

Snapping sensitivity: Very_High

Dimension behavior while dragging the section

☐ Lock modified dimensions

☐ Lock user defined dimensions

Sketcher grid

☑ Show the grid

☑ Snap to grid

Grid angle: 0.000000

Grid type: Polar

Grid spacing type: Static

Spacing along radial lines: .5

Angle between radial lines: 30.000000

Number of radial lines: 12

Figure 14.10(l) Sketcher Grid Settings

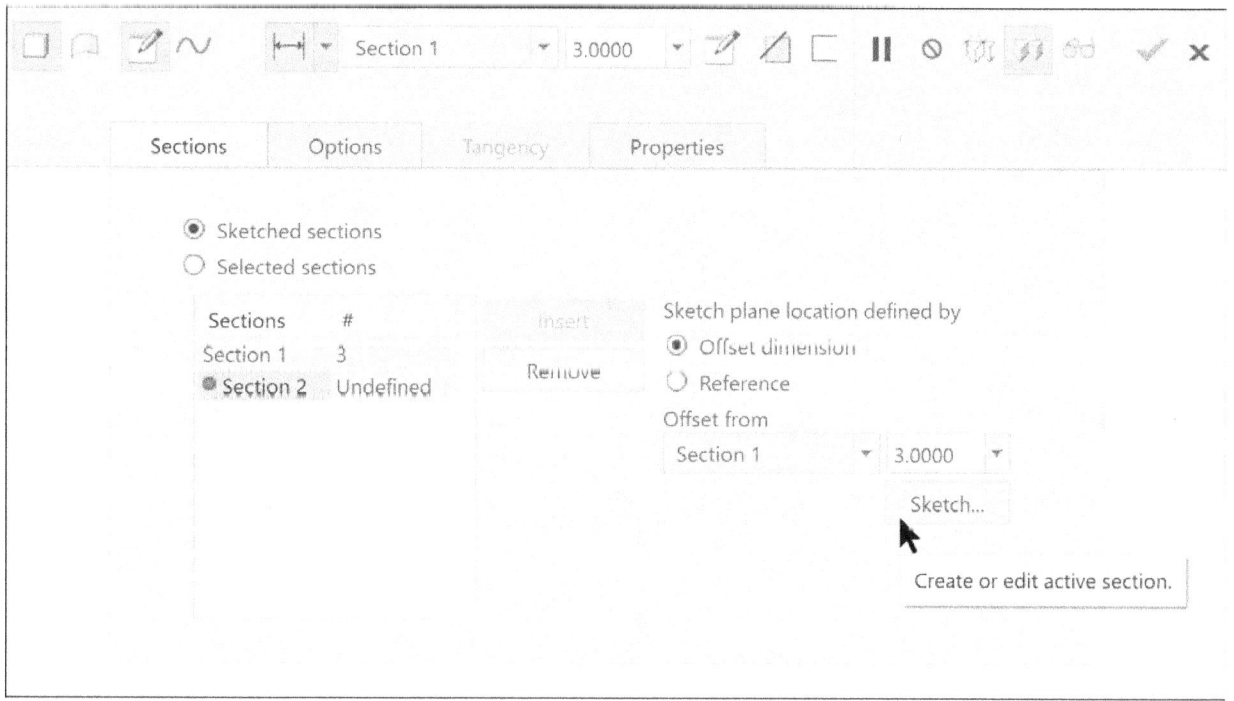

Figure 14.10(m) Section 2 Sketch Setup

639

In the Graphics Window, press **RMB** > ⊥ **Construction Centerline** > add two **30**-degree construction centerlines, start the construction centerlines at the origin > in the Graphics Window, **MMB** to end the current tool > **LMB** to deselect > sketch the second parallel section (the first section is *grayed* out), press **RMB** > ⌃ **Line Chain** > sketch the three lines of the triangle starting at the top so that the *start point* is near the *start point* of the first section and picking points *in the same direction* in which the arcs were created [Fig. 14.10(n)] > **MMB** > **MMB** > ═ **Equal** > select the three lines [Fig. 14.10(o)] > **File** > **Options** > **Sketcher** > turn off the grid display and grid snap > **OK** > **Yes** > **OK** > modify the line length to **3.00** > **LMB** to deselect > move the dimension[Fig. 14.10(p)] > **LMB**

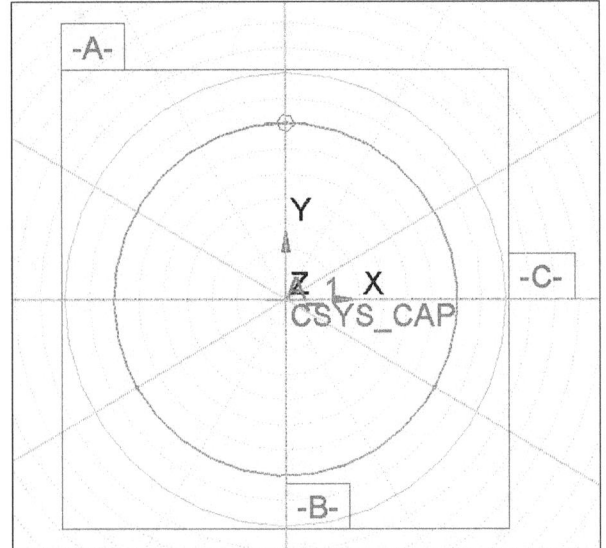

Figure 14.10(n) Second Section Sketch

Figure 14.10(o) Three Length Dimensions

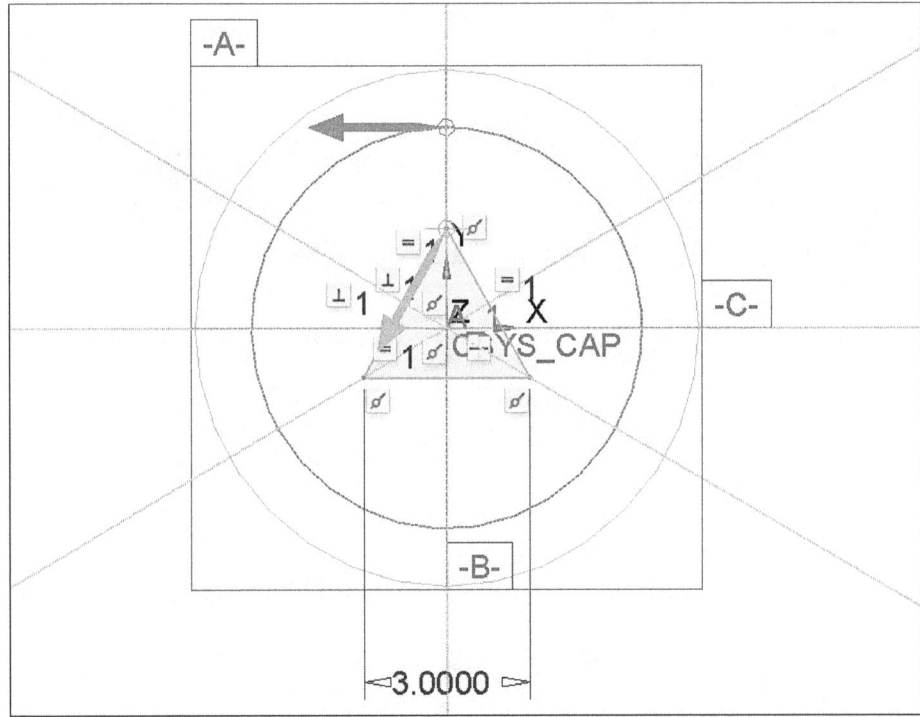

Figure 14.10(p) Dimensioned Sketch

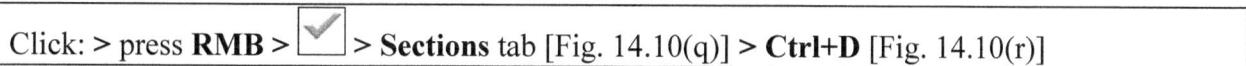

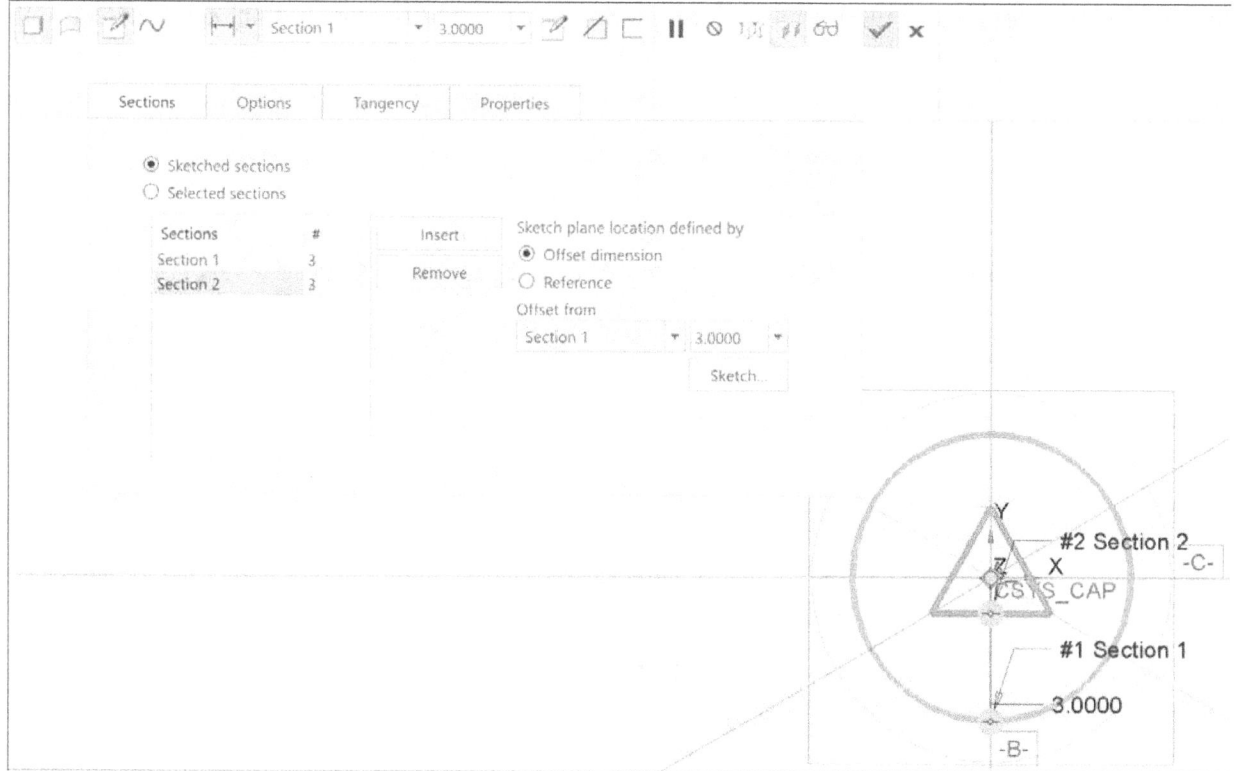

Figure 14.10(q) 120-degree Reference Dimension

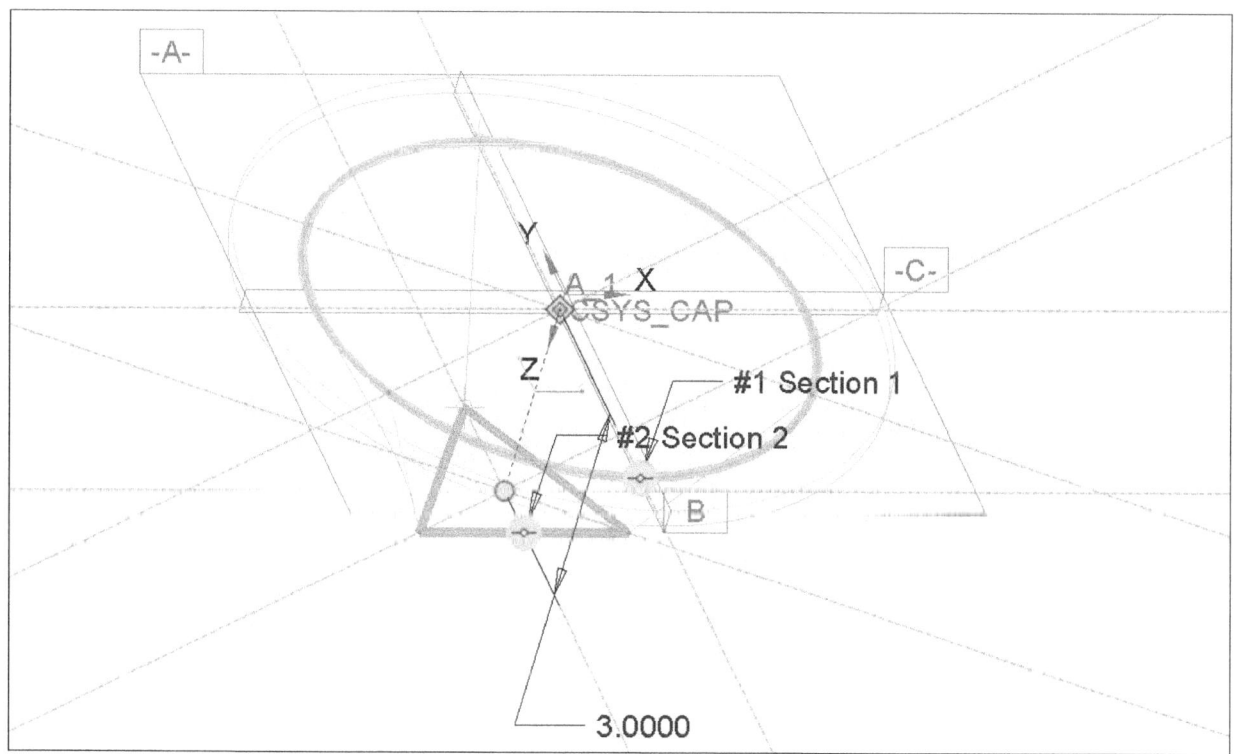

Figure 14.10(r) Section 2

Click: 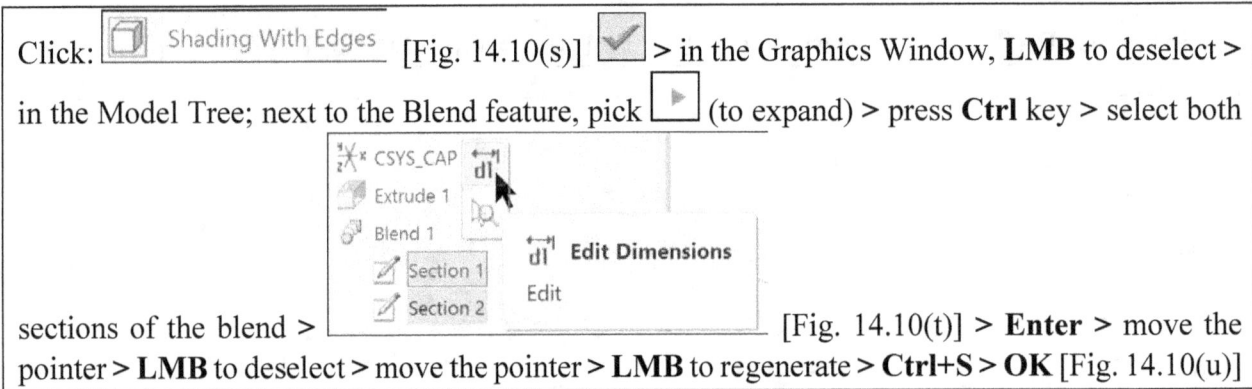 Shading With Edges [Fig. 14.10(s)] ☑ > in the Graphics Window, **LMB** to deselect > in the Model Tree; next to the Blend feature, pick ▶ (to expand) > press **Ctrl** key > select both

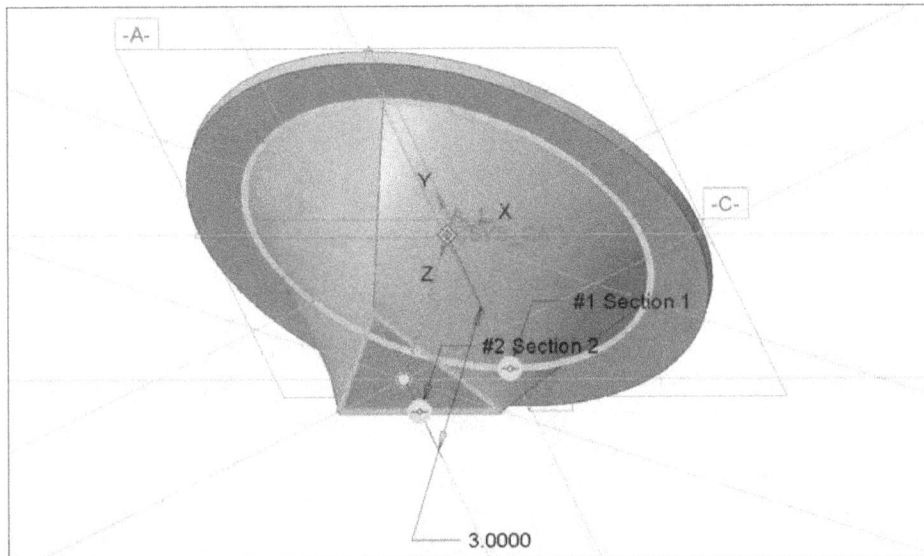

sections of the blend > [Fig. 14.10(t)] > **Enter** > move the pointer > **LMB** to deselect > move the pointer > **LMB** to regenerate > **Ctrl+S** > **OK** [Fig. 14.10(u)]

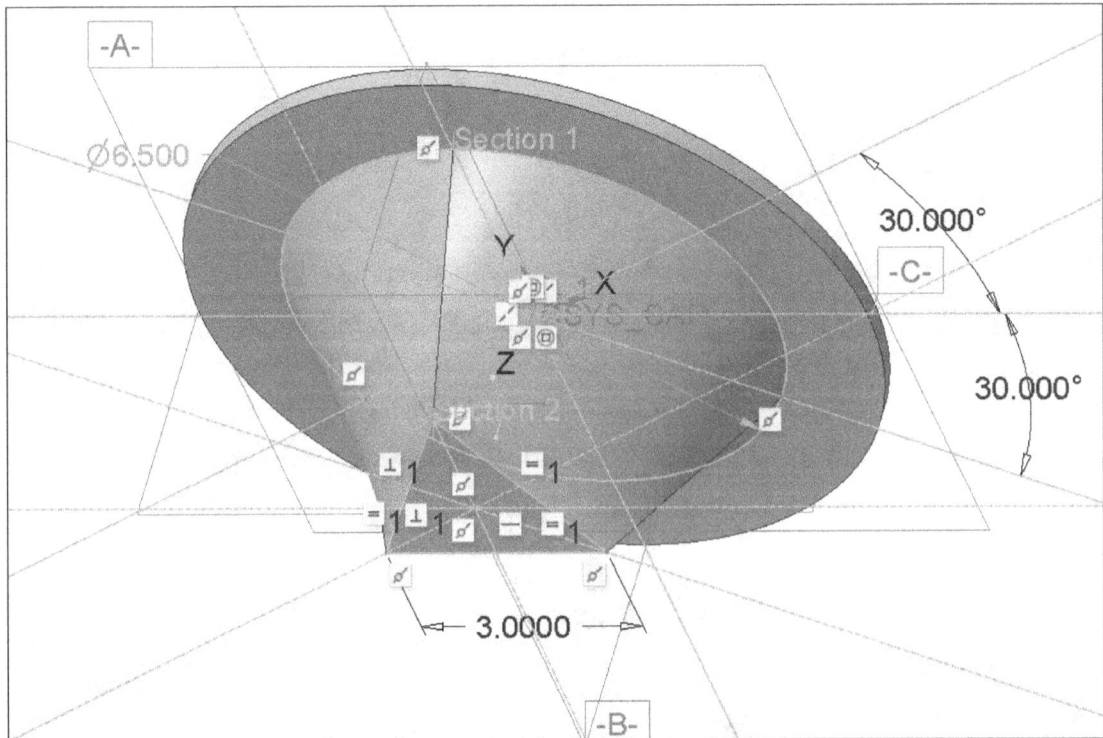

Figure 14.10(s) Blend Preview

Figure 14.10(t) Edit Dimensions

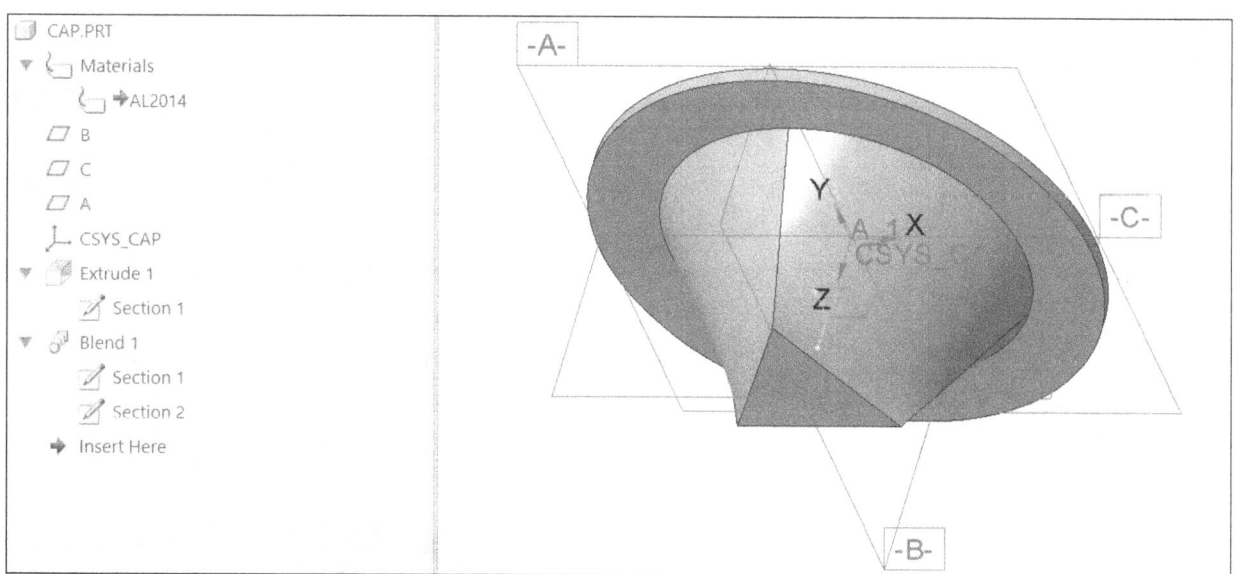

Figure 14.10(u) Completed Blend

Create and pattern the six equally spaced holes ∅.400 on a ∅7.75 bolt circle, click: 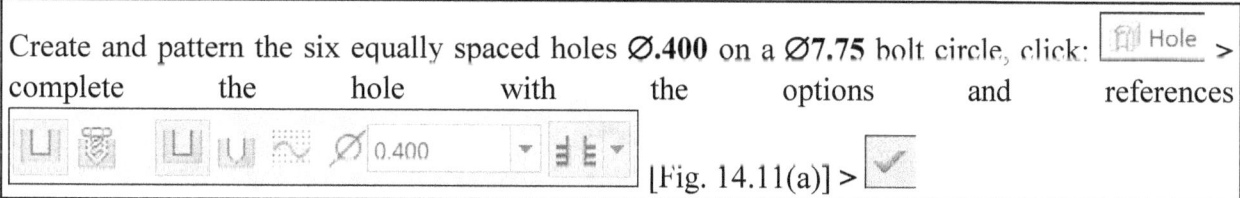 Hole >
complete   the   hole   with   the   options   and   references
[Fig. 14.11(a)] >

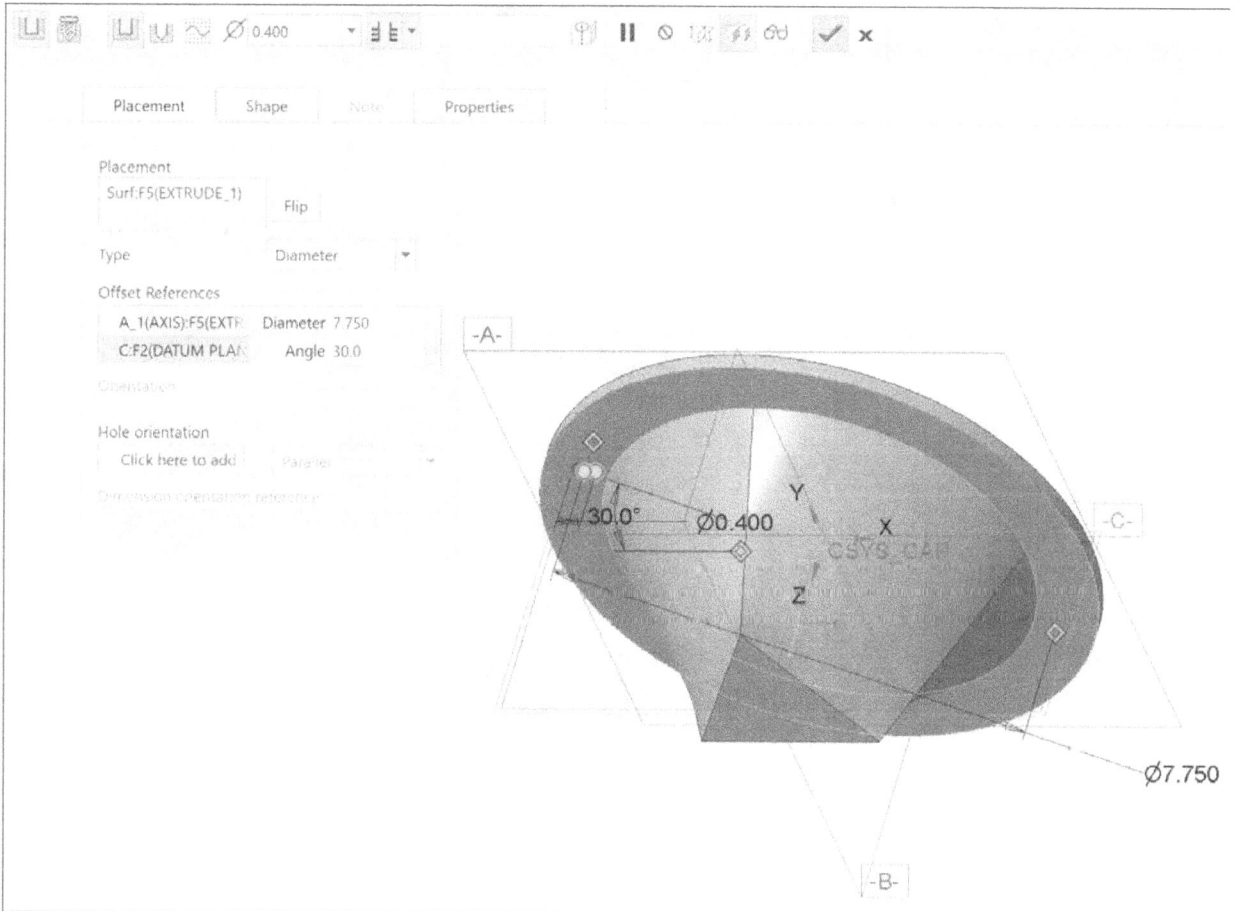

Figure 14.11(a) Hole Options and References (your axis name may be different)

With the hole selected > **RMB** > ⌗ **Pattern** > complete the pattern with the options and references [Fig. 14.11(b)] *(use the **30 degree** dimension to pattern)* [Fig. 14.11(c)] > **LMB** to deselect

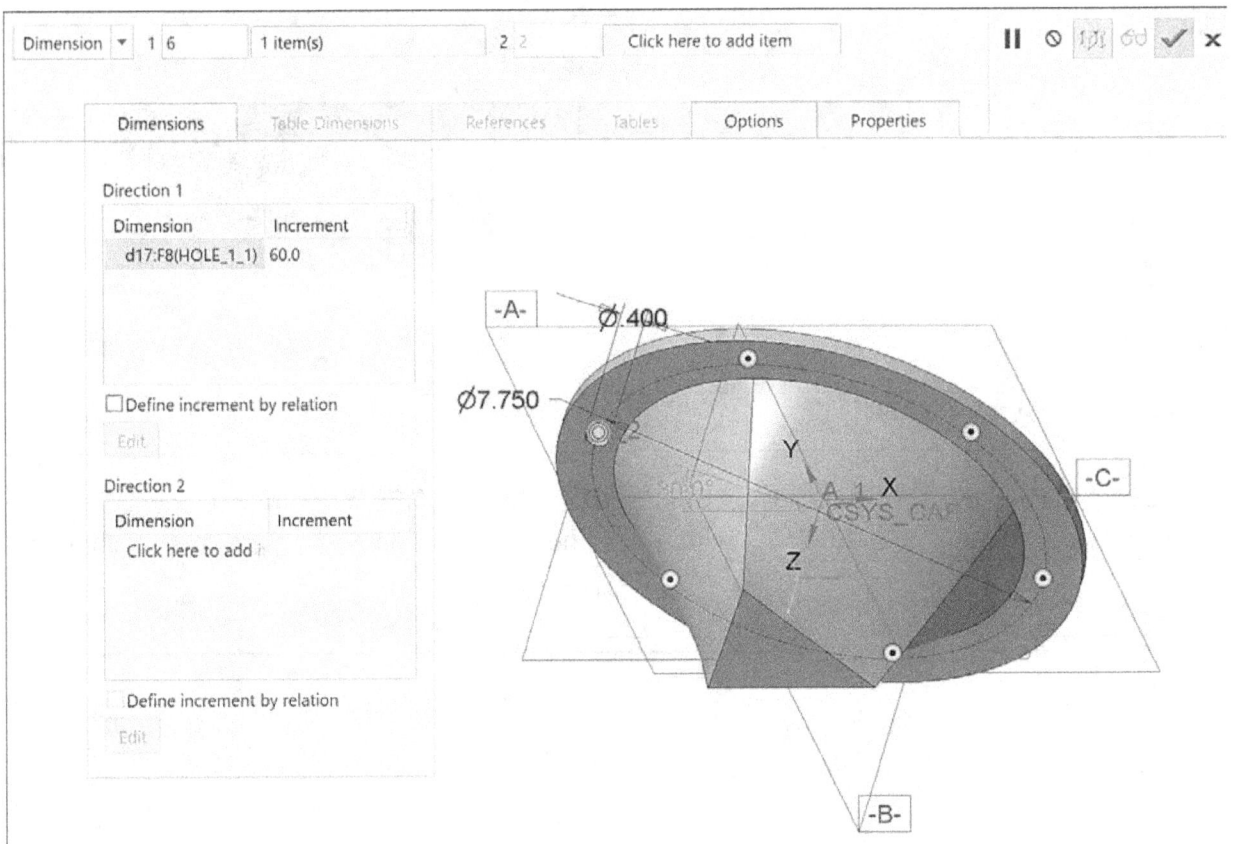

Figure 14.11(b) Patterning the Hole

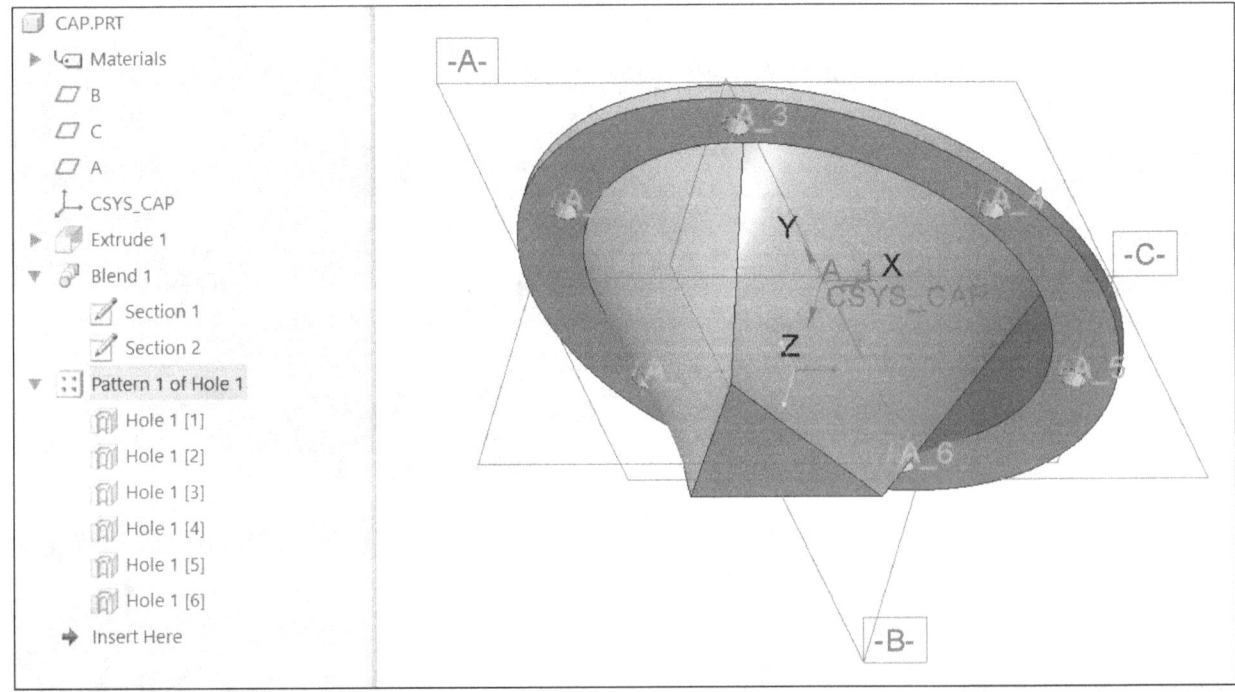

Figure 14.11(c) Completed Hole Pattern

Click: 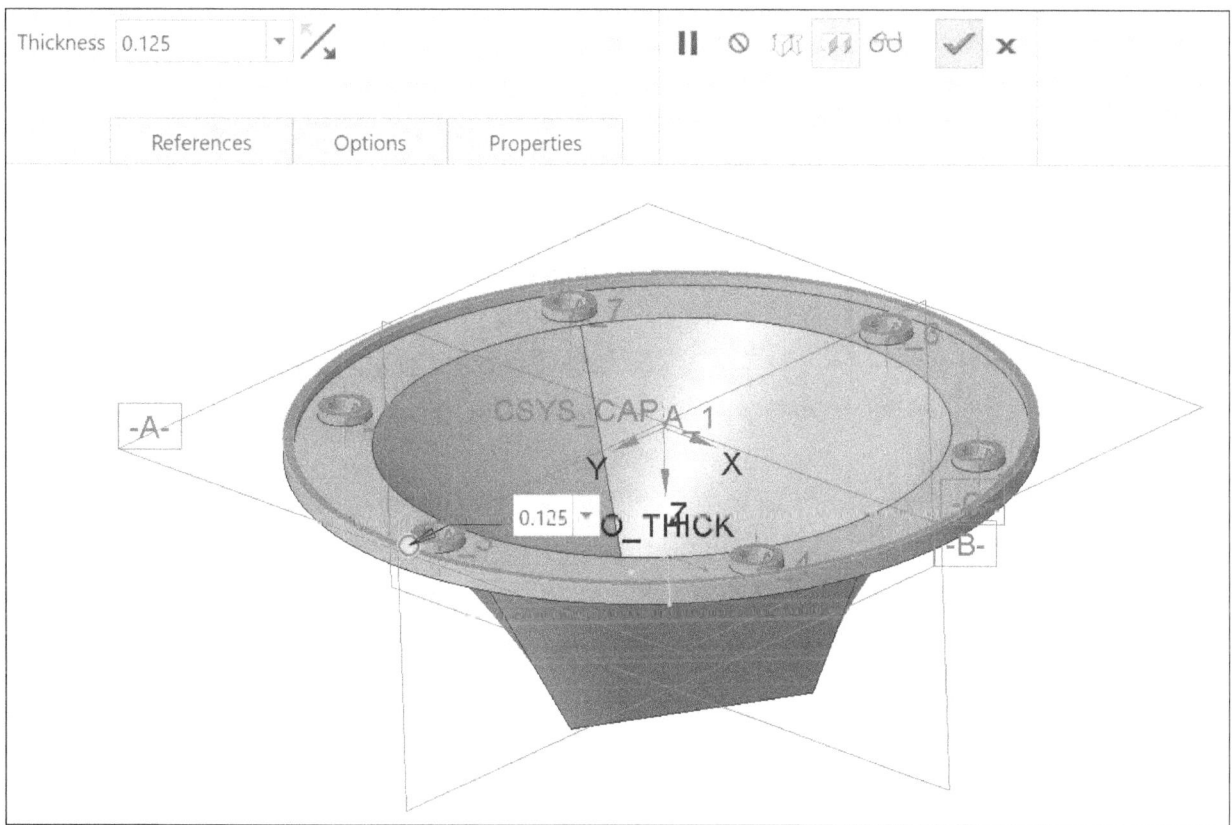 **Shell** > spin the model > select the bottom surface of the part as the surface to remove [Fig. 14.12(a)] > Thickness **.125** > **Enter** > ✓ [Fig. 14.12(b)] > **LMB** > **Ctrl+S**

Figure 14.12(a) Shell Tool

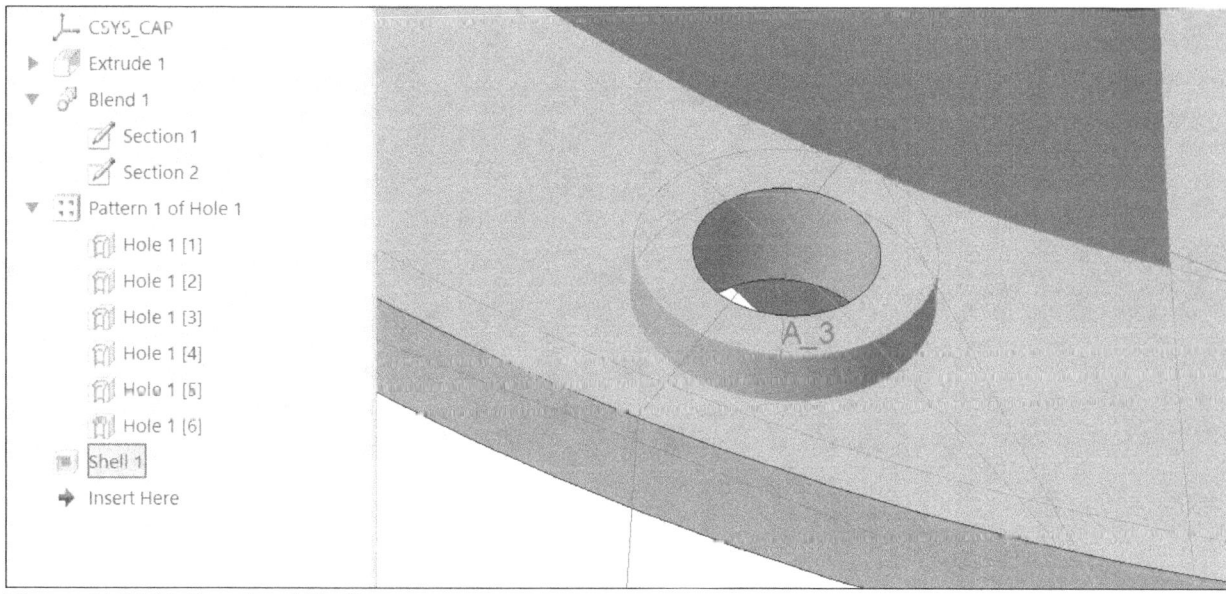

Figure 14.12(b) Completed Shell

In the Model Tree, place the pointer on [Shell 1] > **RMB** > ▣ **Suppress** [Figs. 14.13(a)] > **OK** > **Operations** > **Resume** > **Resume All** [Fig. 14.13(b)] > **Ctrl+D** > **Ctrl+S** > **File** > **Close**

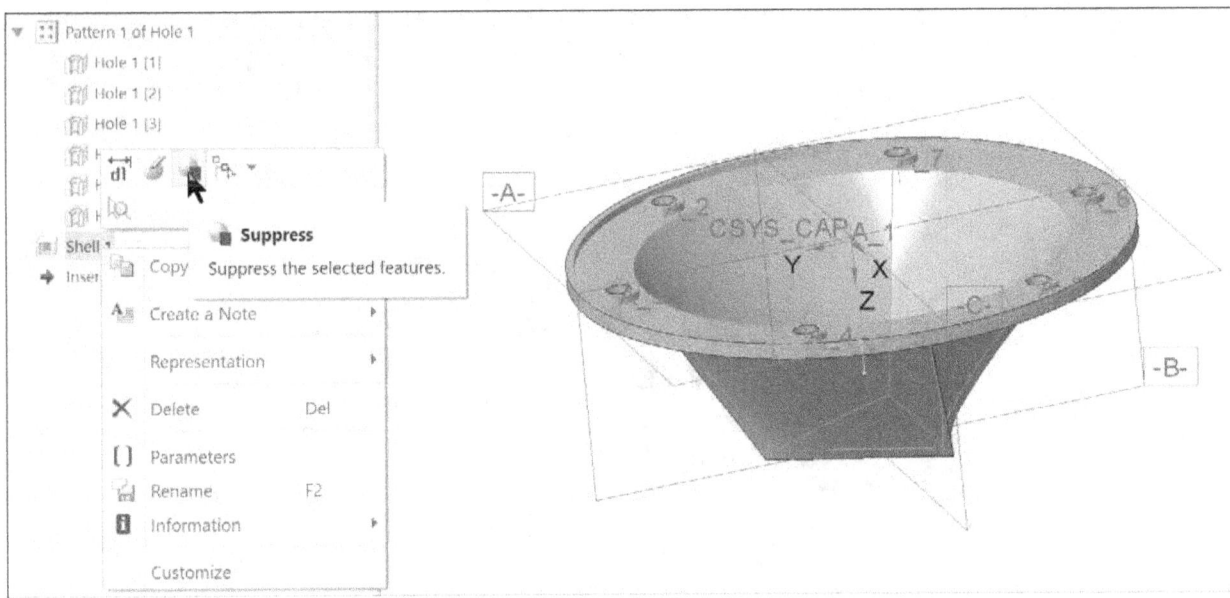

Figure 14.13(a) Suppress

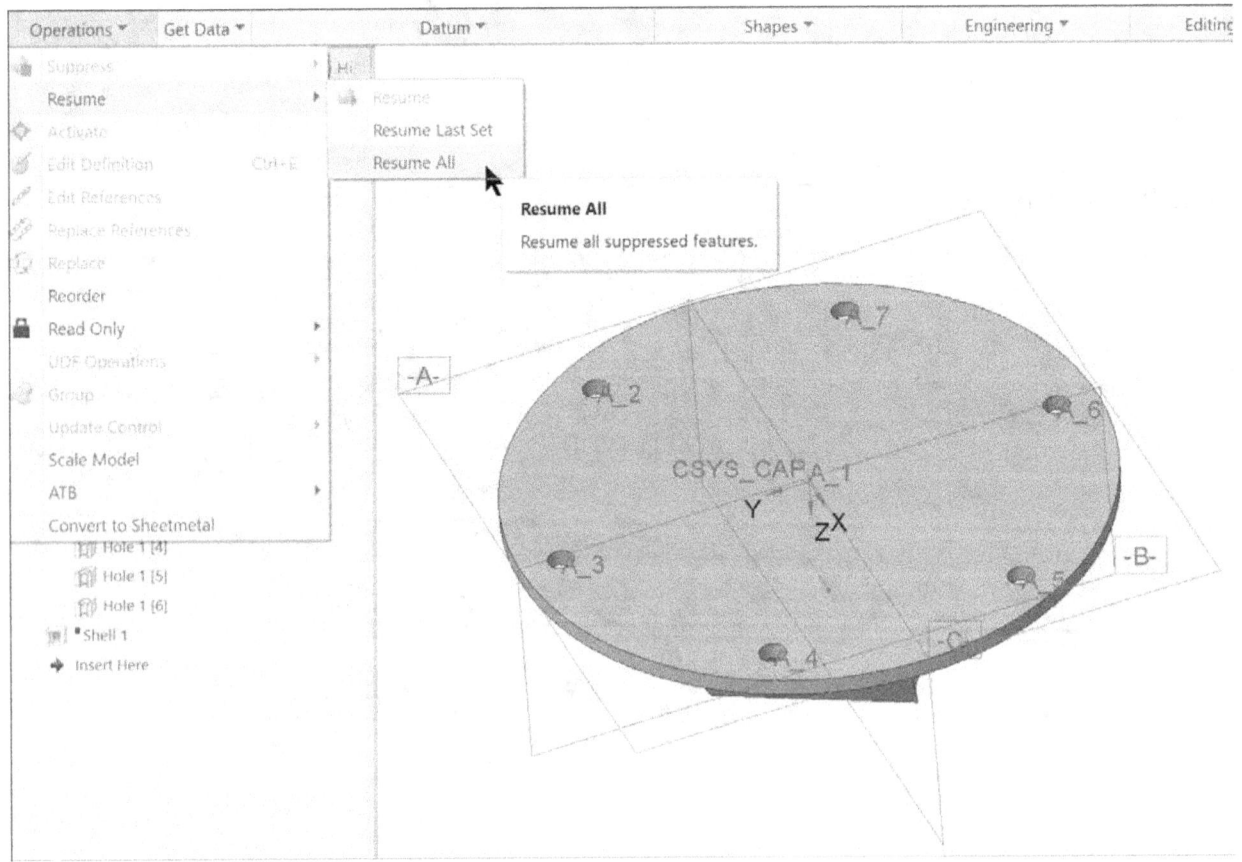

Figure 14.13(b) Resume All

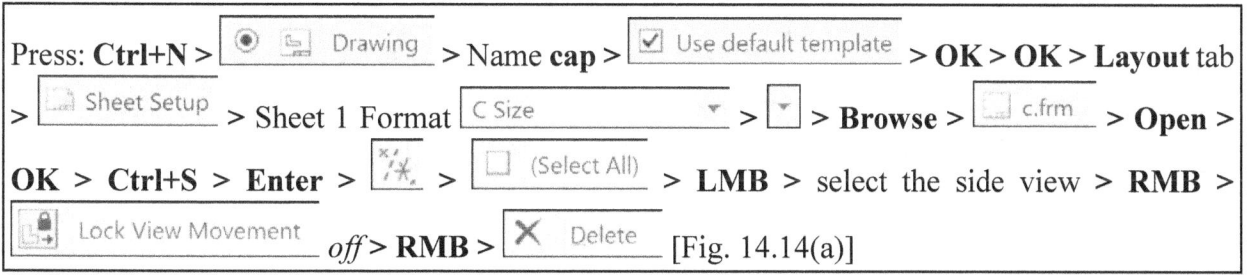

Press: **Ctrl+N** > ⊙ ⬚ Drawing > Name **cap** > ☑ Use default template > **OK** > **OK** > **Layout** tab > ⬚ Sheet Setup > Sheet 1 Format C Size ▼ > ▼ > **Browse** > ⬚ c.frm > **Open** > **OK** > **Ctrl+S** > **Enter** > ⬚ > ⬚ (Select All) > **LMB** > select the side view > **RMB** > ⬚ Lock View Movement *off* > **RMB** > ✕ Delete [Fig. 14.14(a)]

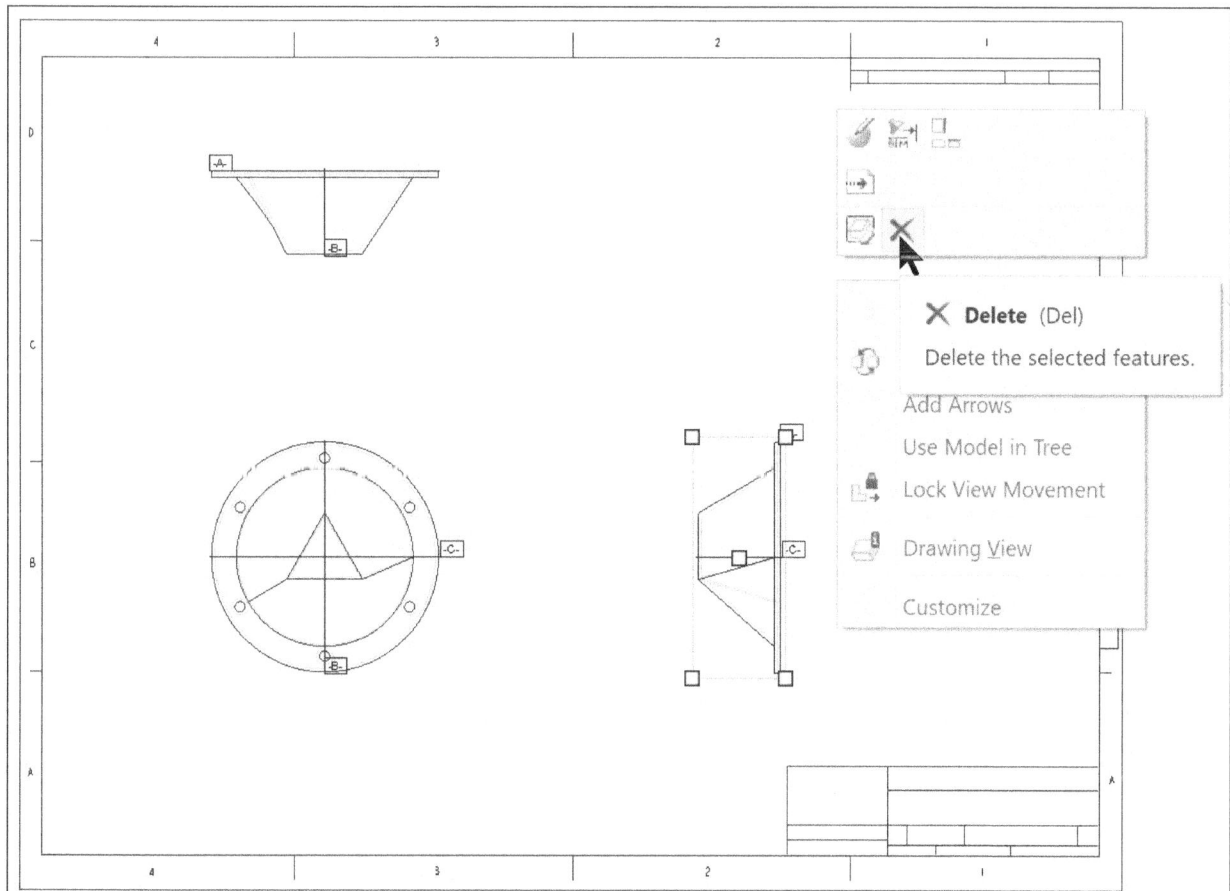

Figure 14.14(a) Cap Drawing

Click: **File** > **Prepare** > **Drawing Properties** > Detail Options **change** > Option: *type* **gtol** > click in Value: field > ▼ > **std_asme** [Fig. 14.14(b)] > **Add/Change** > **Apply** > **Close** > **Close**

radial_pattern_axis_circle | std_ansi *
▽ These options control geometric tolerancing informa | std_ansi_mm
asme_dtm_on_dia_dim_gtol | std_iso_jis
gtol_datum_placement_default | std_iso
gtol_datums | std_asme
 | std_jis
 | std_din   Select or enter a value for config option
Option: | std_ansi_dashed
gtol_datums | std_ansi *

Figure 14.14(b) gtol_datums set to std_asme

Click: **Annotate** tab > 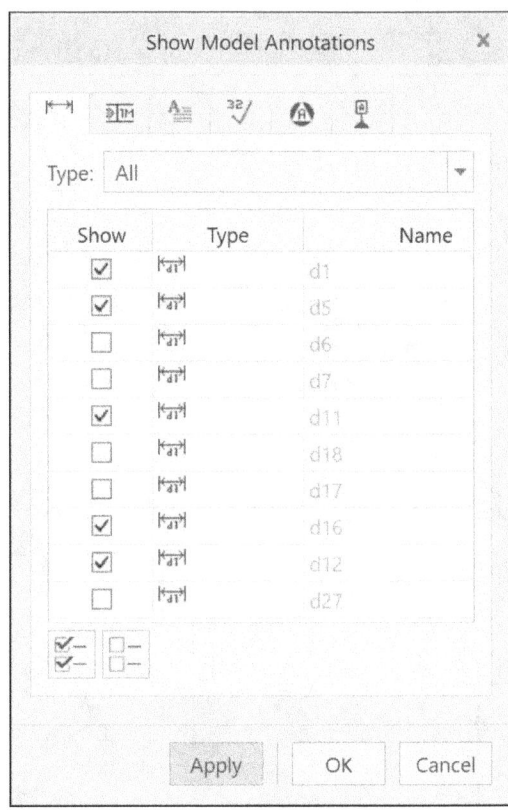 Show Model Annotations > select the front view > press **Ctrl** key > select the right side view > release the **Ctrl** key > check the dimensions required for the detail [Figs. 14.14(c-e)] > **Apply**

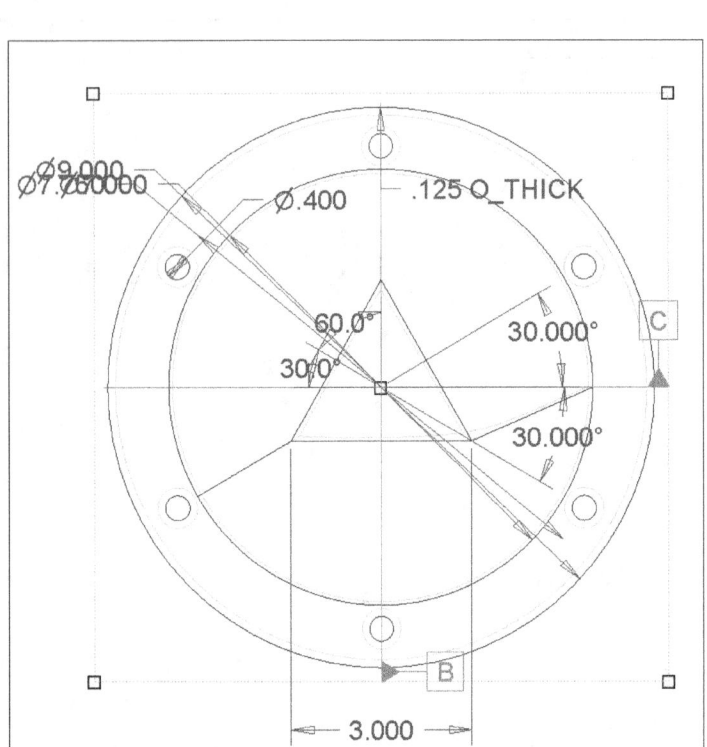

Figure 14.14(c) Model Annotations
(your dimensions may differently)

Figure 14.14(d) Checked Dimensions

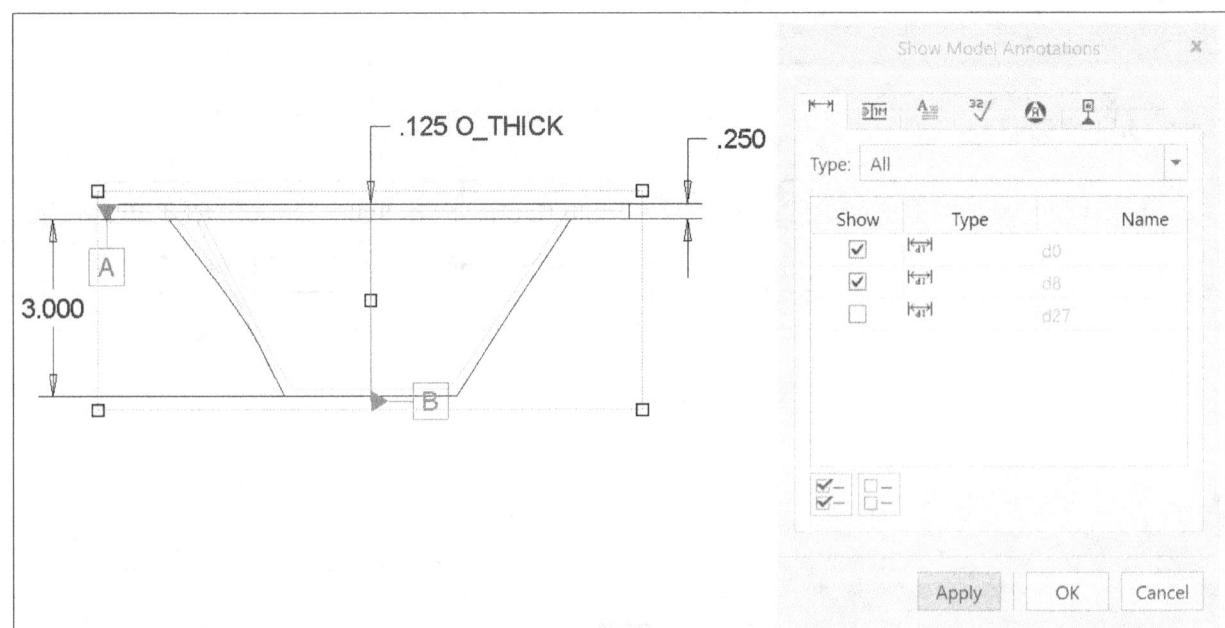

Figure 14.14(e) Top View *(your dimensions may differently)*

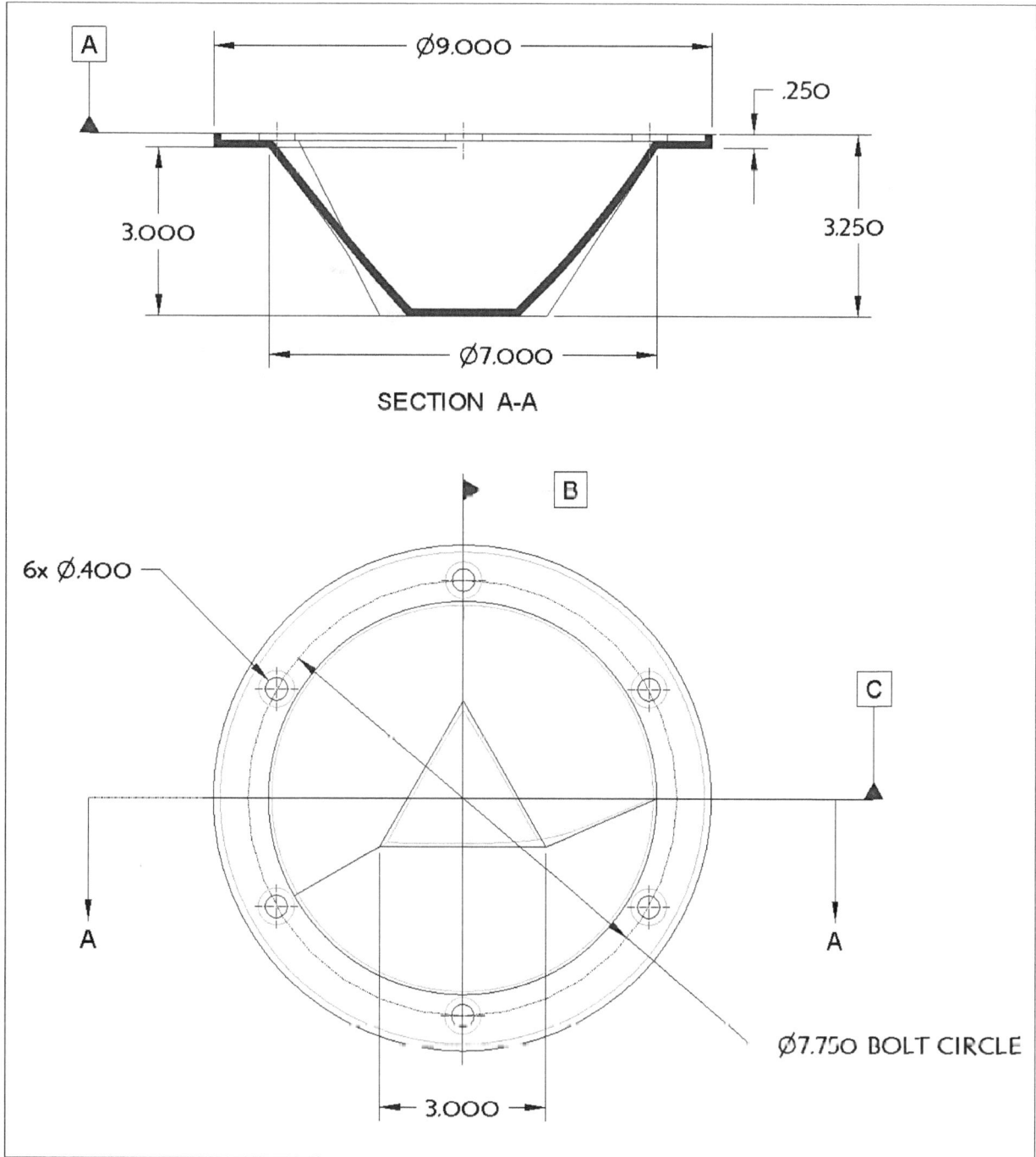

Figure 14.14(f) Drawing

Click: [icon] > **Ctrl+S** > **OK** [Fig. 14.14(g)] > **File** > **Manage File** > **Delete Old Versions** > **Enter** > **File** > **Save As** > Type [icon] > **Zip File (*.zip)** > **OK** > **upload** the zip file to your course interface or attach to an email and send to your instructor and/or yourself > **File** > **Close** > **File** > **Close** > **File** > **Exit** > **Yes**

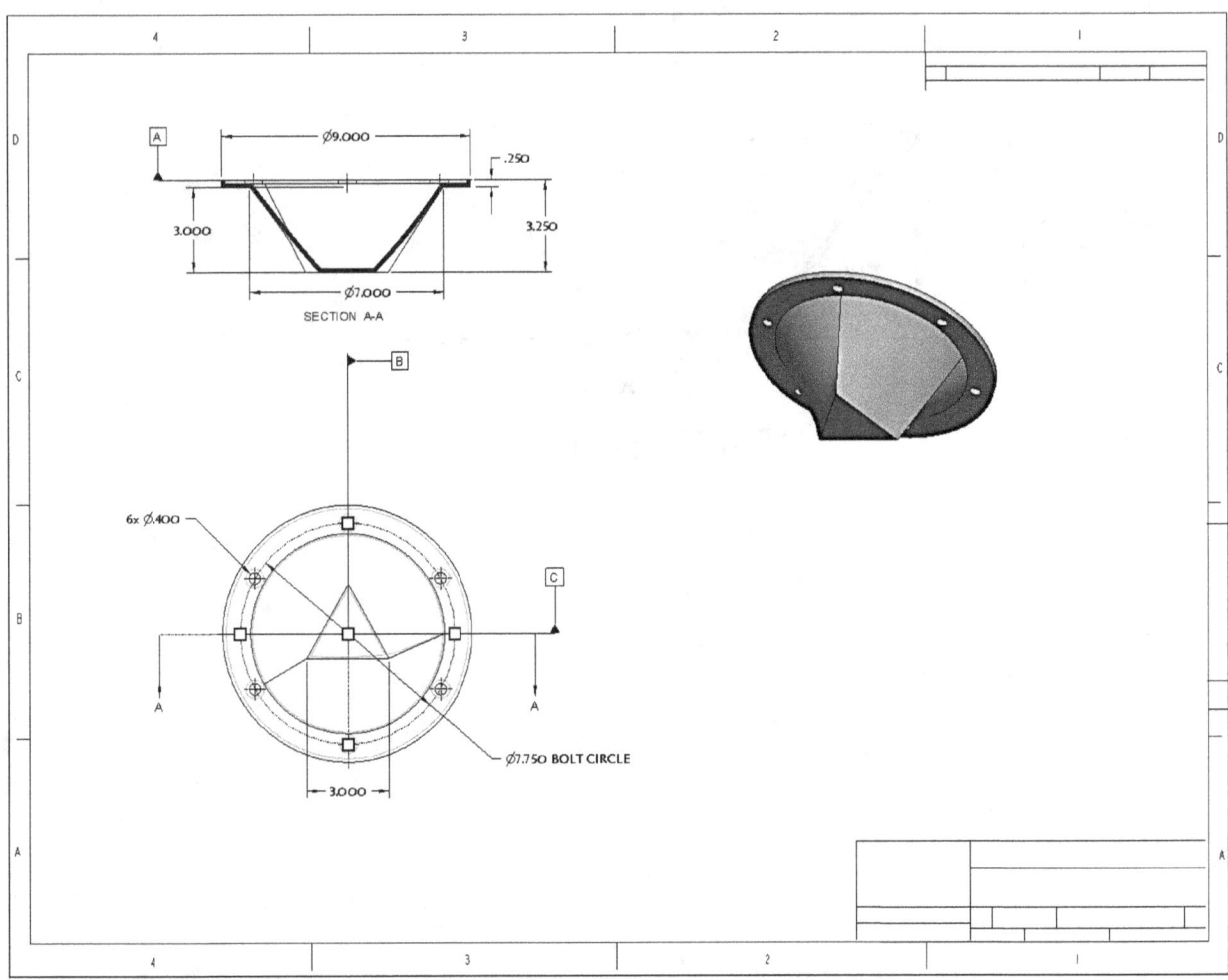

Figure 14.14(g) Completed Detail Drawing

For additional projects, see *www.cad-resources.com* > *click on the image of your book cover.*

Lesson 15 Sweeps

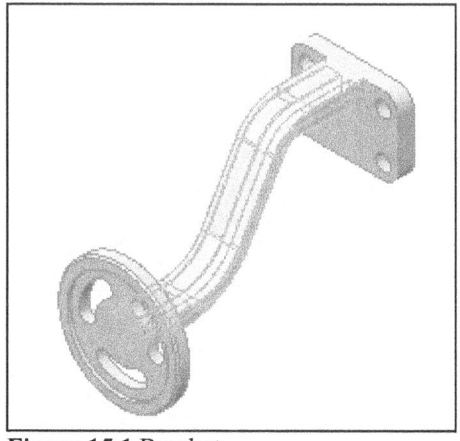
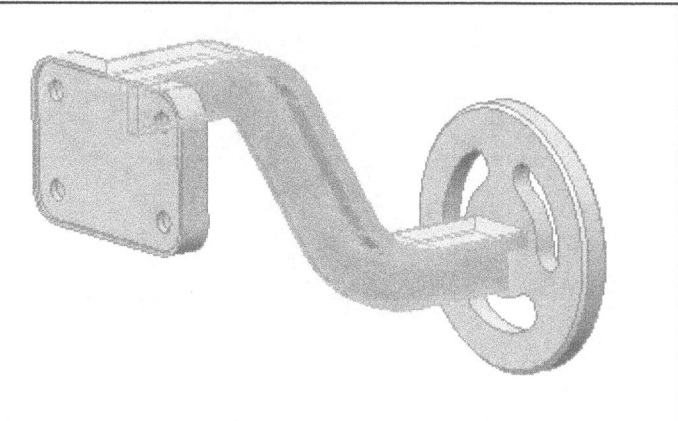

Figure 15.1 Bracket

OBJECTIVES

- Create a **sweep** feature
- Sketch a **Trajectory** for a sweep
- Sketch and locate a **Sweep section**
- **Render** a model

REFERENCES AND RESOURCES

For **Resources** go to **www.cad-resources.com** > click on the PTC Creo Parametric 5.0 Book cover

- Lesson 15 Lecture at **YouTube Creo Parametric Lecture Videos**

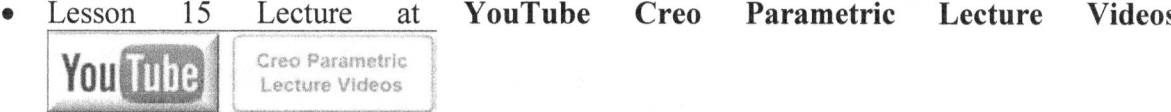

- Lesson 15 3D PDF models embedded in a PDF

SWEEPS

A Sweep is created by sketching or selecting a *trajectory* and then sketching a *section* to follow along it. The Bracket, shown in Figure 15.1, uses a simple sweep in its design. A *constant-section sweep* (Fig. 15.2) can use either trajectory geometry sketched at the time of feature creation or a trajectory made up of selected datum curves or edges. The trajectory [Figs. 15.3 (a-c)] must have adjacent reference surfaces or be planar. When defining a sweep, Creo Parametric checks the specified trajectory for validity and establishes normal surfaces.

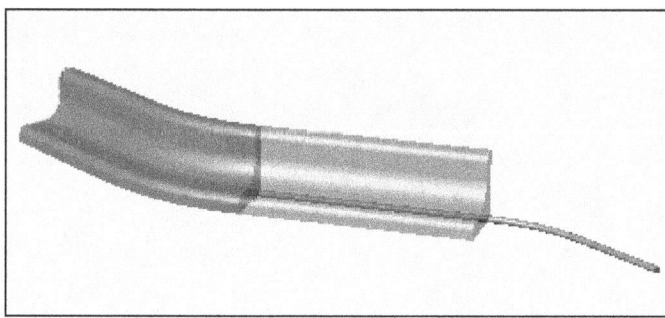

Figure 15.2 Sweep

651

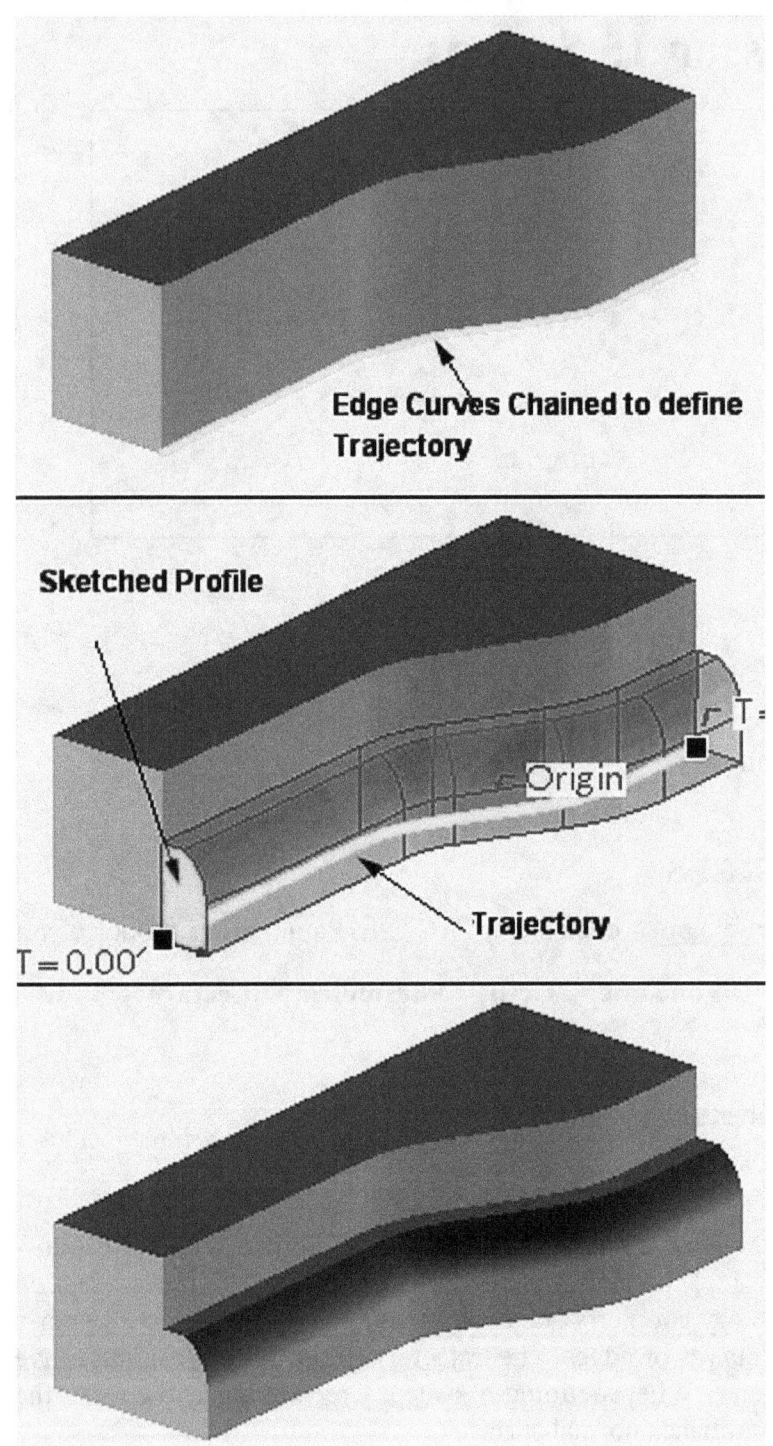

Figures 15.3(a-c) Sweep Trajectory and Section

Lesson 15 STEPS

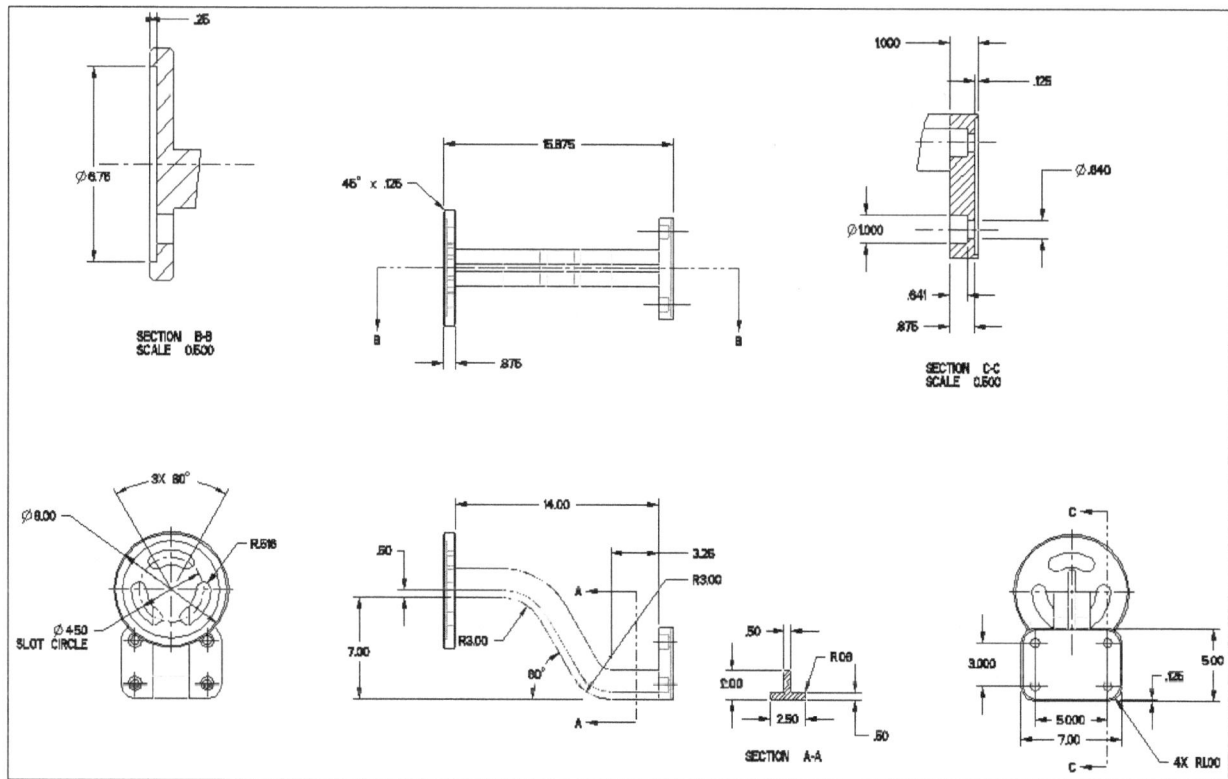

Figure 15.4 Bracket Detail

Bracket

The Bracket (Fig. 15.4) requires the use of the Sweep command. The T-shaped section is swept along the sketched *trajectory.*

Start a new part. Press: **Ctrl+N** > Name **bracket** > **Enter** > **File** > **Prepare** > **Model Properties**

- **Material** = al6061.mtl
- **Units** = Inch lbm Second

Set Datum ⟨-A-⟩ and **Rename** the default datum planes and coordinate system:
- Datum **TOP** = **C**
- Datum **FRONT** = **A**
- Datum **RIGHT** = **B**
- Coordinate System = **CSYS_SWEEP**

In the Model Tree, click on **BRACKET.PRT** > **RMB** > **Info** > **Model** [Fig. 15.5(a)] > ⟨icon⟩ close the browser panel > **LMB** to deselect

Color: set the model color as desired [Fig. 15.5(b)]

Model Info : BRACKET

PART NAME :   BRACKET

MATERIAL FILENAME: AL6061

Units:		Length:	Mass: Force:		Time: Temperature:
Inch lbm Second (Creo Parametric Default) in			lbm	in lbm / sec^2 sec	F

Feature List

No.	ID	Name	Type	Actions	Sup Order
1	1	B	DATUM PLANE		---
2	3	C	DATUM PLANE		---
3	5	A	DATUM PLANE		---
4	7	CSYS_SWEEP	COORDINATE SYSTEM		---

Figure 15.5(a) Bracket Information (*your Browser Window information may appear differently*)

Figure 15.5(b) Bracket Color (your display may appear differently)

The extrusions on both sides of the swept feature are to be created with the dimensions shown in Figures 15.6(a-j).

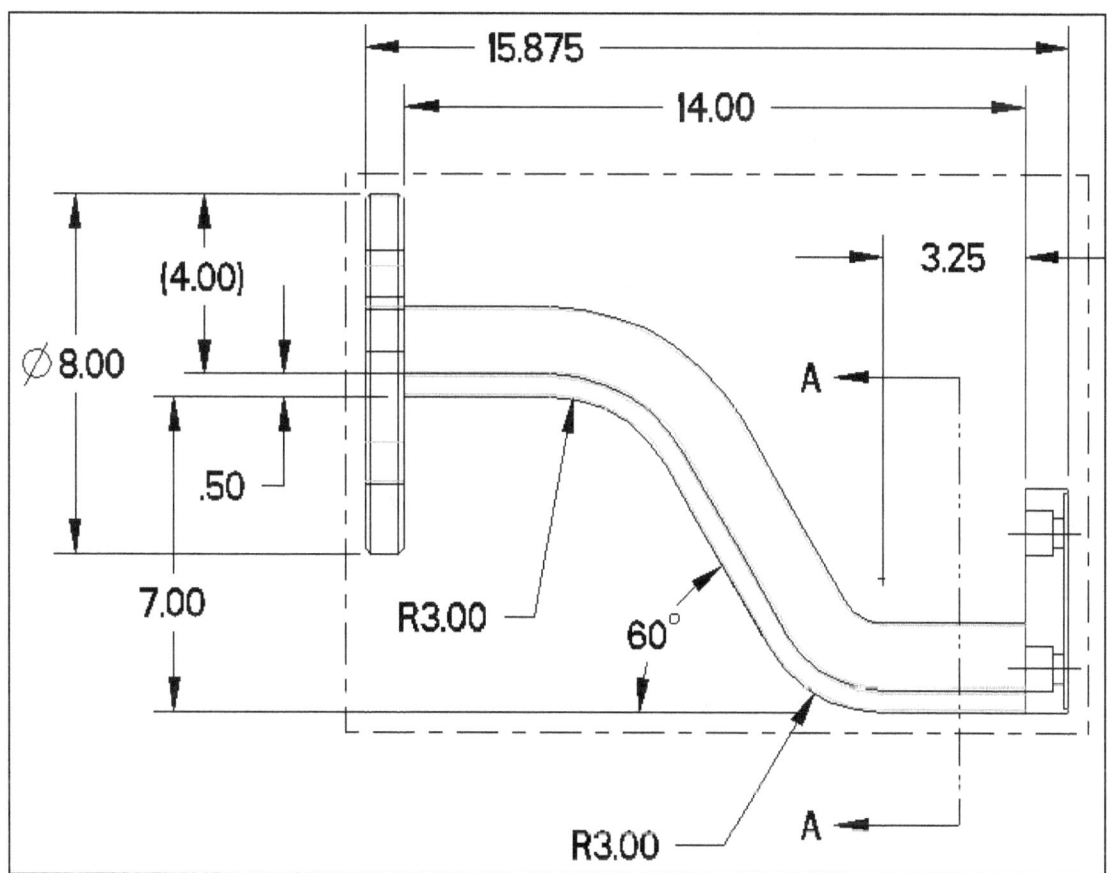

Figure 15.6(a) Bracket Drawing, Front View

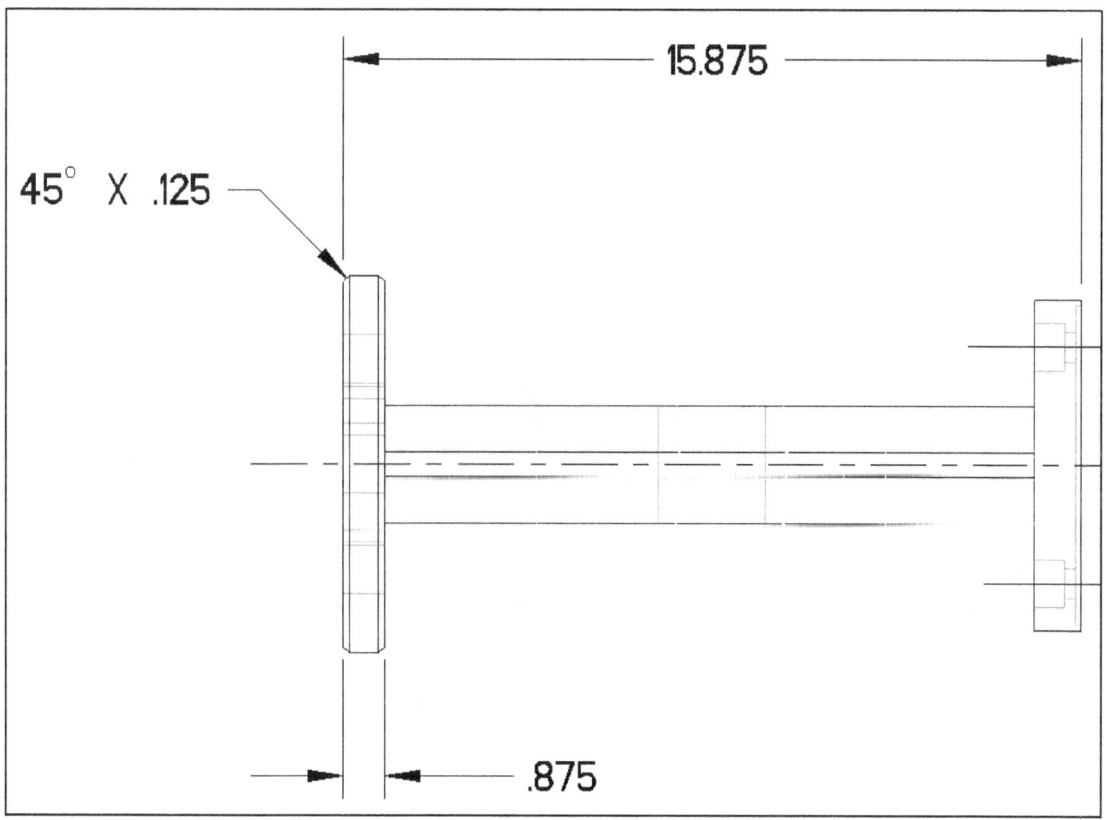

Figure 15.6(b) Bracket Drawing, Top View

655

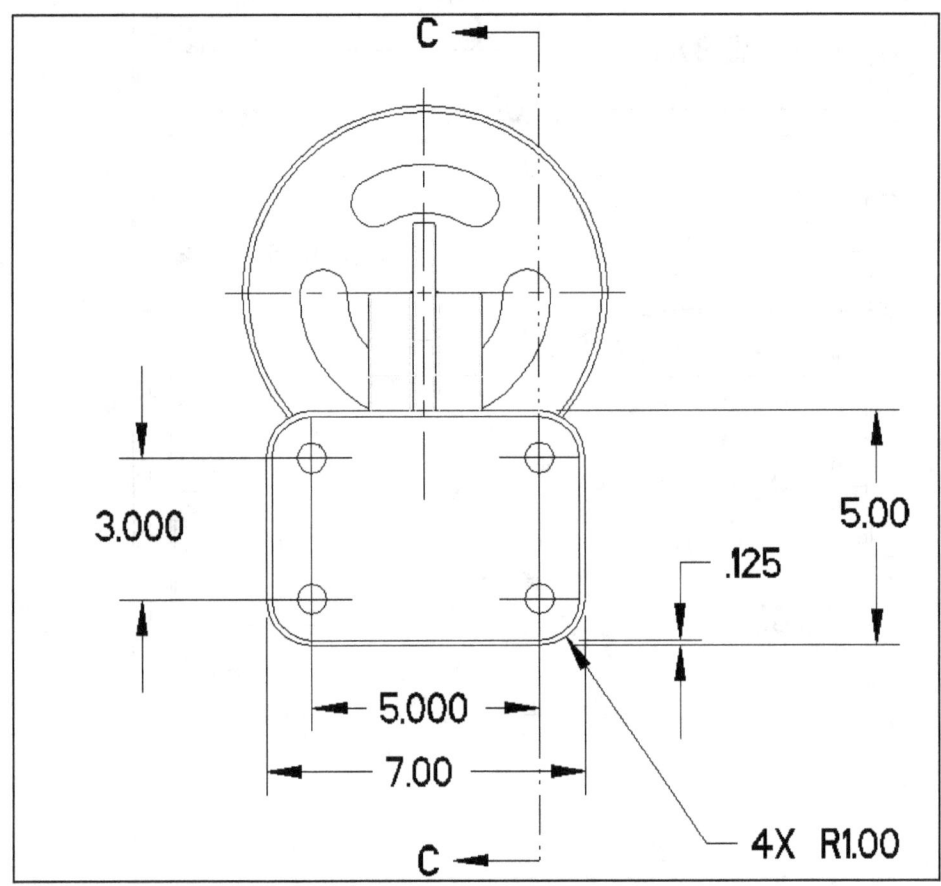

Figure 15.6(c) Bracket Drawing, Right Side View

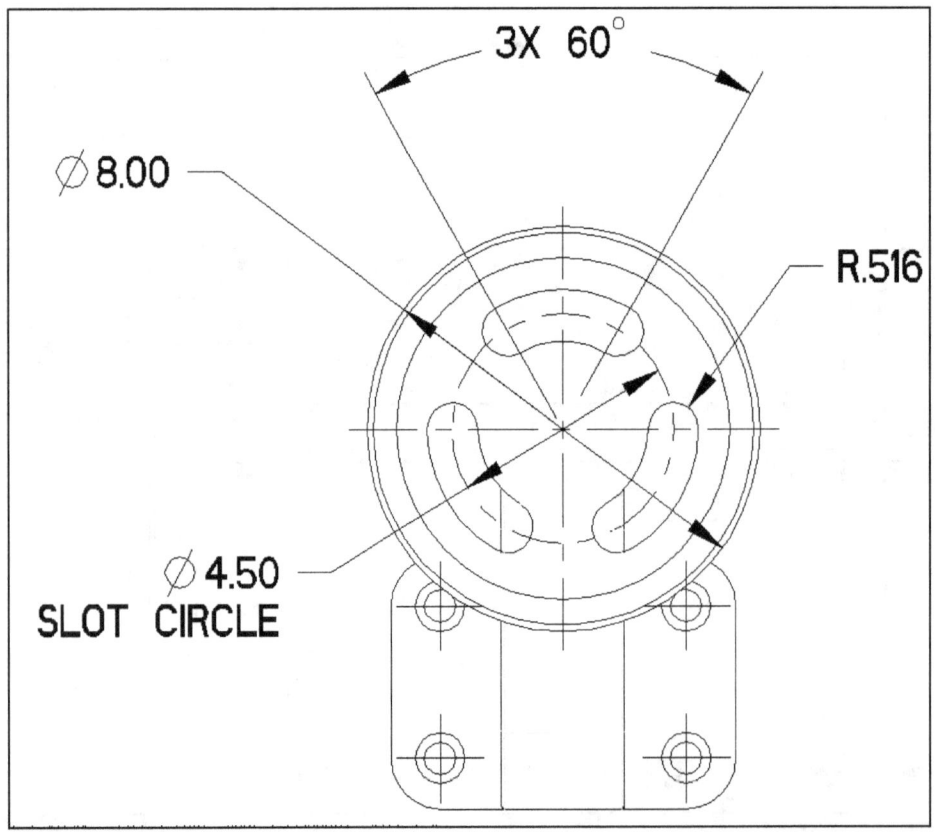

Figure 15.6(d) Bracket Drawing, Left Side View

656

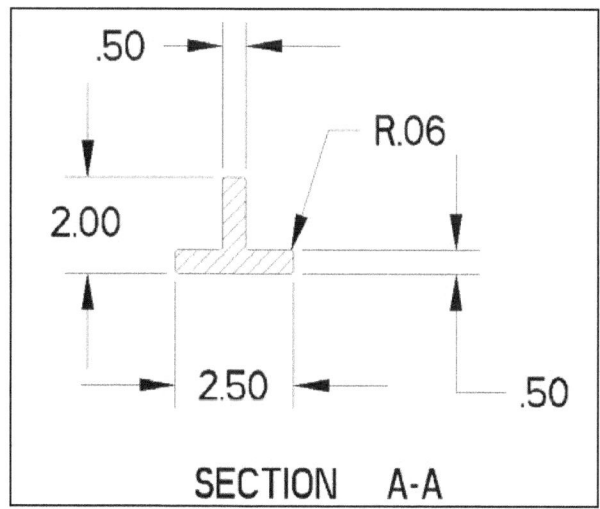

Figure 15.6(e) SECTION A-A

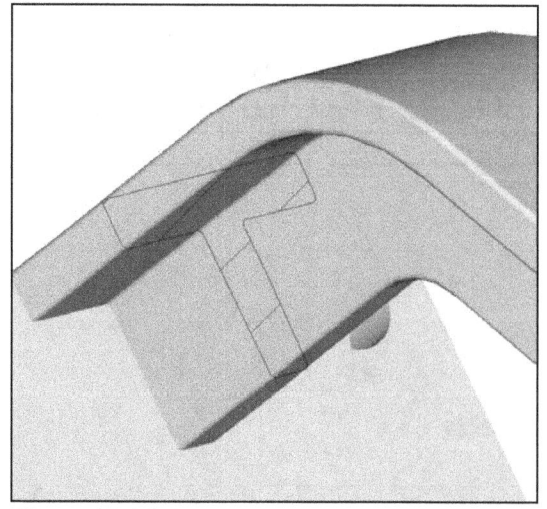

Figure 15.6(f) Swept Arm

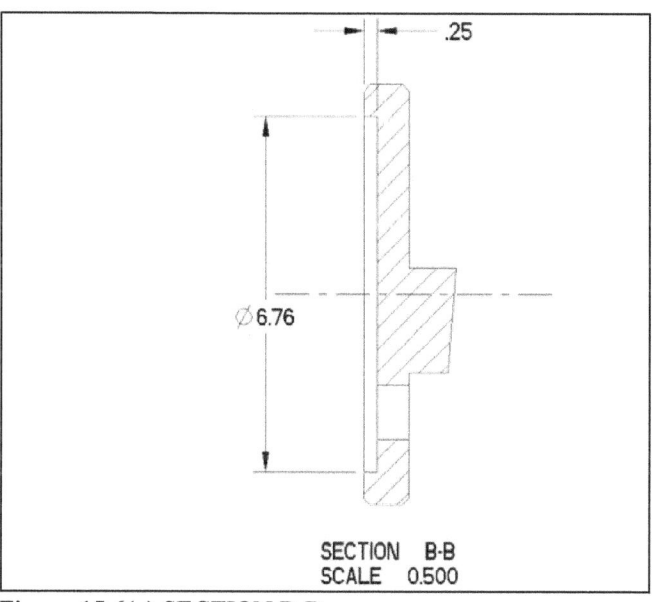

Figure 15.6(g) SECTION B-B

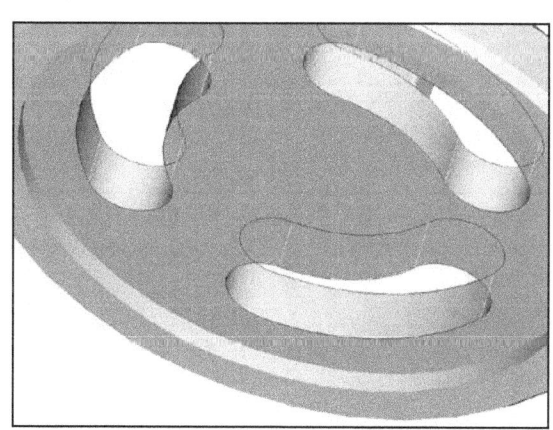

Figure 15.6(h) Cut

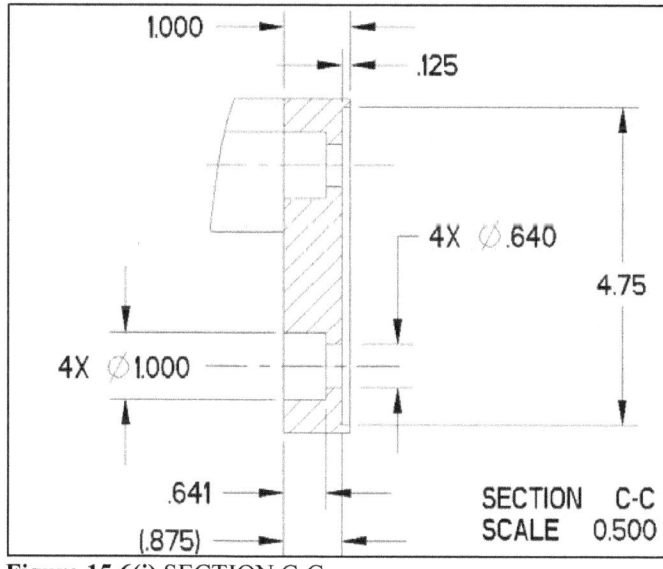

Figure 15.6(i) SECTION C-C

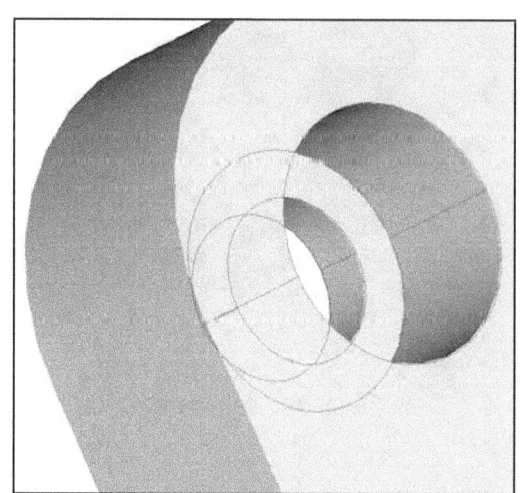

Figure 15.6(j) Counterbore Hole

657

Start by modeling the first feature [Fig. 15.7(a)], it will be used to establish the sweep's position in space. Sketch the extrusion on datum **A** (FRONT) [Fig. 15.7(b)] and centered on **B** (RIGHT) and **C** (TOP). **> Ctrl+D > LMB > Ctrl+S > OK**

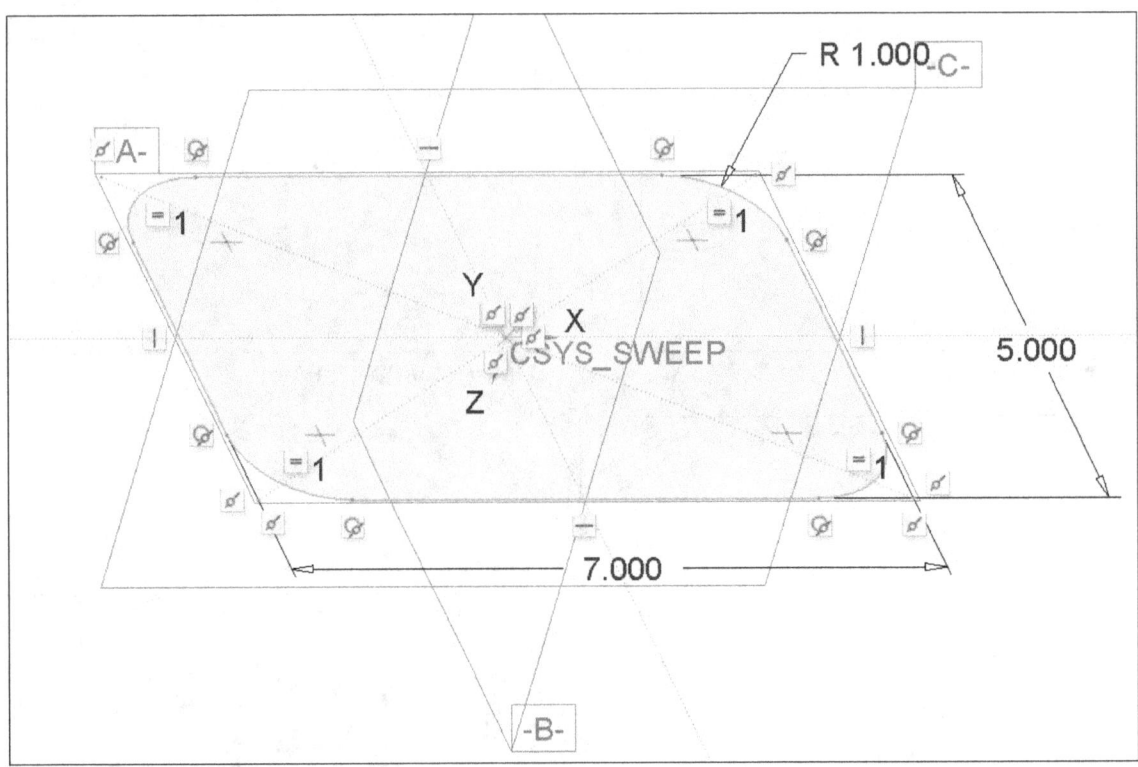

Figure 15.7(a) Bracket's First Section

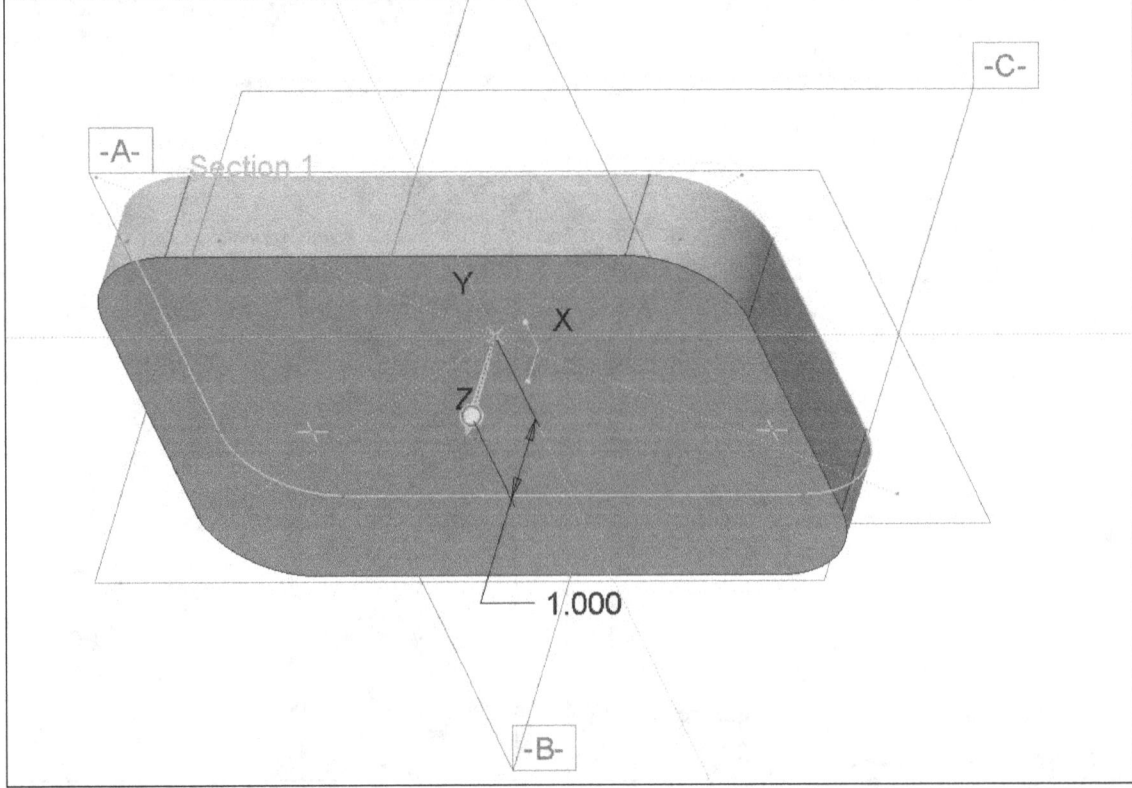

Figure 15.7(b) Completed Extrusion

Click: [Sweep] > [Datum] > [≈] **Sketch** > select datum **B** as the sketching plane [Fig. 15.8(a)] > Orientation **Top** > **Sketch** > [icon] **Sketch View** > **RMB** > **References** > in the References dialog box, delete datums **A** and **C** and add the front and top faces of the first extrusion > **Solve** > Fully Placed [Fig. 15.8(b)] > **Close**

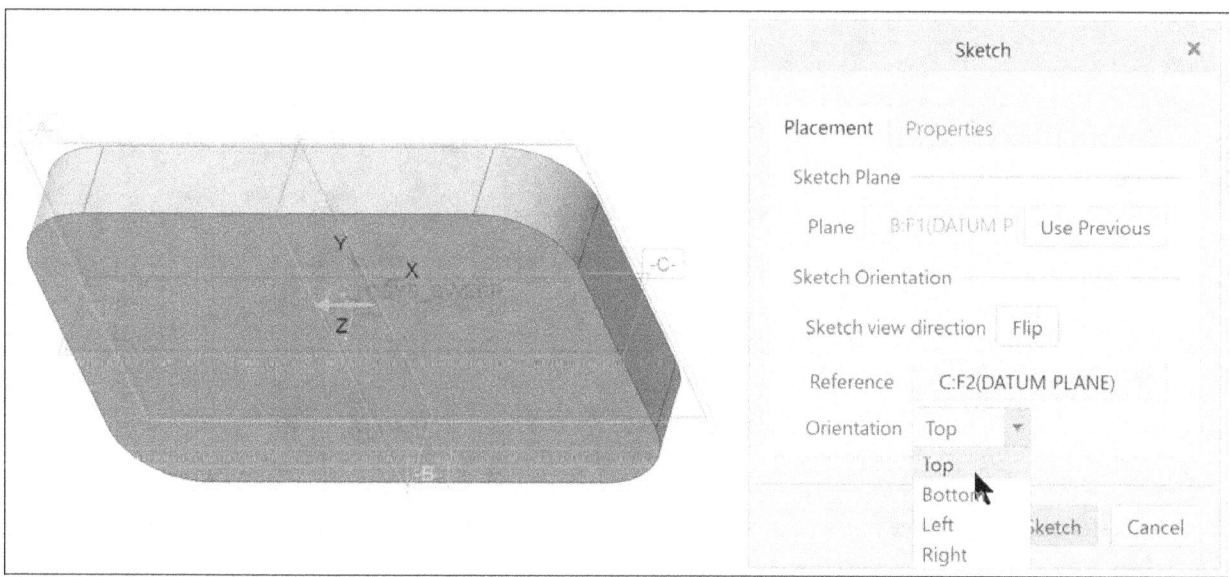

Figure 15.8(a) Select Datum B as the Trajectory Sketching Plane, Orientation is Top

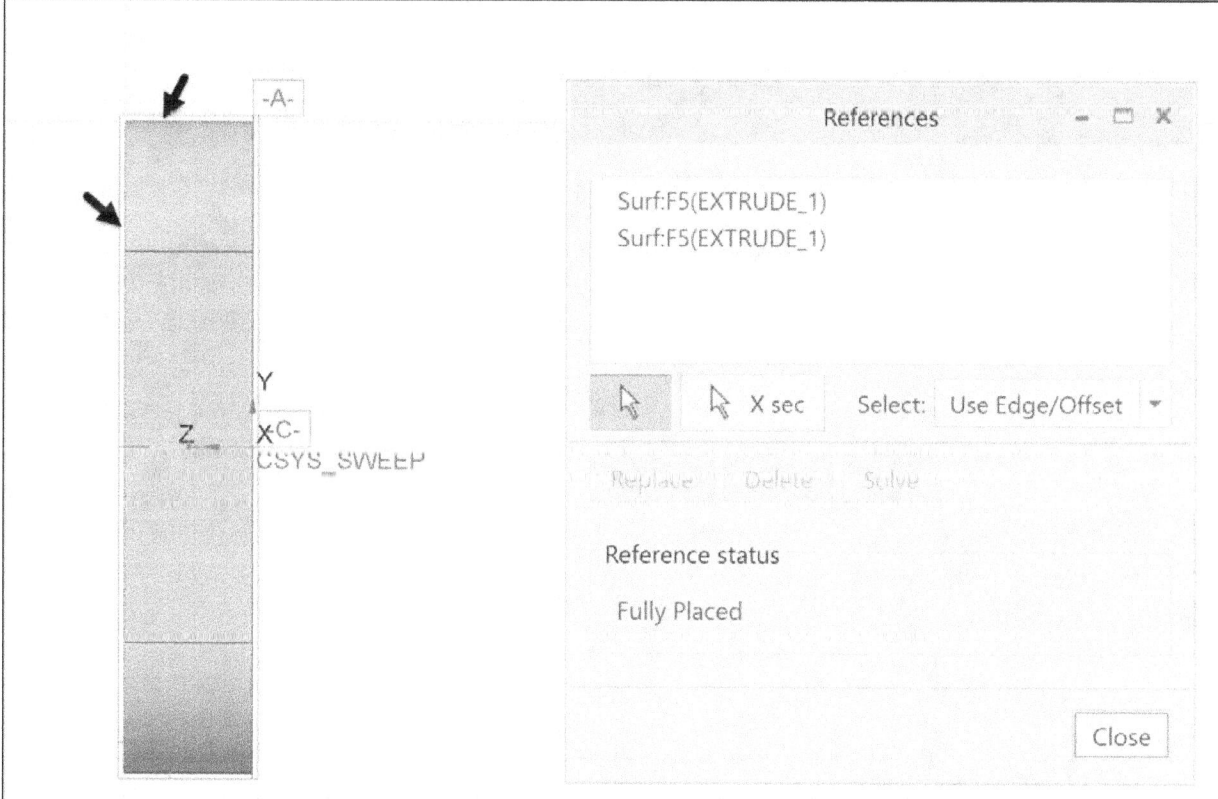

Figure 15.8(b) Delete Datums A and C and add the Front and Top Surfaces as References

Sketch [Fig. 15.8(c)], add fillets, dimension, and modify the trajectory [Fig. 15.8(d)] > ✔

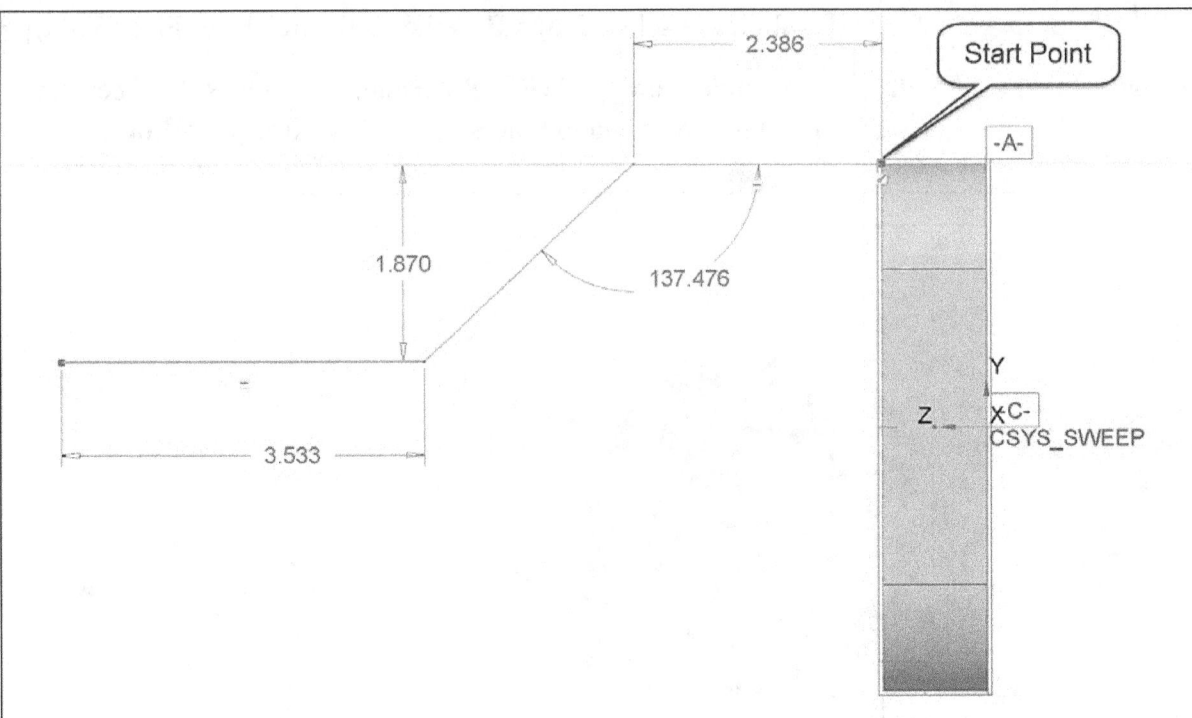

Figure 15.8(c) Sketch the Three Lines. Start the trajectory by sketching a horizontal line from this position.

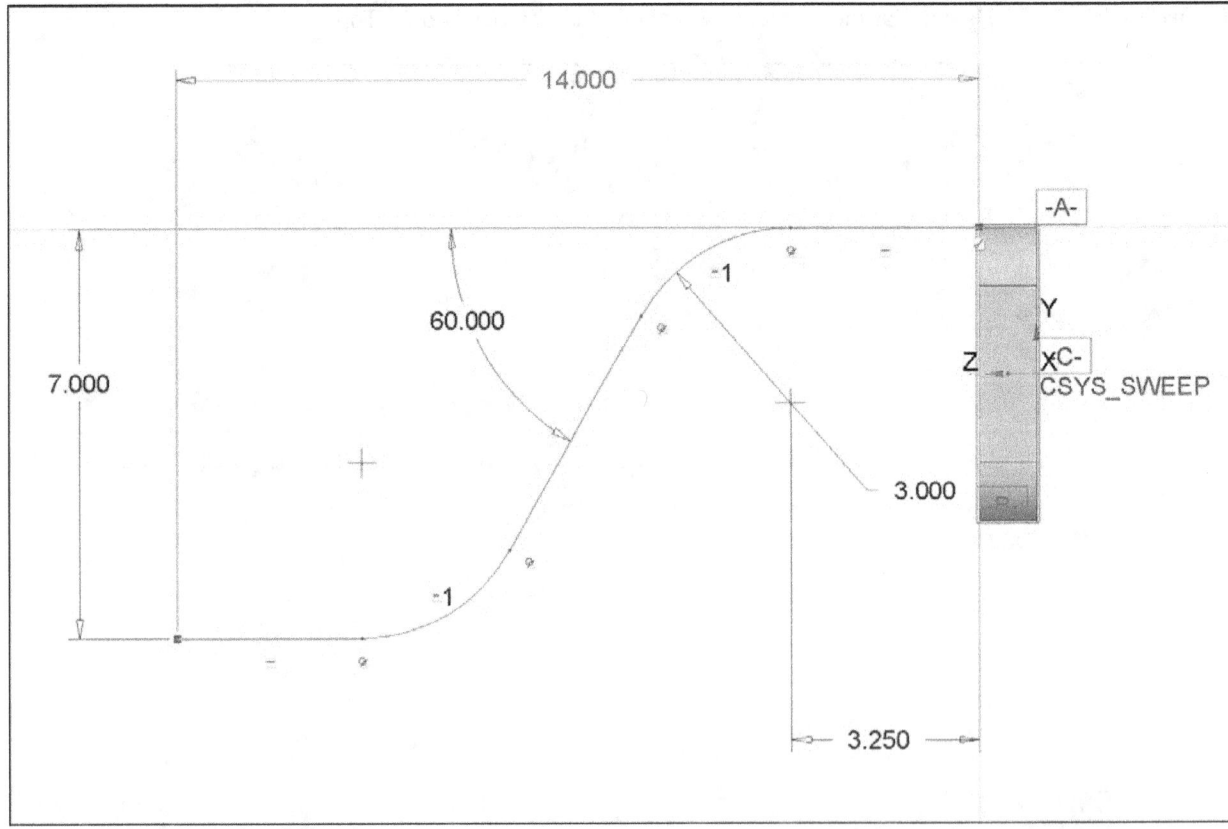

Figure 15.8(d) Completed Sketch

Press: **MMB** to spin the model [Fig. 15.8(e)] > ▶ **Resumes the previously paused tool** > ⬚
[Fig. 15.8(f)]

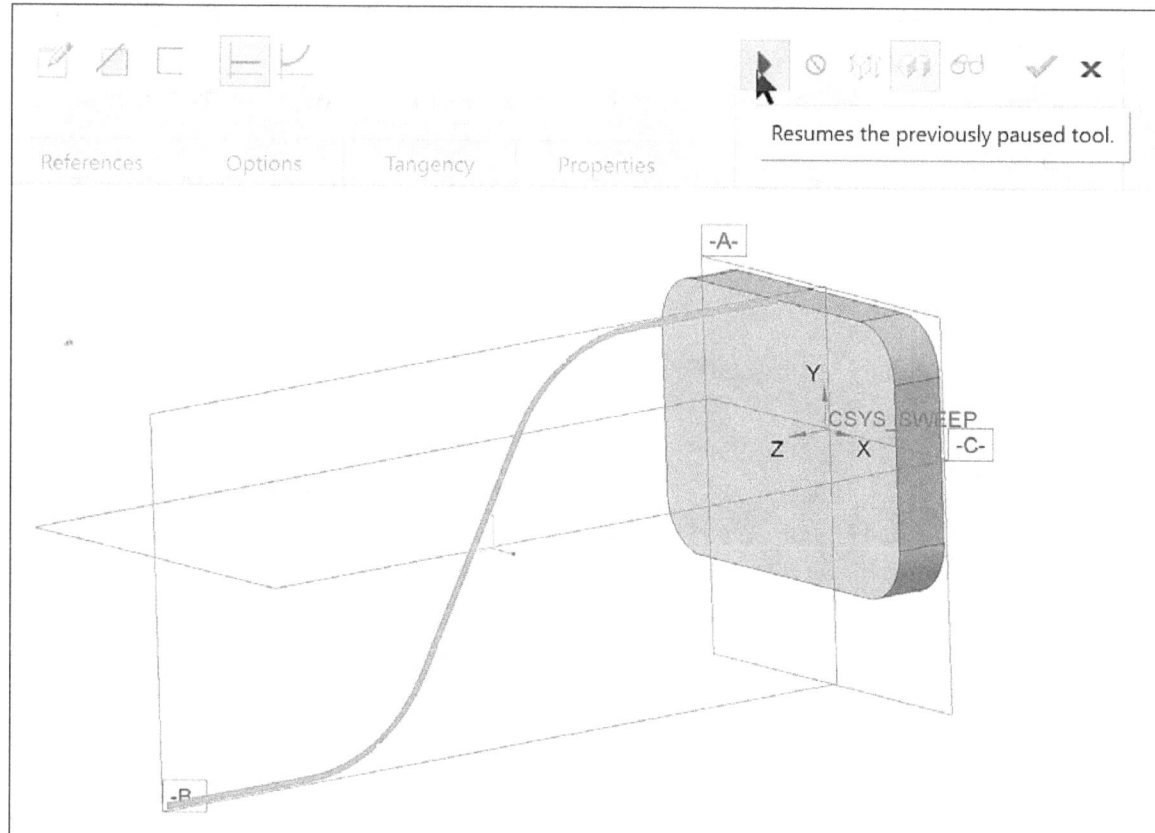

Figure 15.8(e) Resumes the previously paused tool

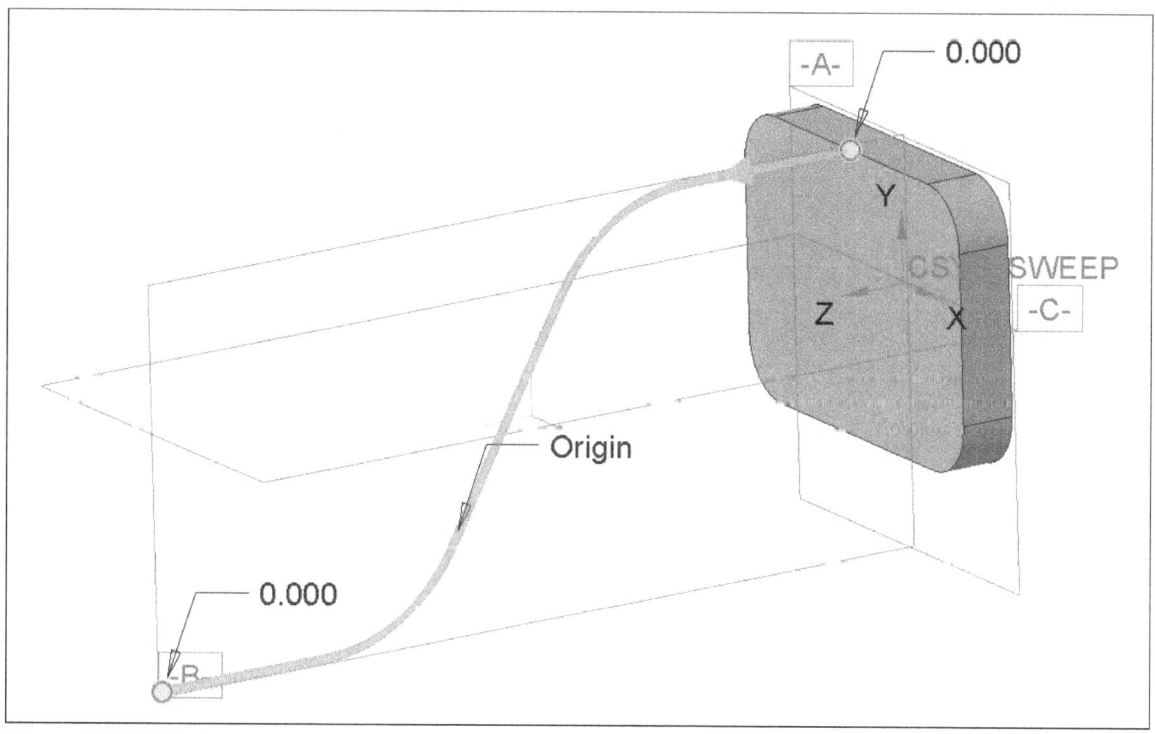

Figure 15.8(f) Trajectory

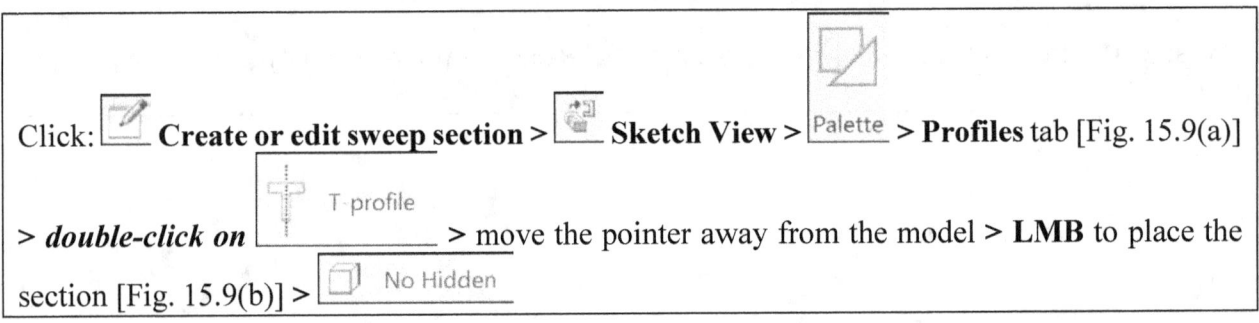

Click: **Create or edit sweep section** > **Sketch View** > Palette > **Profiles** tab [Fig. 15.9(a)]

> *double-click on* T-profile > move the pointer away from the model > **LMB** to place the section [Fig. 15.9(b)] > No Hidden

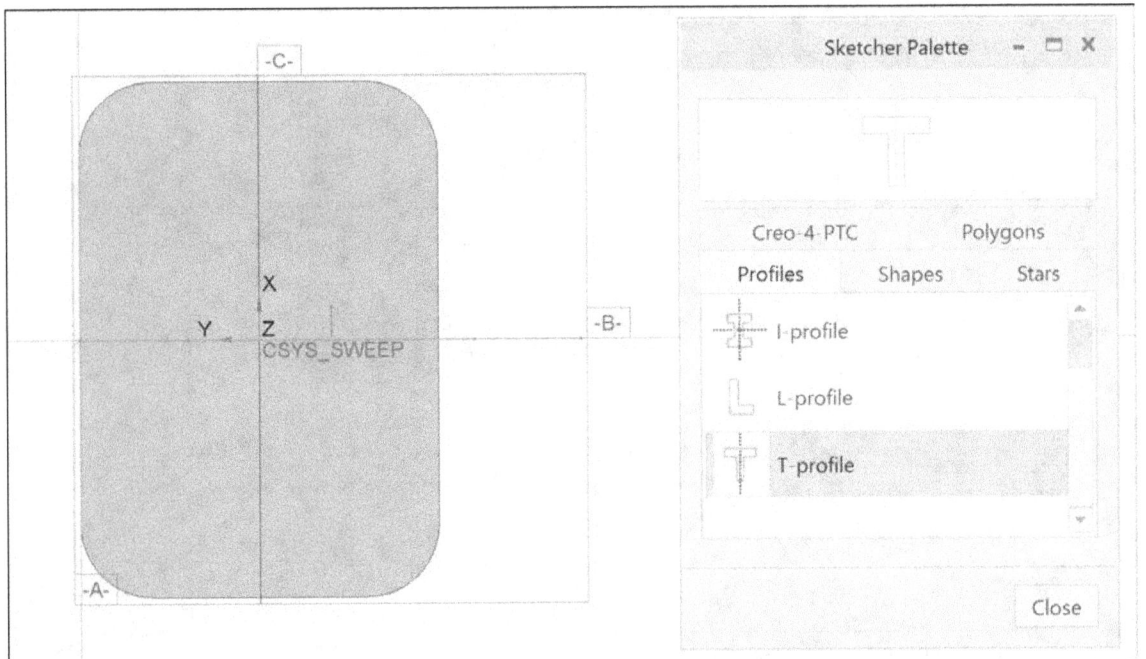

Figure 15.9(a) Sketcher Palette

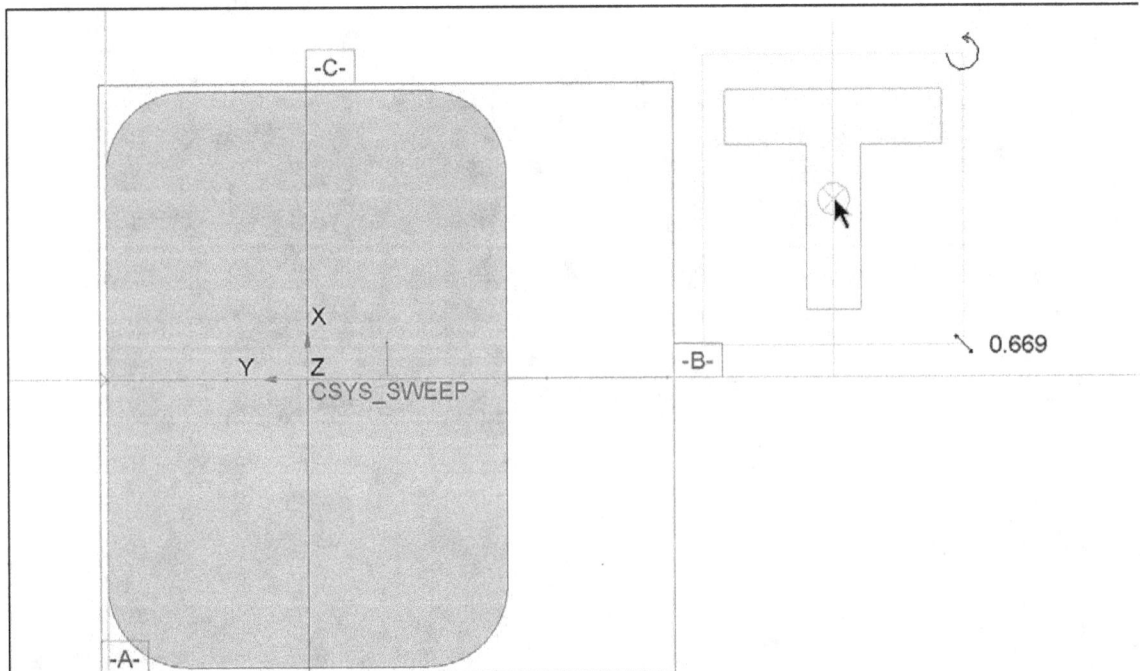

Figure 15.9(b) Place the Section

Place the pointer on the position handle [Fig. 15.9(c)] > press and hold down the **RMB** > move the position handle [Fig. 15.9(d)] (Note the **M**idpoint constraint) > drop the handle (release the **RMB**) in the new position [Fig. 15.9(e)] > place the pointer on the rotate handle [Fig. 15.9(f)] > press and hold down the **LMB** > move the pointer to rotate the section 90 degrees [Fig. 15.9(g)] > drop the handle (release the **LMB**) in the new position [Fig. 15.9(h)] > place the pointer on the position handle [Fig. 15.9(i)] > press and hold down the **LMB** > move the section to the start point of the trajectory [Fig. 15.9(j)] > drop the section in the new position [Fig. 15.9(k)] >  from the **Rotate Resize** ribbon > **Close** the Sketcher Palette dialog box

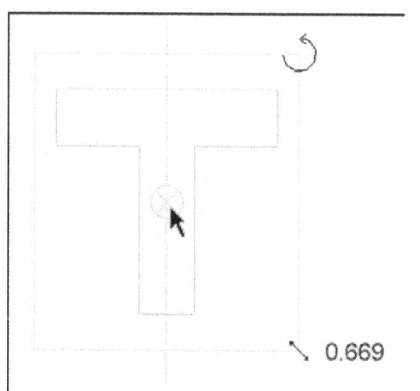
Figure 15.9(c) RMB on Move Handle

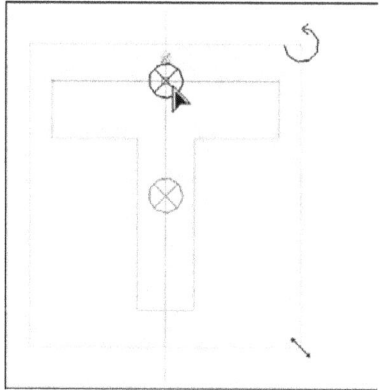
Figure 15.9(d) Move the Handle

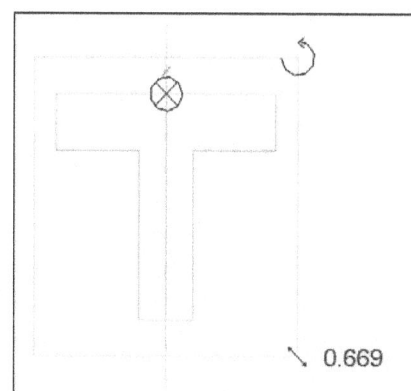
Figure 15.9(e) Drop the Handle

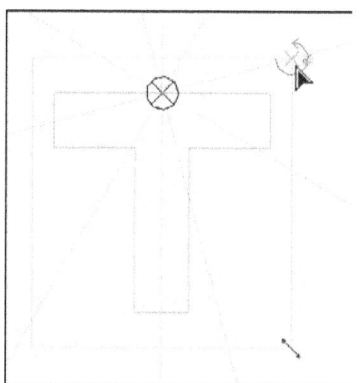
Figure 15.9(f) LMB on Rotate Handle

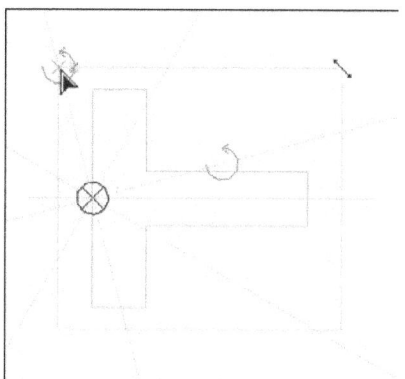
Figure 15.9(g) Rotate **90** Degrees

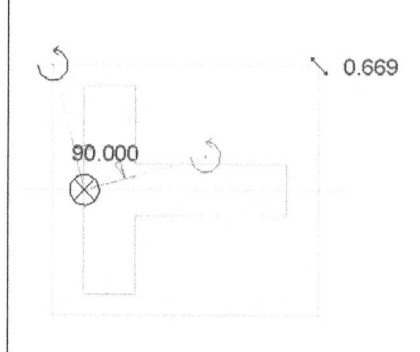
Figure 15.9(h) Rotated Section

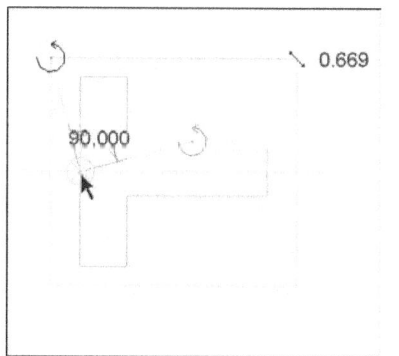
Figure 15.9(i) LMB on Move Handle

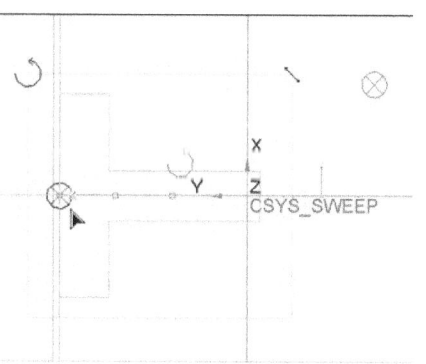
Figure 15.9(j) Place the Section

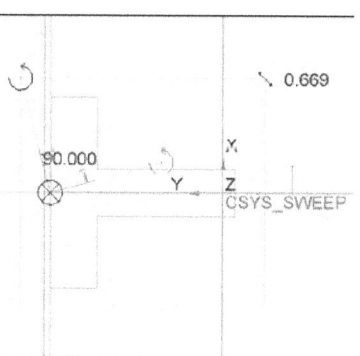
Figure 15.9(k) Drop Section

Add *eight* equal fillets > add constraints and dimensions [Fig. 15.9(l)] > using the dimensions of the detail drawing [Fig. 15.9(m)], modify the section (eliminate the extra **.50** dimension my making them equal) [Fig. 15.9(n)] > ☑ > spin the part

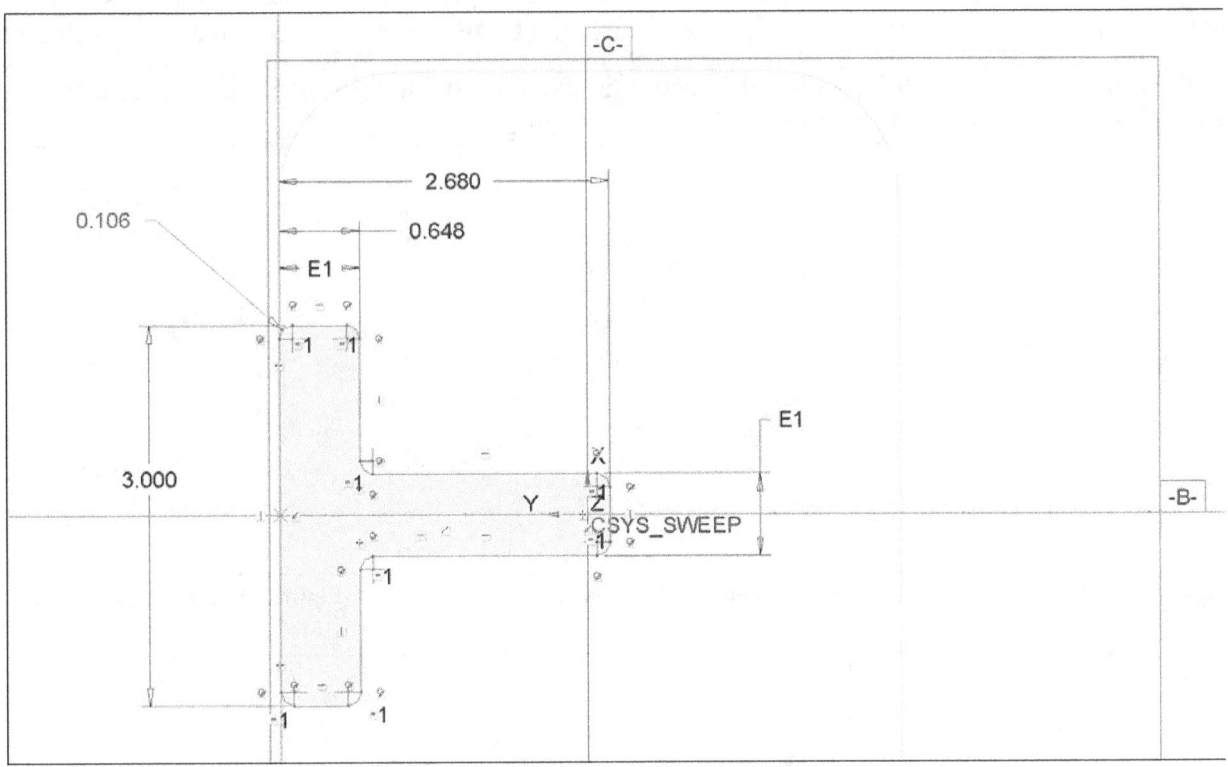

Figure 15.9(l) Add Eight Fillets, Constraints, and the Dimensioning Scheme

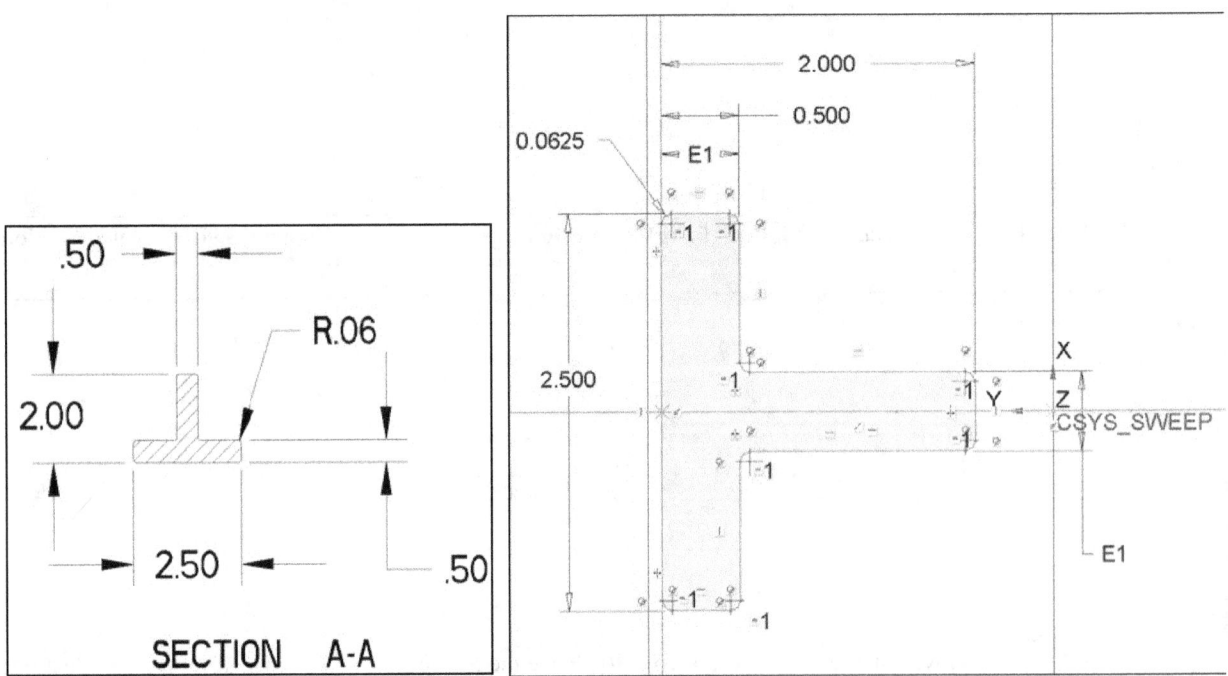

Figure 15.9(m) Section AA from Detail Drawing

Figure 15.9(n) Modified Sketch

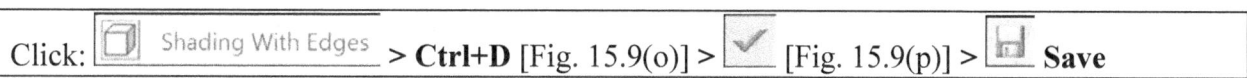

Click: [icon] Shading With Edges **> Ctrl+D** [Fig. 15.9(o)] > [✓] [Fig. 15.9(p)] > [💾] **Save**

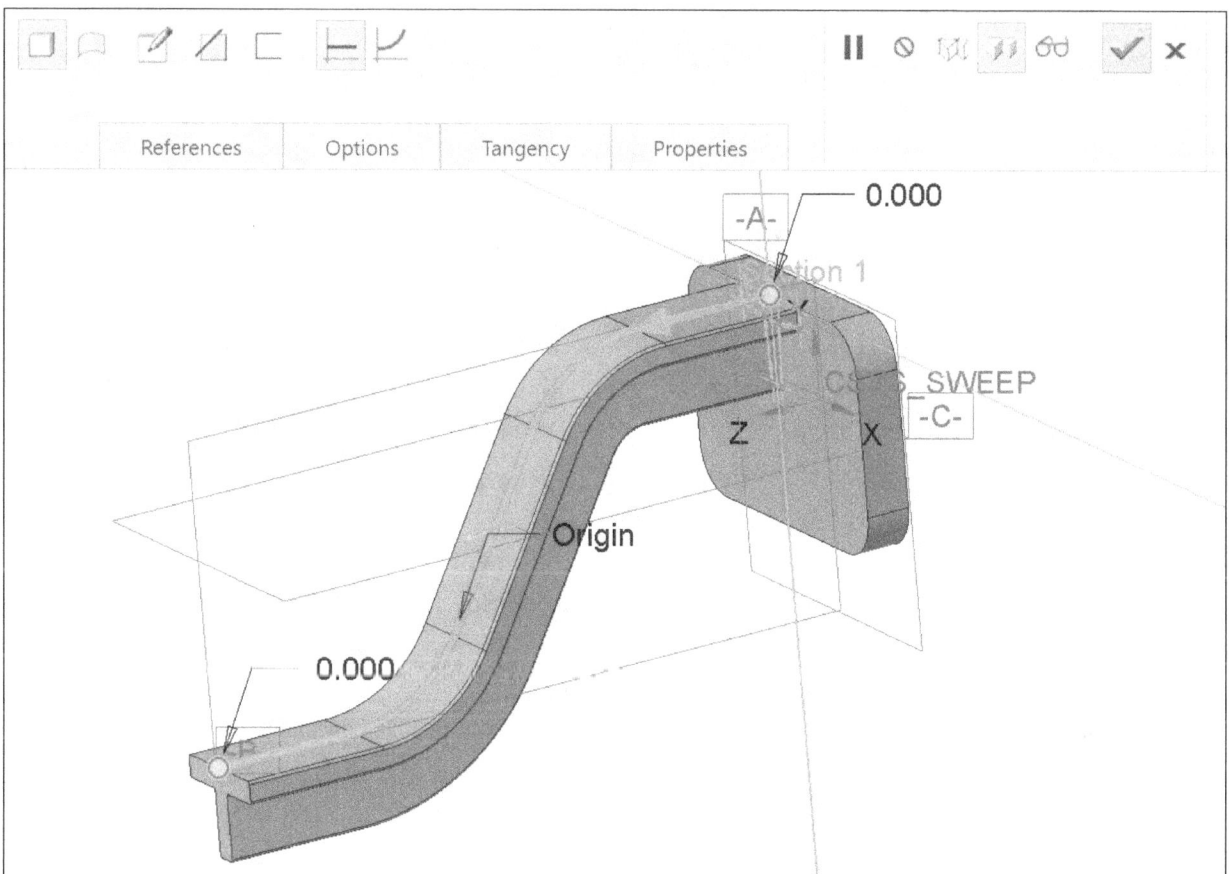

Figure 15.9(o) Shaded Sweep Preview

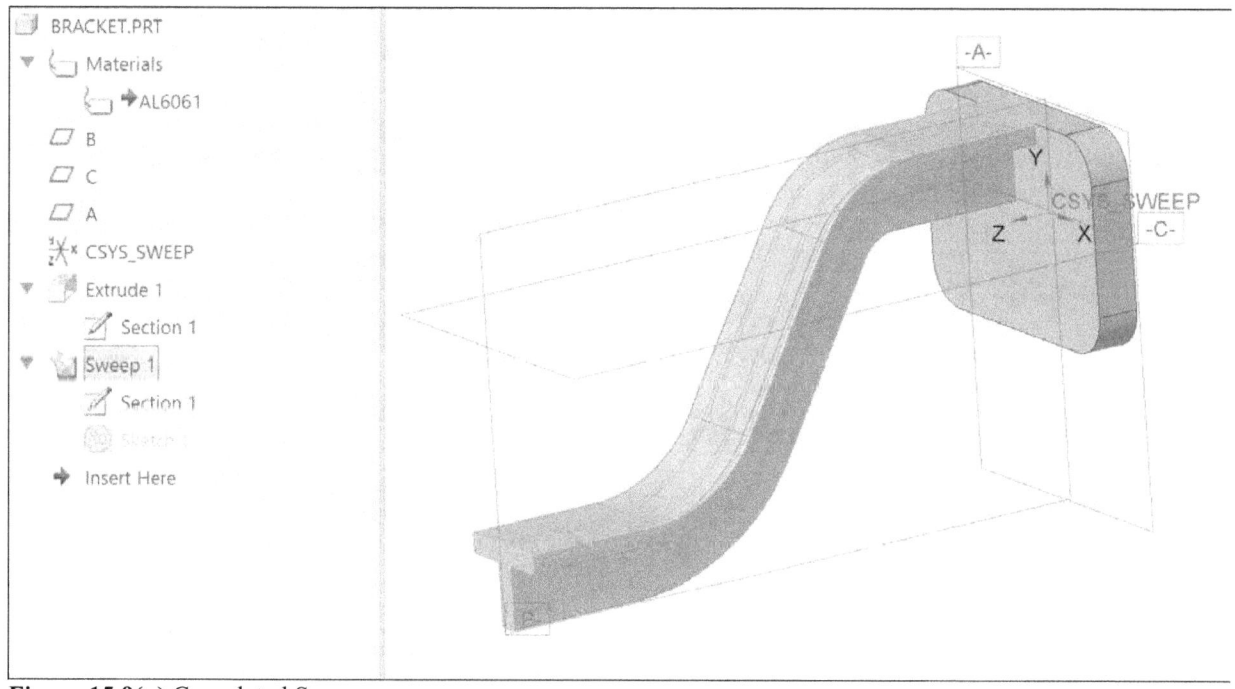

Figure 15.9(p) Completed Sweep

Add the next extrusion. Add a Reference on bottom of the **T**. (∅**8.00** by **.875**) [Figs. 15.10(a-b)] > model the cut feature (∅**6.76** by **.250** deep) [Figs. 15.10(c-d)]

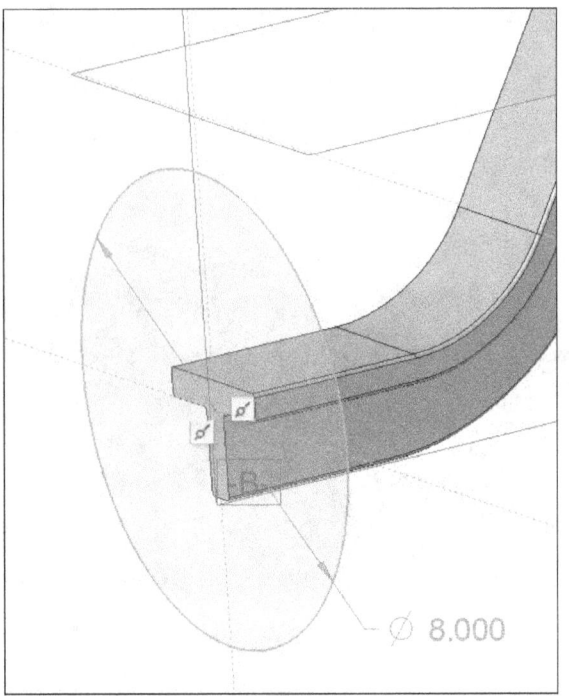

Figure 15.10(a) 8.00 Diameter

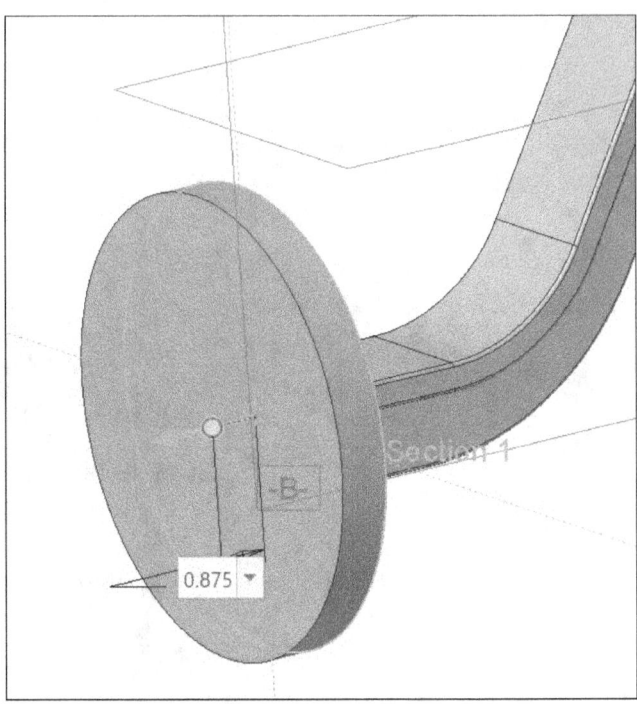

Figure 15.10(b) .875 Thickness

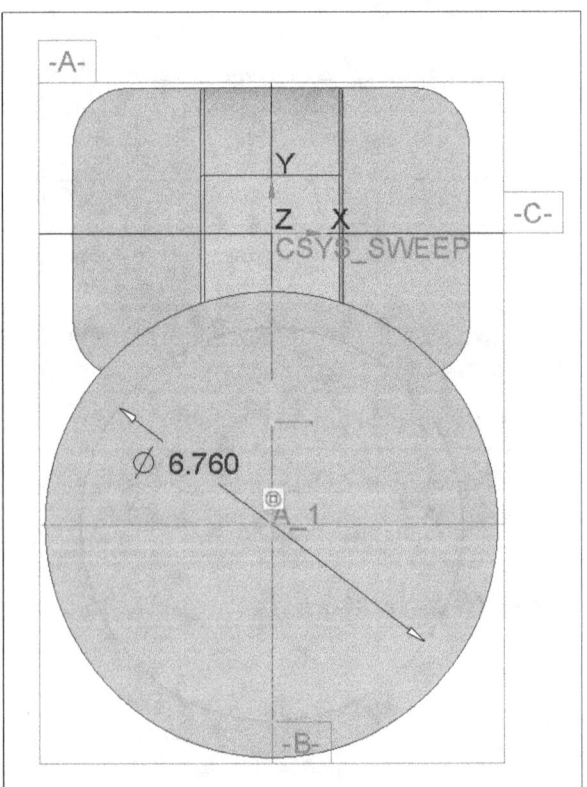

Figure 15.10(c) 6.76 Diameter

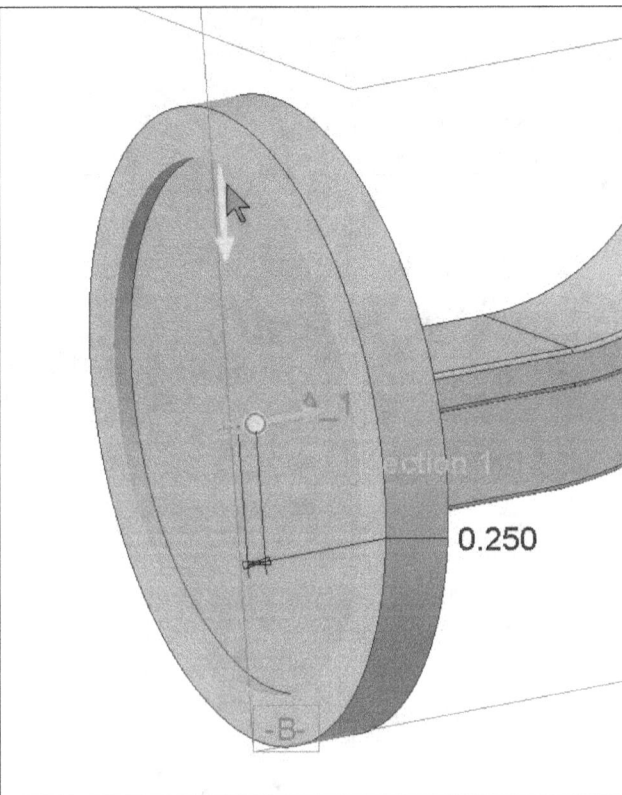

Figure 15.10(d) .250 Cut

Add chamfers (**45° X .125**) [Fig. 15.10(e)] > **Ctrl+S** > the next feature will be the slot [Fig. 15.11(a)]

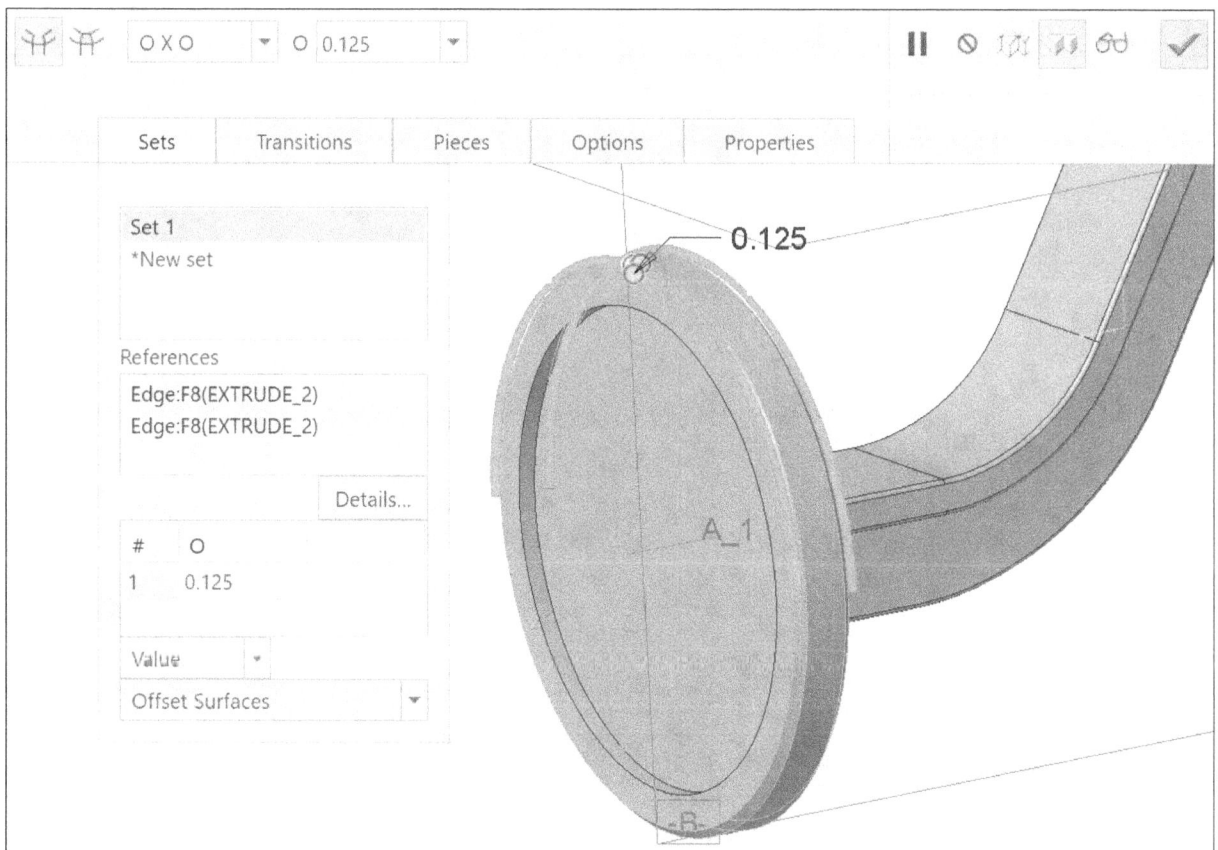

Figure 15.10(e) Chamfer

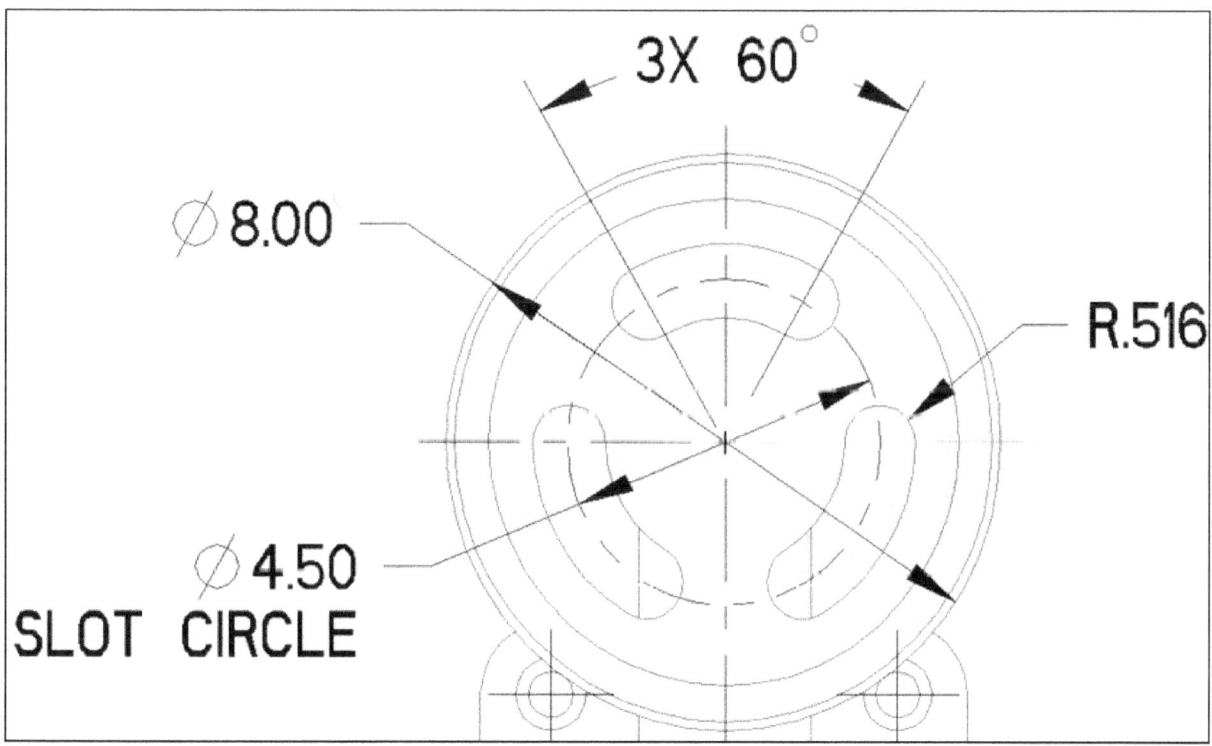

Figure 15.11(a) Patterned Slot

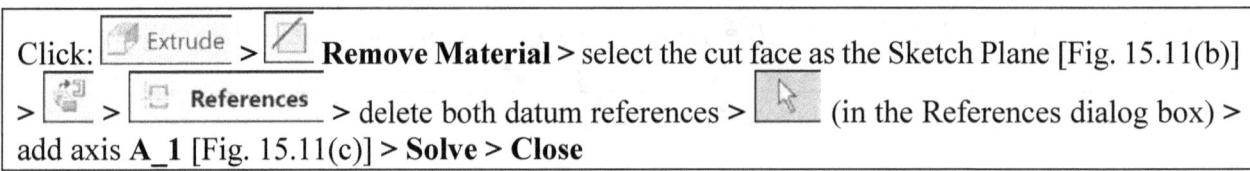

Click: Extrude > Remove Material > select the cut face as the Sketch Plane [Fig. 15.11(b)] > > References > delete both datum references > (in the References dialog box) > add axis **A_1** [Fig. 15.11(c)] > **Solve** > **Close**

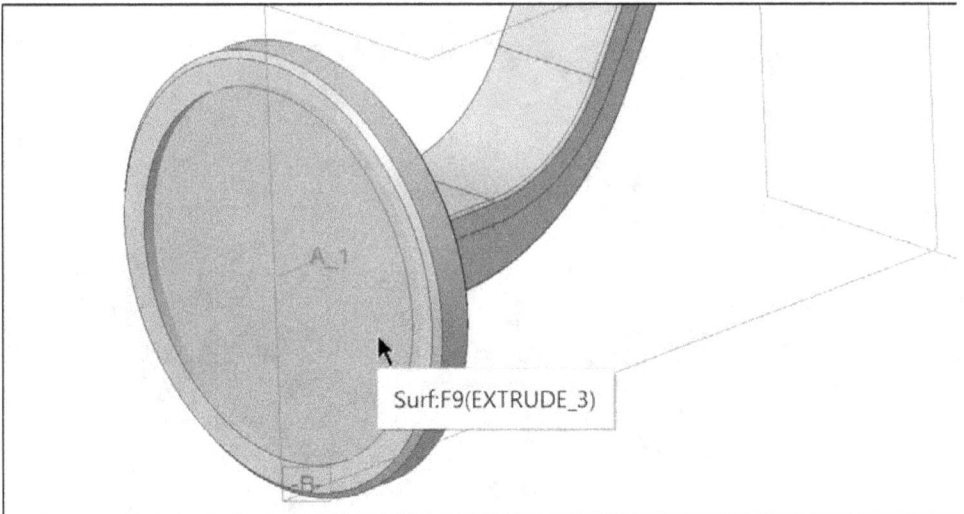

Surf:F9(EXTRUDE_3)

Figure 15.11(b) Sketch Plane

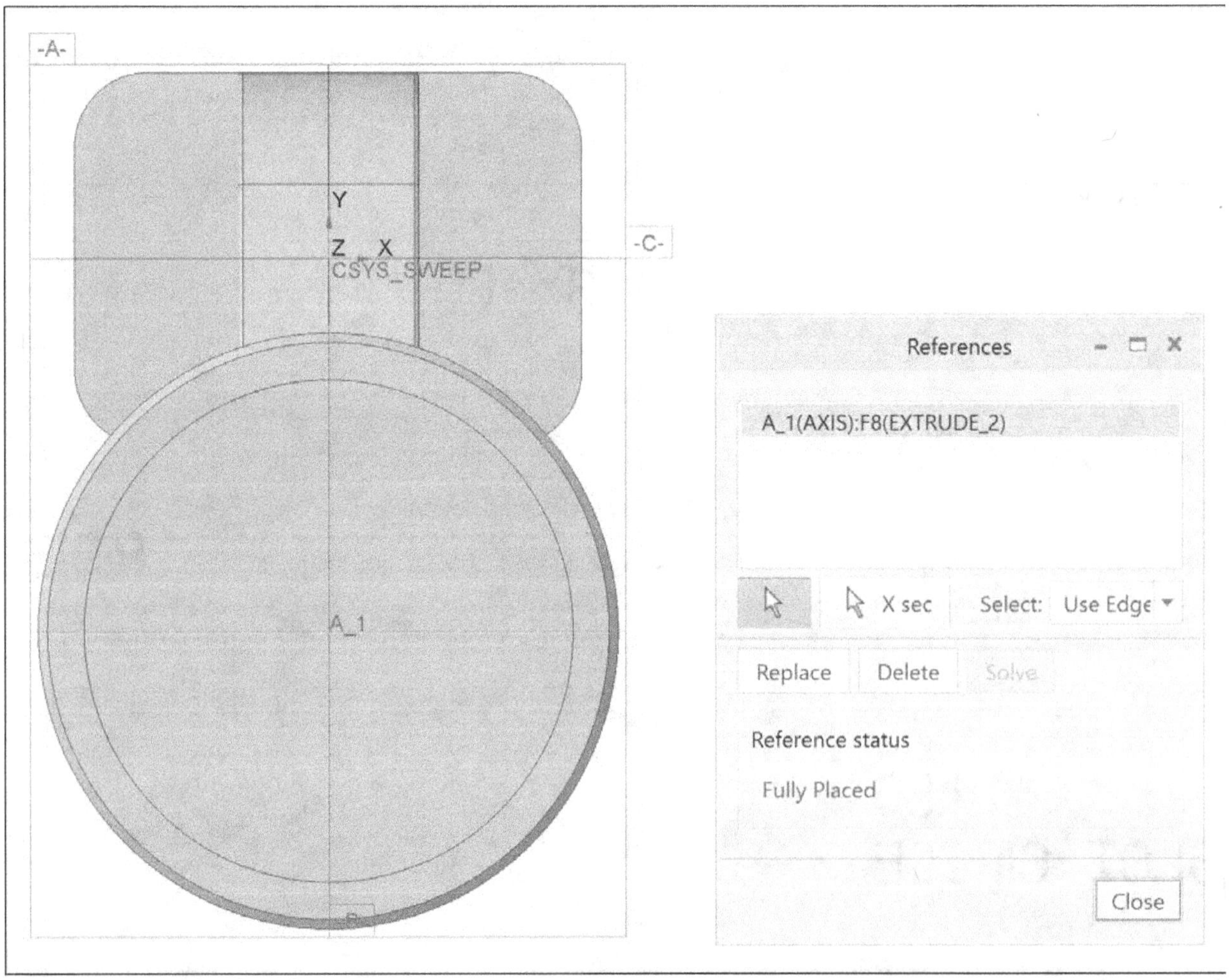

-A-

Y

Z X
CSYS_SWEEP

-C-

A_1

References — ☐ X

A_1(AXIS):F8(EXTRUDE_2)

X sec   Select: Use Edge ▼

Replace   Delete   Solve

Reference status

Fully Placed

Close

Figure 15.11(c) Sketch References

Add a construction etcher point at the center of the round extrusion, click: 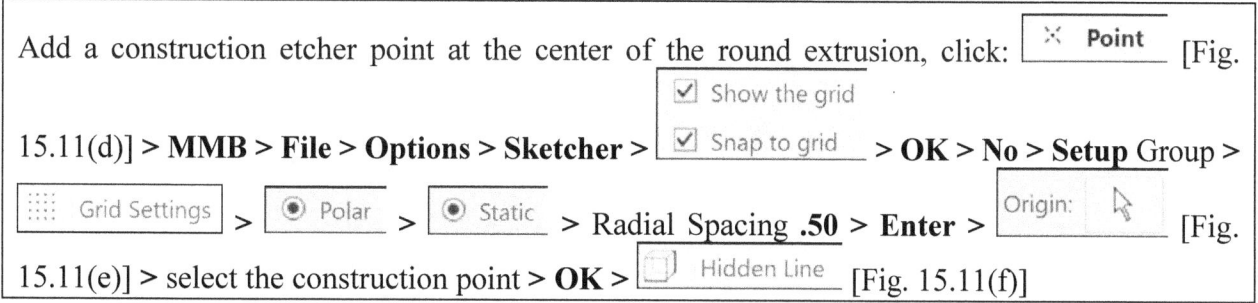 [Fig.

15.11(d)] > **MMB** > **File** > **Options** > **Sketcher** > ☑ Show the grid ☑ Snap to grid > **OK** > **No** > **Setup** Group >

▦ Grid Settings > ⦿ Polar > ⦿ Static > Radial Spacing **.50** > **Enter** > Origin: ⃕ [Fig.

15.11(e)] > select the construction point > **OK** > ▢ Hidden Line [Fig. 15.11(f)]

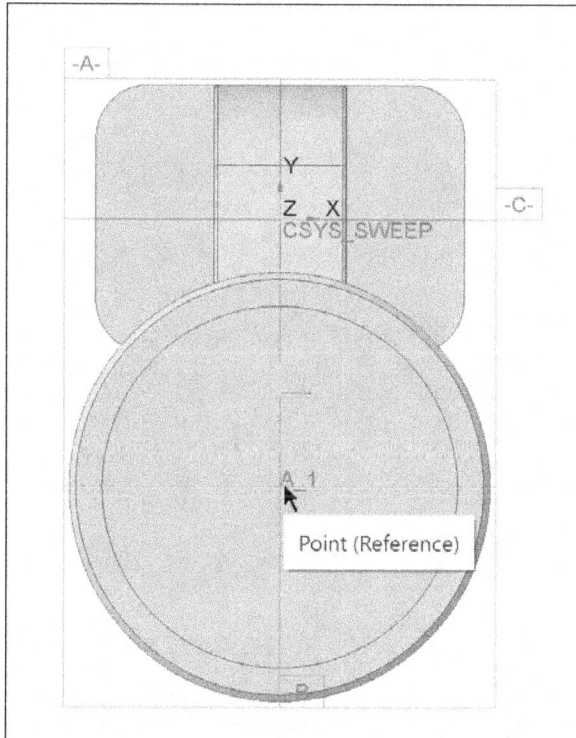

Figure 15.11(d) Add a Construction Point

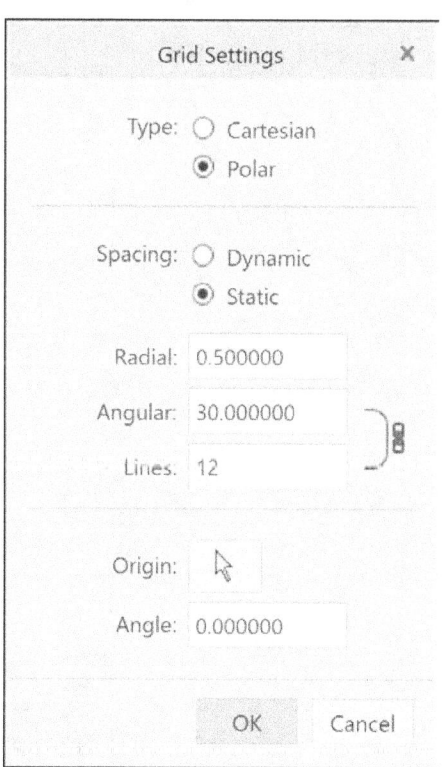

Figure 15.11(e) Grid Settings Dialog Box

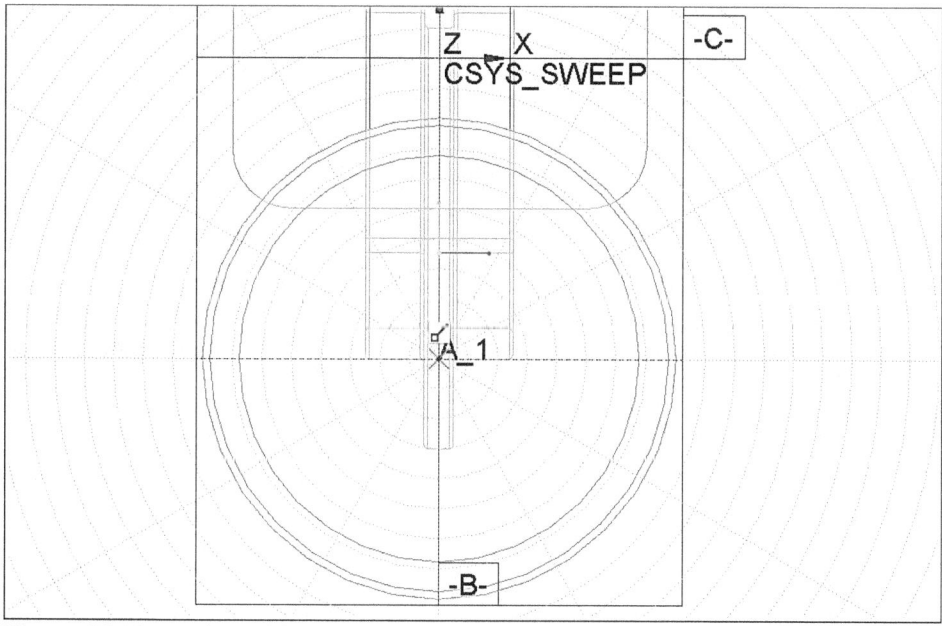

Figure 15.11(f) Polar Grid (note: older illustration that displays the grid clearly)

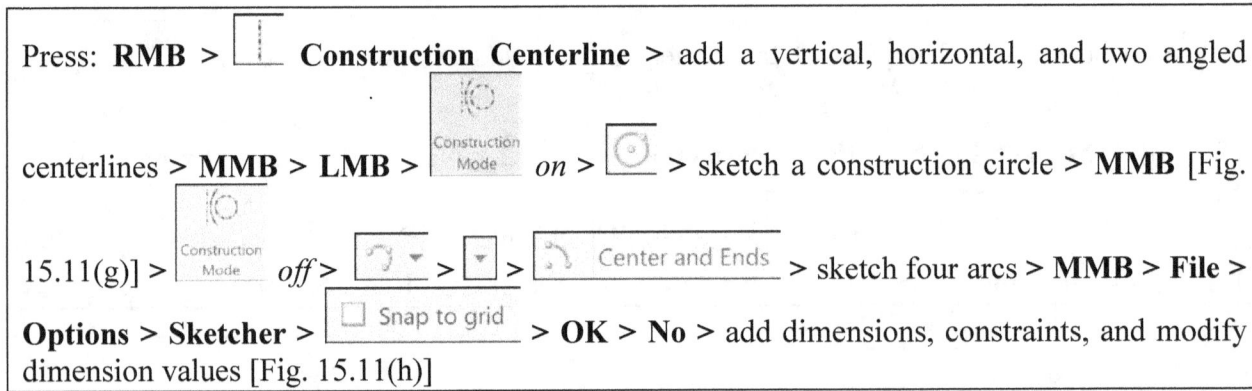

Press: **RMB** > ⬜ **Construction Centerline** > add a vertical, horizontal, and two angled centerlines > **MMB** > **LMB** > [Construction Mode] *on* > [⊙] > sketch a construction circle > **MMB** [Fig. 15.11(g)] > [Construction Mode] *off* > [↻ ▾] > [▾] > [Center and Ends] > sketch four arcs > **MMB** > **File** > **Options** > **Sketcher** > [☐ Snap to grid] > **OK** > **No** > add dimensions, constraints, and modify dimension values [Fig. 15.11(h)]

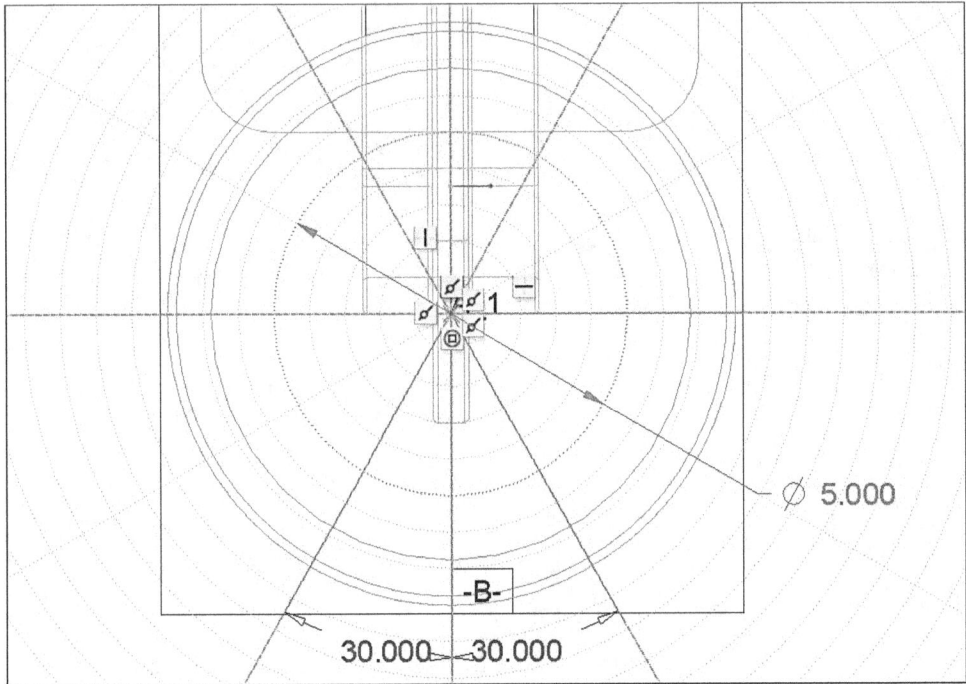

Figure 15.11(g) Construction Circle

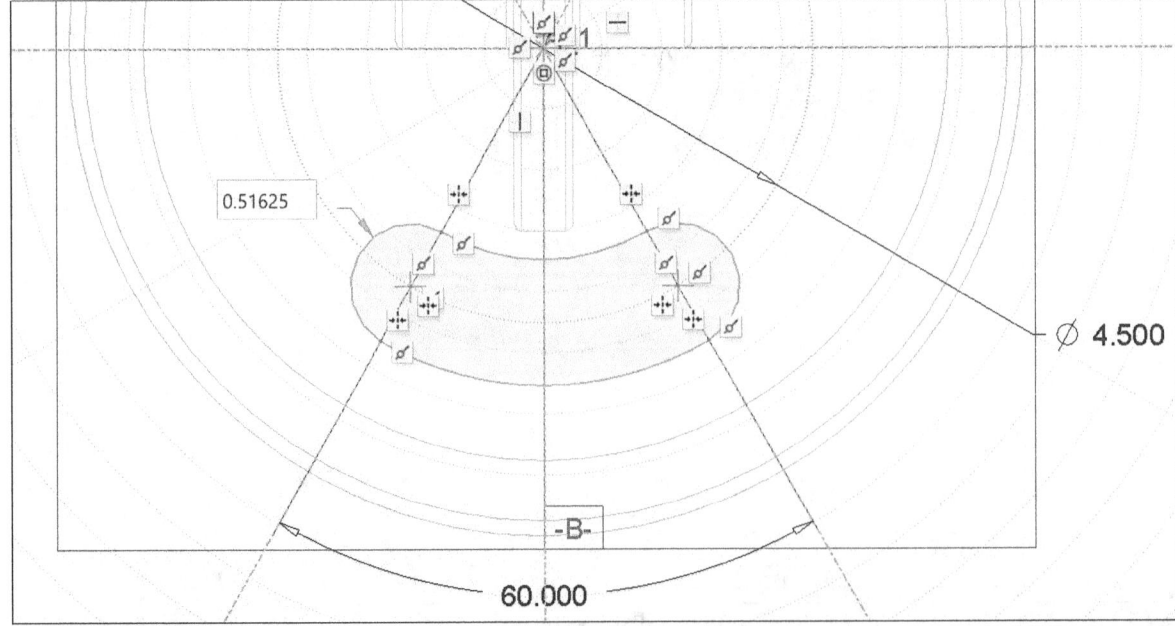

Figure 15.11(h) 0.51625 X 60-Degree Slot on a **4.50** Diameter Bolt Circle

Click: **Ctrl+D** > 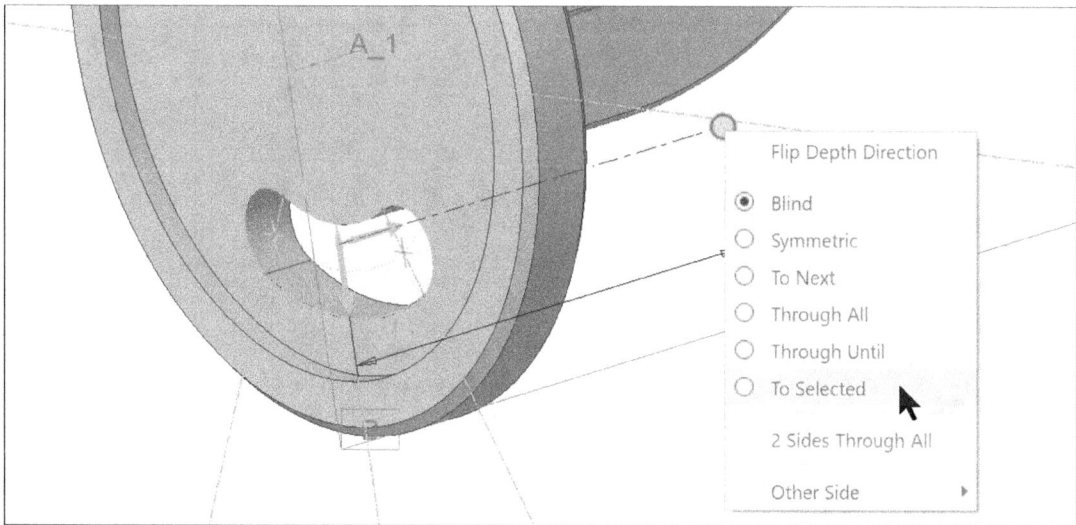 Shading With Edges > **RMB** > ✓ > in the Graphics Window, place the pointer on ➤ > **RMB** > **To Selected** [Fig. 15.11(i)] > select the (back) surface of the circular extrusion [Fig. 15.11(j)] > ✓ [Fig. 15.11(k)] > **Ctrl+S**

Figure 15.11(i) To Selected

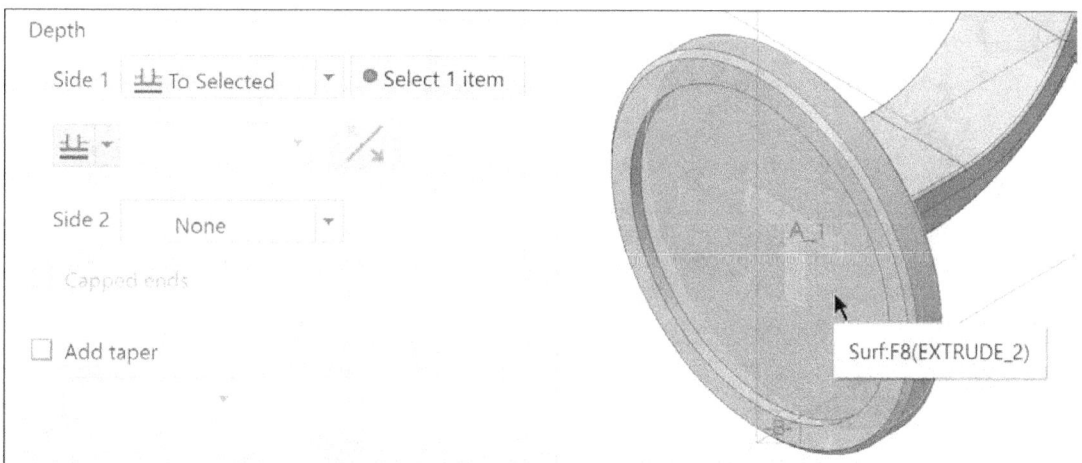

Figure 15.11(j) Select Back Surface of the Circular Extrusion

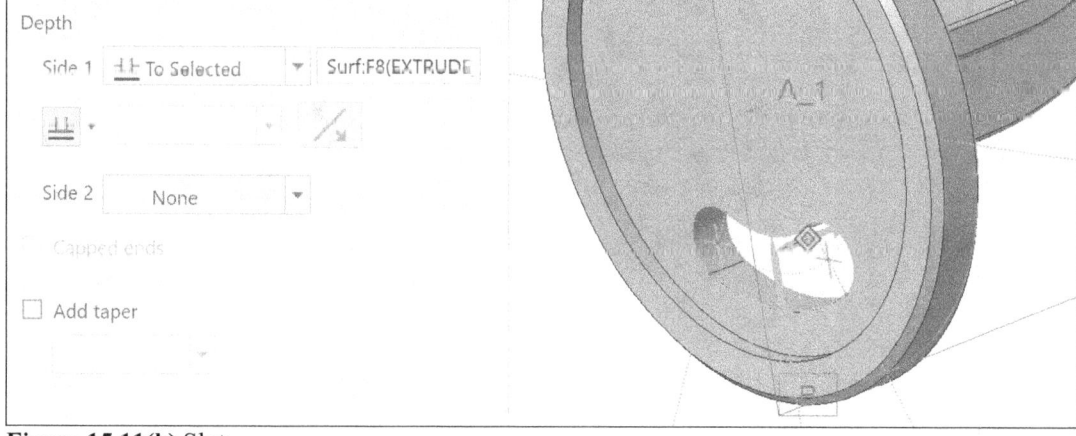

Figure 15.11(k) Slot

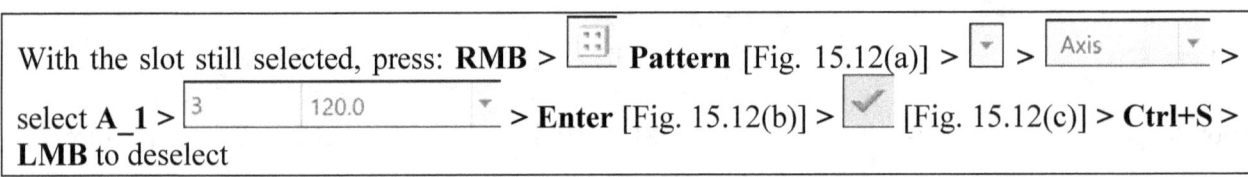

With the slot still selected, press: **RMB** > ⊞ **Pattern** [Fig. 15.12(a)] > ▾ > Axis ▾ > select **A_1** > 3 | 120.0 ▾ > **Enter** [Fig. 15.12(b)] > ✓ [Fig. 15.12(c)] > **Ctrl+S** > **LMB** to deselect

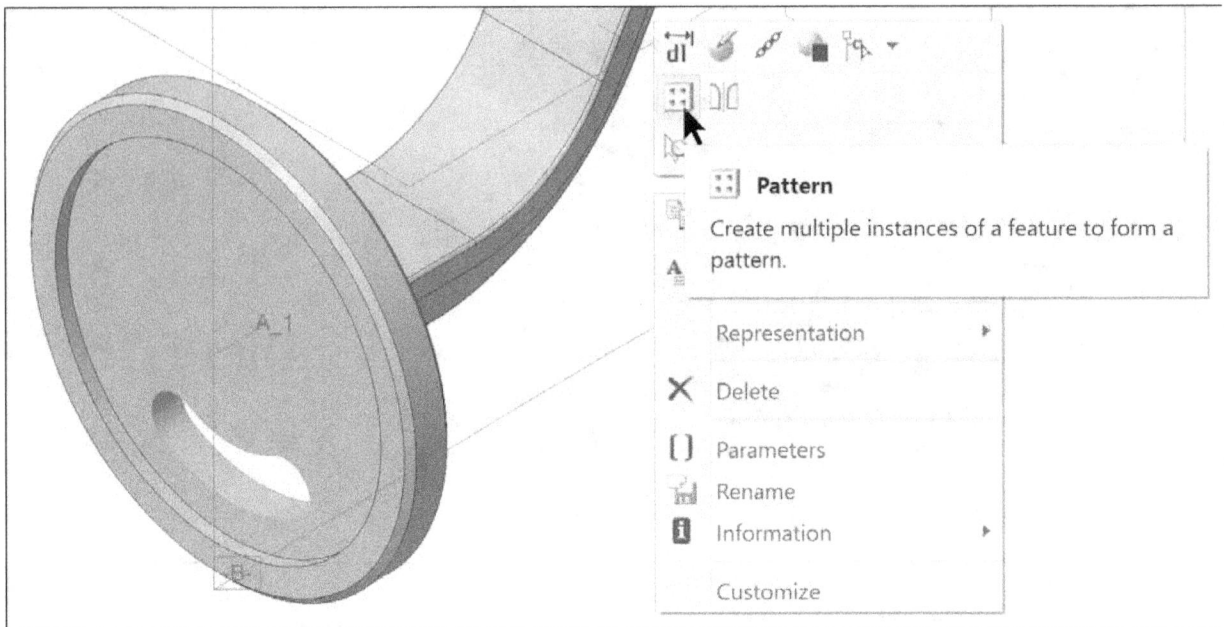

Figure 15.12(a) Press RMB > Pattern

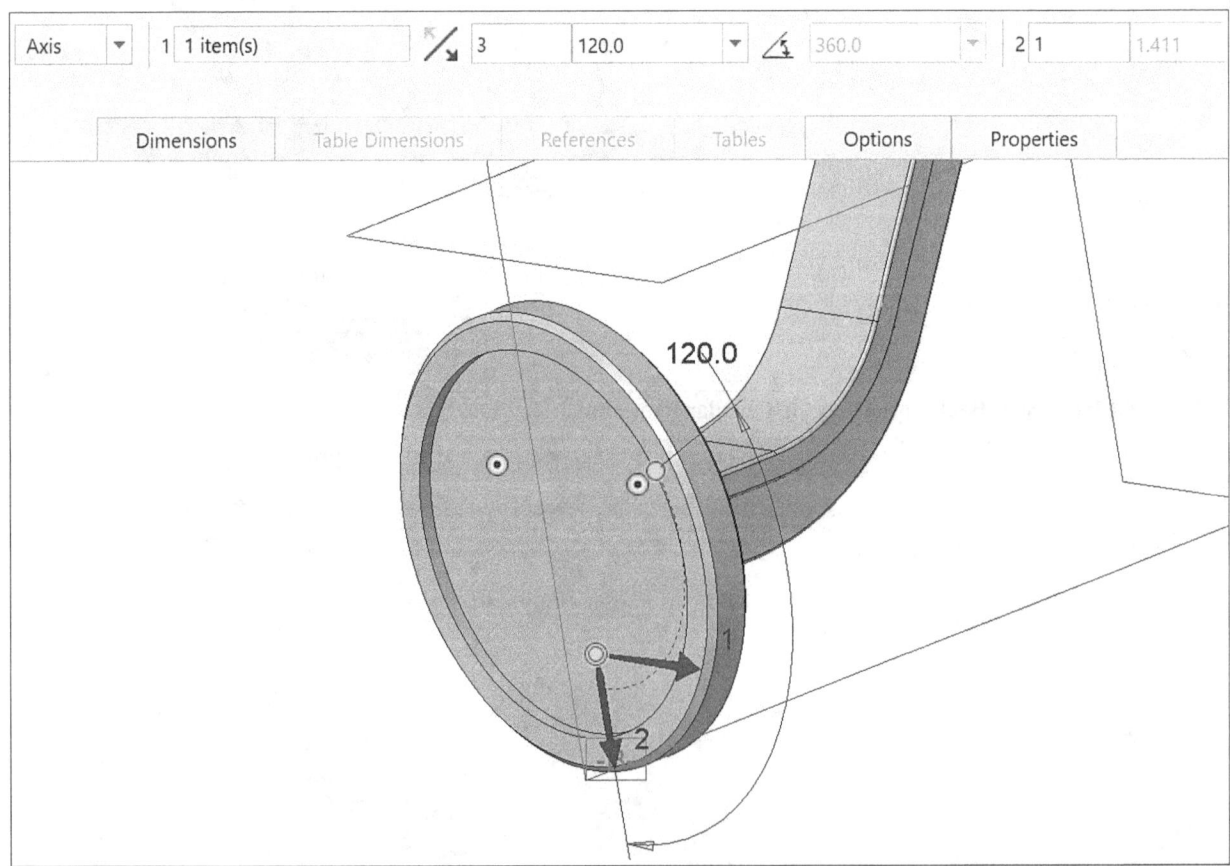

Figure 15.12(b) Pattern the Slot Using the Axis

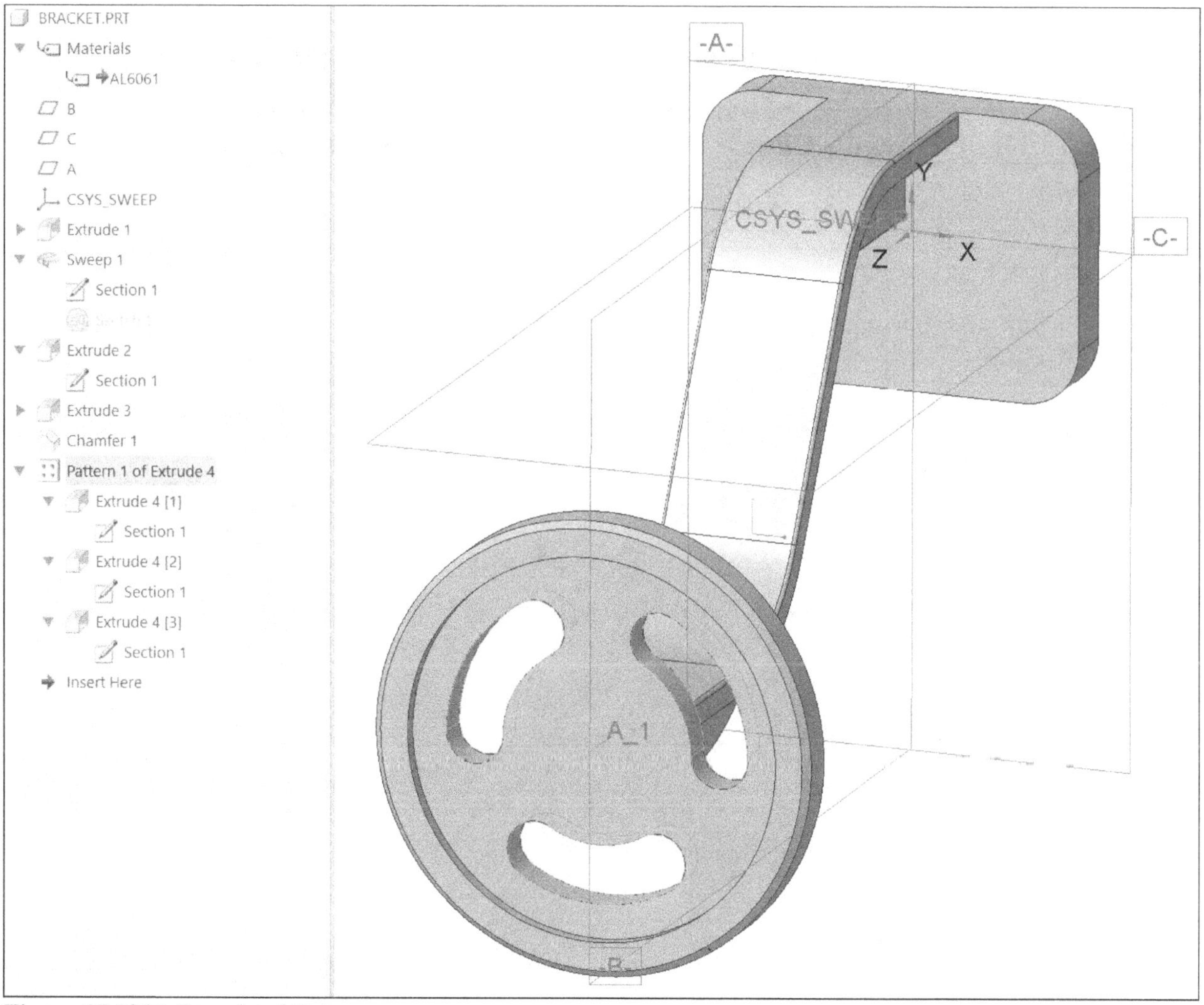

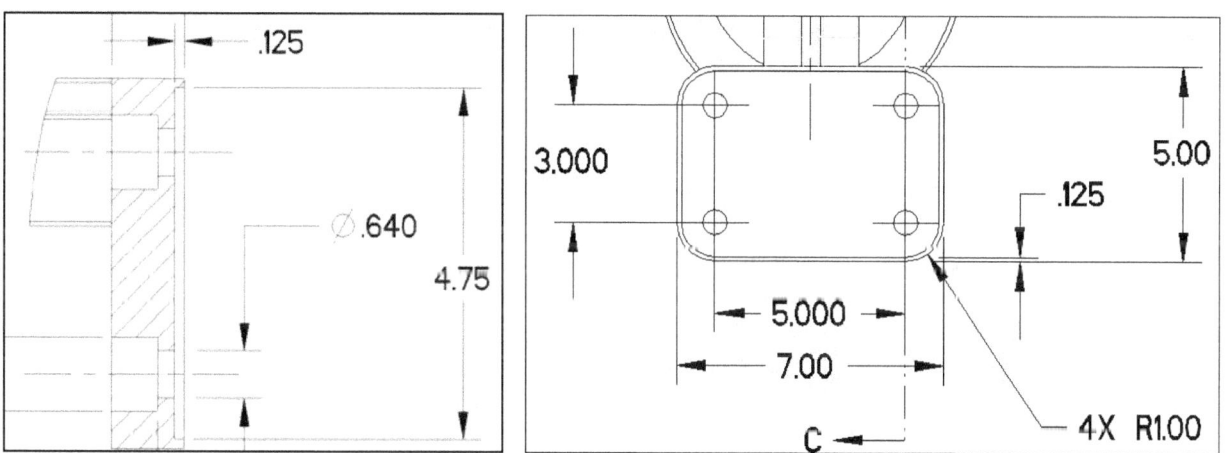

BRACKET.PRT
Materials
AL6061
B
C
A
CSYS_SWEEP
Extrude 1
Sweep 1
Section 1
Extrude 2
Section 1
Extrude 3
Chamfer 1
Pattern 1 of Extrude 4
Extrude 4 [1]
Section 1
Extrude 4 [2]
Section 1
Extrude 4 [3]
Section 1
Insert Here
-A-
-C-
Y
Z
X
A_1

Figure 15.12(c) Completed Pattern

Model the face cut and then create and pattern the counterbore holes [Figs. 15.13(a-b)]

.125
⌀.640
4.75
3.000
5.00
.125
5.000
7.00
4X R1.00
C

Figures 15.13(a-b) Create the Face Cut and Pattern Counterbore Holes

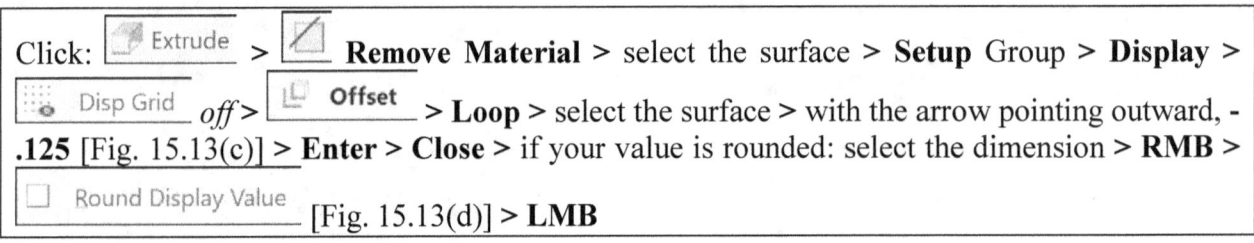

Click: <u>Extrude</u> > <u>Remove Material</u> > select the surface > **Setup** Group > **Display** > <u>Disp Grid</u> *off* > <u>Offset</u> > **Loop** > select the surface > with the arrow pointing outward, -.125 [Fig. 15.13(c)] > **Enter** > **Close** > if your value is rounded: select the dimension > **RMB** > ☐ Round Display Value [Fig. 15.13(d)] > **LMB**

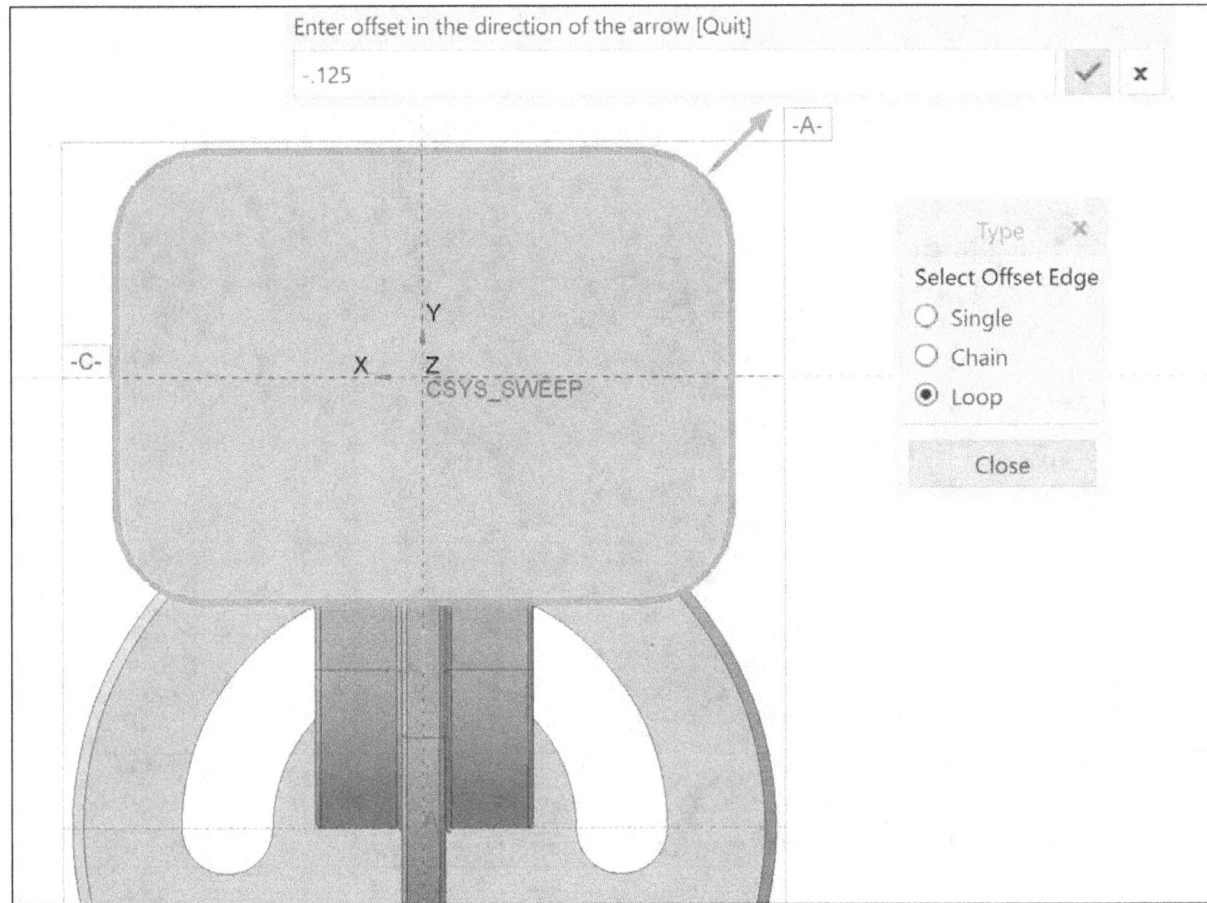

Figure 15.13(c) Use Offset Loop **-.125** (if the arrow points inward, enter .125)

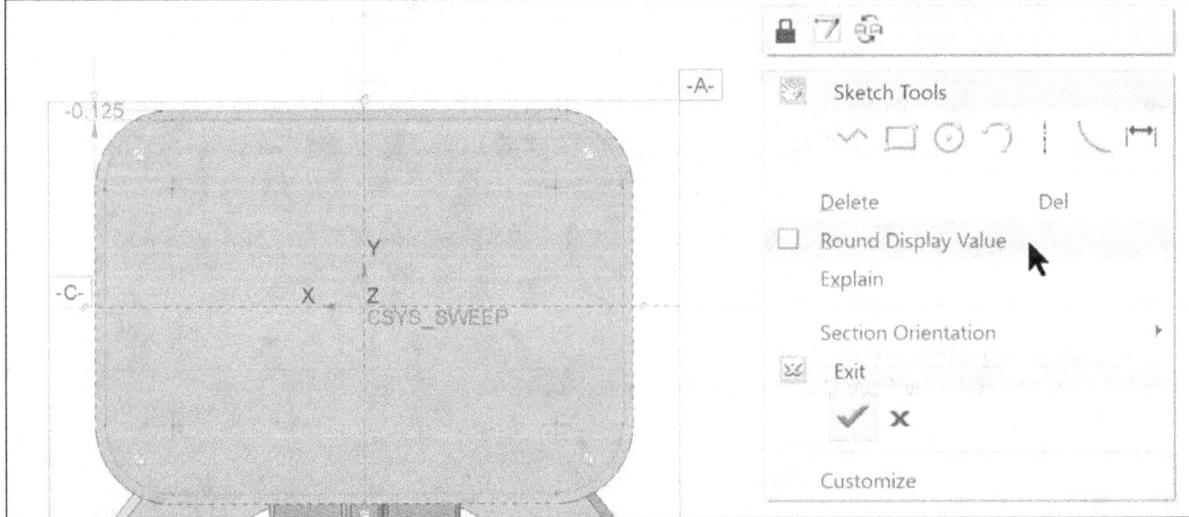

Figure 15.13(d) Select the Dimension > Press RMB > Uncheck Round Display Value

Click: [✓] > for the depth, *type:* **.125 > Enter** [Fig. 15.13(e)] > [✓] > **LMB** [Fig. 15.13(f)] >
Ctrl+S

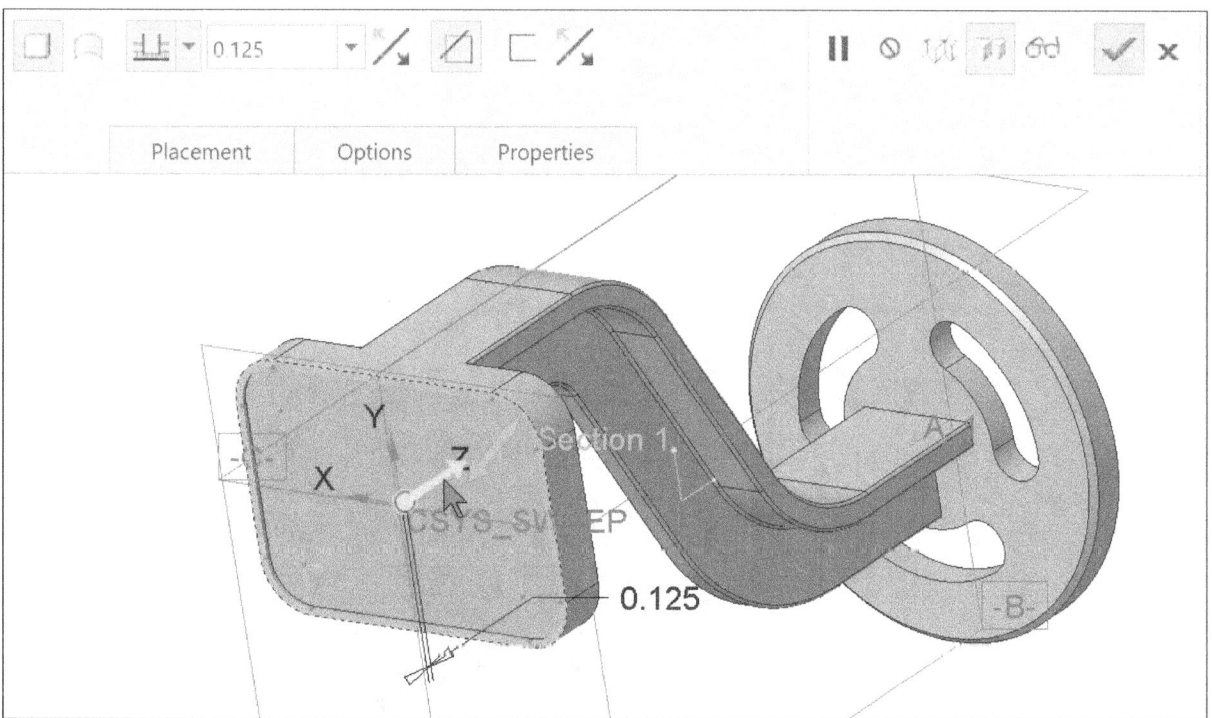

Figure 15.13(e) Cut Preview

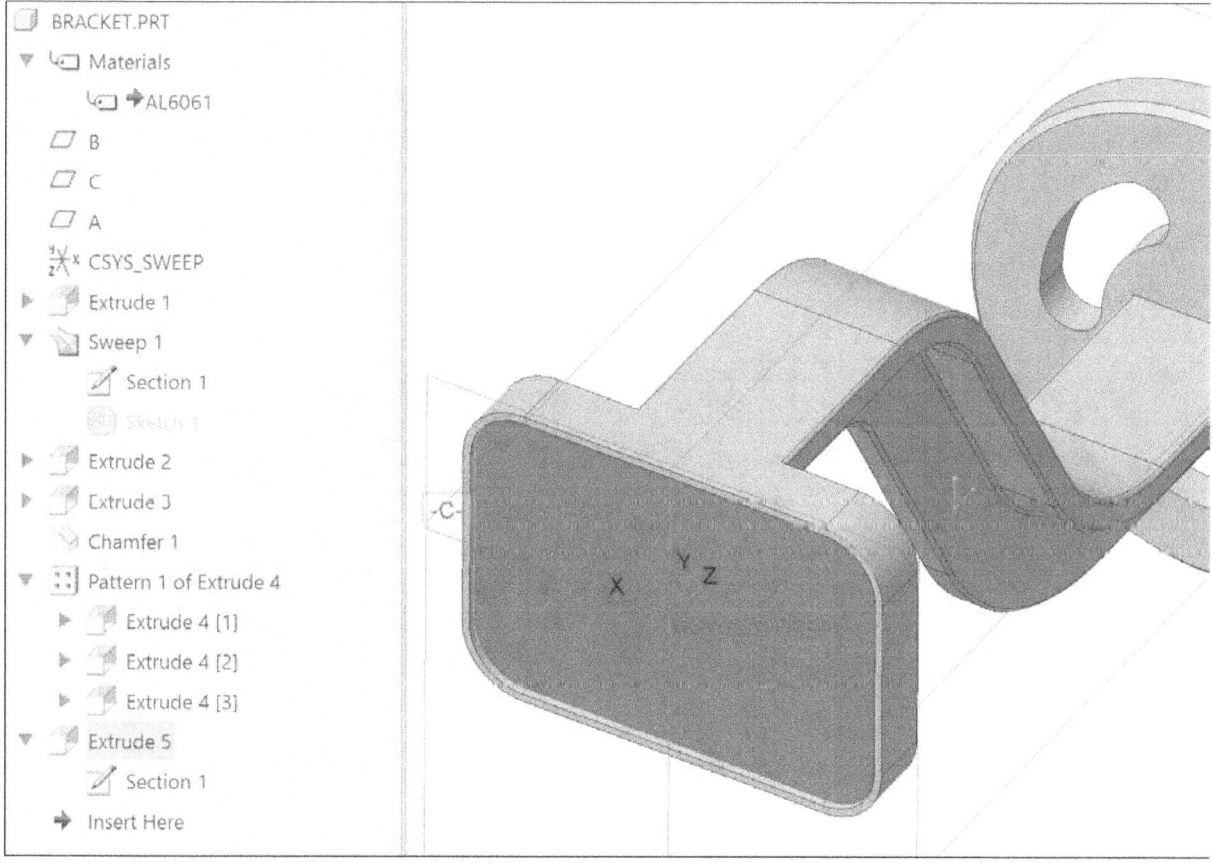

Figure 15.13(f) Completed Cut

Click: **Ctrl+D** > model the hole using the detail dimensions [Fig. 15.14(a)] > 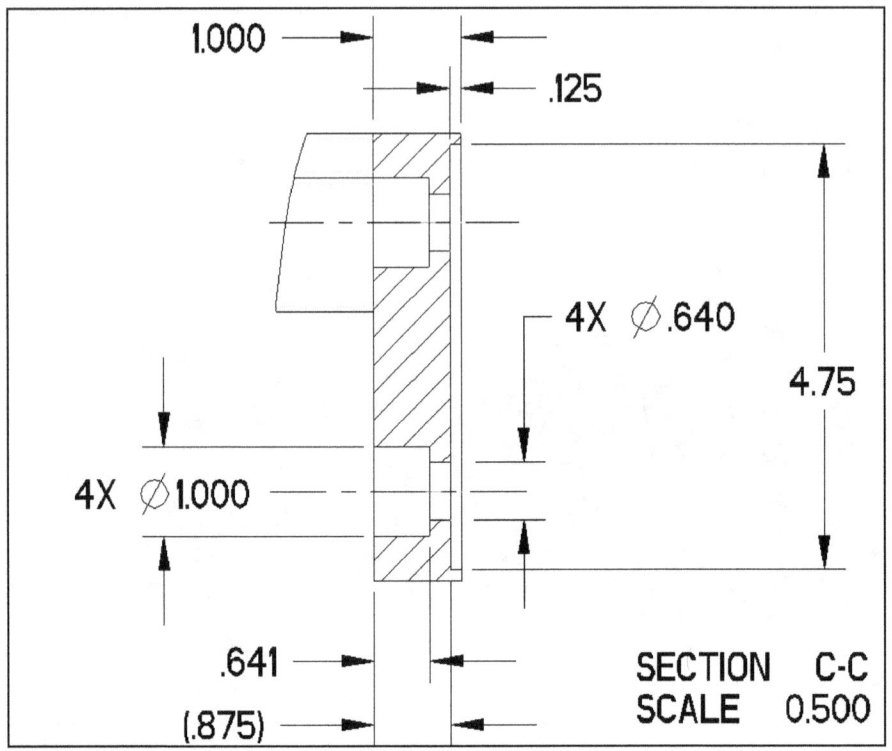 Hole >
Placement tab > place the hole per Placement requirements [Fig. 15.14(b)]

Figure 15.14(a) Counterbore Hole Detail

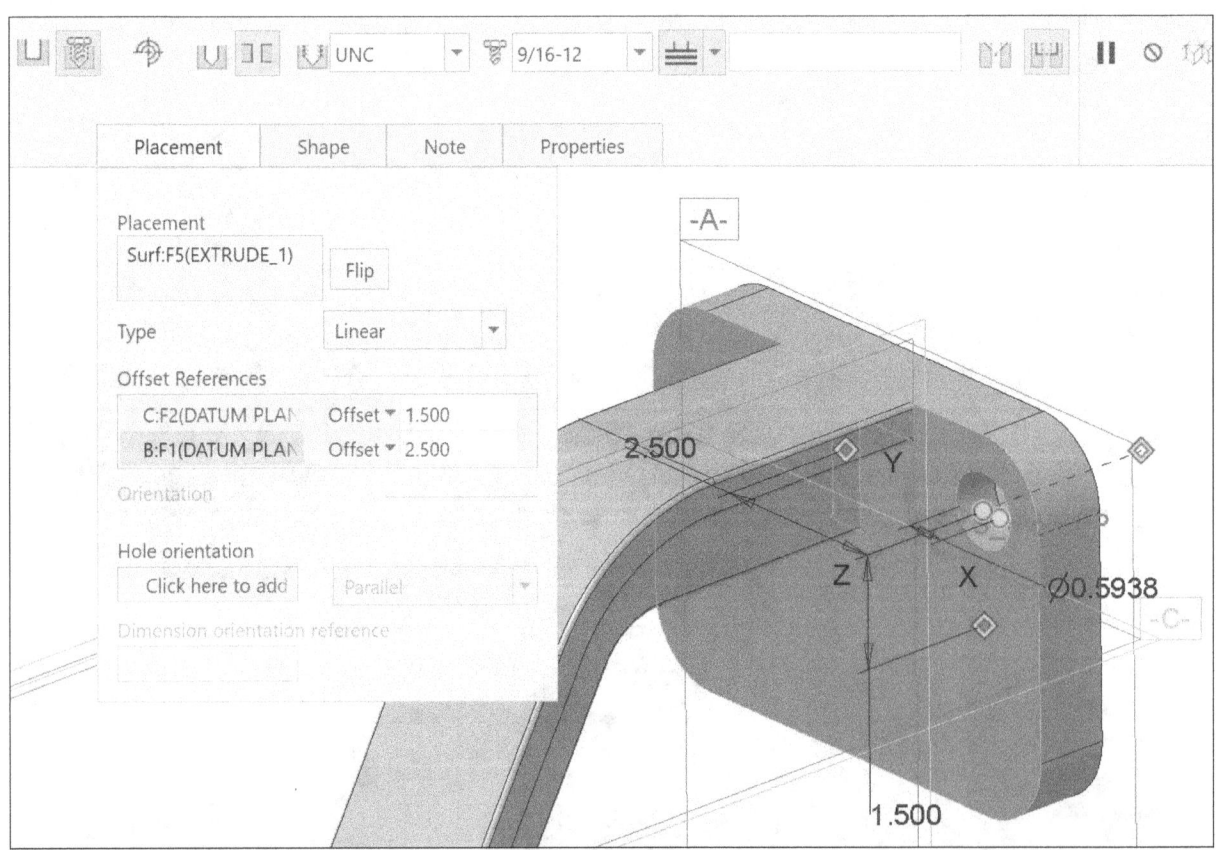

Figure 15.14(b) Hole Placement

Click: **Shape** tab > shape the hole per Shape requirements [Fig. 15.14(c)] > ✓ > [icon] off > **RMB** > [icon] **Pattern** > [icon] > **Dimension** > select for Direction 1 **2.50** > **Enter** > **Dimensions** tab > pick **Click here to** in the Direction 2 field > select for Direction 2 **1.50** > **Enter** > modify Increment *2.50* to **-5.00** > **Enter** > modify Increment *1.50* to **-3.00** > **Enter** [Fig. 15.15(a)]

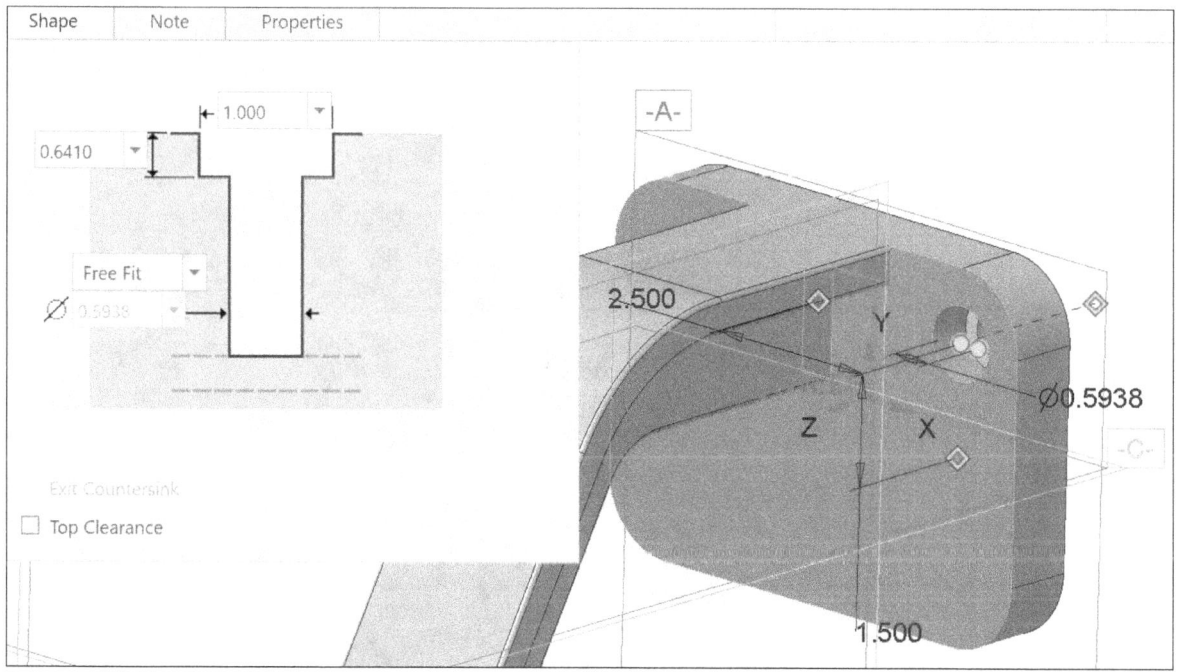

Figure 15.14(c) Hole Shape

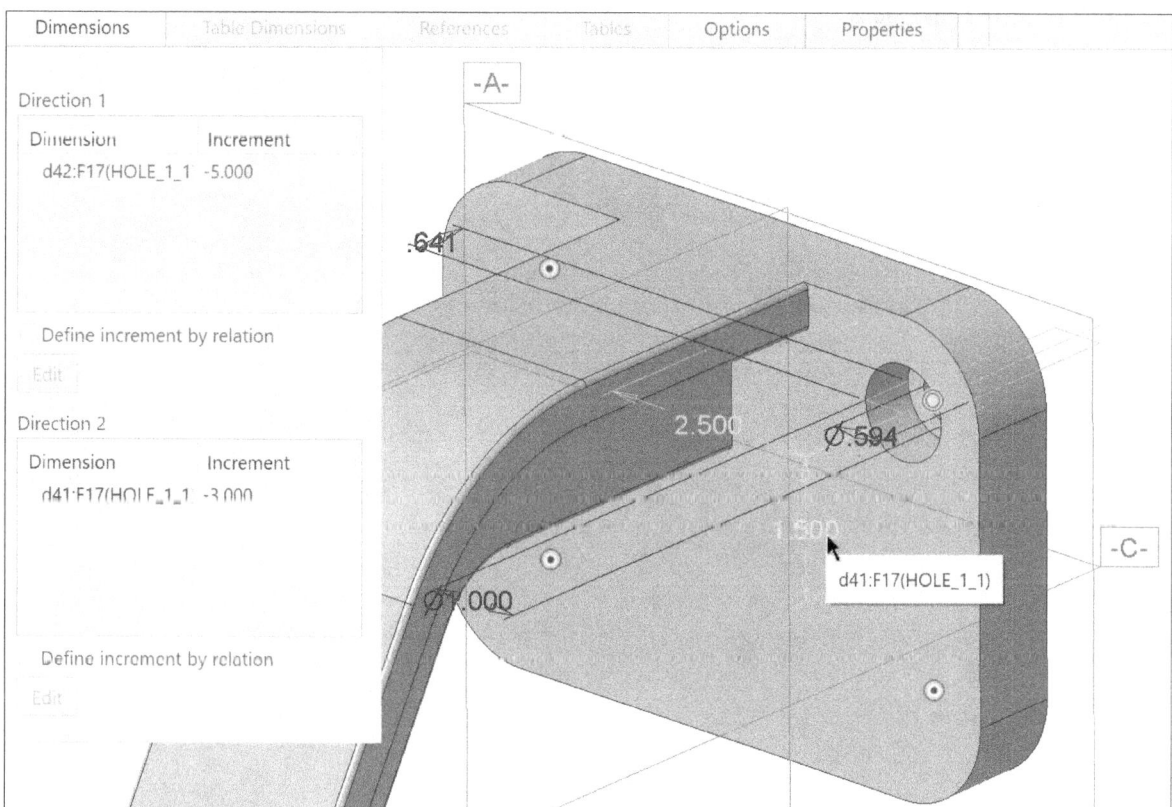

Figure 15.15(a) Pattern Dimensions

677

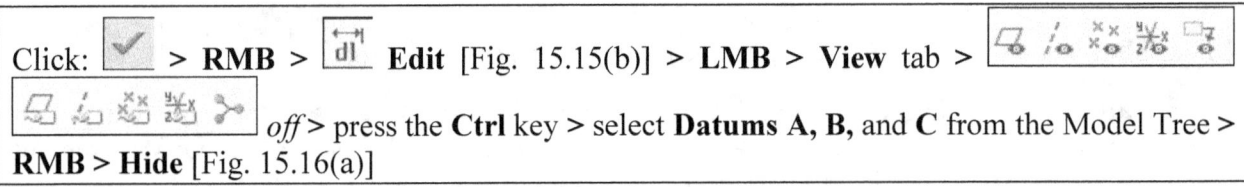

Click: [✓] > **RMB** > [d1] **Edit** [Fig. 15.15(b)] > **LMB** > **View** tab > [icons] [icons] *off* > press the **Ctrl** key > select **Datums A, B,** and **C** from the Model Tree > **RMB** > **Hide** [Fig. 15.16(a)]

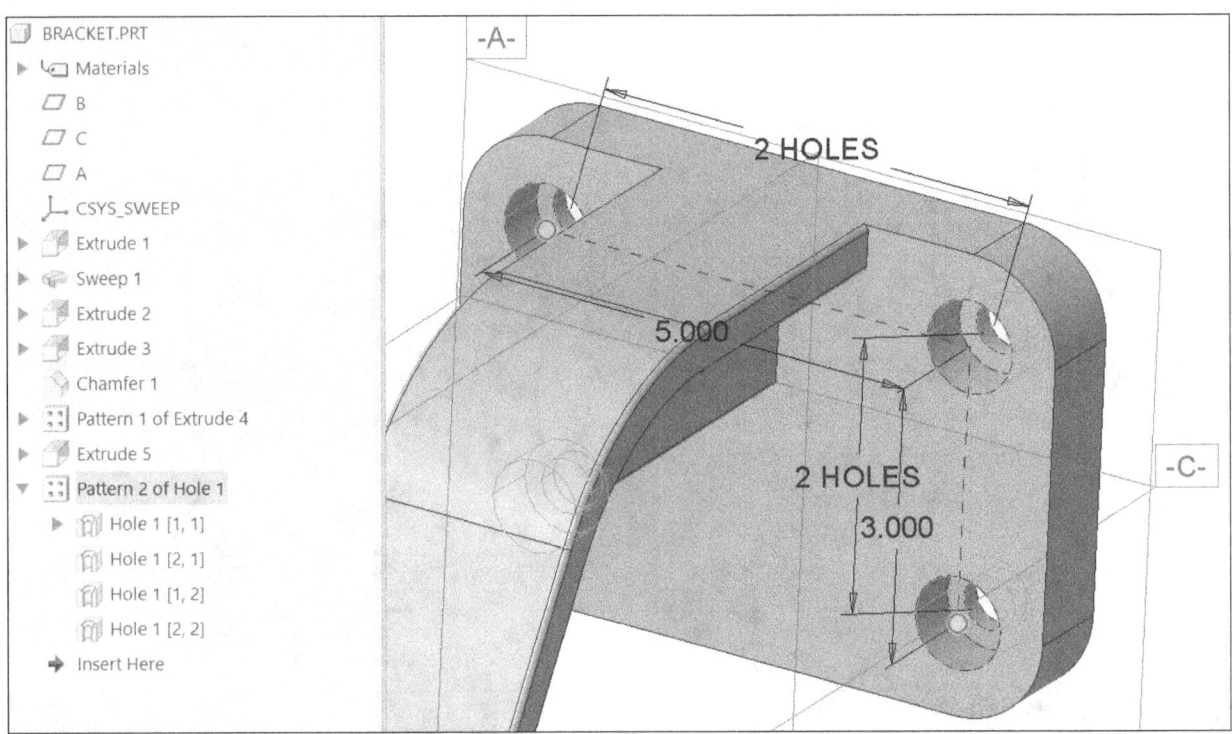

Figure 15.15(b) Pattern Dimensions

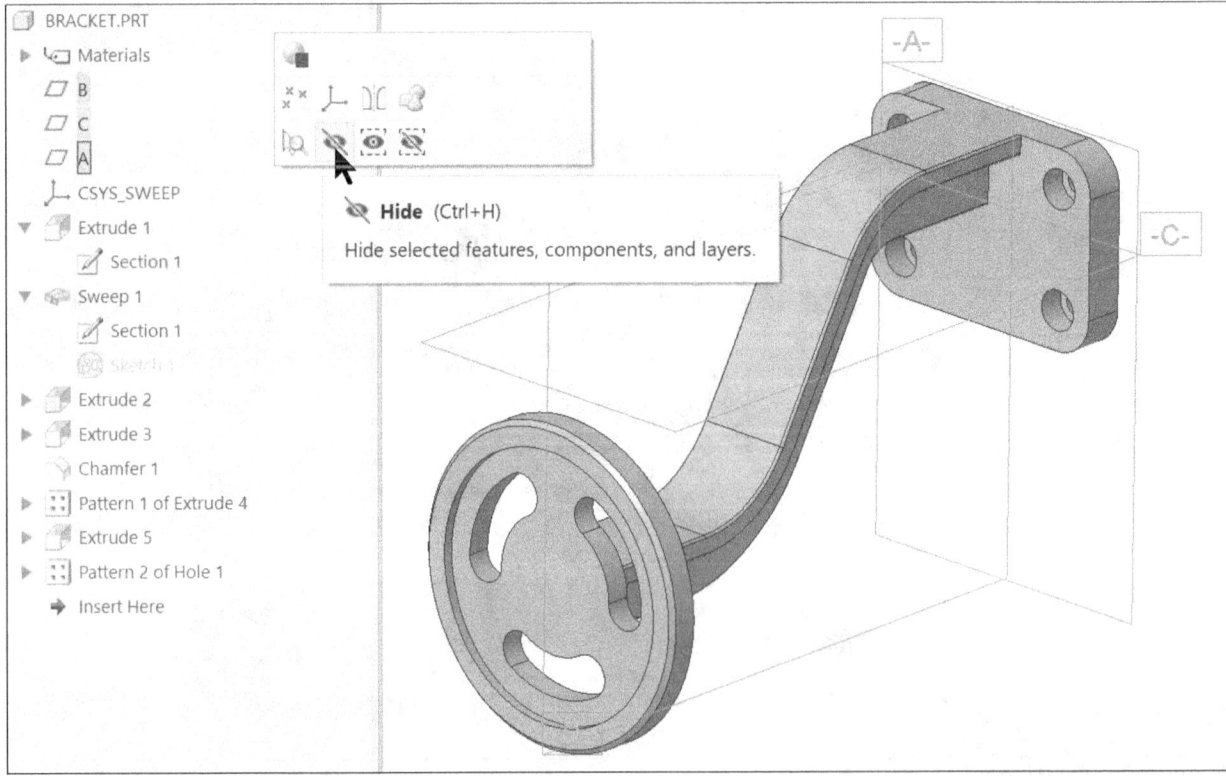

Figure 15.16(a) Hide the Datum Planes and the Sketch

Click: **File > Options > Model Display** > set as shown [Fig. 15.16(b)]

Favorites Environment System Appearance **Model Display** Entity Display Selection Sketcher Assembly Notification Center ECAD Assembly Board Data Exchange Sheetmetal Update Control Additive Manufacturing ▼ Customize 　　Ribbon 　　Quick Access Toolbar 　　Shortcut Menus 　　Keyboard Shortcuts Window Settings Licensing Configuration Editor	Change how the model is displayed. **Model orientation** Default model orientation: ☐ Isometric ▼ **Model display settings when reorienting the model** ☐ Show the orientation center ☐ Show datums ☐ Fast hidden lines removal ☐ Show surface mesh ☐ Show silhouette edges ☐ Use time to draw frames 　　Set frames per seconds rate to ☐ 3 ☐ Show animation 　　Set maximum number of seconds spent animating to: ☐ 1.0 　　Set minimum number of frames displayed for animation to: ☐ 6 **Shaded model display settings** Set shade quality to: 50 ☐ Show surface features ☐ Show datum curve features ☑ Shade very small surfaces ☐ Shade manufacturing reference model ☐ Use level of detail for the shaded model when manipulating the view 　　Set detail percentage to: 50 ☐ Show textures on shaded model ☑ Enable transparency 　　Transparency: ☐ Blended ▼ ☐ Show clipped model as solid

Figure 15.16(b) Model Display Options *(Note: A shade quality of 50 will greatly increase your Models' file size)*

Click: **Entity Display** > set as shown [Fig. 15.16(c)]

Creo Parametric Educational Edition Options

Favorites
Environment
System Appearance
Model Display
Entity Display
Selection
Sketcher
Assembly
Notification Center
ECAD Assembly
Board
Data Exchange
Sheetmetal
Update Control
Additive Manufacturing

▼ Customize
    Ribbon
    Quick Access Toolbar
    Shortcut Menus
    Keyboard Shortcuts

Window Settings

Licensing
Configuration Editor

Change how entities are displayed.

Geometry display settings

Default geometry display: Shade With Edges ▼

Edge display quality: Very High ▼

Tangent edges display style: Dimmed ▼

Anti-Aliasing: Off ▼

Lines Anti-Aliasing: Off ▼

Text Anti-Aliasing: Off ▼

☑ Show colors assigned to model surface

☑ Show silhouette edges

Datum display settings

☑ Show datum planes

☑ Show datum plane tags

☑ Show datum axes

☑ Show datum axis tags

☑ Show datum points

    Show point symbol as: Cross ▼

☑ Show datum point tags

☑ Show datum coordinate system

☑ Show coordinate system tags

☑ Show images

Dimensions, annotations, notes and reference designators display settings

☐ Show dimension tolerances

☐ Show note names instead of note text

☐ Show reference designators of cabling, ECAD and piping components

☑ Show annotations and Annotation Elements

☑ Show annotation orientation grid

    Set grid spacing to: 0.100

Assembly display settings

☐ Show connections

☑ Show animation while exploding the assembly

    Maximum seconds an animation takes between explode states: 1.0

    ☐ Follow explode sequence

☑ Show name for components in Symbolic Representation

☐ Show component interference in cross sections

Figure 15.16(c) Entity Display Options

Click: **Environment** > set as shown [Fig. 15.16(d)] > **OK** > **No** > 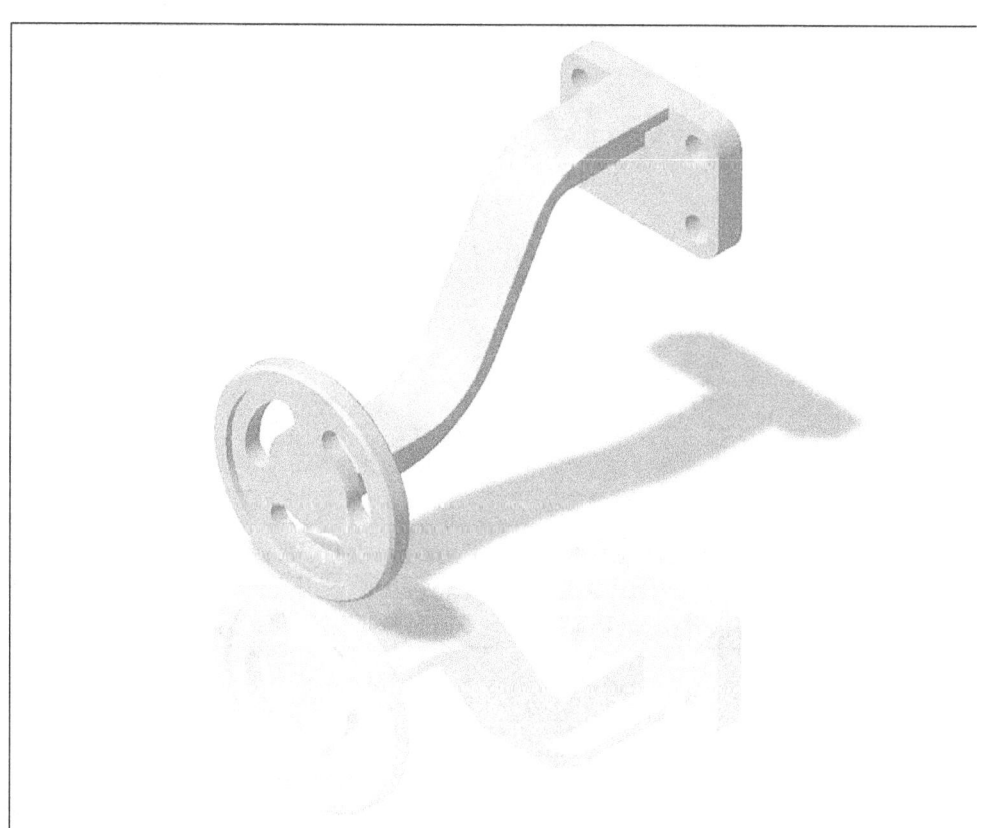 **Shading with Reflections** [Fig. 15.16(e)] > **View** tab > **Model Display** > **Temporary Shade** > **Ctrl+D** > **Ctrl+S**

Favorites
Environment
System Appearance
Model Display
Entity Display
Selection
Sketcher
Assembly
Notification Center
ECAD Assembly
Board
Data Exchange
Sheetmetal
Update Control
Additive Manufacturing

▼ Customize
    Ribbon
    Quick Access Toolbar
    Shortcut Menus
    Keyboard Shortcuts

Change environment options for working with Creo.

General environment options

Working directory: C:\Users\Public\Documents\

☑ Show user help prompts on pointer

☐ Sound bell for prompts and messages

☑ Save display with model

☑ Add to the model datums that are created during information operations

Define, run, and manage mapkeys: Mapkeys Settings...

ModelCHECK settings

Configure ModelCHECK: ModelCHECK Settings...

Instance creation options

☑ Create instance accelerator files upon instance storage

Create files: Saved Objects   ▼

Distributed computing settings

Figure 15.16(d) Environment Options (your Working directory may be different)

Figure 15.16(e) Shading With Reflections
(the quality of your graphics card and graphics settings may prevent this display)

681

If you have the Render License do the following: **Render** tab > **Scene** > **Lights** tab > **OK** (if needed) > 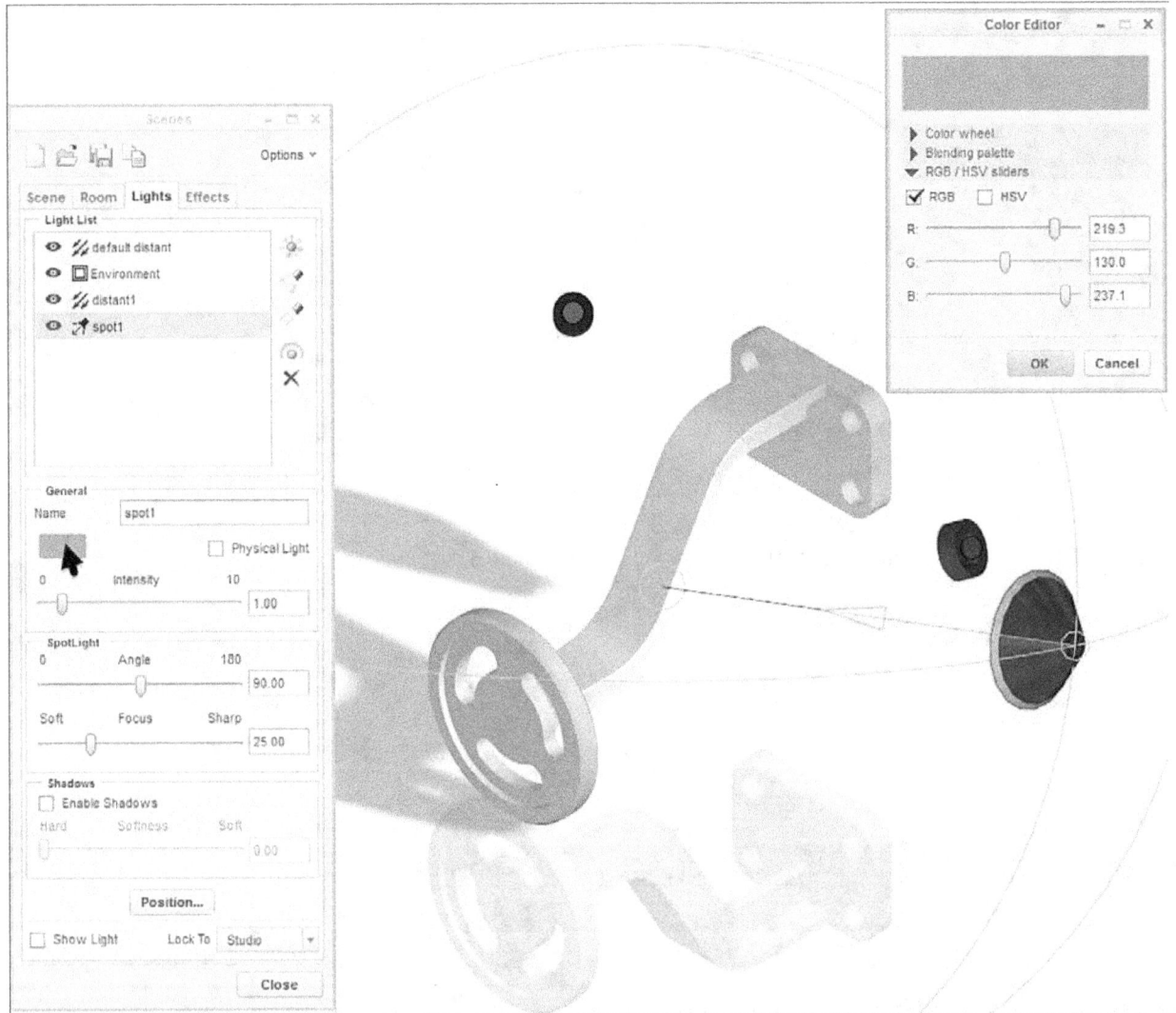 **Add new spotlight** > Name [   ] **Color for lighting** > adjust the slide bars in the Color Editor to the RGB values provided [Fig. 15.17(a)] > **OK** (from the Color Editor dialog box)

If you do not have the Render License see pages 797-804 for steps for using the Render capabilities within basic Creo Parametric 5.0.

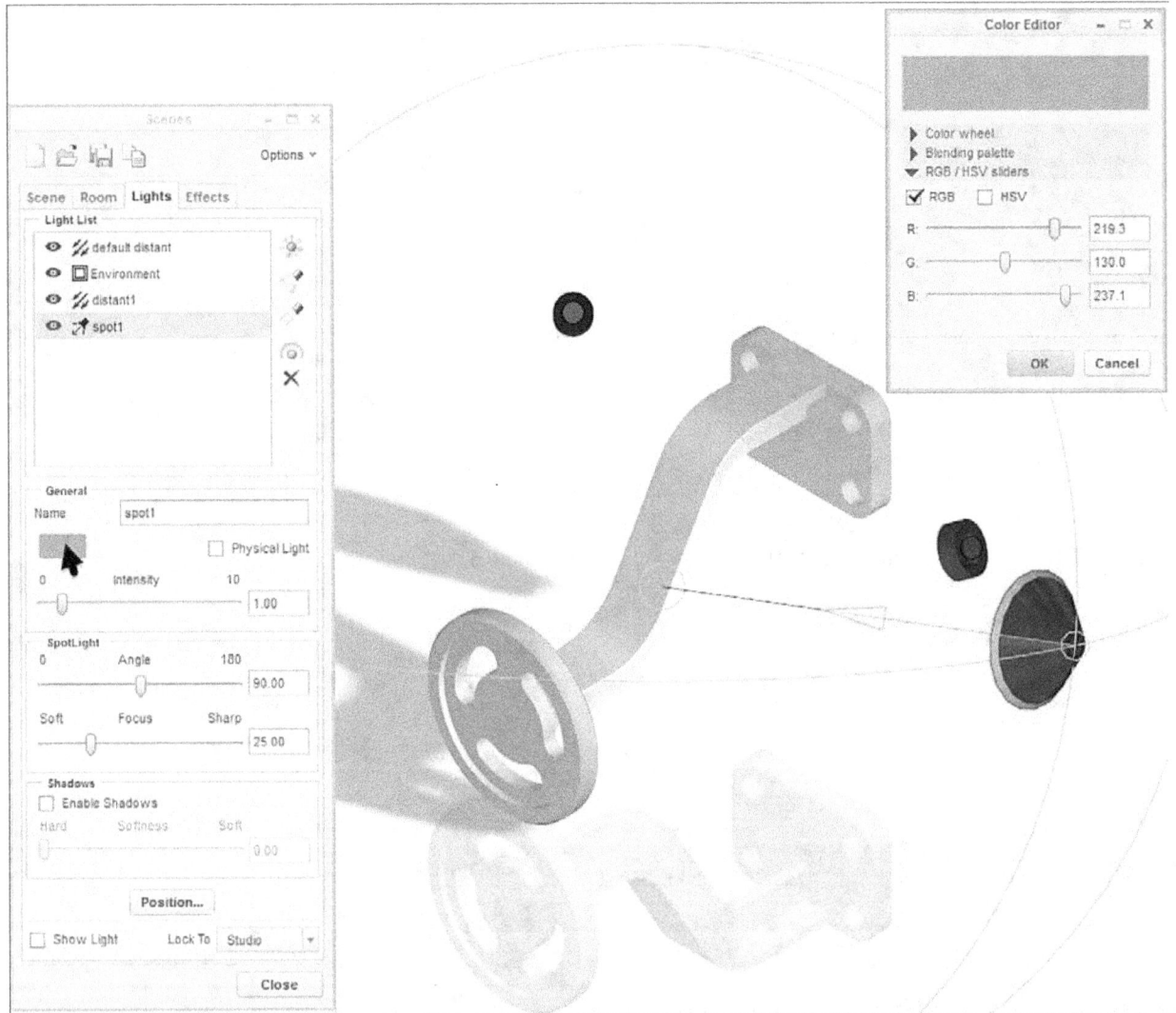

Figure 15.17(a) Light Setup

Move the light from its default position to the other side of the model. Place the pointer on the circle behind the light (highlights) [Fig. 15.17(b)] > **LMB** > move the pointer to the other side of the model [Fig. 15.17(c)] > **Close** (the Scenes dialog box) > **Ctrl+S** > **File** > **Close**

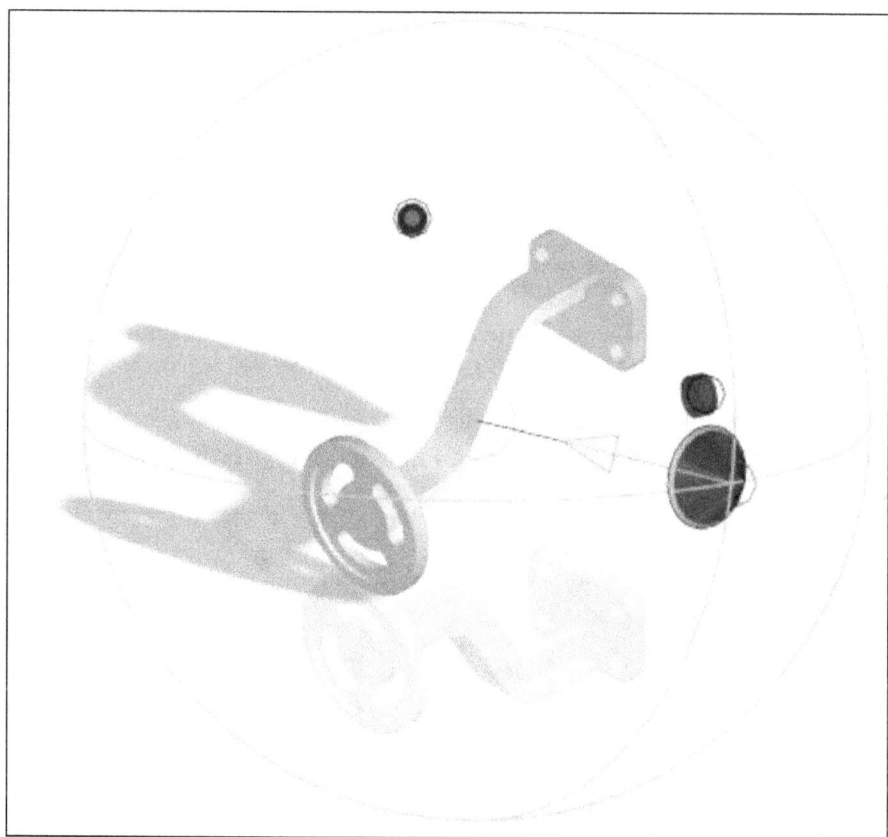

Figure 15.17(b) Move the Light (experiment with different positions)

Figure 15.17(c) New Light Position (experiment with different positions)

683

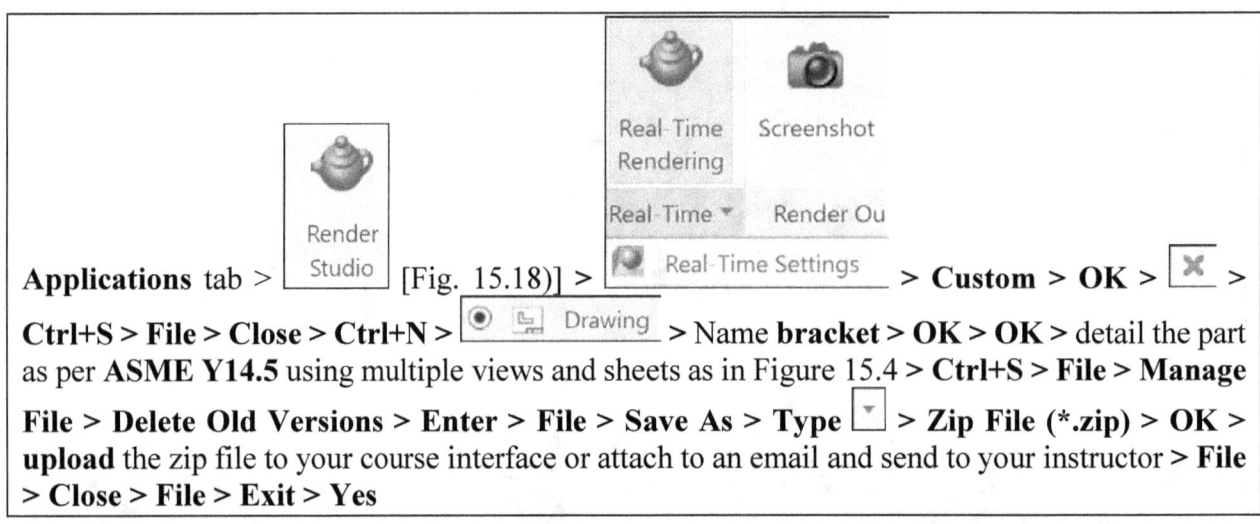

Applications tab > [Render Studio] [Fig. 15.18)] > [Real-Time Rendering / Screenshot / Real-Time Settings] > **Custom > OK >** [X] > **Ctrl+S > File > Close > Ctrl+N >** [Drawing] > Name **bracket > OK > OK >** detail the part as per **ASME Y14.5** using multiple views and sheets as in Figure 15.4 > **Ctrl+S > File > Manage File > Delete Old Versions > Enter > File > Save As > Type** [▼] > **Zip File (*.zip) > OK > upload** the zip file to your course interface or attach to an email and send to your instructor > **File > Close > File > Exit > Yes**

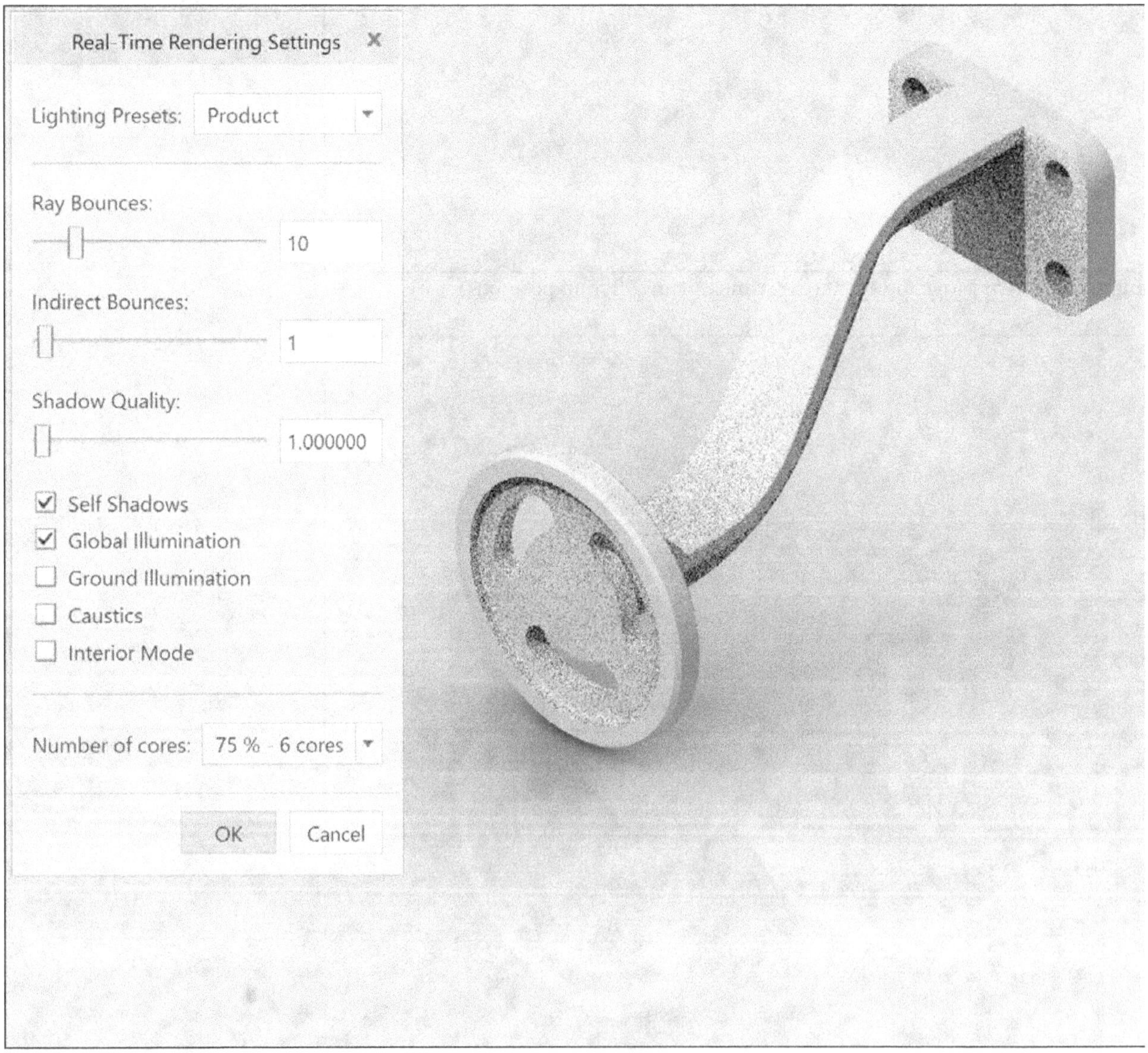

Figure 15.18 Real-Time Rendering

Download additional projects from *www.cad-resources.com*.

Lesson 16 Helical Sweeps and Annotations

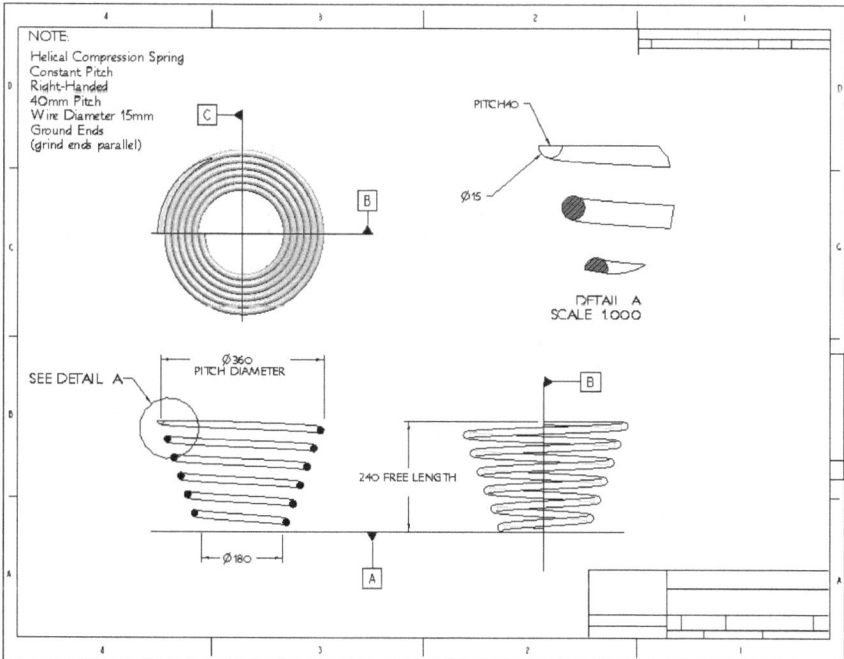

Figure 16.1 Helical Compression Spring Drawing

OBJECTIVES

- Create a **helical compression spring** with a **Helical Sweep**
- Use sweeps to create **hooks** on **extension springs**
- Create **plain ground** or **hook ends** on a spring
- Create **3D Notes** and **Annotation Features**

REFERENCES AND RESOURCES

For **Resources** go to www.cad-resources.com > click on the PTC Creo Parametric 5.0 Book cover

- Lesson 16 Lecture at **YouTube Creo Parametric Lecture Videos**

- Lesson 16 3D PDF models embedded in a PDF

Helical Sweeps and Annotations

A **helical sweep** (Fig. 16.1) is created by sweeping a section along a helical *trajectory*. The trajectory is defined by both the *profile* of the *surface of revolution* (which defines the distance from the section origin of the helical feature to its *axis of revolution*) and the *pitch* (the distance between coils). The trajectory and the surface of revolution are construction tools and do not appear in the resulting geometry.

    Annotation features are data features that you can use to manage the model annotation including surface finish, geometric tolerances, notes, and so on. **Model notes** are pieces of text, which can contain links (URL's) to World Wide Web pages, which you can attach to objects. Model notes, increase the amount of information that you can attach to any entity in your model.

685

Helical Sweeps

The Helical Sweep command is available (Fig. 16.2) for both solid and surface features. You can define the helical sweep feature using the following options:

- **Keep constant section** The pitch is constant
- **Vary section** The pitch is variable and defined by a graph
- **Thru axis of revolution** The section lies in a plane that passes through the axis of revolution
- **Normal To trajectory** The section is oriented normal to the trajectory
- **Normal to projection** The section is oriented normal to the projection
- **Use right handed rule** The trajectory is defined by the right-hand rule
- **Use left handed rule** The trajectory is defined by the left-hand rule

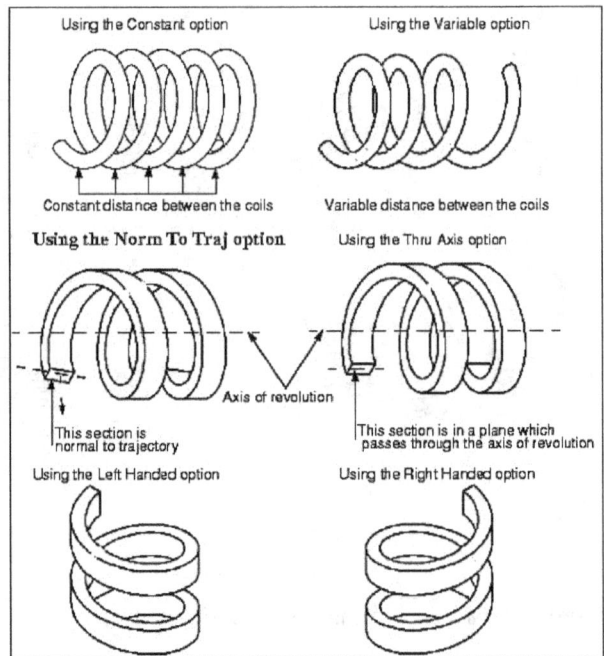

Figure 16.2 Helical Sweeps

Annotations

Model notes are text strings, which can be placed flat to the screen (view plane) in model space (Fig. 16.3). Note(s) can be attached to any entity in your model. When you attach a note to an entity, that entity is considered the parent of the note. If you delete the parent entity, all child note(s) are deleted with it. You can also allocate a URL to each model note. You can use model notes to communicate with members of your workgroup as to how to review or use a model, explain how you approached or solved a design problem when modeling, and explain changes that you have made to the features of a model over time.

Annotation features can also be notes, but also include: symbols, surface finish, geometric tolerance, set datum tags, ordinate baseline dimensions, driven dimension, and so on.

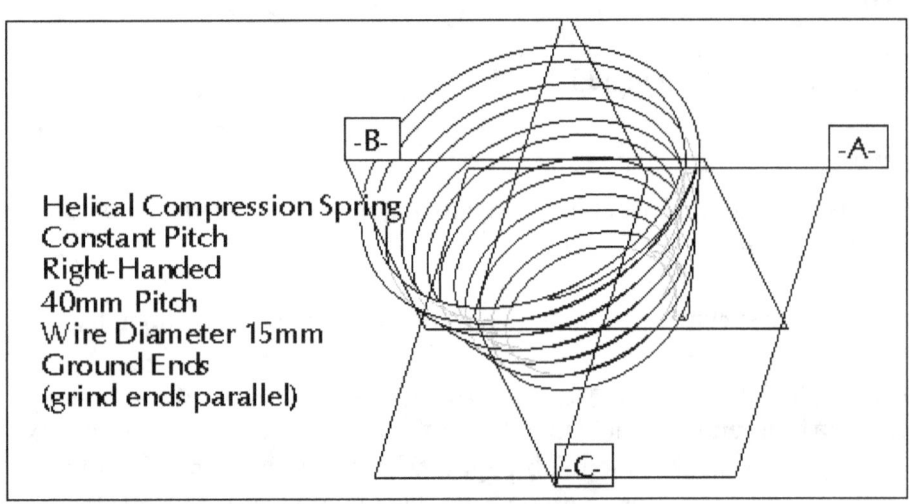

Figure 16.3 Model Notes

Lesson 16 STEPS

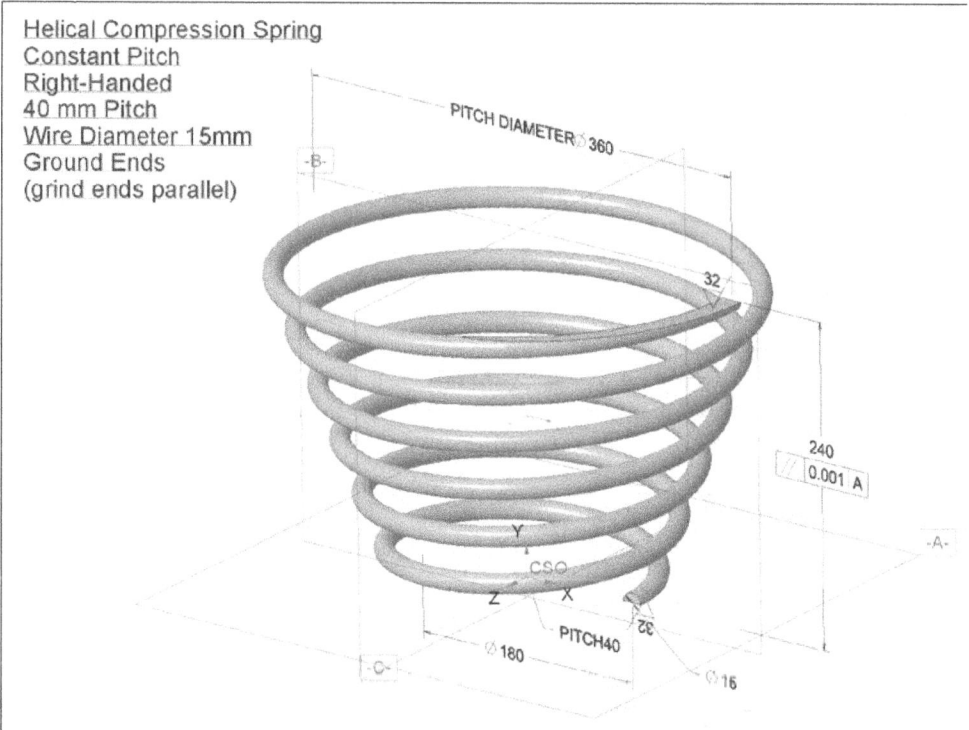

Helical Compression Spring
Constant Pitch
Right-Handed
40 mm Pitch
Wire Diameter 15mm
Ground Ends
(grind ends parallel)

Figure 16.4(a) Helical Compression Spring with Datum Planes and Model Note

Helical Compression Spring

Springs [Fig. 16.4(a)] and other helical features are created with the Helical Sweep command. A helical sweep is created by sweeping a *section* along a *trajectory* that lies in the *surface of revolution:* The trajectory is defined by both the *profile* of the surface of revolution and the distance between coils. The model for this lesson is a *constant-pitch right-handed helical compression spring with ground ends, a pitch of **40 mm**, and a wire diameter of **15 mm*** [Figs. 16.4(b-e)].

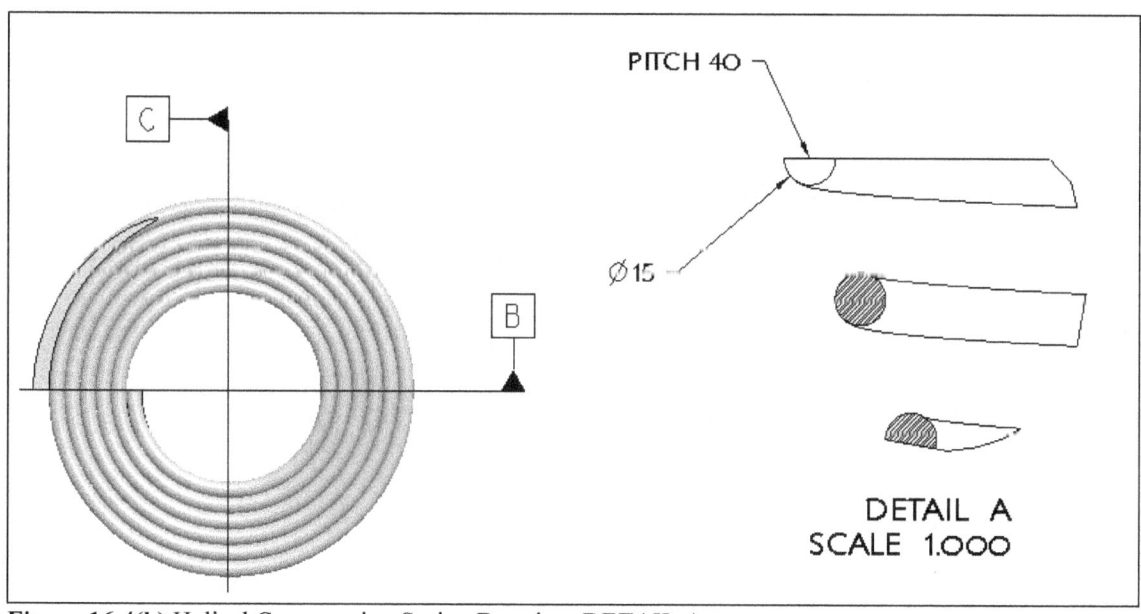

Figure 16.4(b) Helical Compression Spring Drawing: DETAIL A

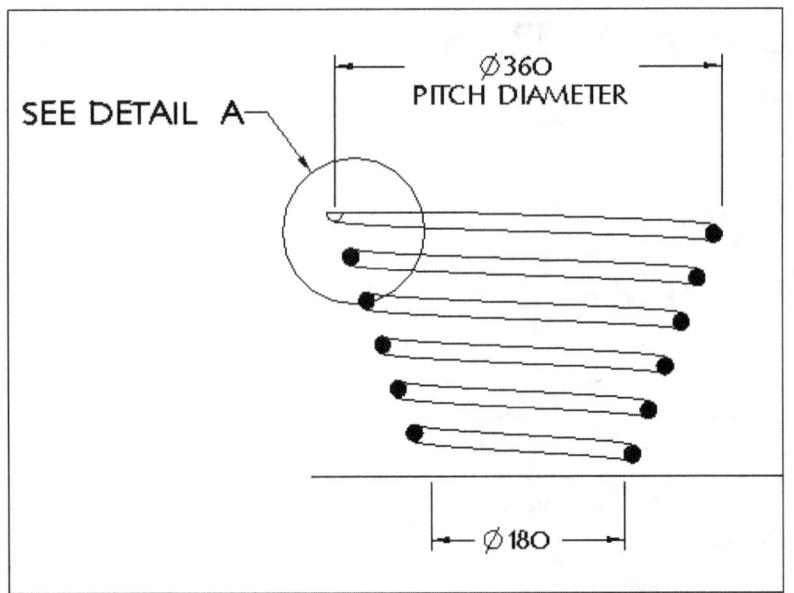

Figure 16.4(c) Helical Compression Spring Drawing, Section

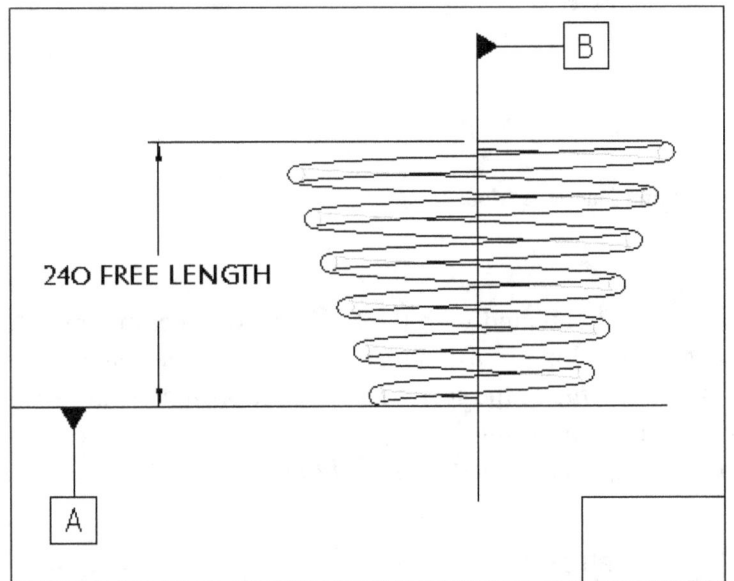

Figure 16.4(d) FREE LENGTH 240

Figure 16.4(e) 3D Model Note

Start a new part. Click: [ ] **New** > (●) [ ] Part > Name **helical_compression_spring** >
[✓] Use default template > **OK** > **File** > **Prepare** > **Model Properties** (set the material and units):

- **Material** = ss.mtl

Set Datum [-A-] and **Rename** the default datum planes and coordinate system:

- Datum TOP = **A**
- Datum FRONT = **B**
- Datum RIGHT = **C**
- Coordinate System = **CS0**

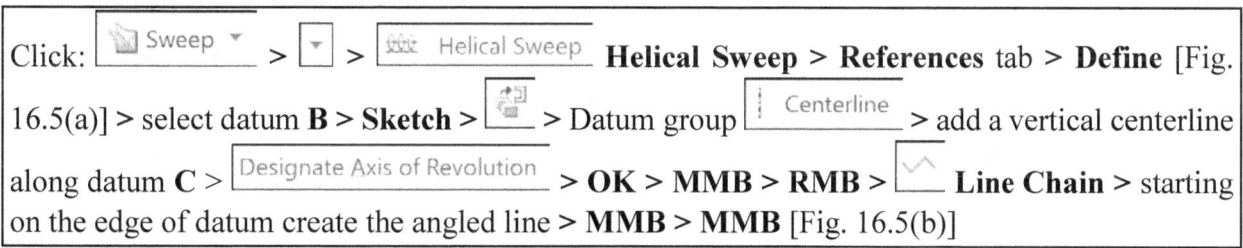

Click: ⬚ Sweep ▾ > ▾ > ⬚ Helical Sweep **Helical Sweep > References** tab > **Define** [Fig. 16.5(a)] > select datum **B** > **Sketch** > ⬚ > Datum group ⎸ Centerline > add a vertical centerline along datum **C** > ⎸Designate Axis of Revolution > **OK** > **MMB** > **RMB** > ⬚ **Line Chain** > starting on the edge of datum create the angled line > **MMB** > **MMB** [Fig. 16.5(b)]

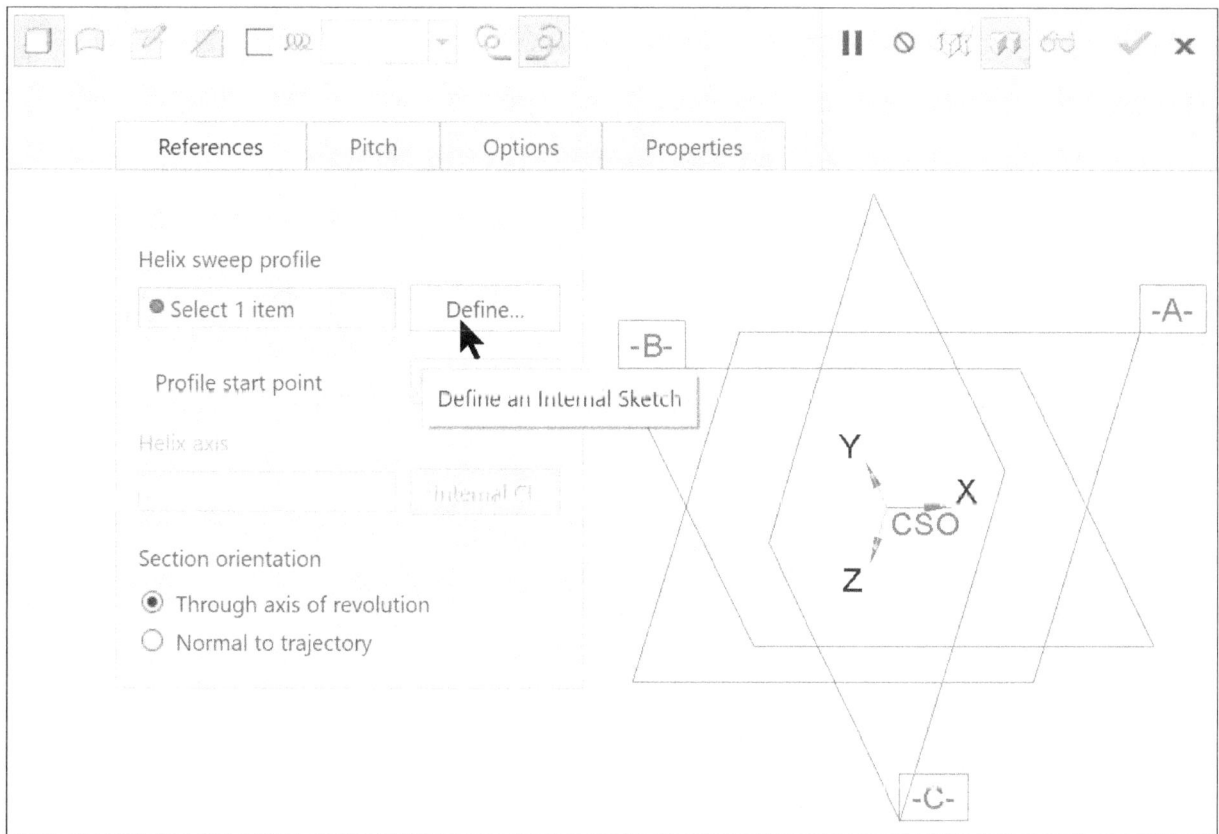

Figures 16.5(a) Helical Sweep Tool

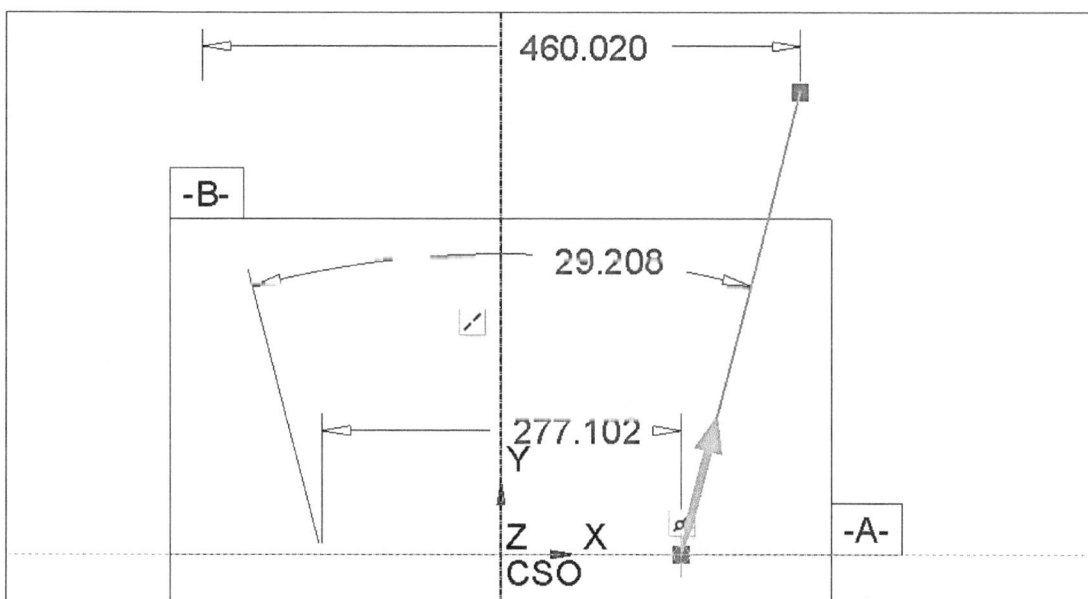

Figure 16.5(b) Helix Sweep Profile Sketch. Note the Start Arrow Direction.

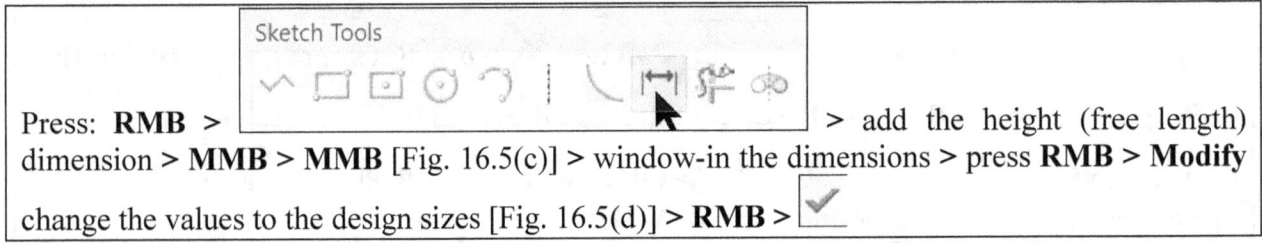

Press: **RMB** > _____ > add the height (free length) dimension > **MMB** > **MMB** [Fig. 16.5(c)] > window-in the dimensions > press **RMB** > **Modify** change the values to the design sizes [Fig. 16.5(d)] > **RMB** > ✓

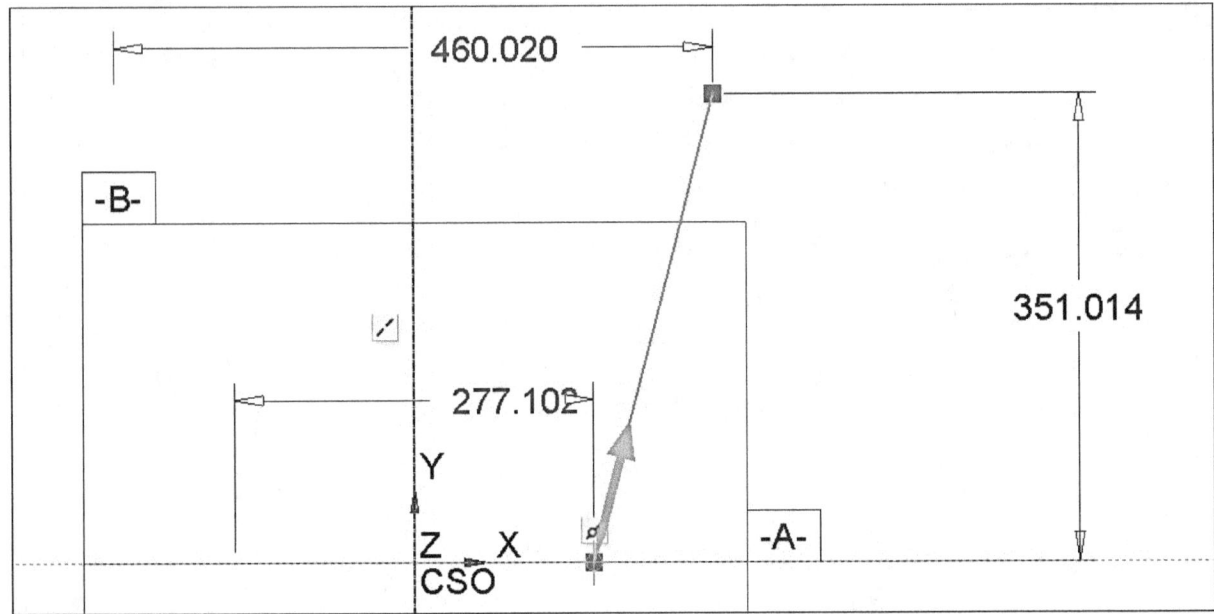

Figure 16.5(c) Dimensioned Sketch

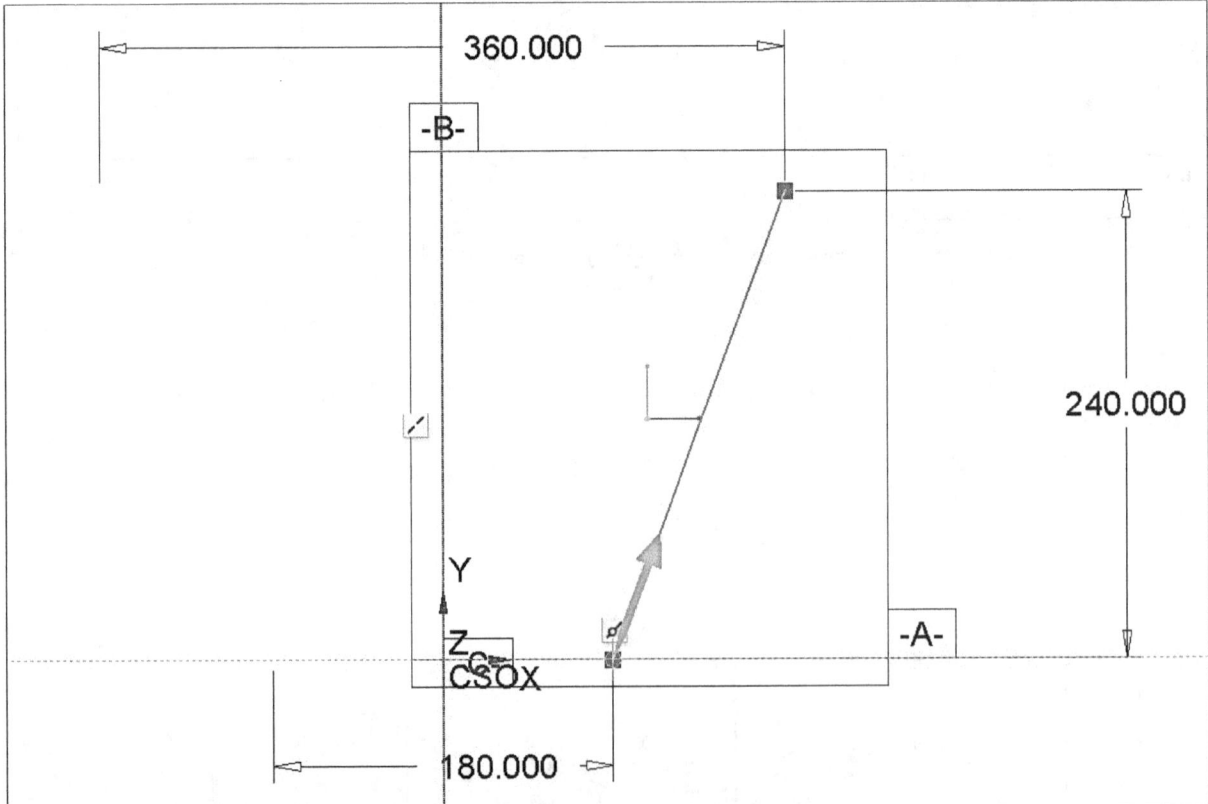

Figure 16.5(d) Modified Dimensions

Enter the pitch value **40** ![40.000] > **Enter** [Fig. 16.5(e)] > ![icon] **Create or edit sweep section** from the Dashboard > ![Center and Point] > sketch the section geometry of the spring at the intersection of the crosshairs [Fig. 16.5(f)] > **MMB** > **LMB** > select the dimension > **RMB** > **Modify** > *type* **15** > **Enter** > **OK** > **LMB** [Fig. 16.5(g)]

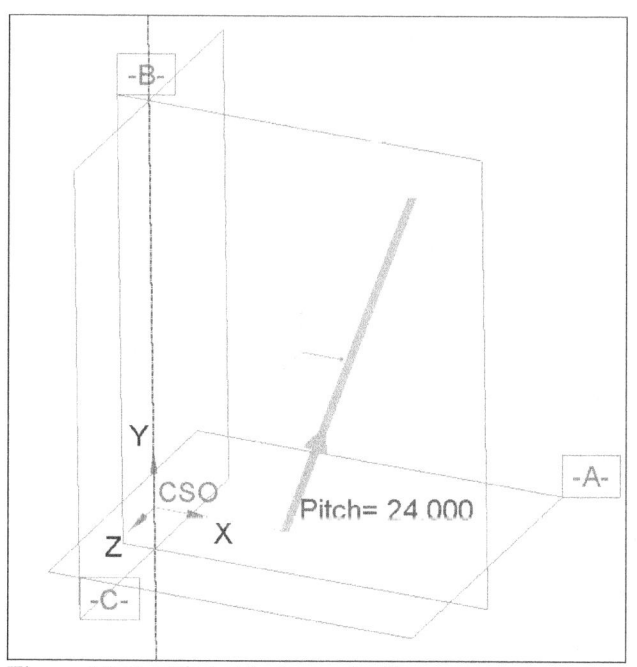

Figure 16.5(e) Pitch **24**, change to **40**

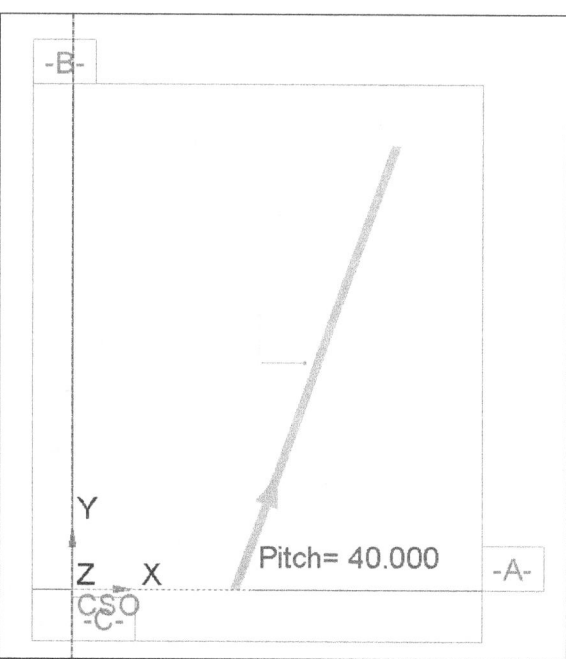

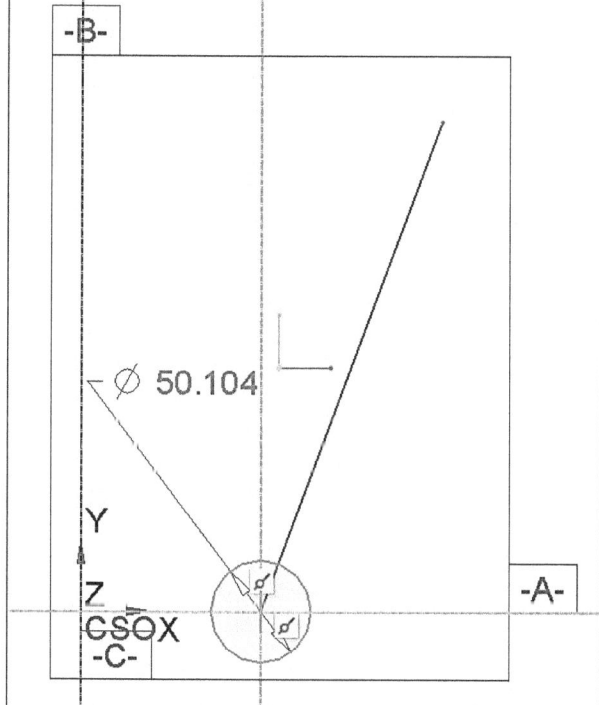

Figure 16.5(f) Sketch a Circle

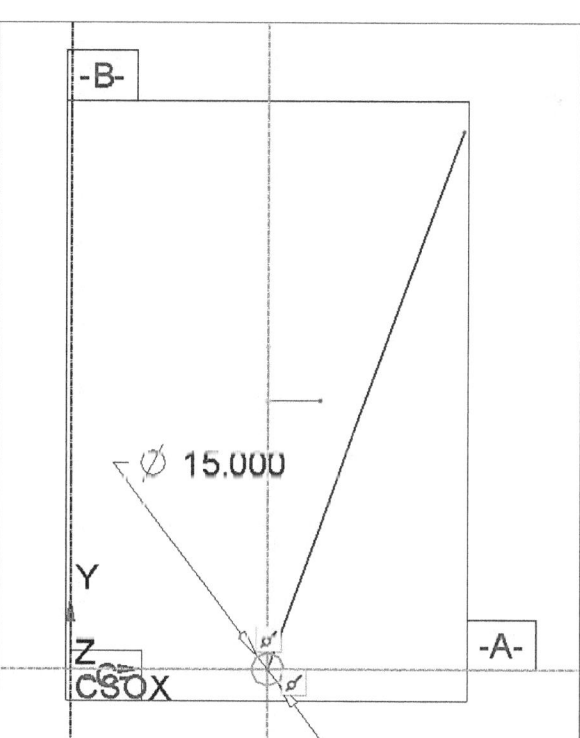

Figure 16.5(g) Wire Diameter **15**

Click: **RMB** > 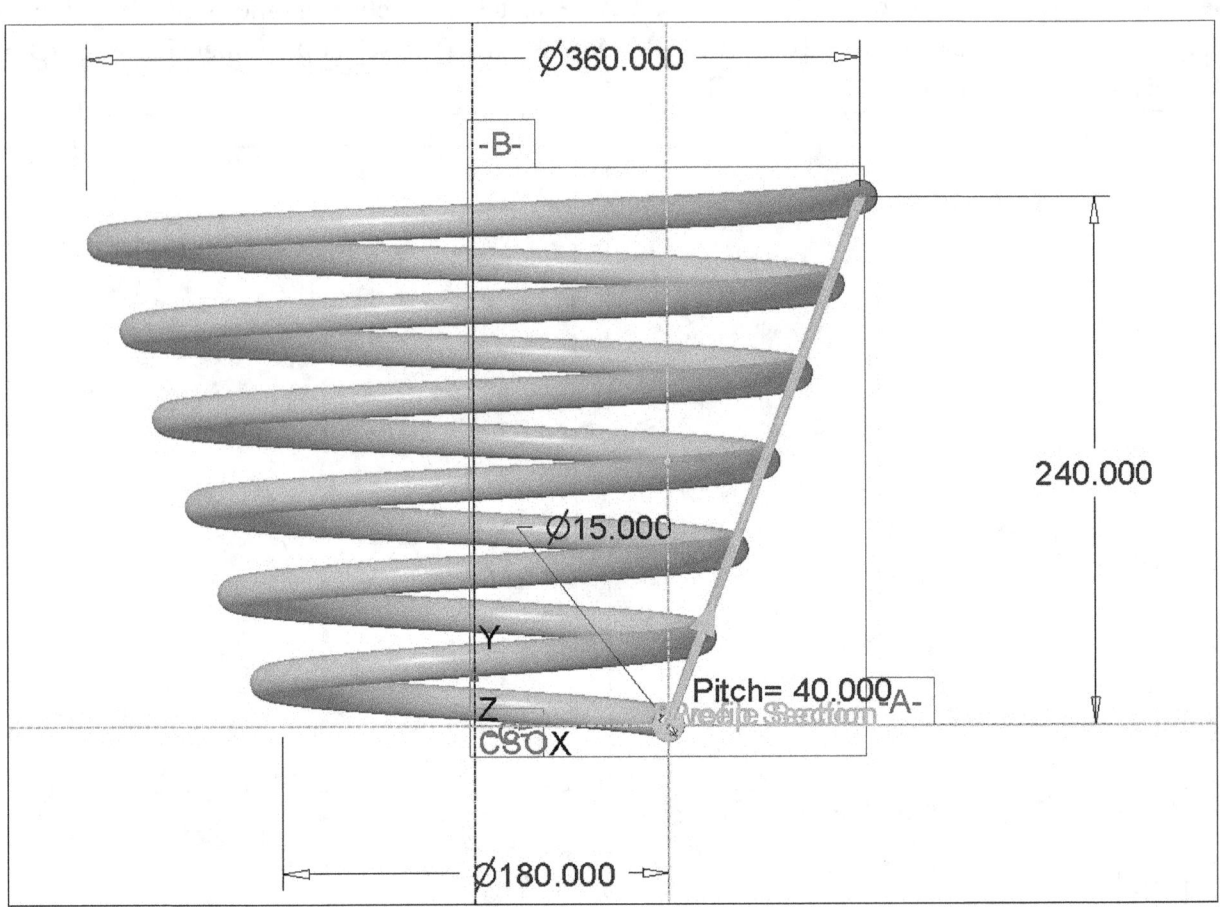 [Fig. 16.5(h)] > **RMB** > **Show Section Dimensions** > **Ctrl+D** > ✔ > **View** tab > **Appearance Gallery** > change the color of the part > **Ctrl+S** > **OK** [Fig. 16.5(i)]

Figure 16.5(h) Helix Preview

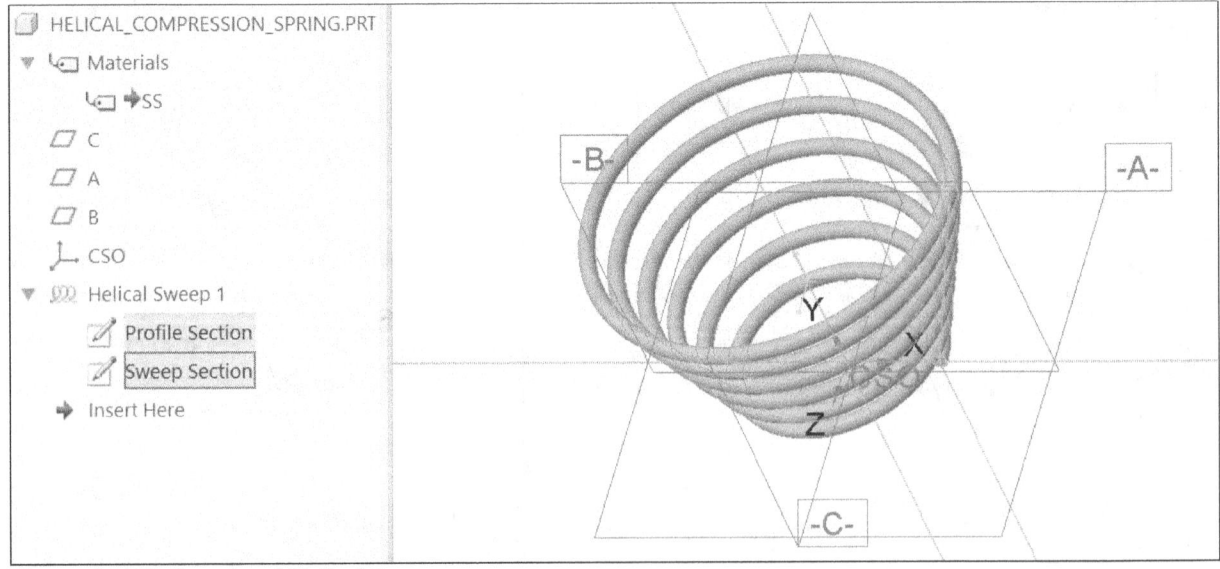

Figure 16.5(i) Completed Helical Sweep

Create the *ground ends*, click: **Model** tab > [Extrude] > [⊥] expand depth options by opening slide-up panel > [⊟] **Extrude on both sides** > [◹] **Remove Material** > in the Graphics Window, press **RMB > Define Internal Sketch** > Sketch Plane- pick datum **C** > Reference- pick datum **A** > Orientation- **Bottom** [Fig. 16.6(a)] > **Sketch** > [⟳] > **RMB** > [⌇] **Line Chain** > draw a horizontal line > **MMB > MMB > LMB** > modify the dimension [Fig. 16.6(b)] > spin the model as needed > [✓]

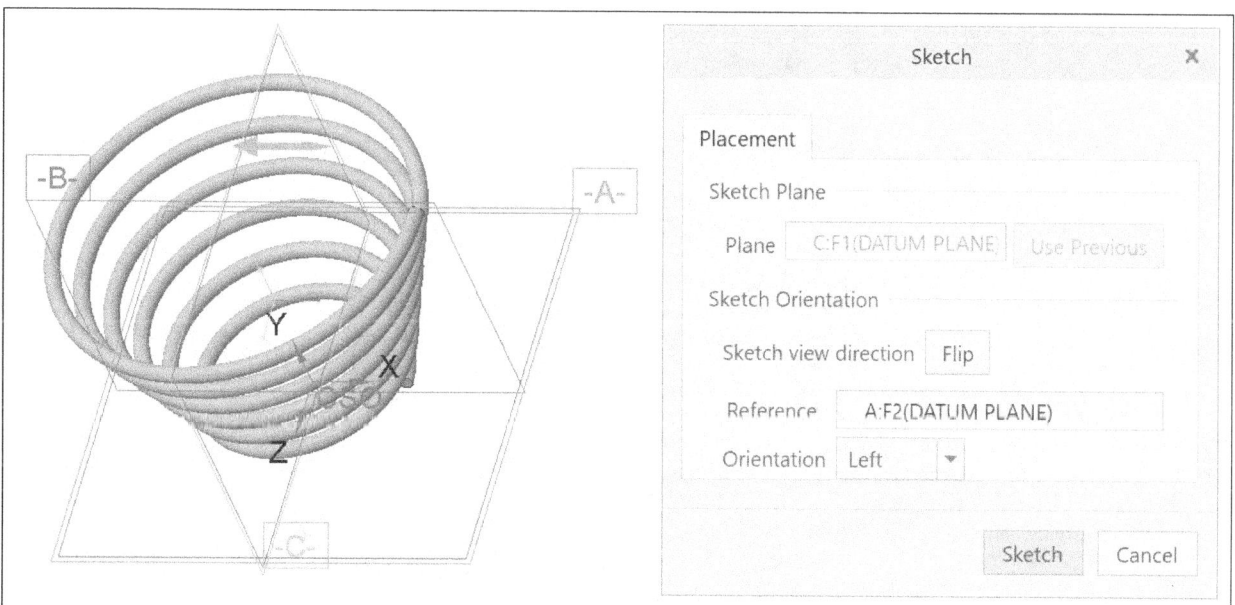

Figure 16.6(a) Cut Sketch Orientation

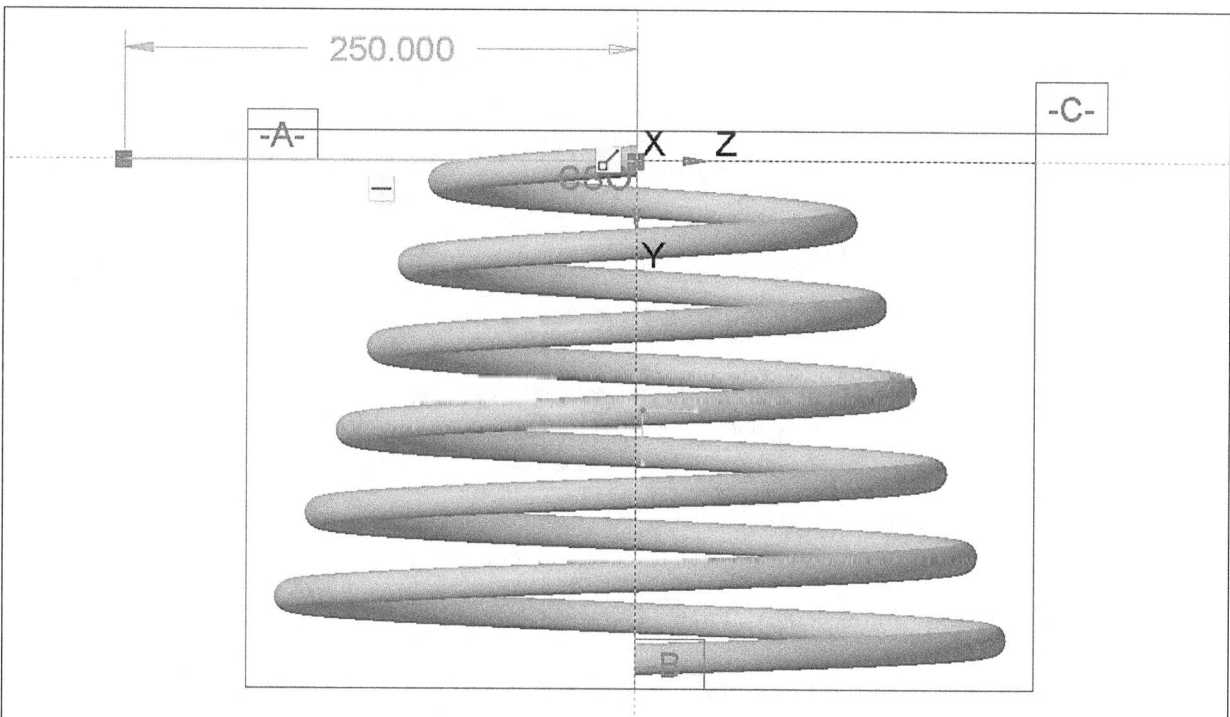

Figure 16.6(b) Creating Ground Ends (any length will work greater than the spring diameter)

693

Extend a depth handle to **300** so as to include the full spring > in the Graphics Window, press **RMB > Flip Material Side** [Fig. 16.6(c)] > 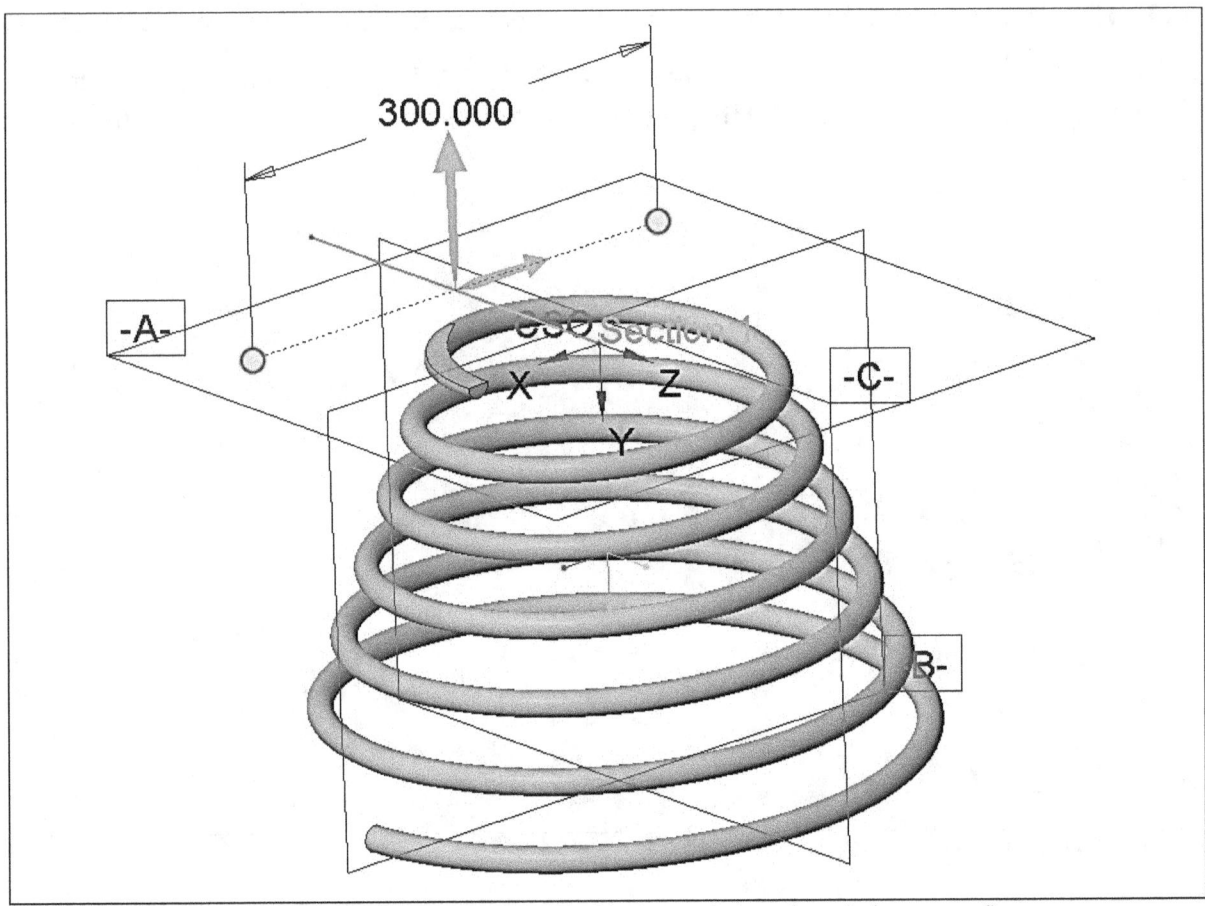 [Fig. 16.6(d)] > **LMB**

Figure 16.6(c) Depth Handles (Squares) and Material Side Arrow (currently pointing upward)

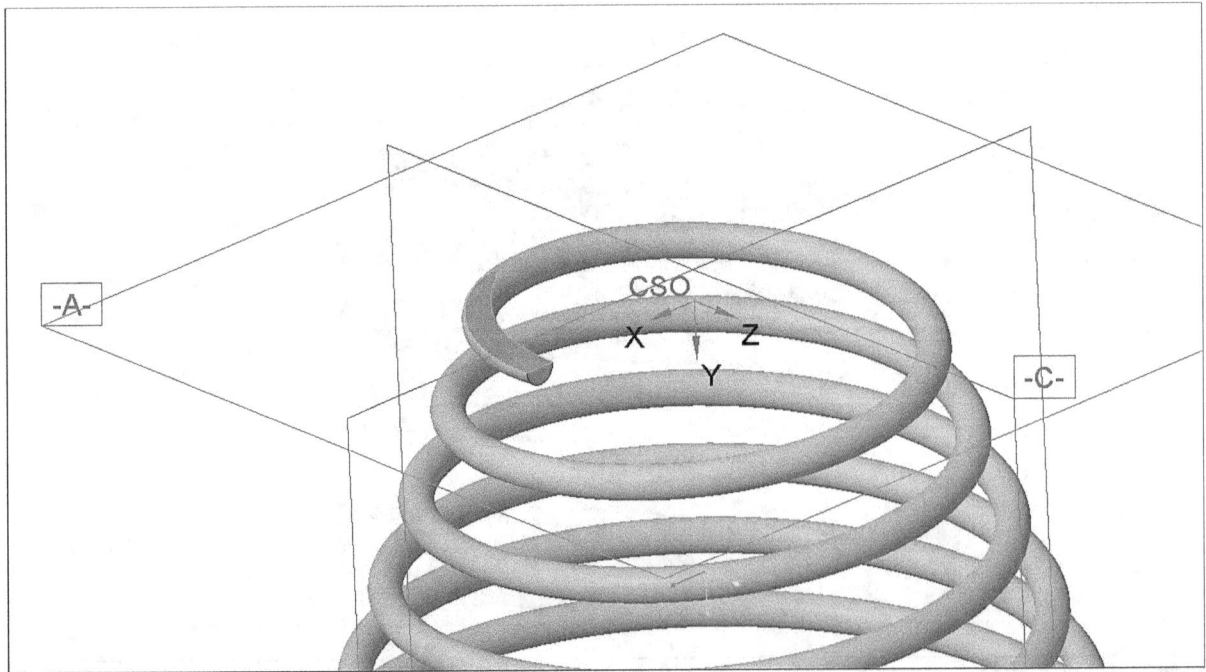

Figure 16.6(d) Completed Cut for One Ground End

The spring is a metric size. Change units to metric. **File > Prepare > Model Properties > change > Units >** 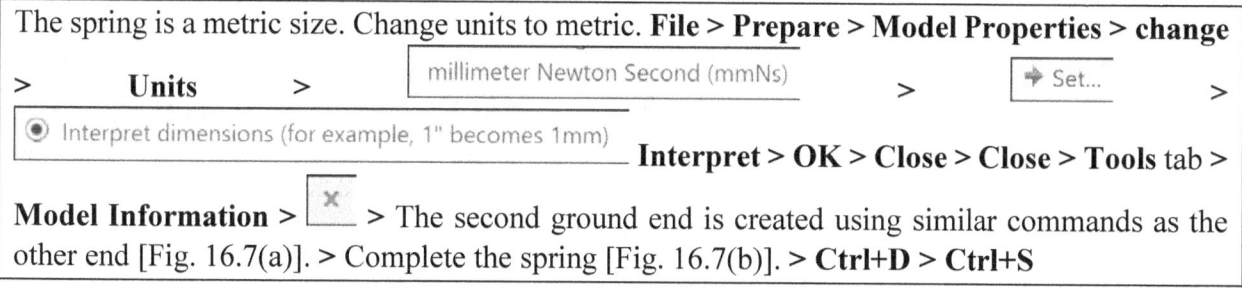 millimeter Newton Second (mmNs) **> | ↦ Set... | > | ◉ Interpret dimensions (for example, 1" becomes 1mm) | Interpret > OK > Close > Close > Tools** tab > **Model Information >** | × | **>** The second ground end is created using similar commands as the other end [Fig. 16.7(a)]. > Complete the spring [Fig. 16.7(b)]. > **Ctrl+D > Ctrl+S**

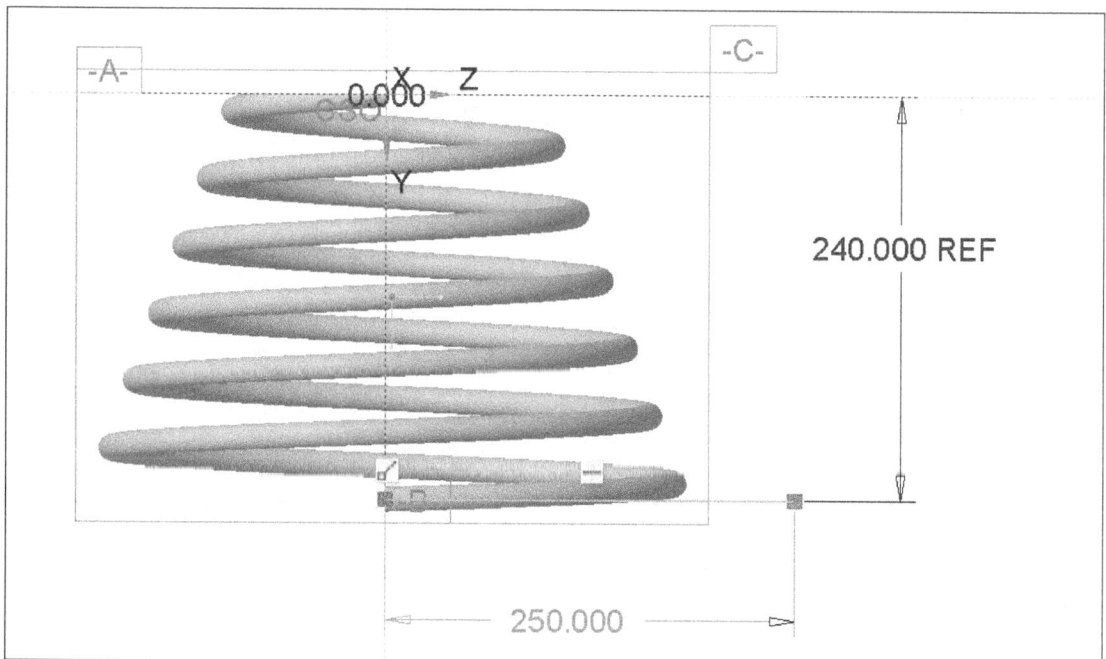

Figure 16.7(a) Creating the Second Ground End

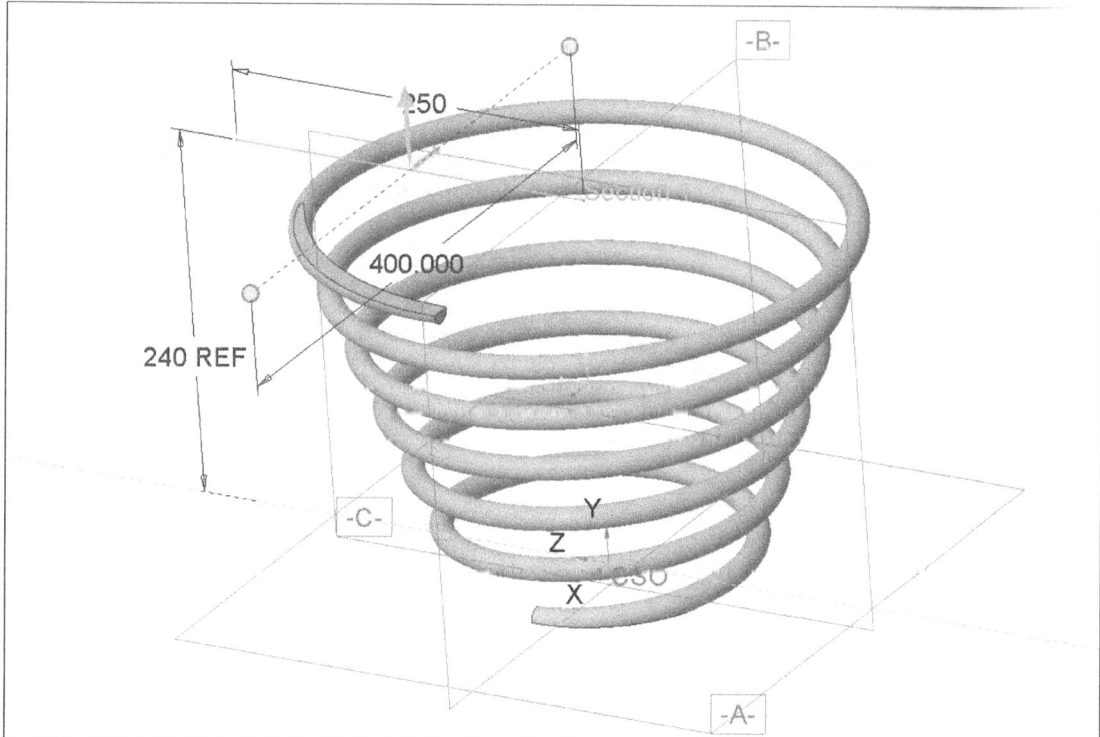

Figure 16.7(b) Preview of the Completed Cut for the Second Ground End

Figure 16.8 provides an **ECO** (Engineering Change Order) for a new spring. Copy the file you are working on by clicking: **File > Save As > Save a Copy > HELICAL_EXTENSION_SPRING > OK > File > Close > File > Open > In Session > helical_extension_spring.prt > Open >** delete the existing ground ends **>** modify the pitch to **10 mm >** change the wire diameter to **7.5 mm >** complete the extension spring [Figs. 16.9(a) through 16.10(d)]. The free length is to be **120 mm**. The large diameter will now be **180 mm**, and the small diameter will be **120 mm**. **> Ctrl+S > OK > File > Close**

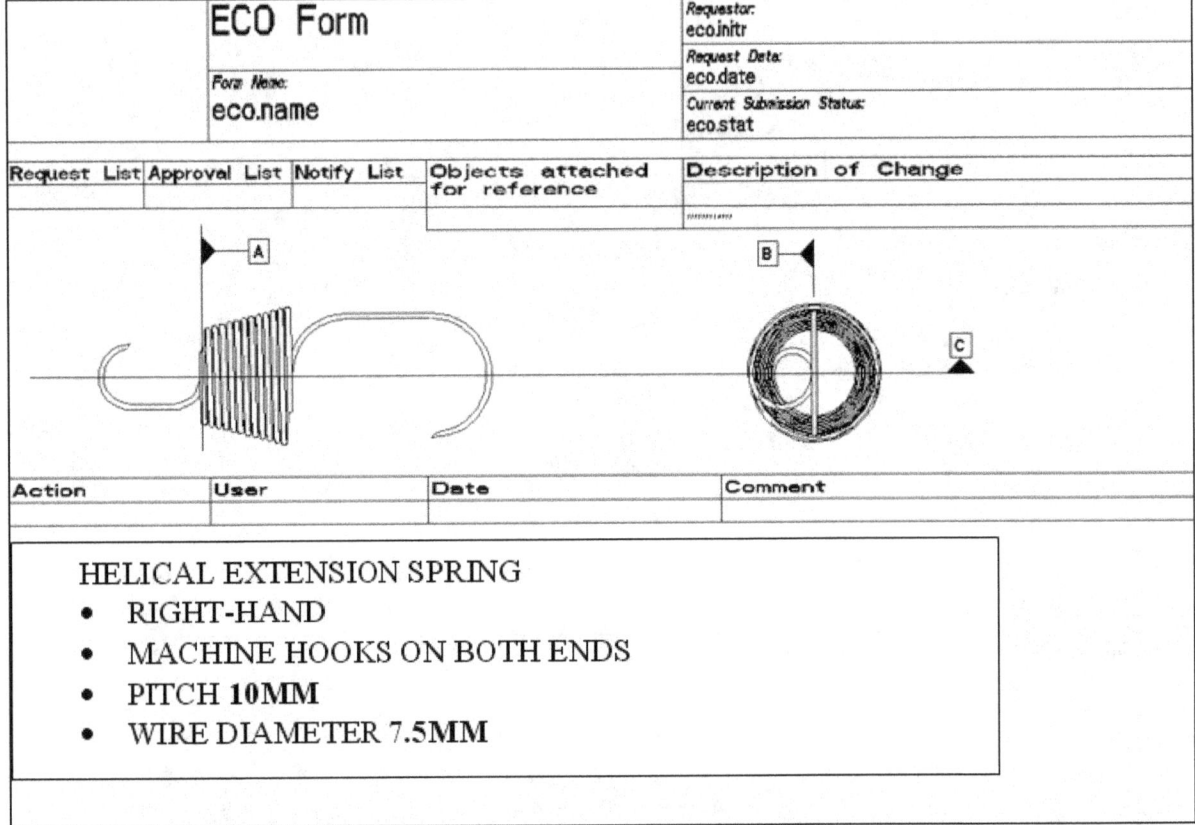

Figure 16.8 ECO to Create a Helical Extension Spring [**You are not creating this ECO drawing; you are making a new part from an existing part (copied) using different dimensions and features**]

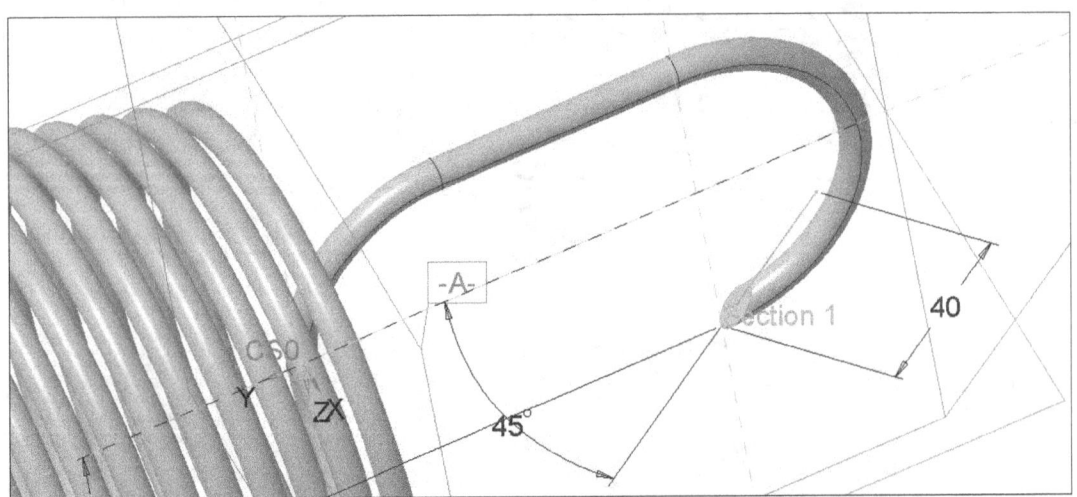

Figure 16.9(a) Ground End

696

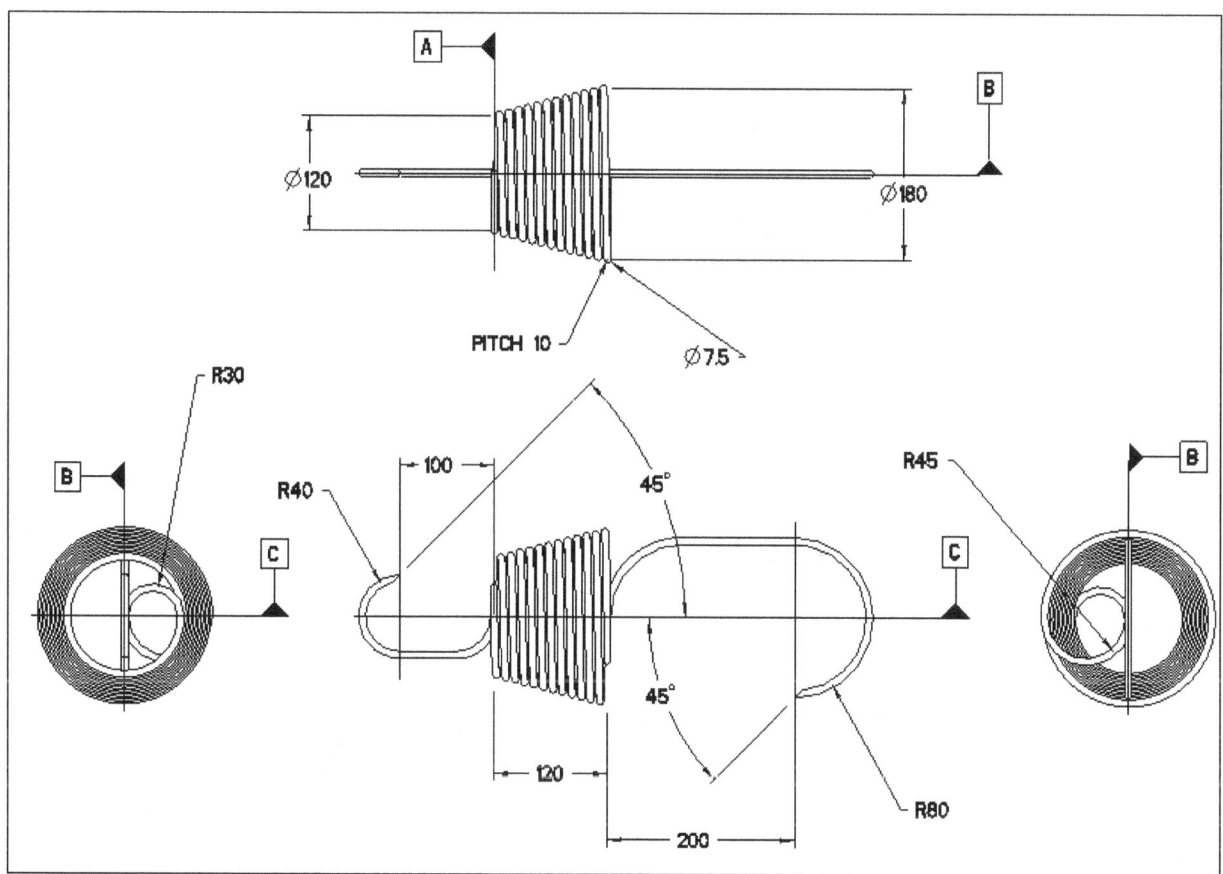

Figure 16.9(b) Detail Drawing of Helical Extension Spring with Machine Hook Ends

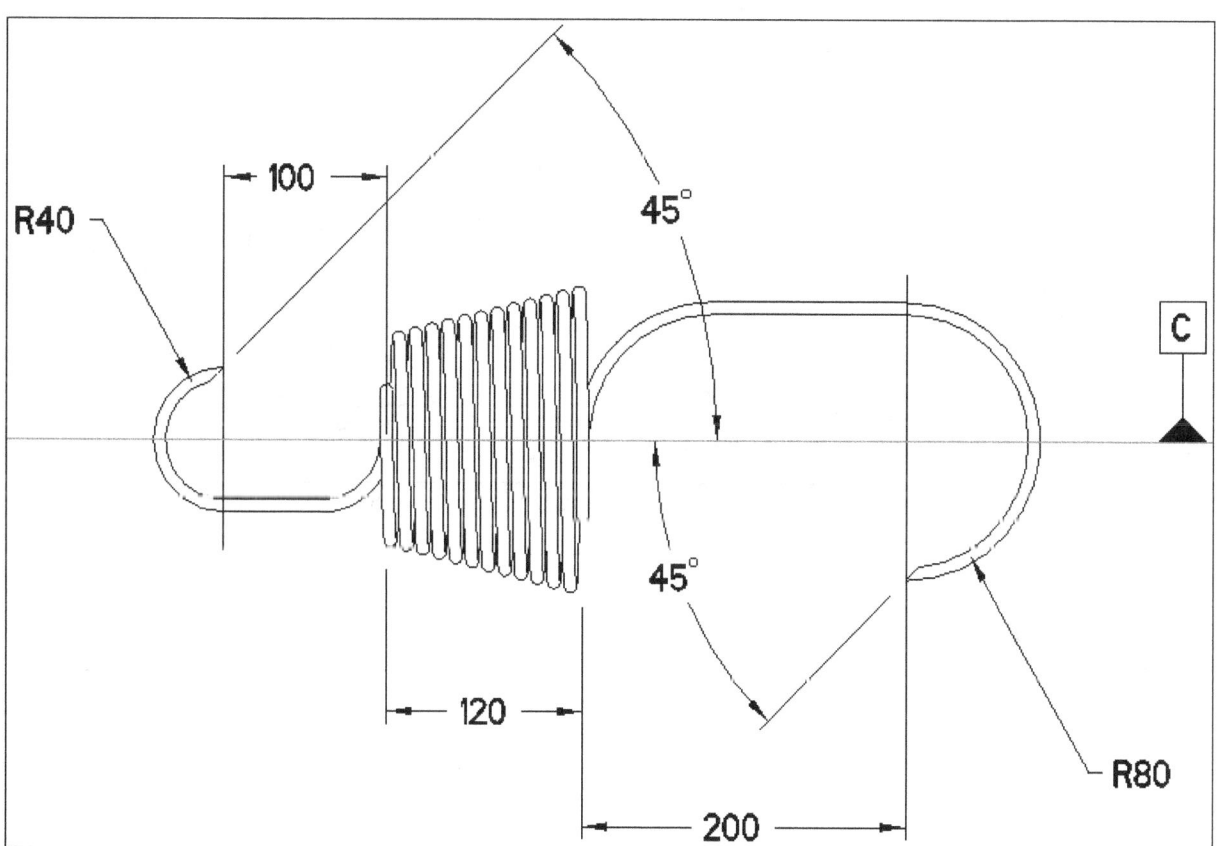

Figure 16.9(c) Front View

697

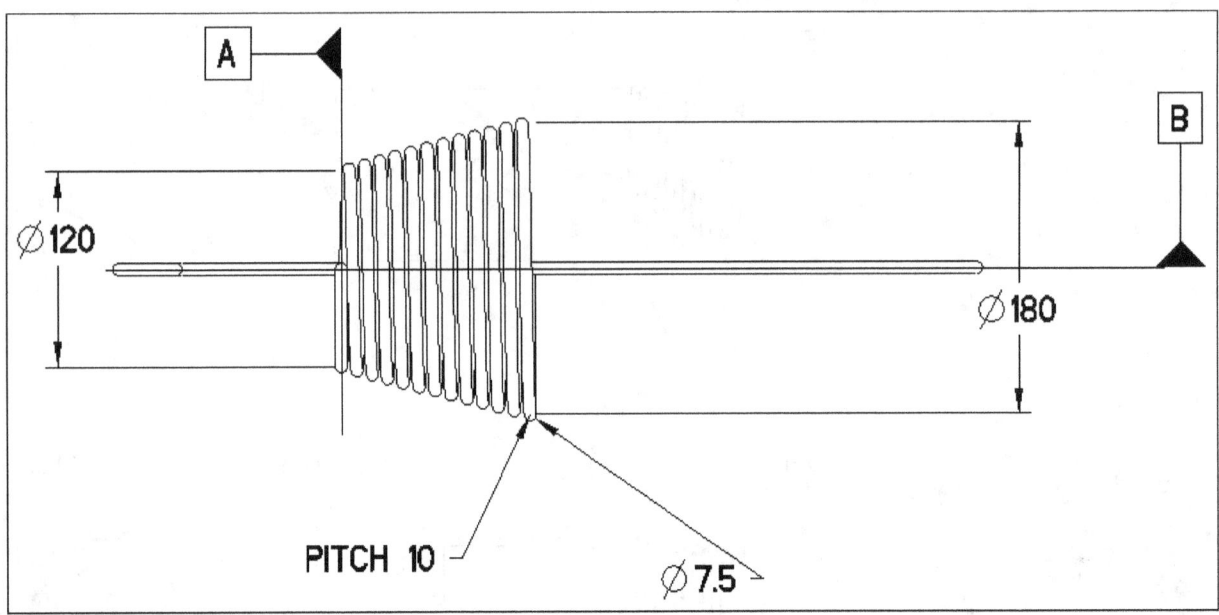

Figure 16.9(d) Top View

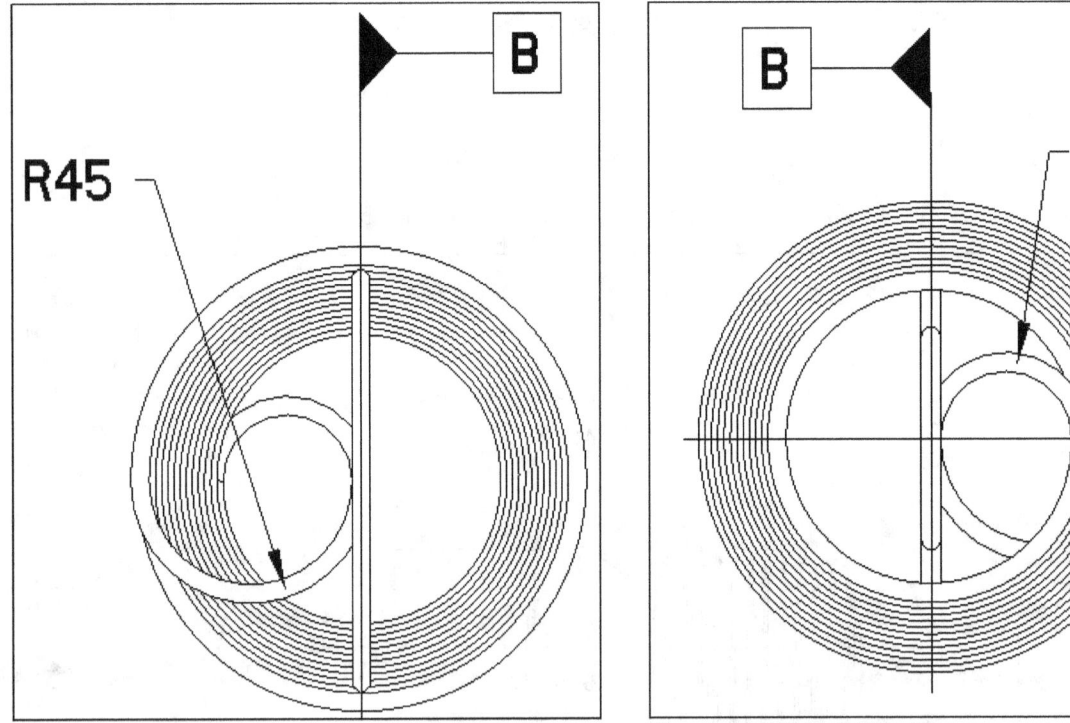

Figure 16.9(e) Right Side View

Figure 16.9(f) Left Side View

Watch the lecture!

698

Create the machine hooks using simple sweeps and cuts, as shown in Figures 16.10(a) through 16.10(d).

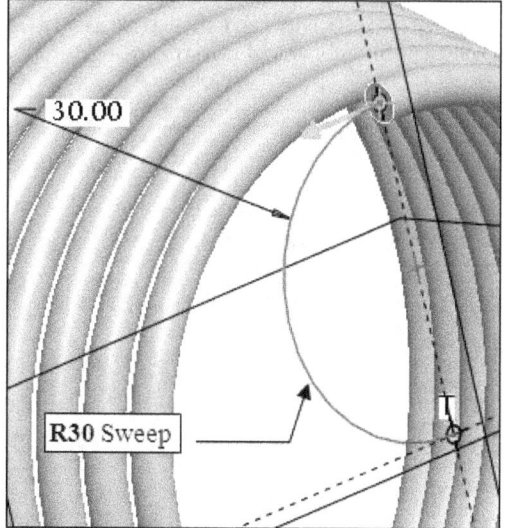

Figure 16.10(a) Sweep **R30**

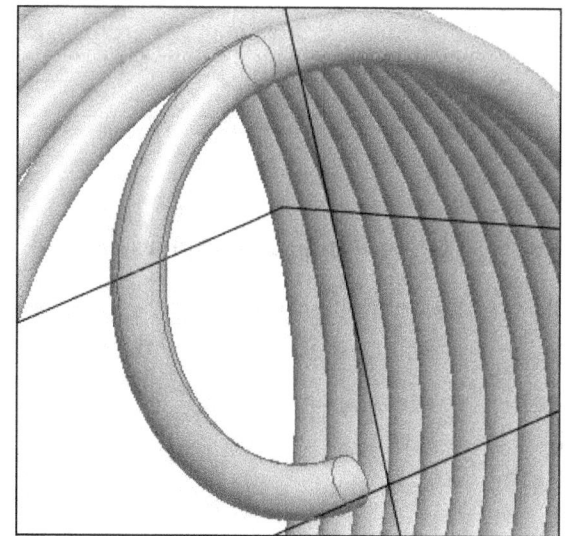

Figure 16.10(b) Completed Sweep

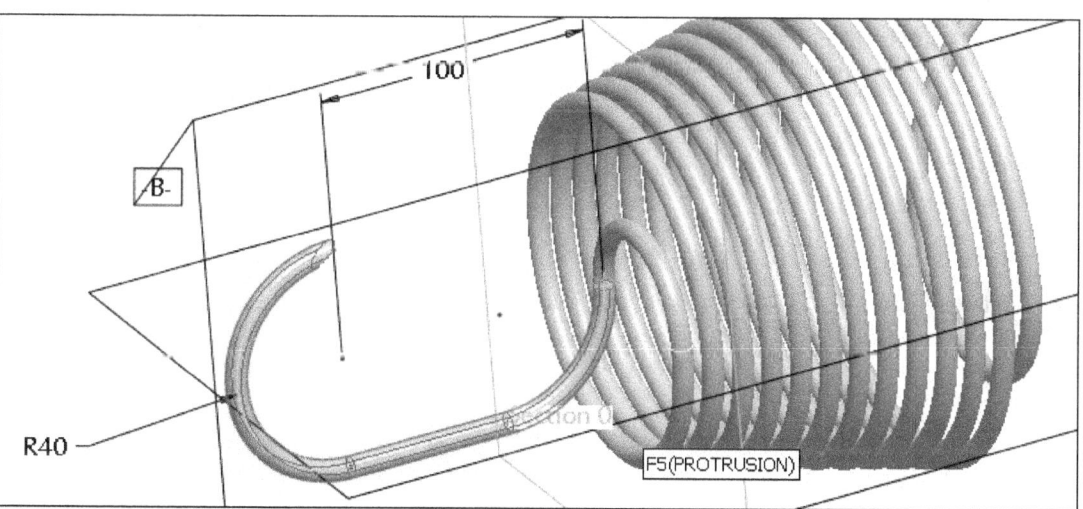

Figure 16.10(c) Small Hook End Sweep

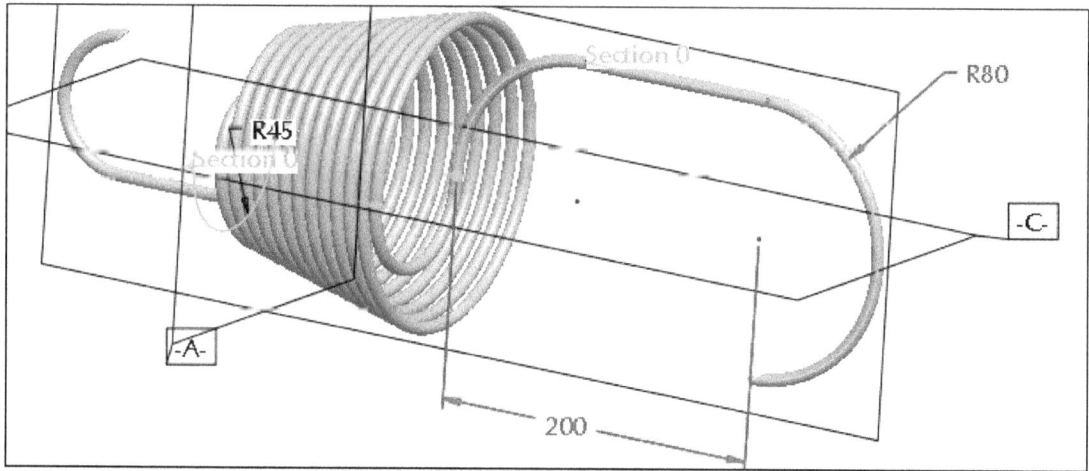

Figure 16.10(d) Large Hook End Sweep

Annotations

When you attach a note to an entity, that entity is considered the "parent" of the note. Deleting the parent deletes all of the notes of the parent. You can attach model notes anywhere in the model; they do not have to be attached to a parent. Here we will add a note to the part and describe the spring.

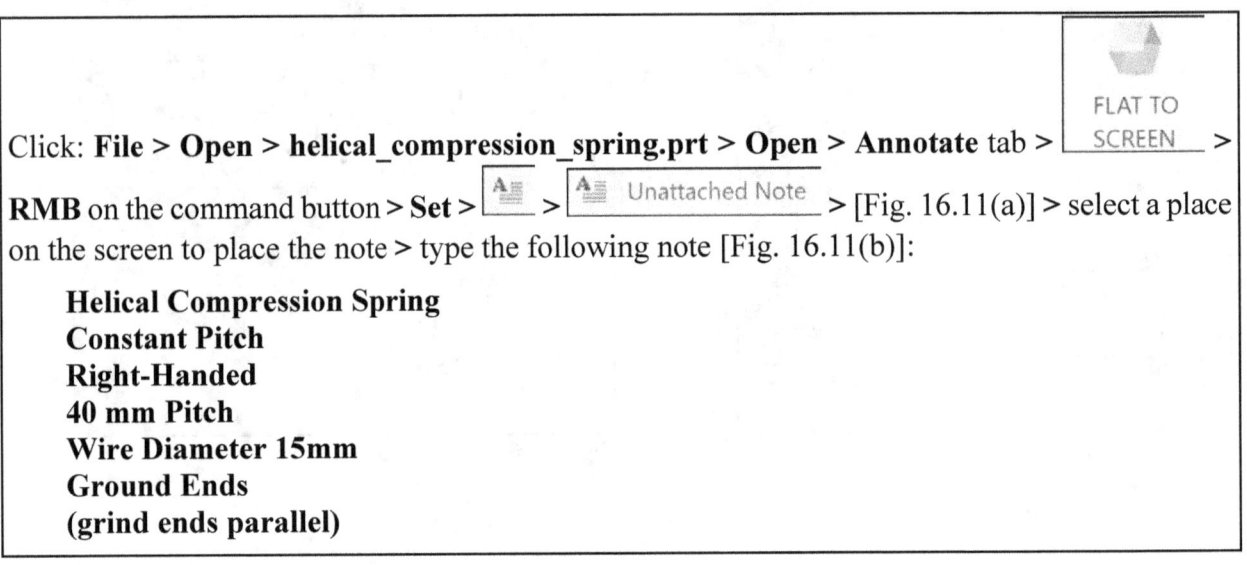

Click: **File > Open > helical_compression_spring.prt > Open > Annotate** tab > FLAT TO SCREEN >
RMB on the command button > **Set** > | A≡ | > | A≡ Unattached Note | > [Fig. 16.11(a)] > select a place on the screen to place the note > type the following note [Fig. 16.11(b)]:

Helical Compression Spring
Constant Pitch
Right-Handed
40 mm Pitch
Wire Diameter 15mm
Ground Ends
(grind ends parallel)

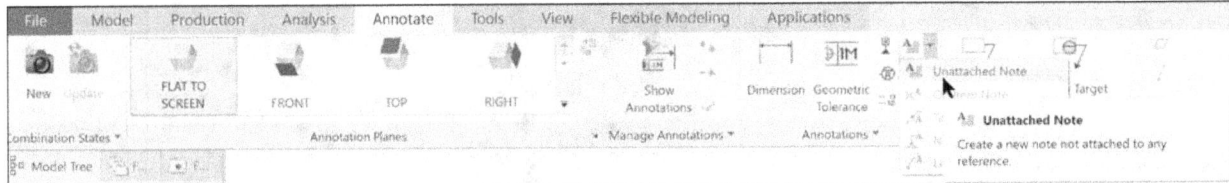

Figure 16.11(a) Unattached Note

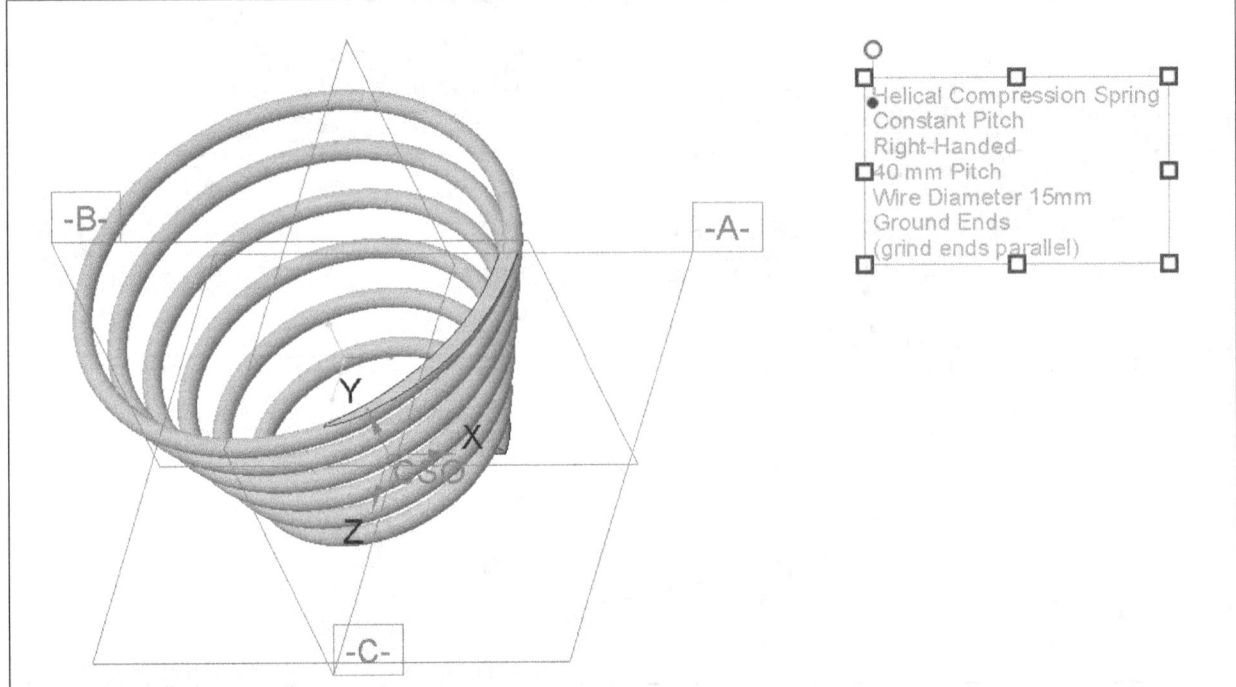

Figure 16.11(b) Note

Drag 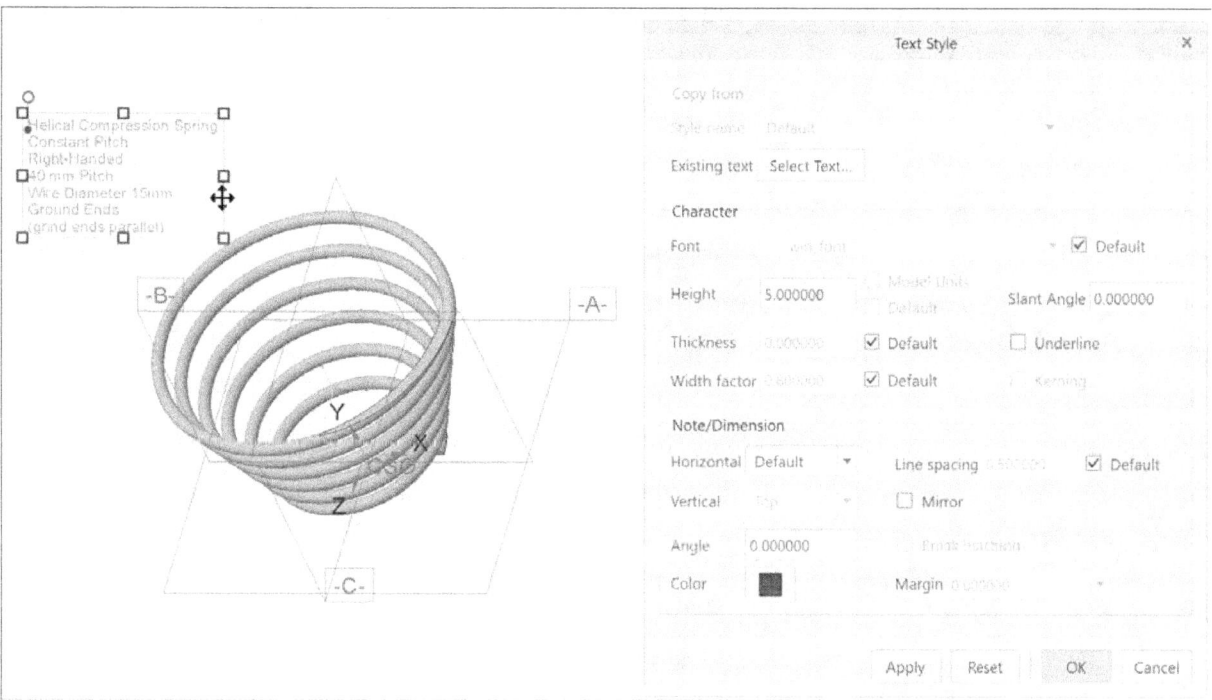 > move the note as needed > with the note still selected (highlighted); **RMB > Text Style** [Fig. 16.11(c)] > ☐ Default > Height **10** (or as needed) > **Enter** [Fig. 16.11(d)] > **OK > LMB > Ctrl+S**

Figure 16.11(c) Placing the Note

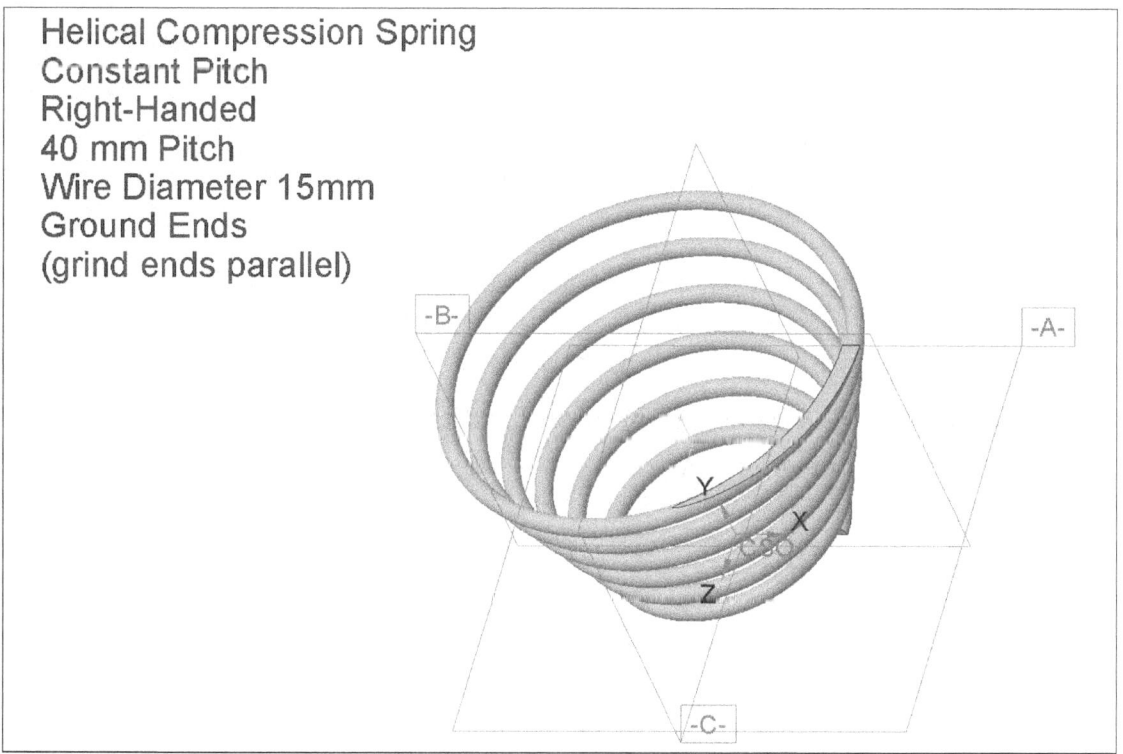

Figure 16.11(d) Completed Note

701

You can toggle model annotations on and off using ⬛ **Turn on or off 3D annotations and annotation elements** > toggle the annotations *off* and *on* > display the note in the Model Tree by clicking: 📐 ▾ > 🔽 Tree Filters... > toggle all *on* [Fig. 16.12(a)] > **Apply** > **OK** > select 🅰 Note_0 from **Model Tree or Drawing Tree** > **RMB** > 🖌 **Properties** > **OK** [Fig. 16.12(b)]

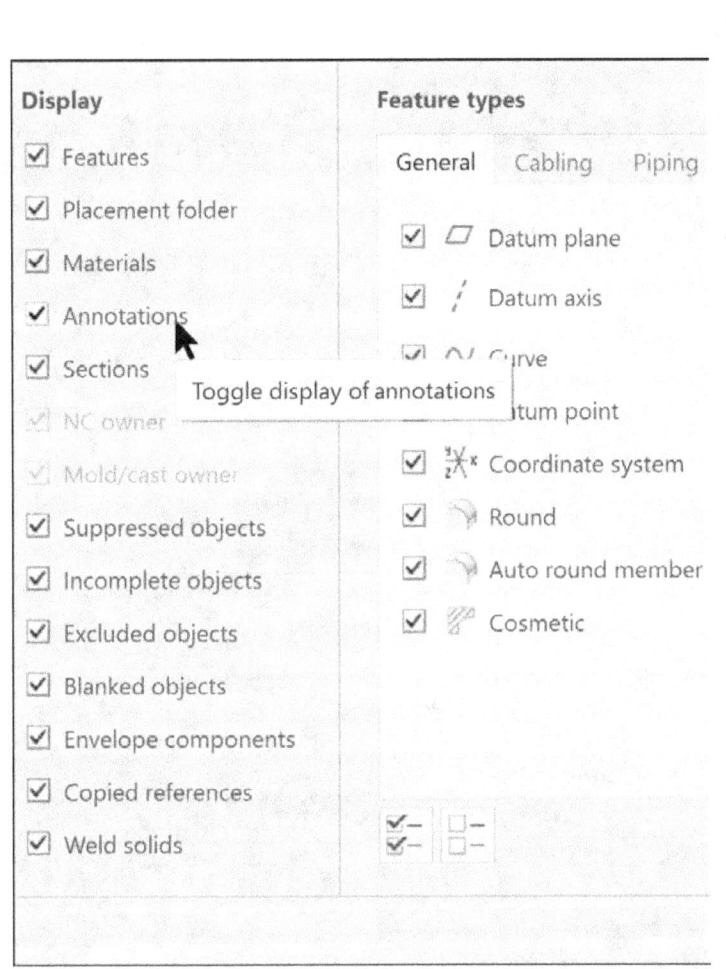

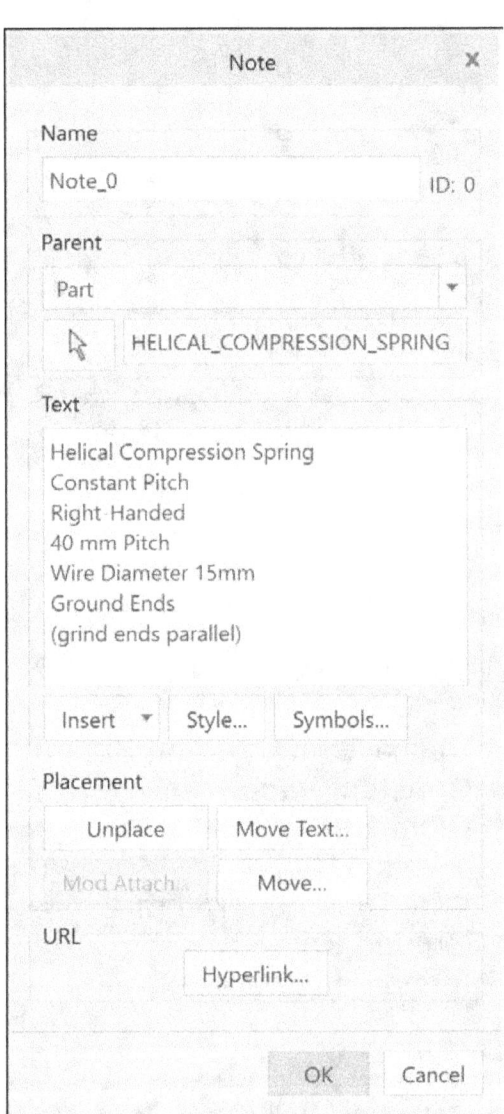

Figure 16.12(a) Displaying 3D Notes (Annotations) in Model Tree   **Figure 16.12(b)** Detail Tree and Model Tree

Click on 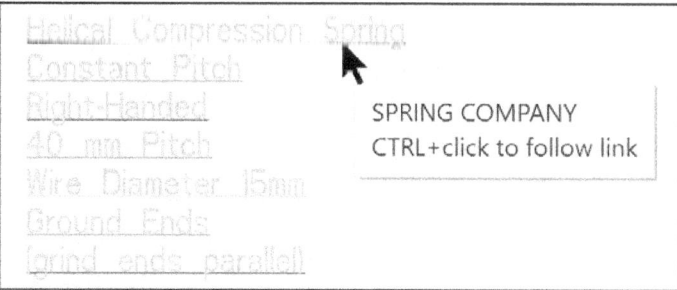 Note_0 in the Model Tree > **RMB** > **Rename** > type: **Compression_Spring** > **Enter** > **RMB** > **Add Link** [Fig. 16.13(a)] > type: http://www.americanprecspring.com [Fig. 16.13(b)] > ScreenTip... > **SPRING COMPANY** [Fig. 16.13(c)] > **OK** > **OK** > **LMB** [Fig. 16.13(d)] > place your pointer over the note in the graphics area to see the Screen Tip [Fig. 16.13(e)] > **Ctrl+S**

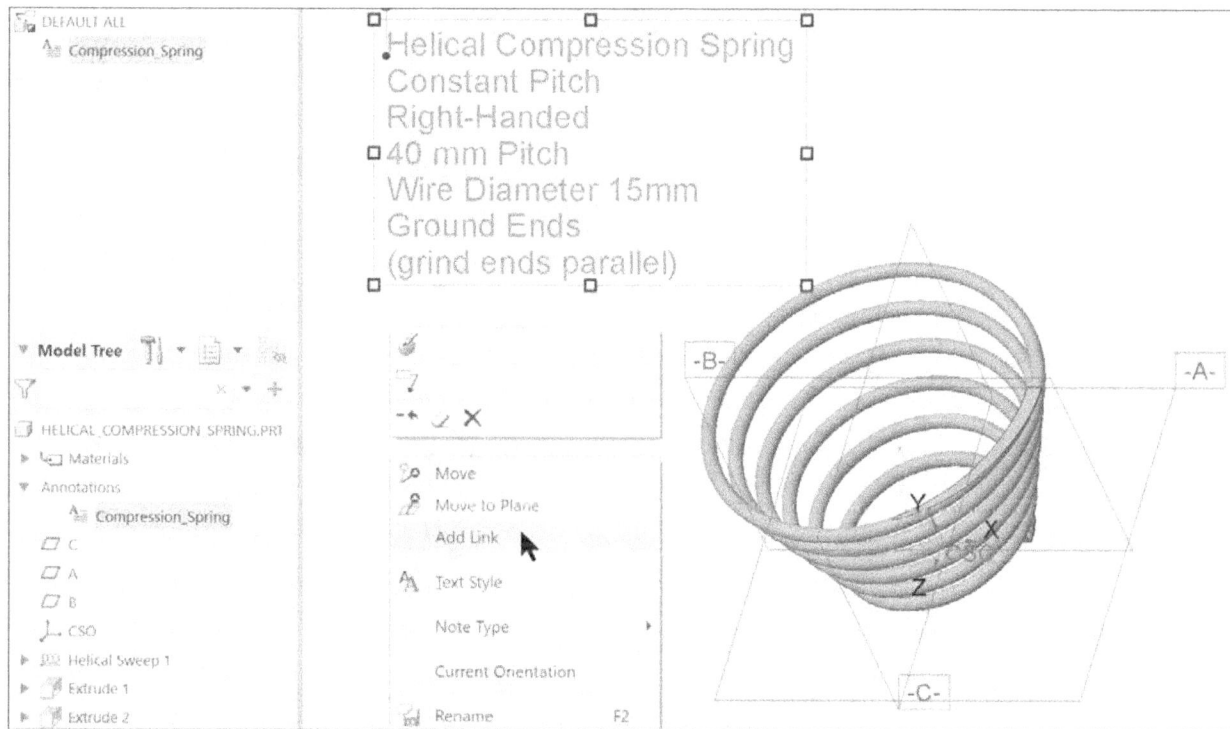

Figure 16.13(a) Press RMB > Properties (your options list may appear differently)

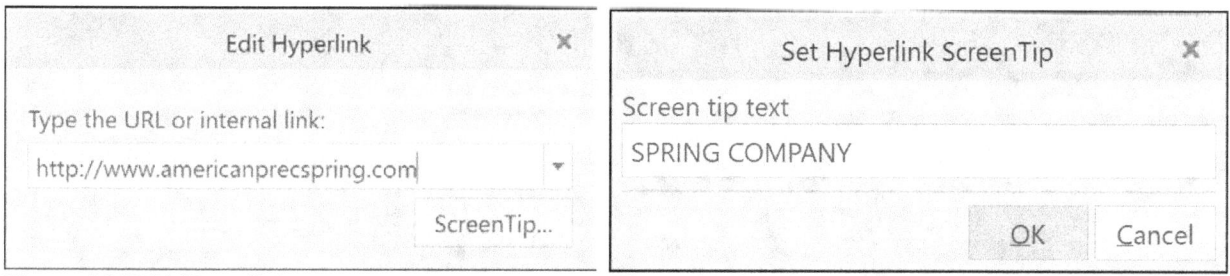

Figure 16.13(b) Hyperlink   **Figure 16.13(c)** Screen Tip

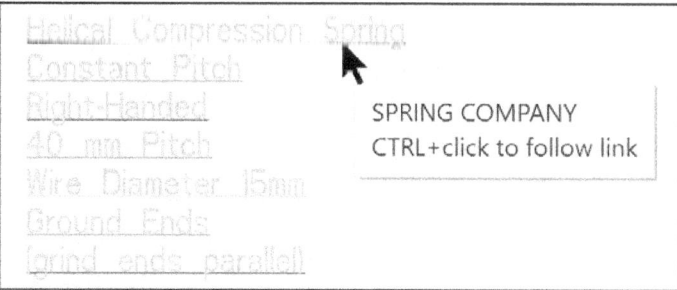

Figure 16.13(d) Note   **Figure 16.13(e)** Screen Tip Displayed

Open the URL, click: [⊞ Compression_Spring] **> RMB > Open URL** [Fig. 16.13(f)] *(URL opens in the browser window)* [Fig. 16.13(g) >* [⊞] **> LMB** to deselect **> Ctrl+D > Ctrl+S > File > Manage File > Delete Old Versions > Enter**

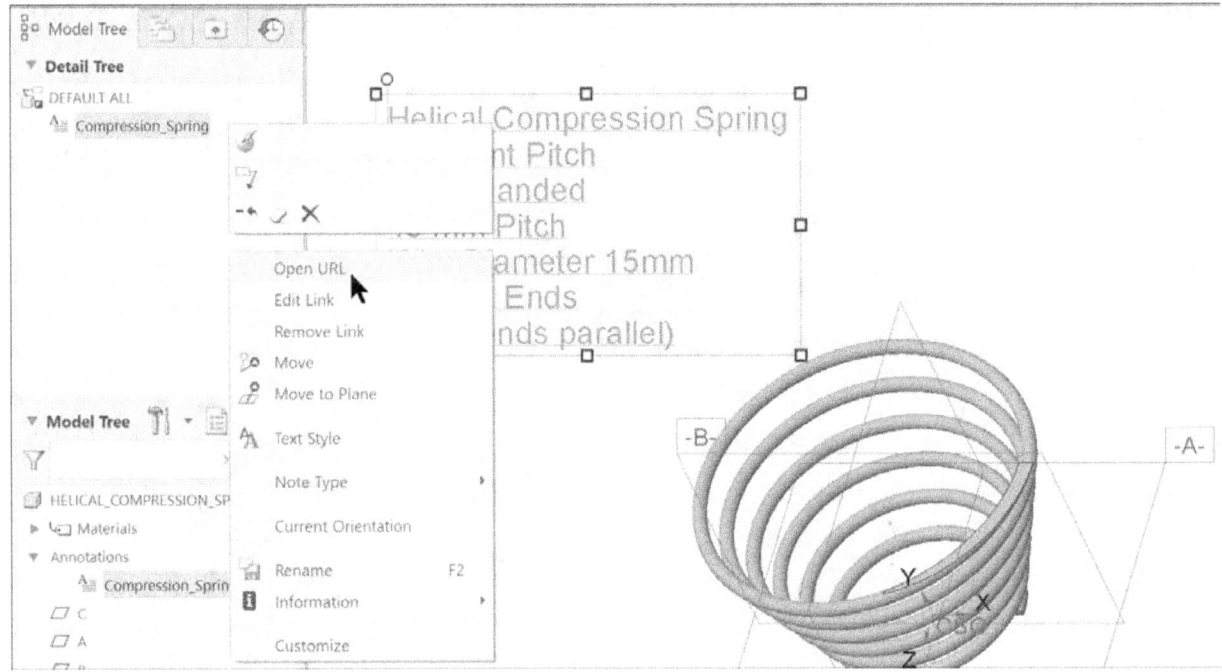

Figure 16.13(f) Open URL

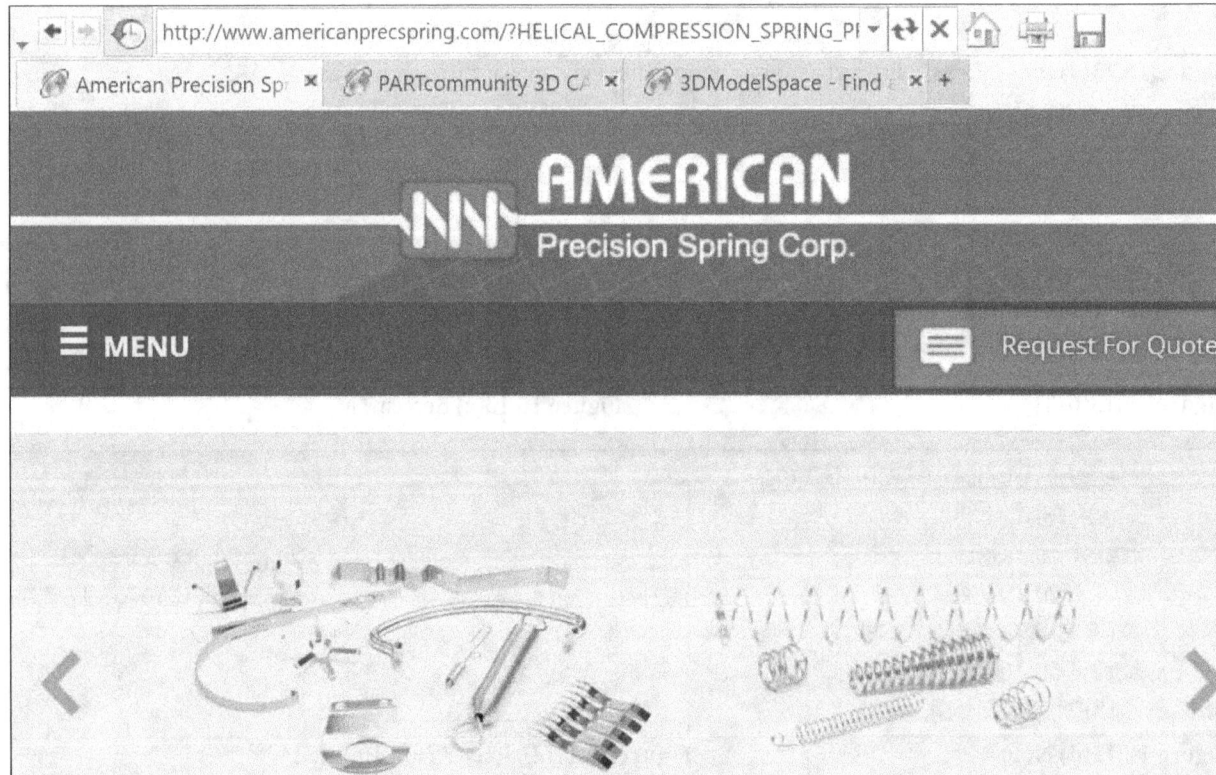

Figure 16.13(g) American Precision Spring Website (this web page may have since been updated)

Model Based Definition

Creo enables designers to successfully implement **MBD** and increase efficiency in product development by reducing dependency on 2D drawings. Creo enables designers to reduce the errors that result from incorrect, incomplete, or misinterpreted information by guiding and educating designers in the proper application of Geometric Dimensioning and Tolerance (GD&T) information. Creo also validates that the GD&T is captured in the 3D CAD model in a fully semantic way, that the model is compliant with ASME and ISO standards, and that it constrains model geometry to enable efficient and error-free downstream use in manufacturing and inspection.

Annotation Features

3D Notes can also be added to an entity using **Annotation Features**. Annotation features are data features that you can use to manage the model annotation and propagate model information to other models, or to manufacturing processes. The Annotation Feature Tool options correspond to the **ASME Y14.41 Digital Product Definition Data Practices.**

An Annotation feature consists of one or more Annotation Elements. Each Annotation Element (AE) can contain one annotation item, along with associated references and parameters. You can include the following types of annotations in an Annotation Element: Note, Symbol, Surface Finish, Geometric Tolerance, Set Datum Tag, Ordinate Baseline, Driven Dimension, Ordinate Driven Dimension, Reference Dimension, Ordinate Reference Dimension, and Existing Annotation.

Digital Product Definition Data Practices

"ASME Y14.41 establishes requirements for preparing, organizing and interpreting 3-dimensional digital product images (Fig. 16.14). Digital Product Definition Data Practices, which represents an extension of the popular Y14.5 standard for 2-dimensional drawings, reflects the growing need for a uniform method of documenting the data created in today's computer-aided design (CAD) environments. The standard provides a guide for CAD software developers working on improved modeling and annotation practices for the engineering community. ASME Y14.41 sets forth the requirements for tolerances, dimensional data, and other annotations. ASME Y14.41 advances the capabilities of Y14.5, Dimensioning and Tolerancing, the standard pertaining to 2-D engineering drawings".

In the following steps you will create a single-view 3D definition of the model for manufacturing, instead of a traditional multi-view drawing.

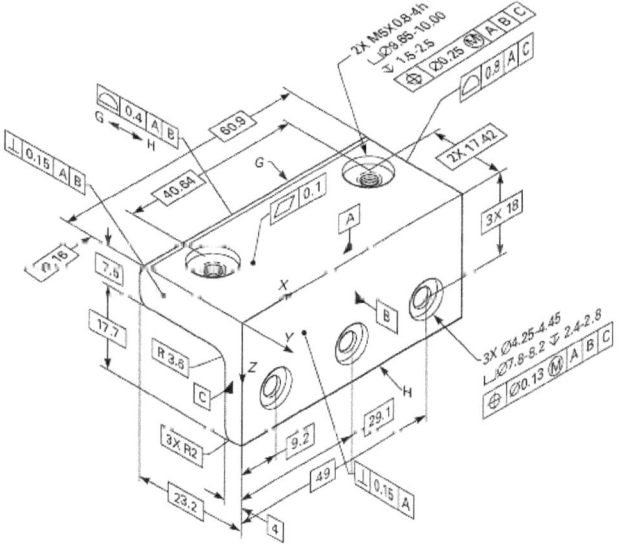

Figure 16.14 Digital Product Definition, ASME Y14.41

Click: 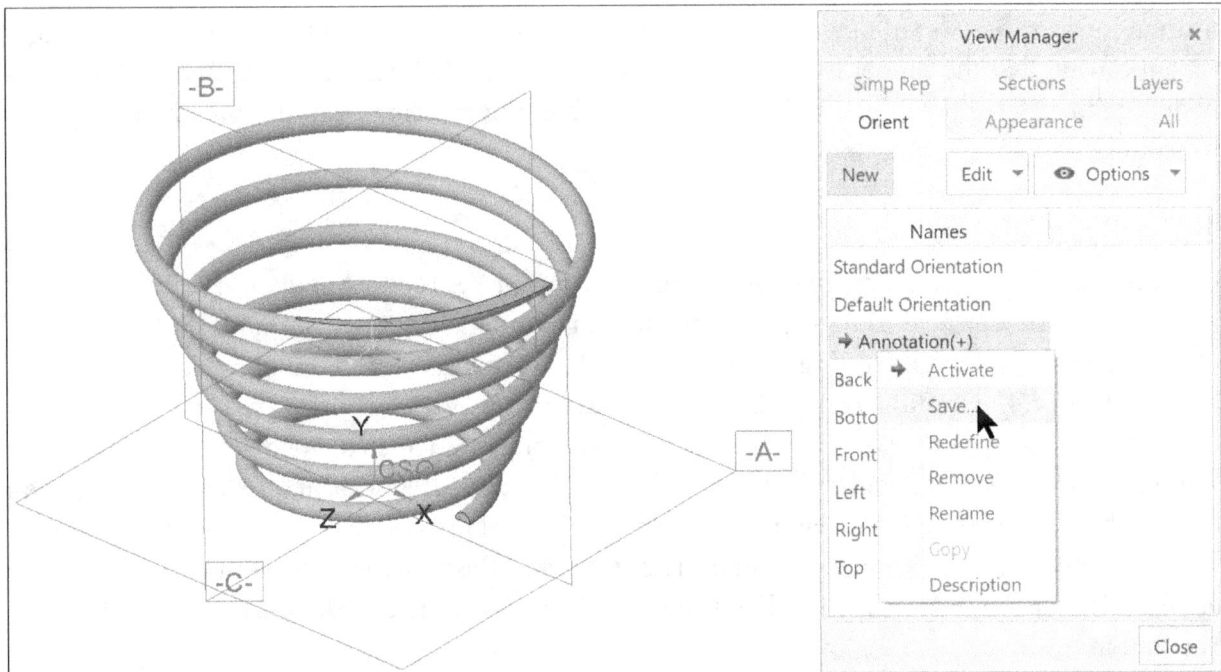 **Open the View Manager > Orient** tab **> New >** *type* **Annotation > Enter >** rotate the view **>** click on **>** ➜ Annotation(+) **> RMB > Save** (Fig. 16.15) **> OK** *(+ sign disappears)* **> Close > Ctrl+S**

Figure 16.15 Reorient the Model

Click: Geometry ▼ **>** Annotation ▼ **>** select the 3D note [Fig. 16.16(a)] **>** ✥ select a new note position [Fig. 16.16(b)] **> LMB** to deselect **> Ctrl+S**

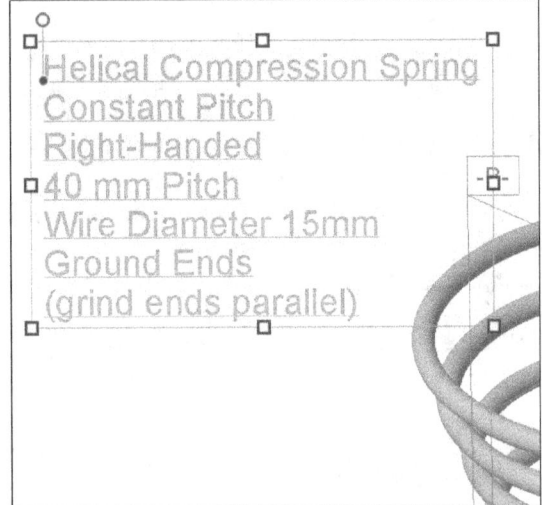

Figure 16.16(a) Move

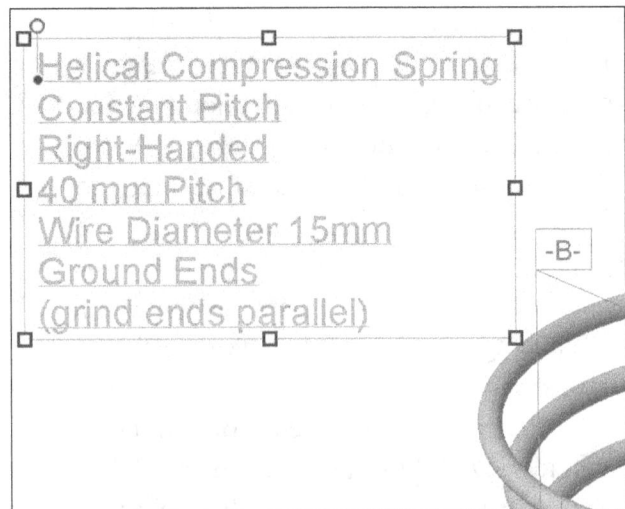

Figure 16.16(b) Place the Note

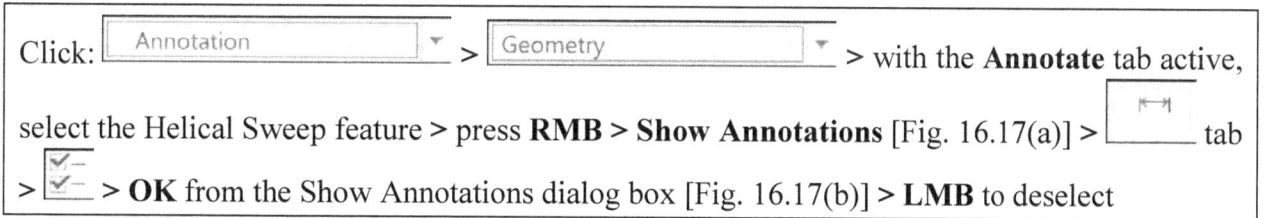

Click: [Annotation ▾] > [Geometry ▾] > with the **Annotate** tab active, select the Helical Sweep feature > press **RMB** > **Show Annotations** [Fig. 16.17(a)] > [↦⊣] tab > [☑] > **OK** from the Show Annotations dialog box [Fig. 16.17(b)] > **LMB** to deselect

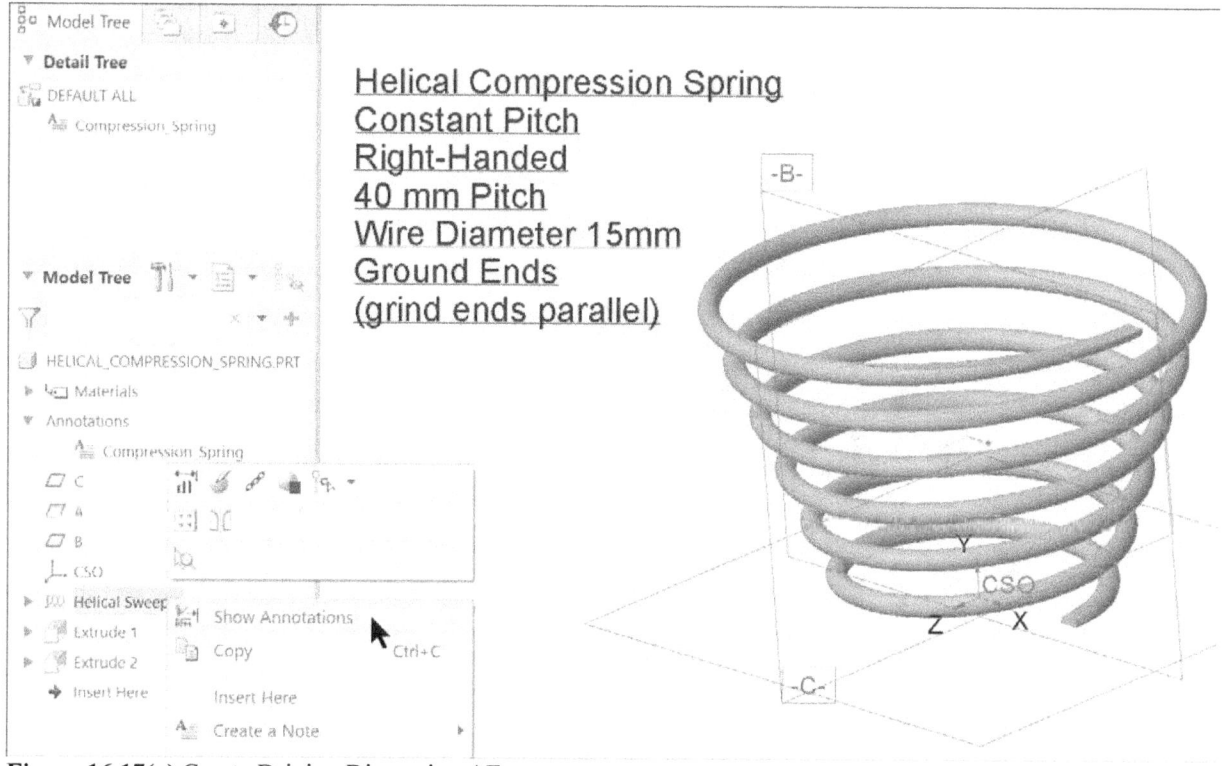

Figure 16.17(a) Create Driving Dimension AE

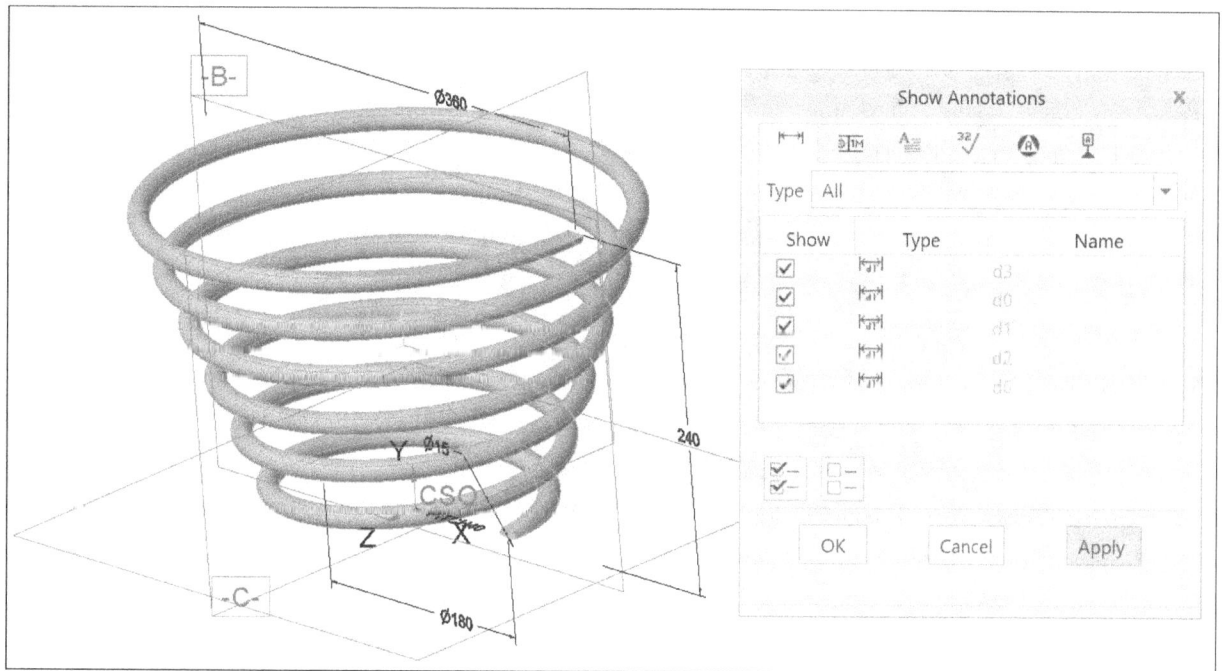

Figure 16.17(b) Displayed Driving Dimensions

Select a dimension > move the pointer [Fig. 16.18(a)] > move the dimension as desired > select a different dimension > move annotation to a better location > continue moving the annotations using either method so that the 3D view displays all dimensions appropriately Fig. 16.18(b)] > **Ctrl+S**

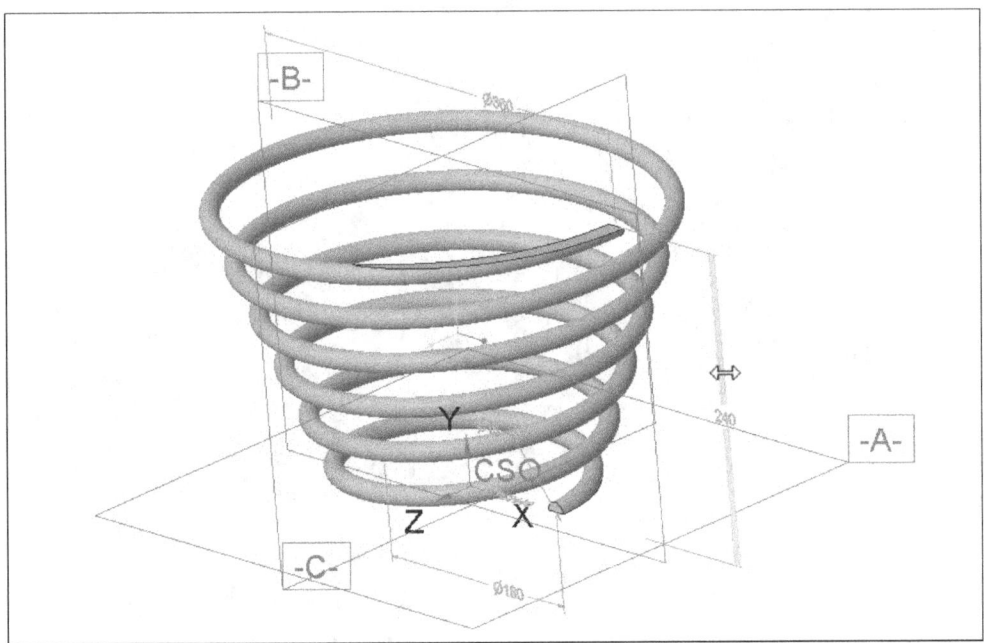

Figure 16.18(a) Flip **Datum_Tag_B**

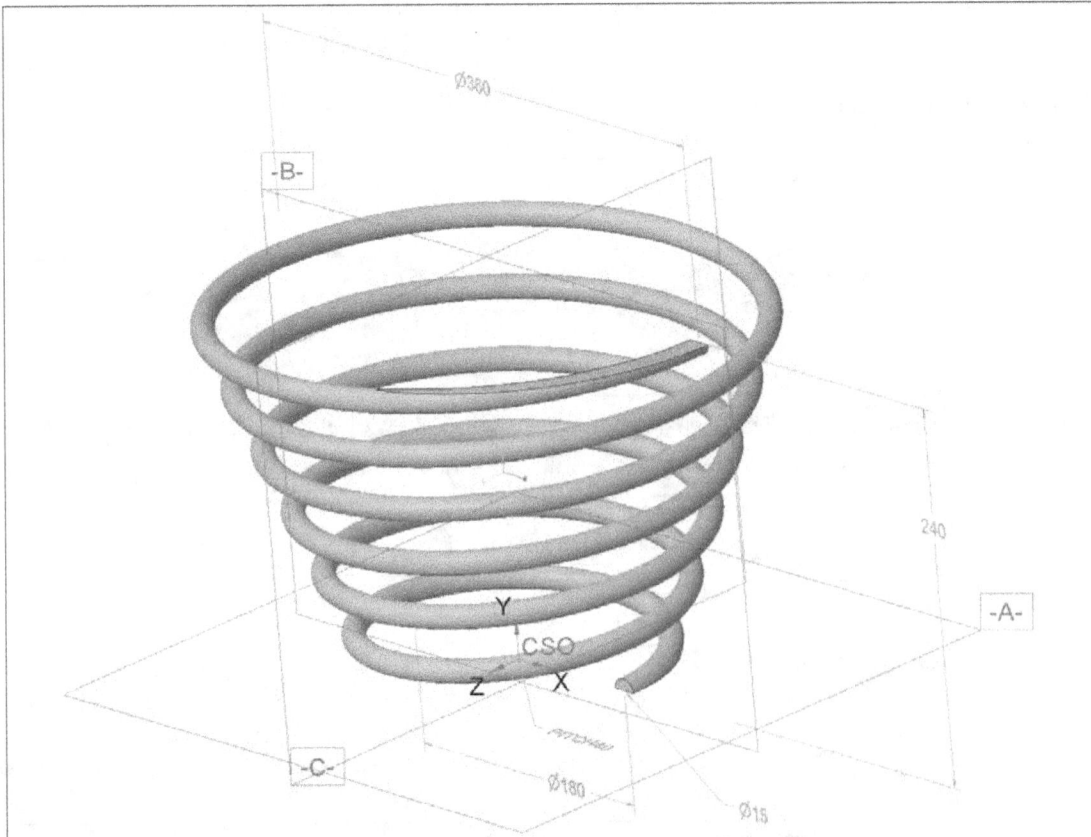

Figure 16.18(b) Current Orientation (your options list may appear differently)

Select the **PITCH 40** annotation > press **RMB** > **Current Orientation** [Figs. 16.18(a-b)] Click: ⊙ Named Model Orientation > ▾ > **TOP** [Fig. 16.18(c)] > **Flip** [Fig. 16.18(d)] > **OK** > move the **PITCH40** annotation as needed > **RMB** > 𝔍 **Flip Text** [Fig. 16.18(e)]

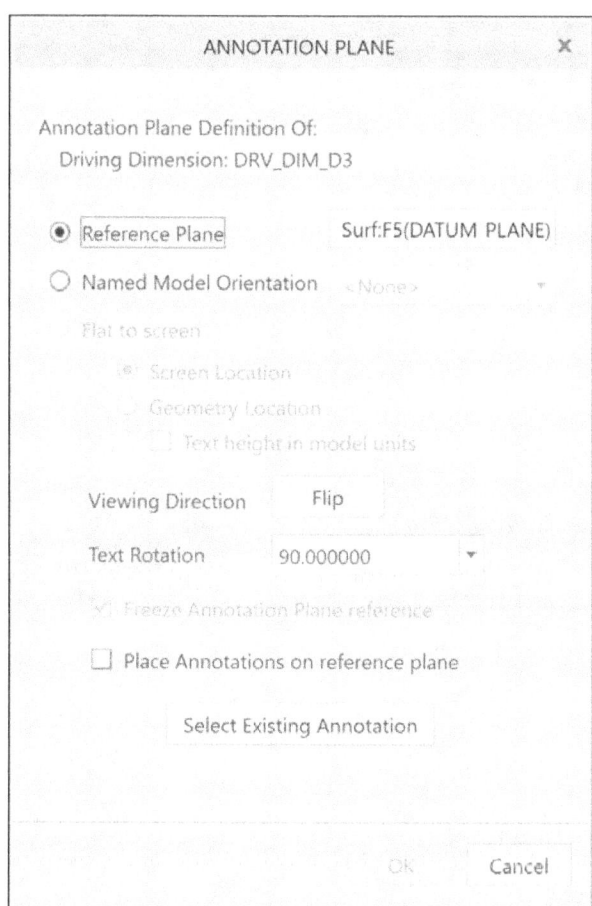

Figure 16.18(c) Annotation Plane Dialog Box   **Figure 16.18(d)** Named Model Orientation: TOP

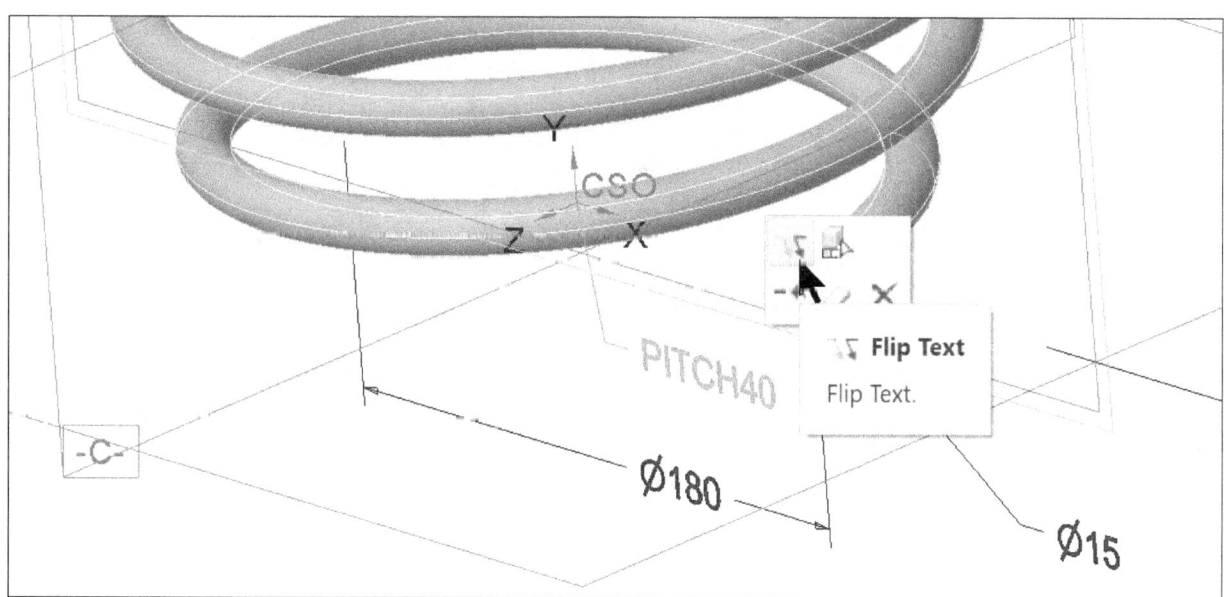

Figure 16.18(e) Moved Dimensions

Click: 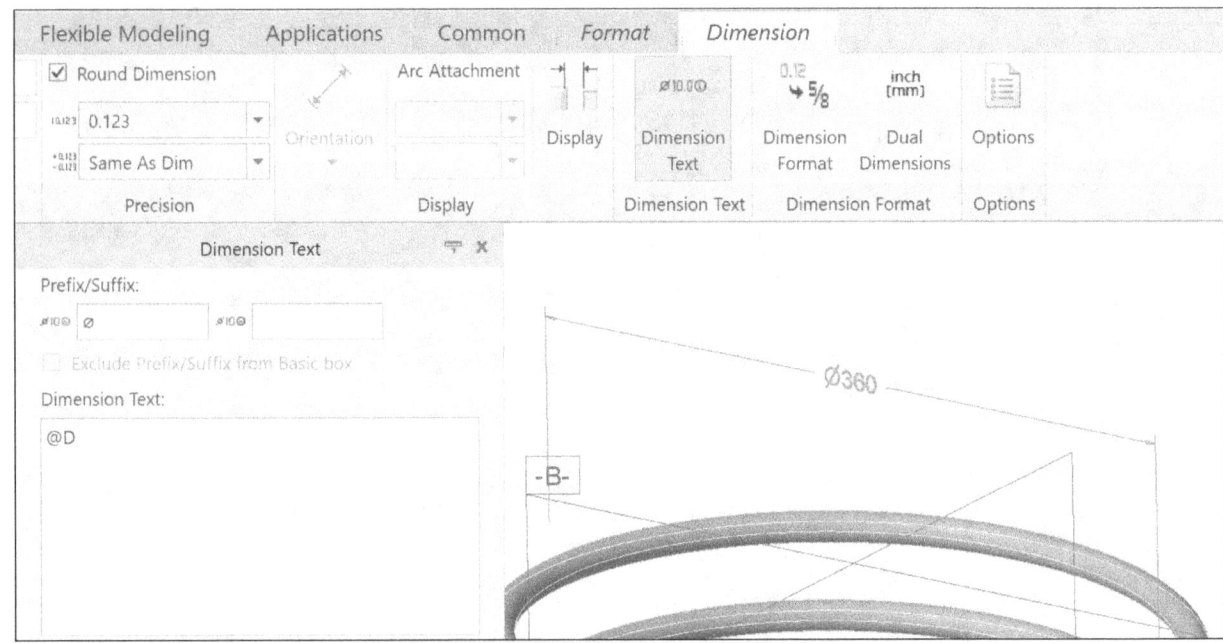 **Open the View Manager > Orient** tab > if (+) click on **Annotation(+) > RMB > Save > OK > Close > Annotate** tab > select the **360** dimension [Fig. 16.19(a)] > **Display Text** [Fig. 16.19(b)] > *type* **PITCH DIAMETER > LMB** to deselect > **Ctrl+S** [Fig. 16.19(c)]

Figure 16.19(a) Dimension Properties (your options list may appear differently)

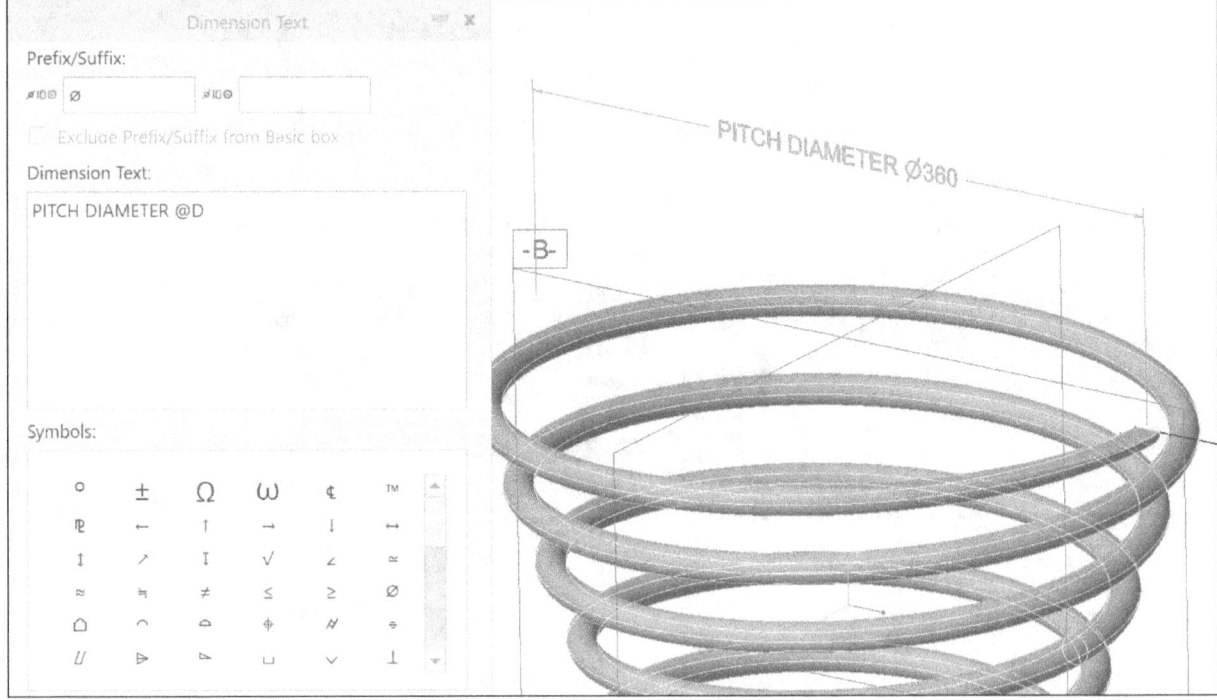

Figure 16.19(b) Dimension Properties Dialog Box, Display Tab *(type in added text)*

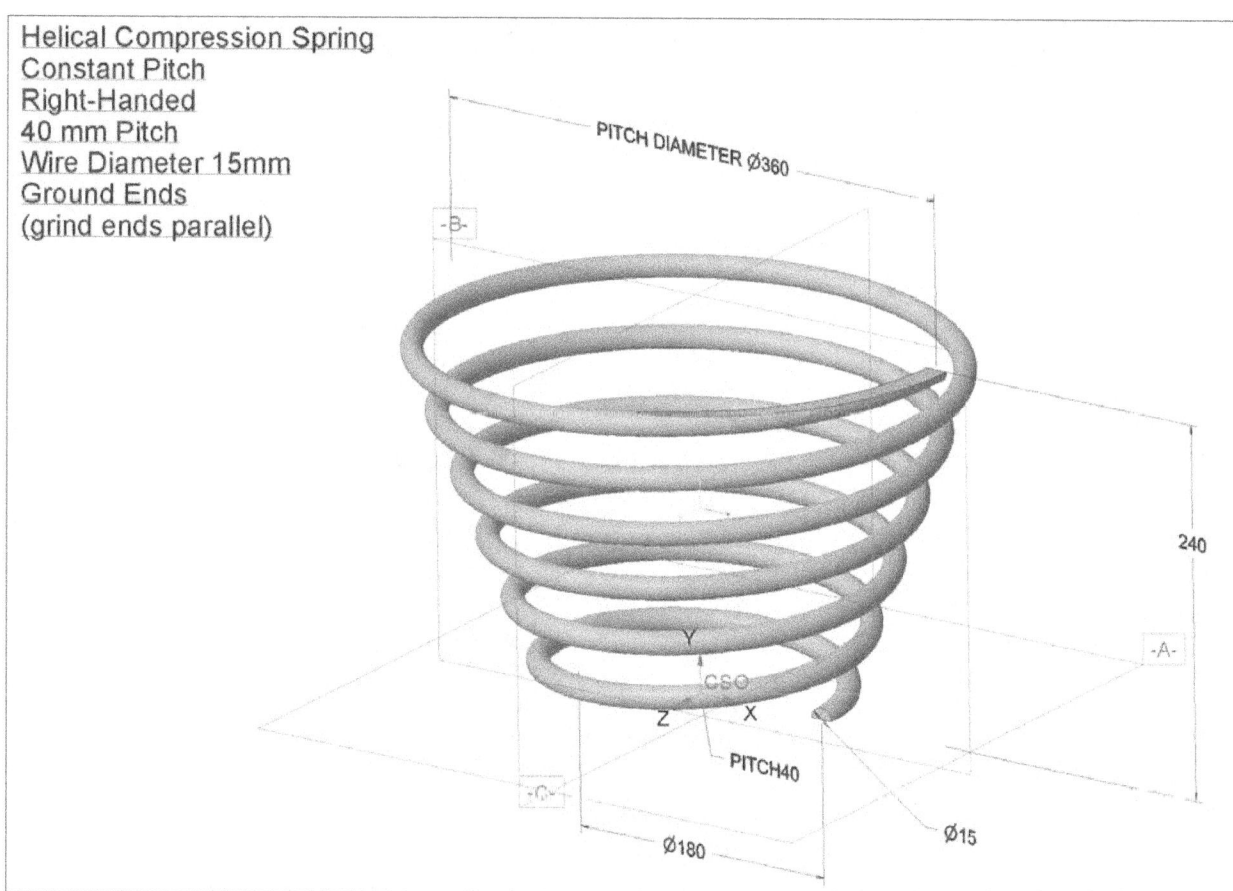

Figure 16.19(c) Annotated Part

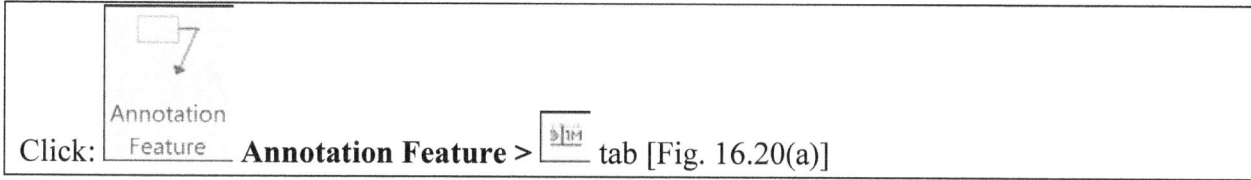

Click: **Annotation Feature > [⊙|1M]** tab [Fig. 16.20(a)]

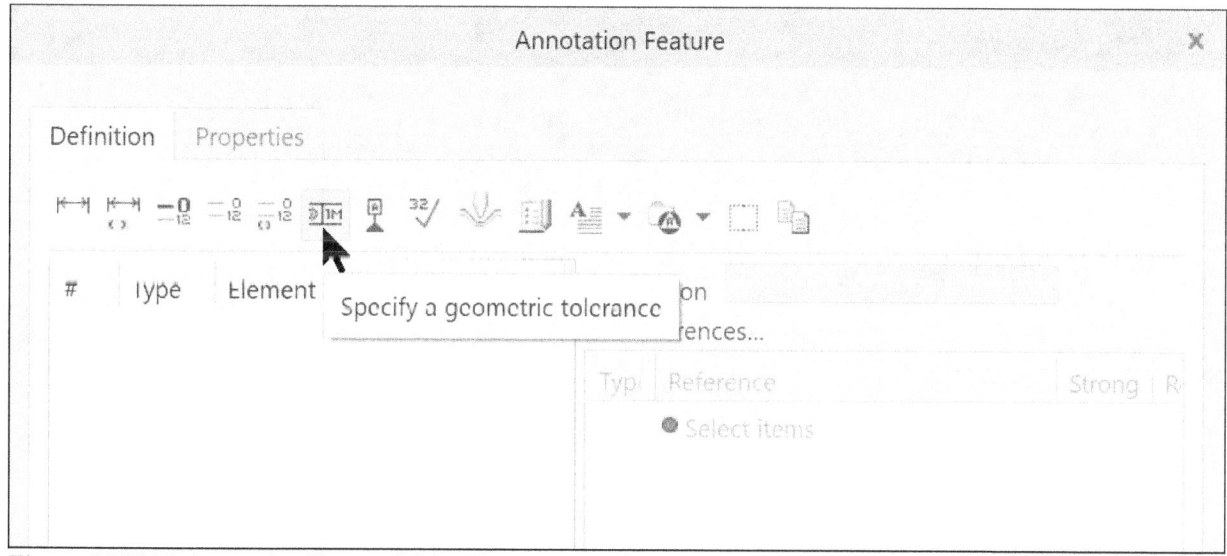

Figure 16.20(a) Annotation Feature Dialog Box, Geometric Tolerance

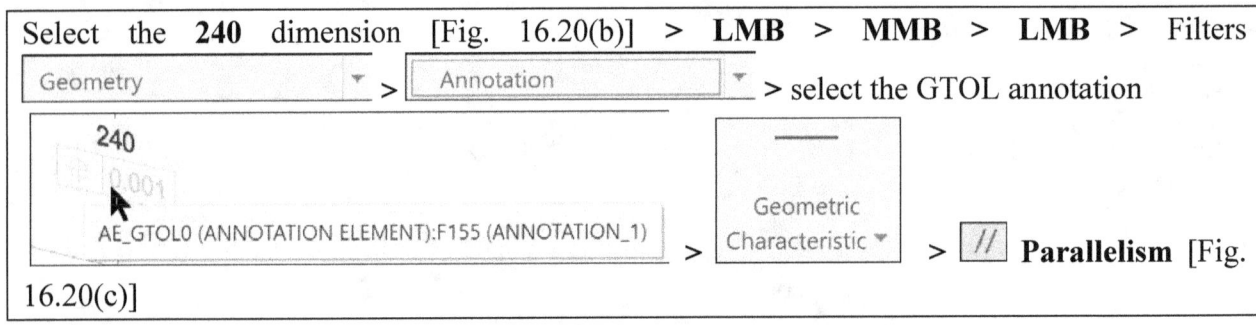

Select the **240** dimension [Fig. 16.20(b)] > **LMB** > **MMB** > **LMB** > Filters Geometry > Annotation > select the GTOL annotation > Geometric Characteristic > // **Parallelism** [Fig. 16.20(c)]

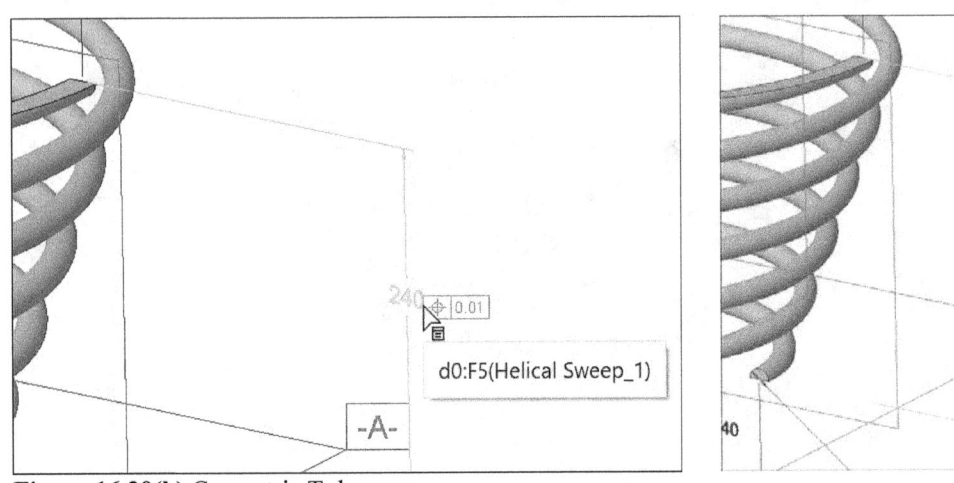
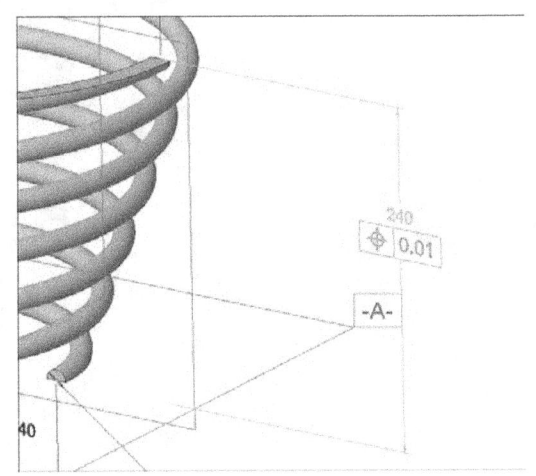

Figure 16.20(b) Geometric Tolerance

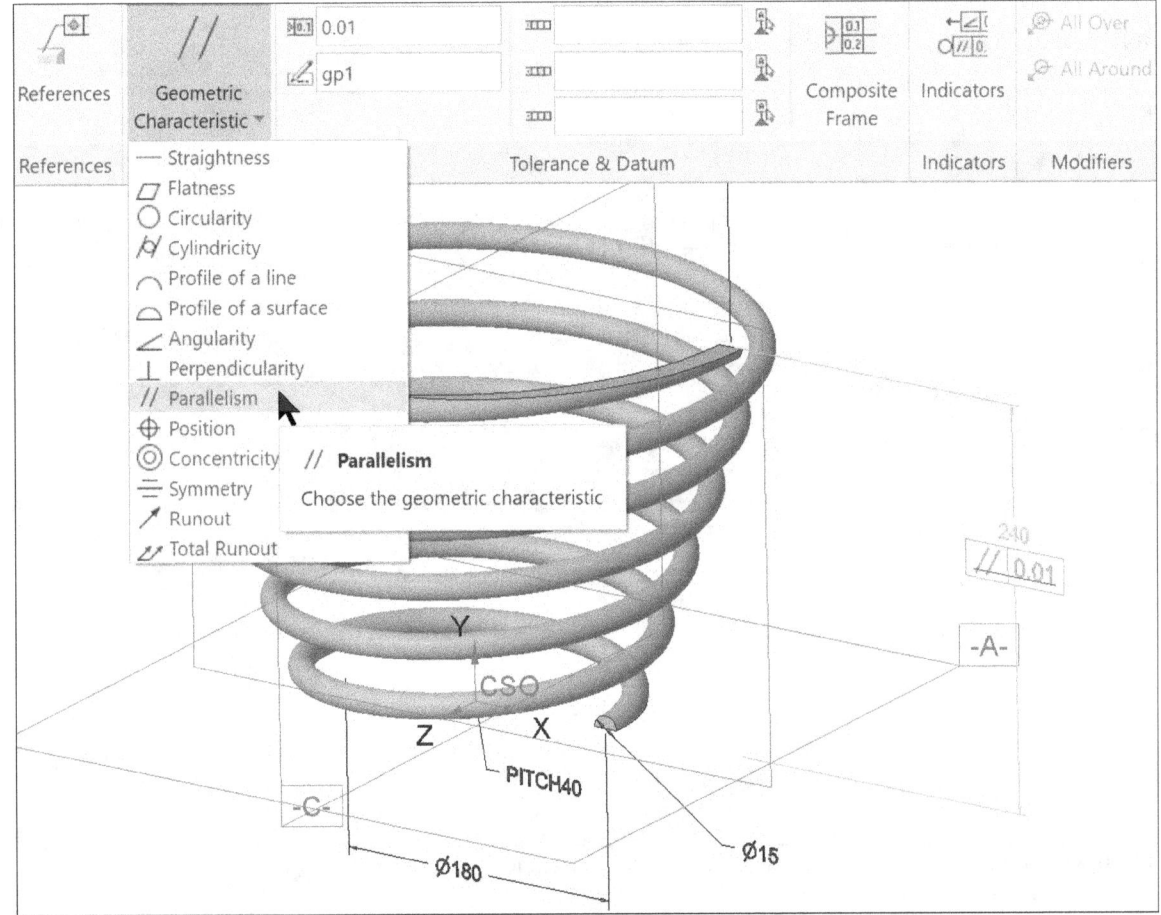

Figure 16.20(c) Parallelism

| Type **A** as the Primary Datum Reference | ▥ A | > **LMB** |

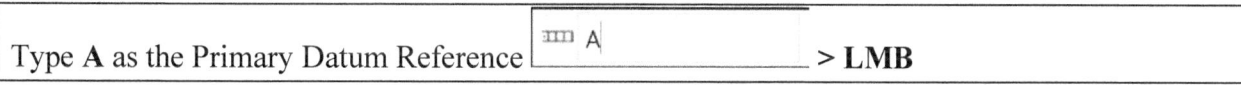

| References | Geometric Characteristic ▾ | ▣ 0.01 ⬚ gp1 | ▥ A ▥ ▥ | 🔒 🔒 🔒 | Composite Frame | Indicators | All Over All Around | Additional Text |
| References | Symbol | | Tolerance & Datum | | | Indicators | Modifiers | Additional Text |

Helical Compression Spring
Constant Pitch
Right-Handed
40 mm Pitch
Wire Diameter 15mm
Ground Ends
(grind ends parallel)

PITCH DIAMETER ⌀360

-B-

240

// 0.01 A

-A-

Y

CSO

Z   X

PITOH 40

-C-

⌀180

⌀15

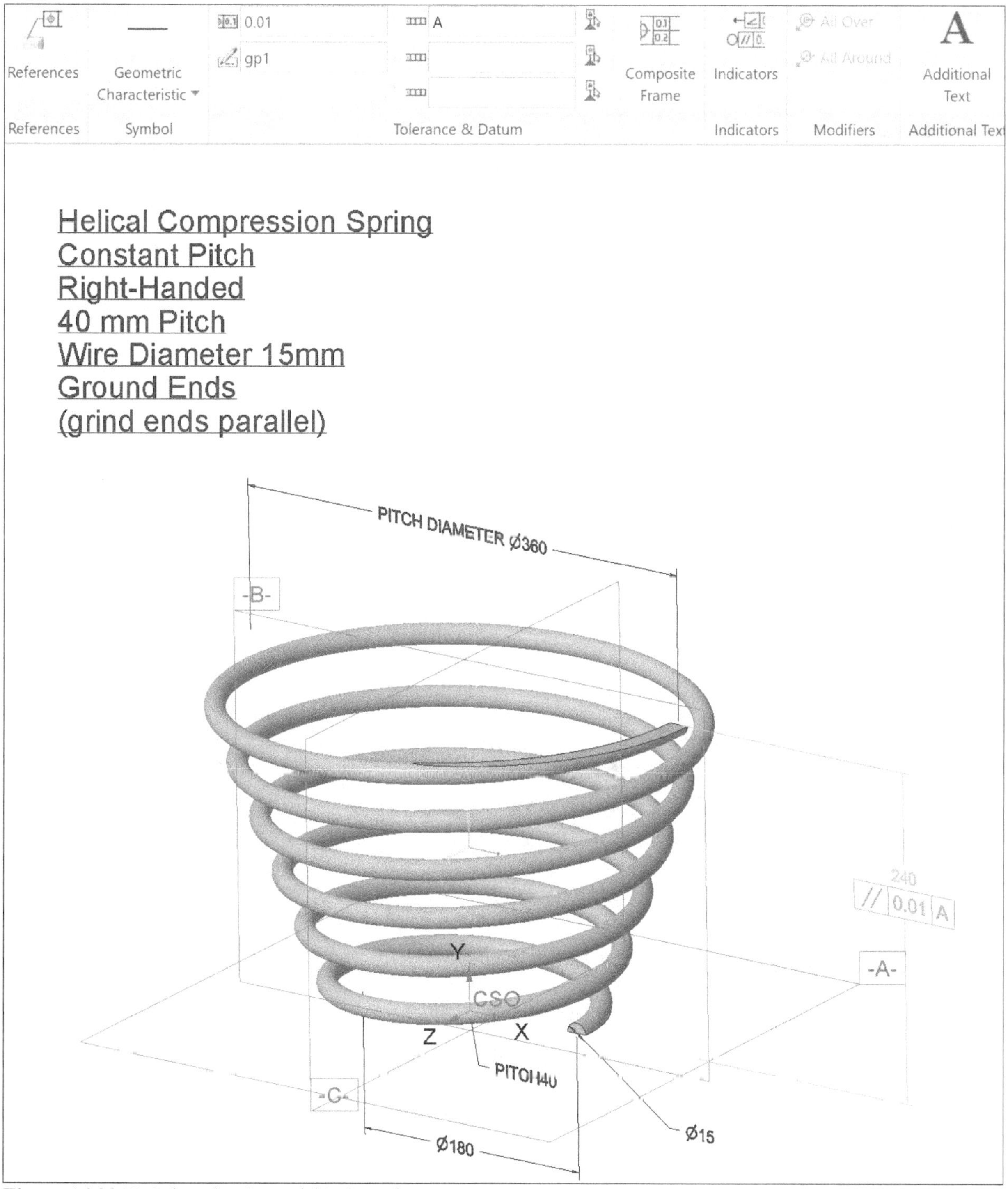

Figure 16.20(d) Select the Ground (Cut) Surface

Click on the GTOL > **Format** tab

240

0.001 A

AE_GTOL0 (ANNOTATION ELEMENT):F155 (ANNOTATION_1)

[Fig. 16.20(e)] > **LMB > Ctrl+S >**

Annotate tab

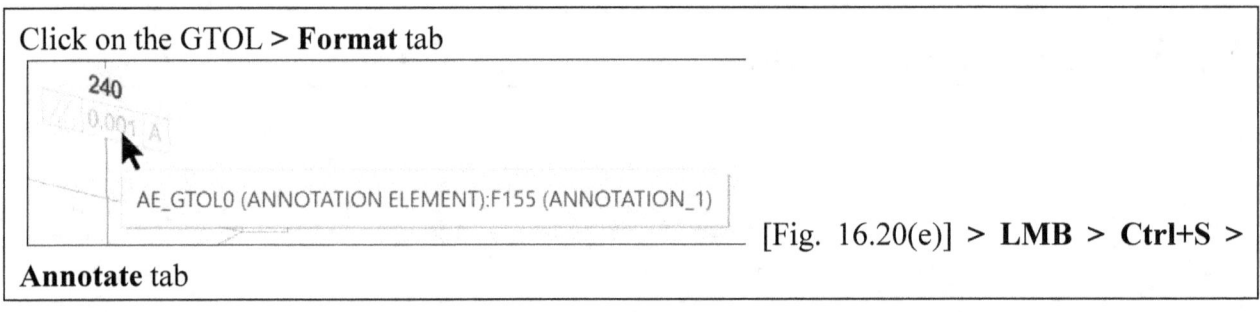

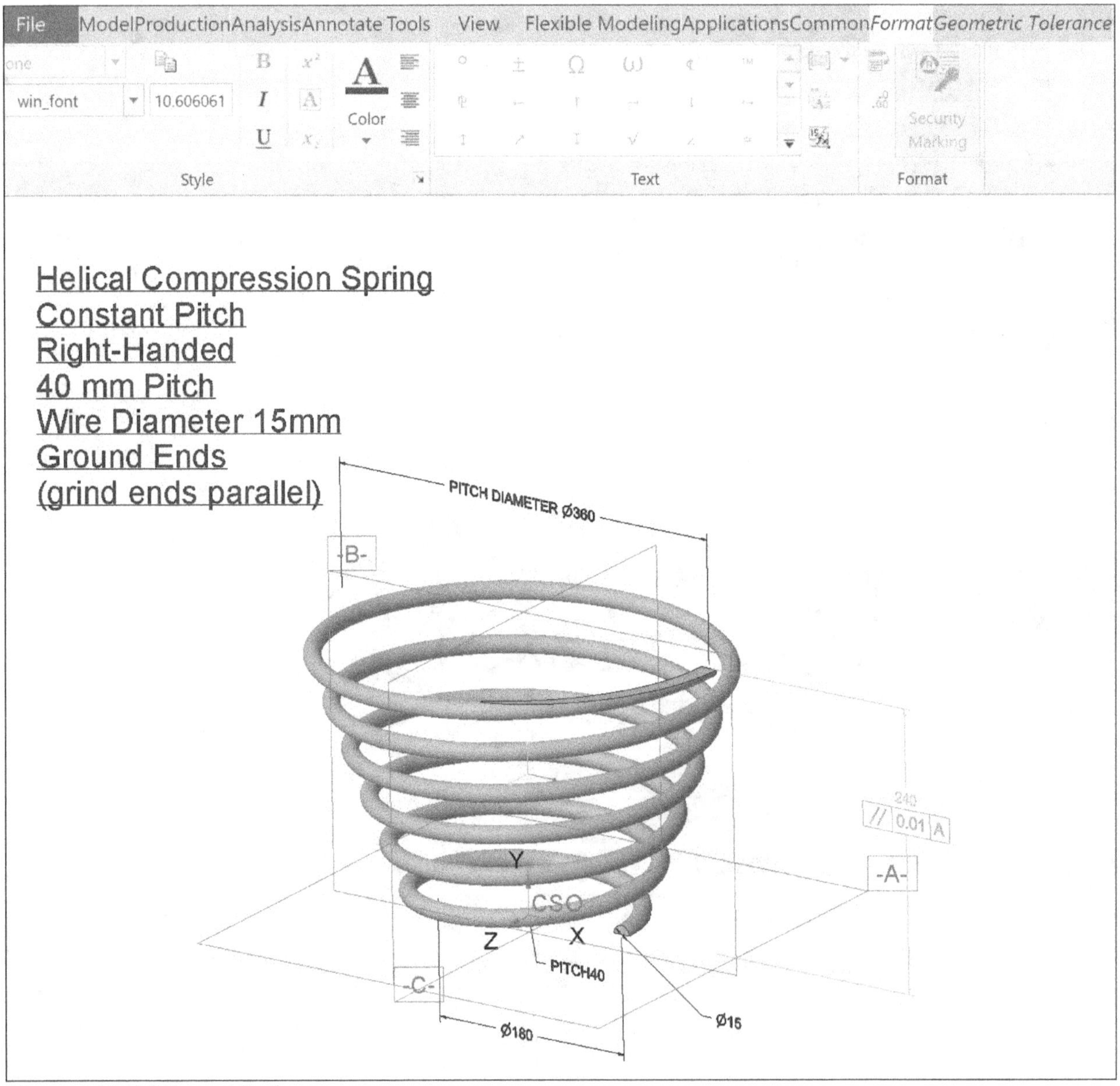

File   ModelProductionAnalysisAnnotate Tools   View   Flexible ModelingApplicationsCommon*FormatGeometric Tolerance*

Helical Compression Spring
Constant Pitch
Right-Handed
40 mm Pitch
Wire Diameter 15mm
Ground Ends
(grind ends parallel)

PITCH DIAMETER Ø360

-B-

240

// 0.01 A

-A-

Y

CSO

Z   X

-C-

PITCH40

Ø180   Ø15

Figure 16.20(e) Format Options

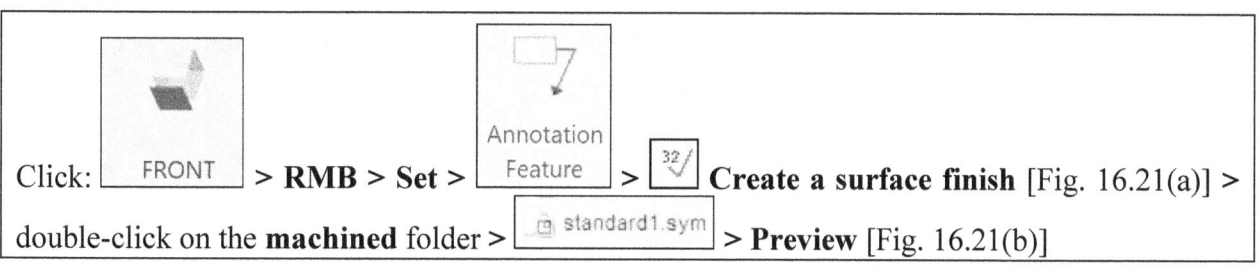

Click: FRONT > **RMB** > **Set** > Annotation Feature > 32/ **Create a surface finish** [Fig. 16.21(a)] > double-click on the **machined** folder > standard1.sym > **Preview** [Fig. 16.21(b)]

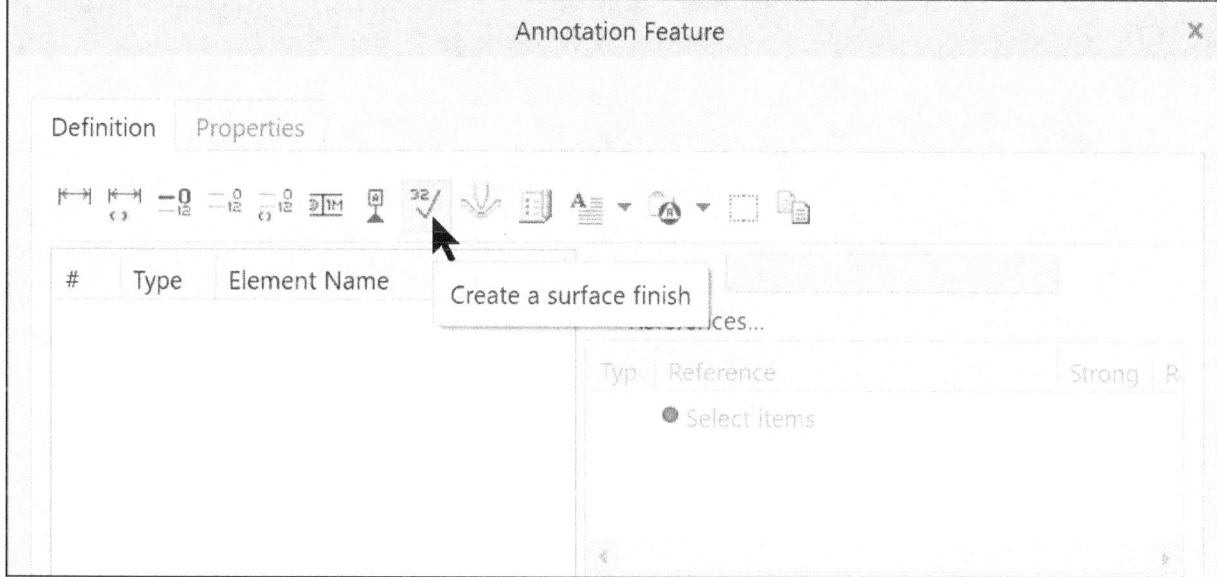

Figure 16.21(a) Create a Surface Finish

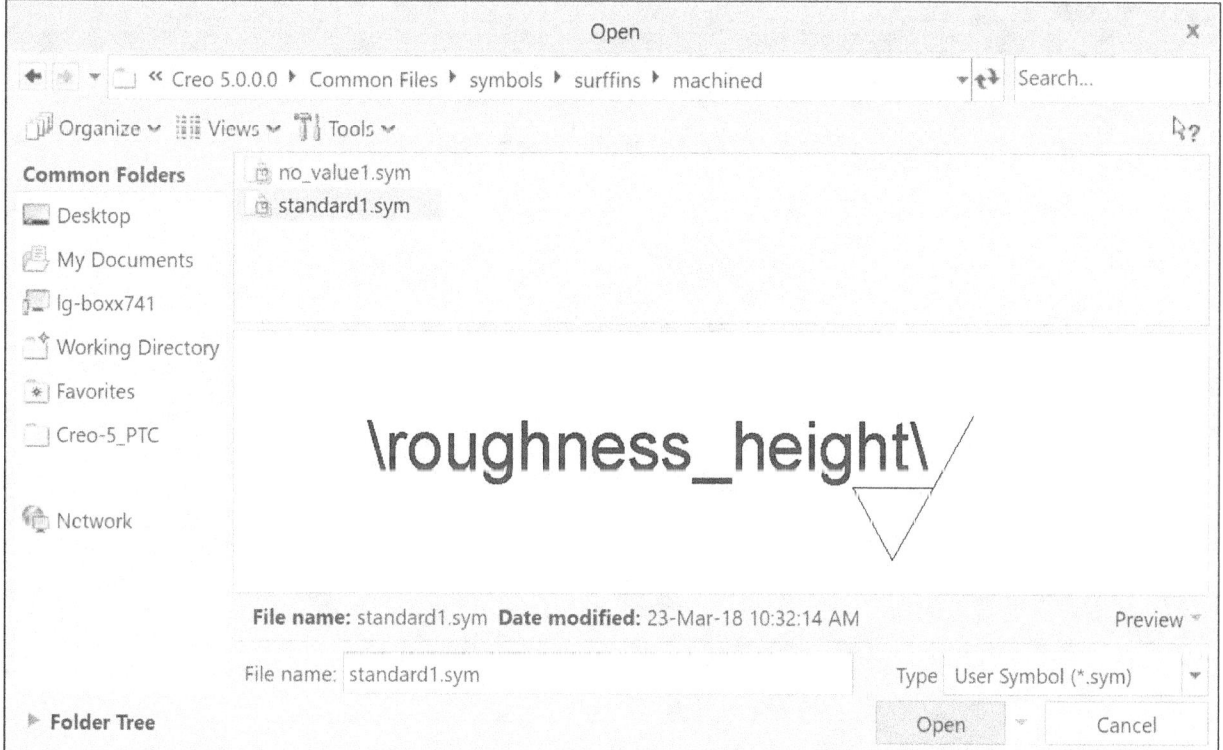

Figure 16.21(b) Preview of Surface Symbol **standard1.sym**

Click: **Open** [Fig. 16.21(c)] and the Surface Finish dialog box opens with its References collector active > select the ground surface [Fig. 16.21(d)]

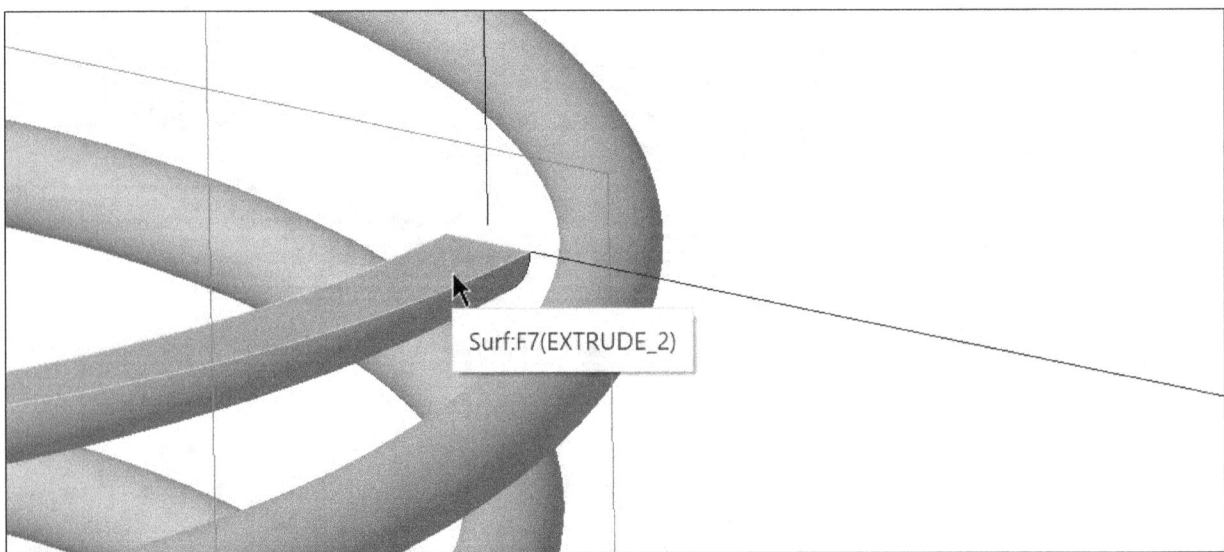

Figure 16.21(c) Surface Finish Dialog Box

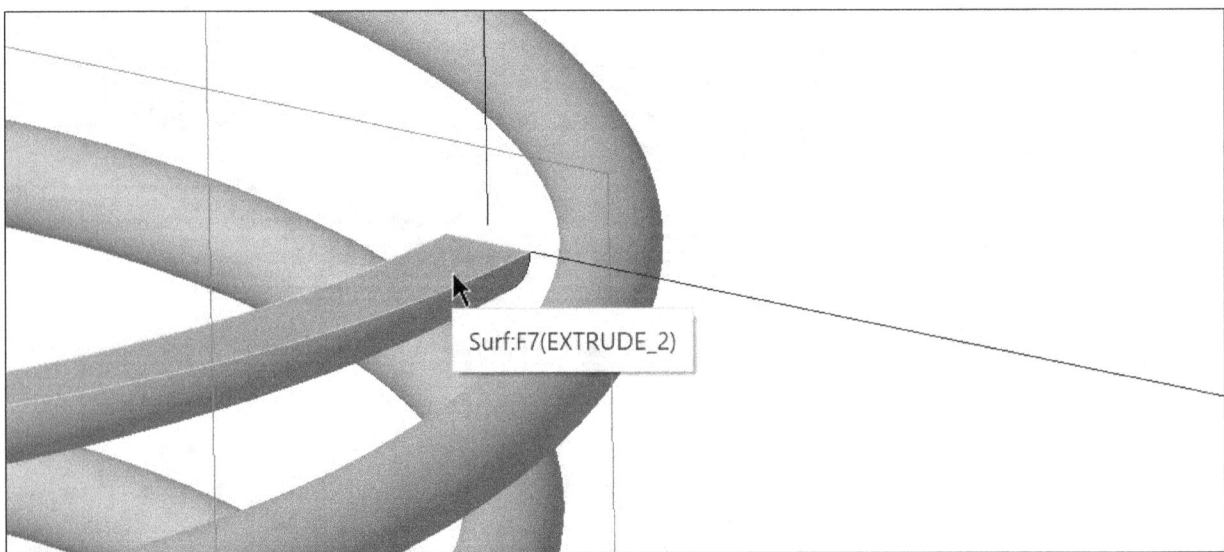

Figure 16.21(d) Select the Reference Surface

Click inside the Placement collector for Attachment references [Fig. 16.21(e)] > select the symbol position on the cut surface [Figs. 16.21(f-g)] > **MMB** [Fig. 16.21(h)]

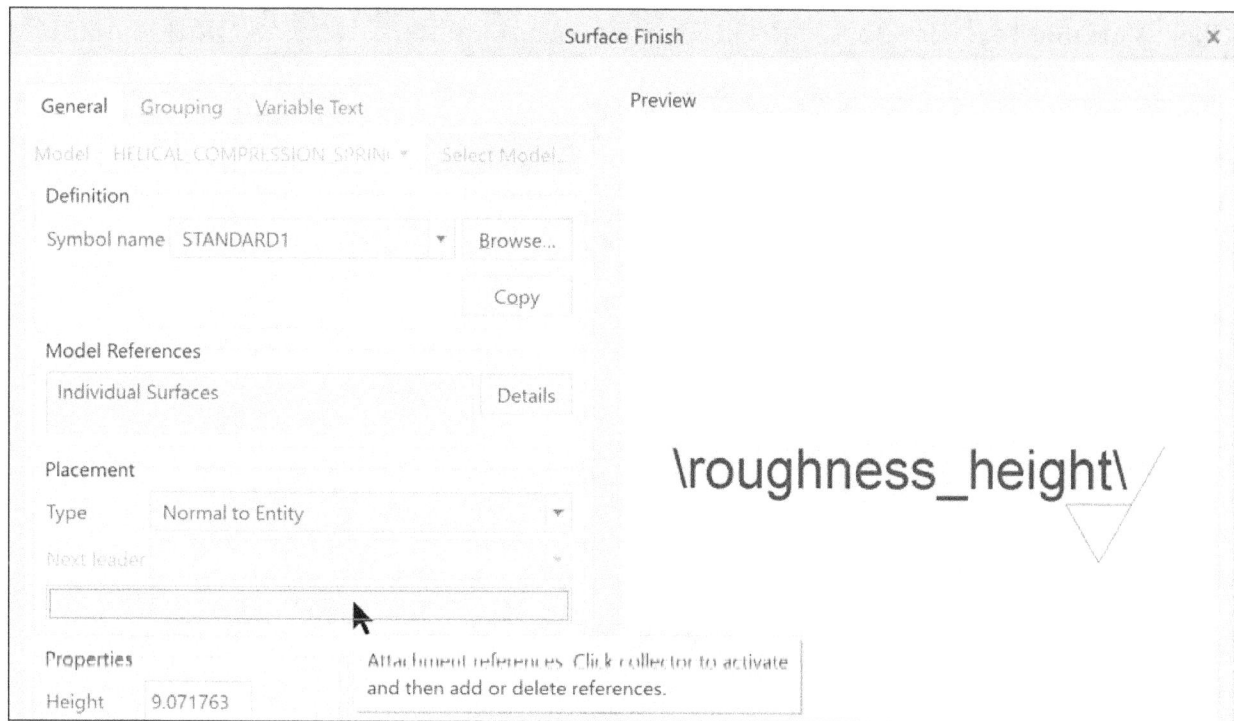

Figure 16.21(e) Placement Collector

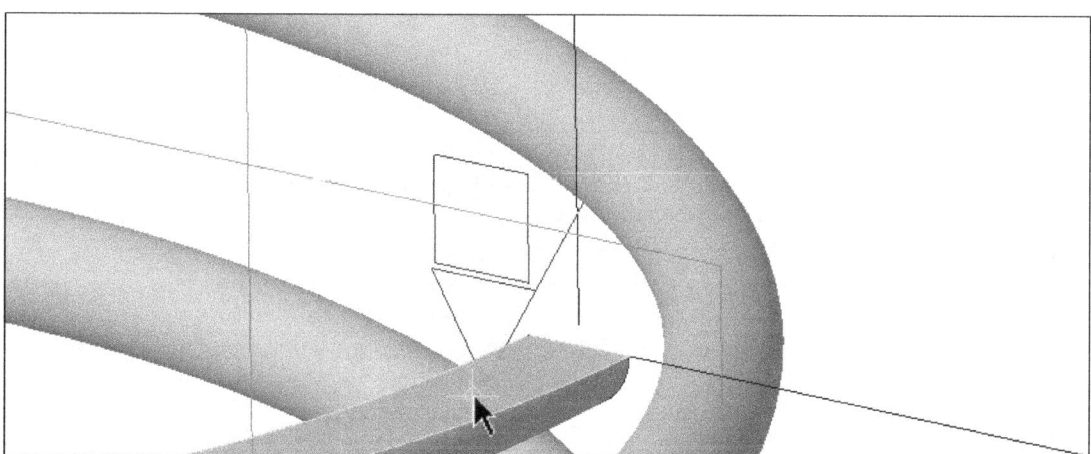

Figure 16.21(f) Select the Surface Finish Symbol Position

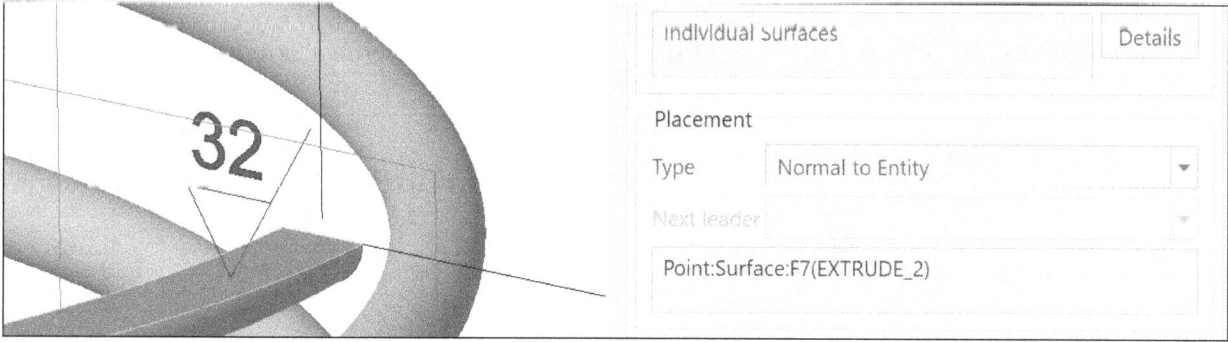

Figure 16.21(g) Completed Symbol Placement

Click: **Variable Text** tab **> 32 > MMB** in Graphics Area 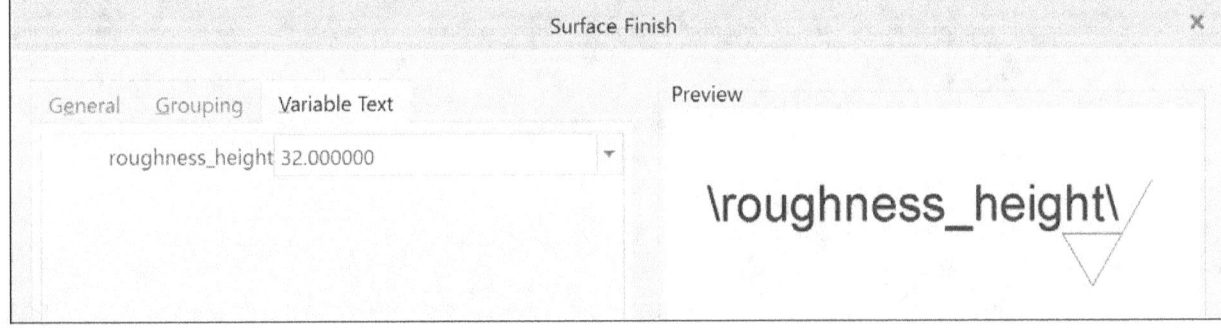 **> OK** [Fig. 16.21(i)]

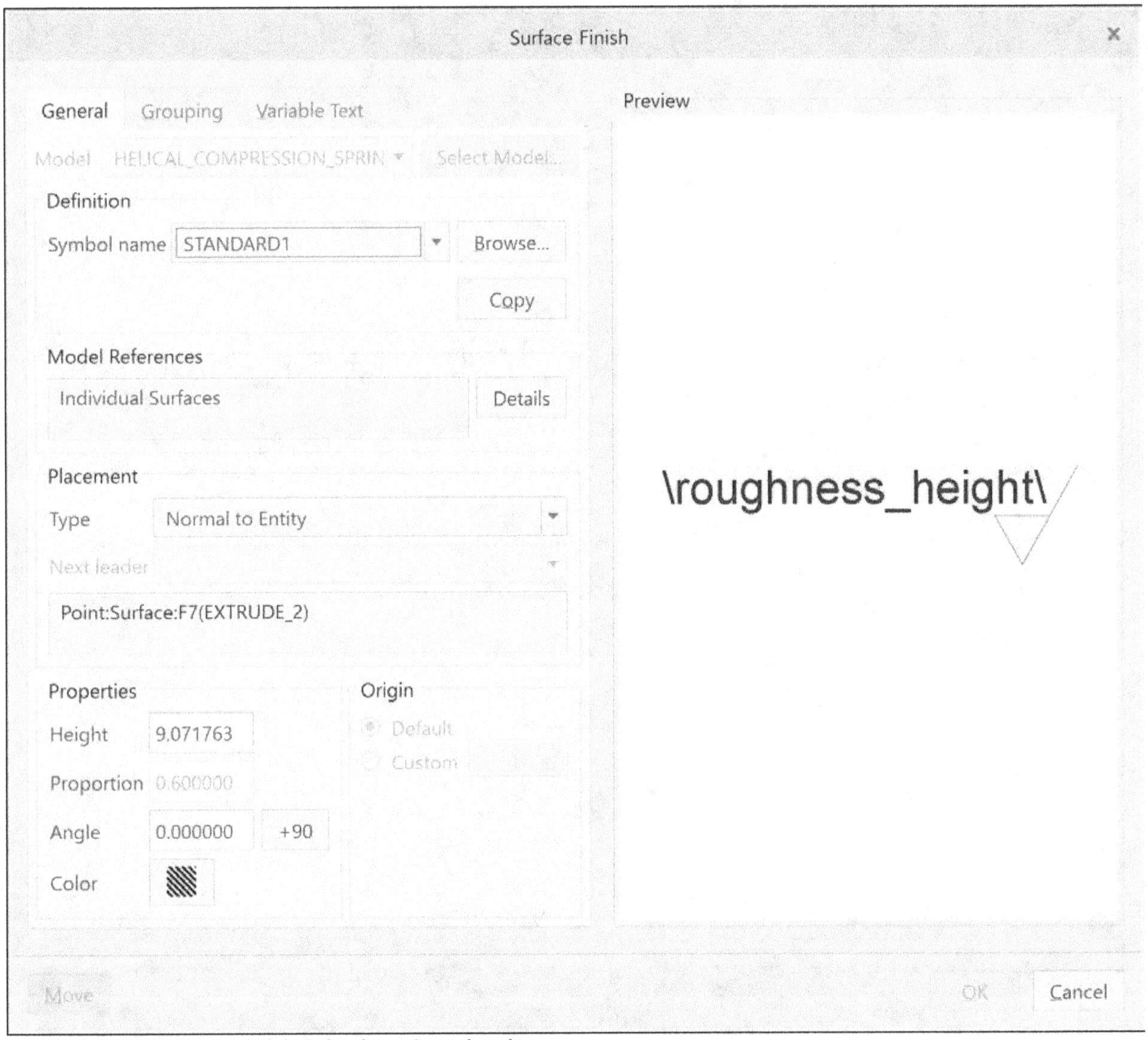

Surface Finish ✕

General   Grouping   Variable Text

Model   HELICAL_COMPRESSION_SPRIN ▼   Select Model...

Definition

Symbol name   STANDARD1   ▼   Browse...

Copy

Model References

Individual Surfaces   Details

Placement

Type   Normal to Entity   ▼

Next leader   ▼

Point:Surface:F7(EXTRUDE_2)

Properties   Origin

Height   9.071763   ⦿ Default
                          ○ Custom
Proportion 0.600000

Angle   0.000000   +90

Color   ▨

Preview

\roughness_height\

Move   OK   Cancel

Figure 16.21(h) General Tab Selections Completed

Surface Finish ✕

General   Grouping   **Variable Text**

roughness_height 32.000000   ▼

Preview

\roughness_height\

Figure 16.21(i) Variable Text Tab

Click: [32✓] tab [Fig. 16.21(j)] > repeat the process to create an annotation feature finish symbol on the opposite end of the spring [Fig. 16.21(k)] > **MMB** > **OK** from the Annotation Feature dialog box > **OK** > **LMB**

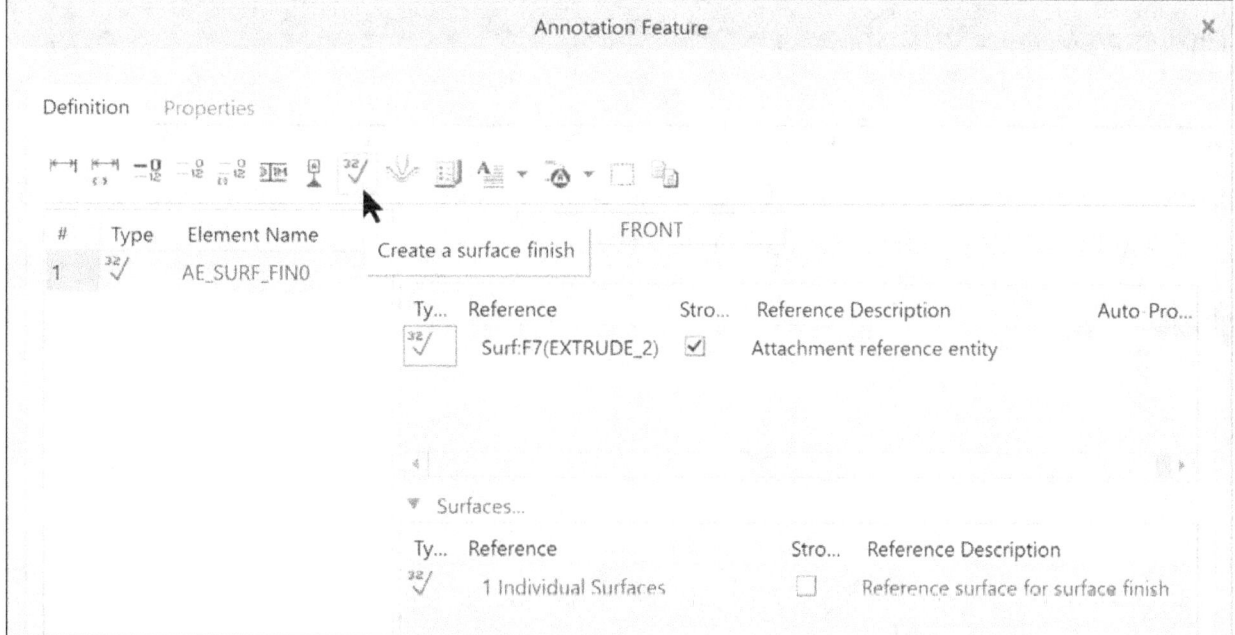

Figure 16.21(j) Annotation Feature Dialog Box

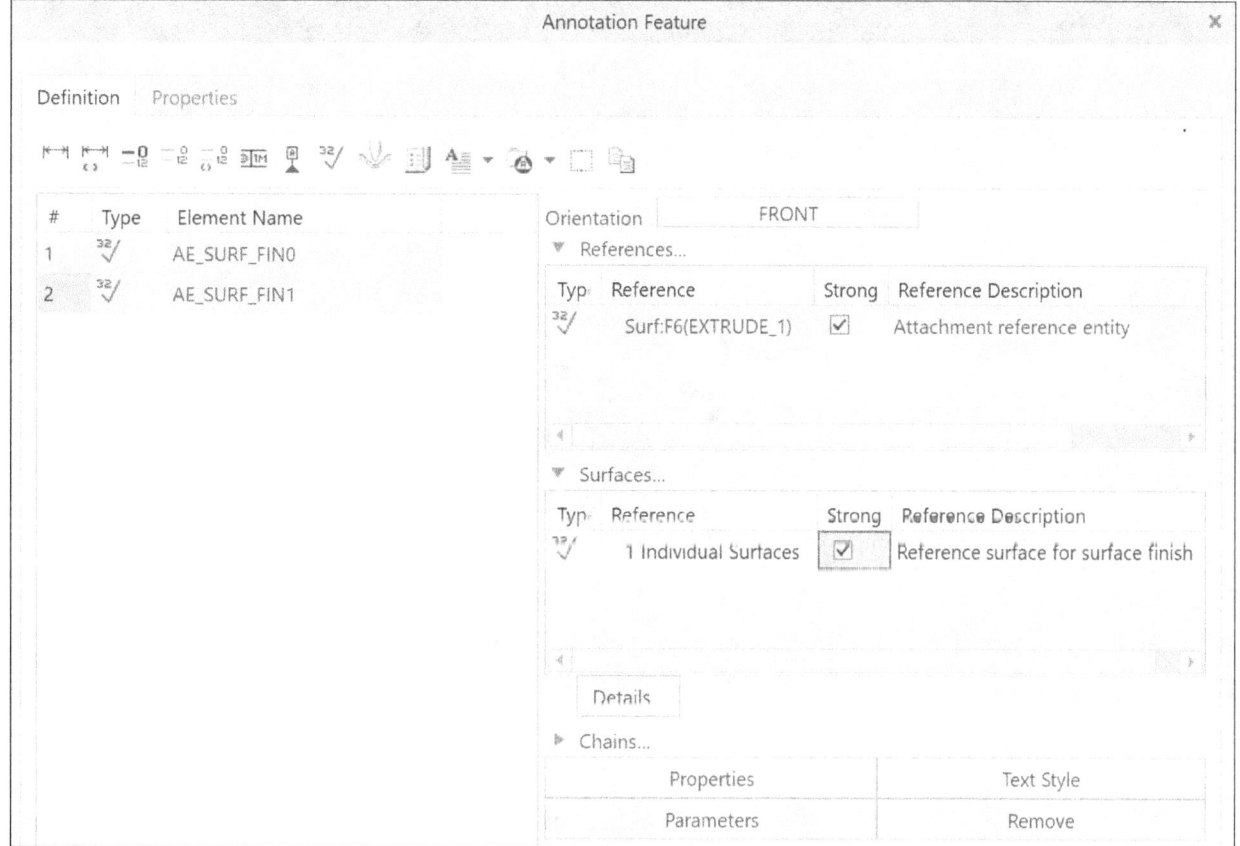

Figure 16.21(k) Second Surface Finish Annotation

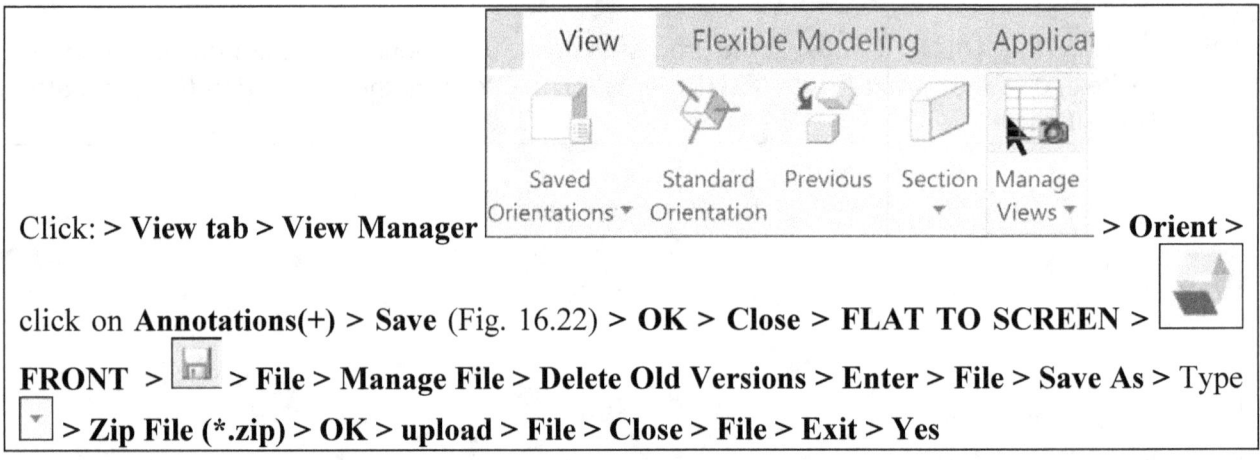

Click: > **View tab** > **View Manager** > **Orient** >

click on **Annotations(+)** > **Save** (Fig. 16.22) > **OK** > **Close** > **FLAT TO SCREEN** >

FRONT > 🖫 > **File** > **Manage File** > **Delete Old Versions** > **Enter** > **File** > **Save As** > Type

▾ > **Zip File (*.zip)** > **OK** > **upload** > **File** > **Close** > **File** > **Exit** > **Yes**

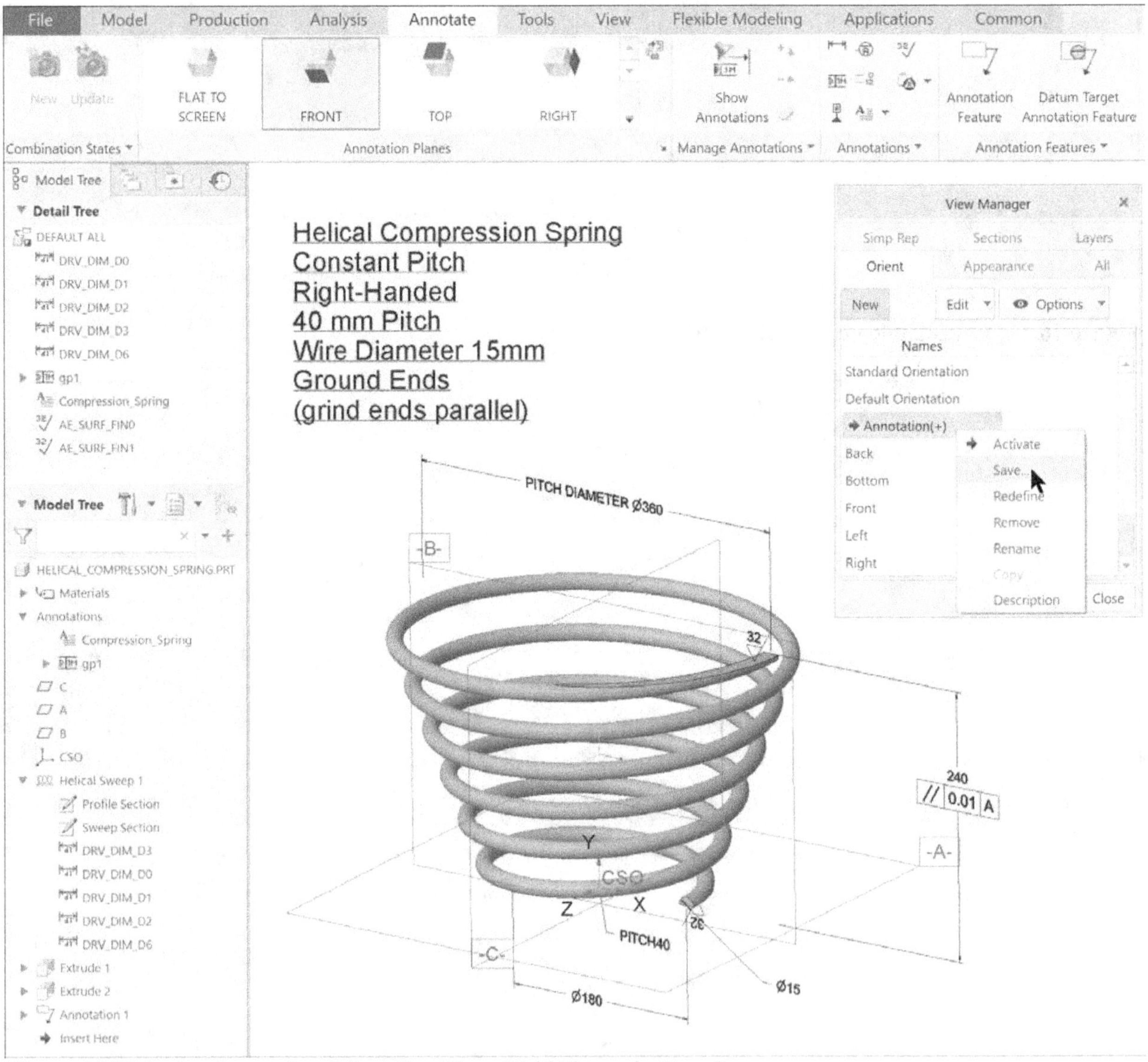

Helical Compression Spring
Constant Pitch
Right-Handed
40 mm Pitch
Wire Diameter 15mm
Ground Ends
(grind ends parallel)

Figure 16.22 Active Annotation Orientation Plane (Grid Shown in Green)

Download a different spring project from *www.cad-resources.com*.

Lesson 17 Shell, Reorder, and Insert Mode

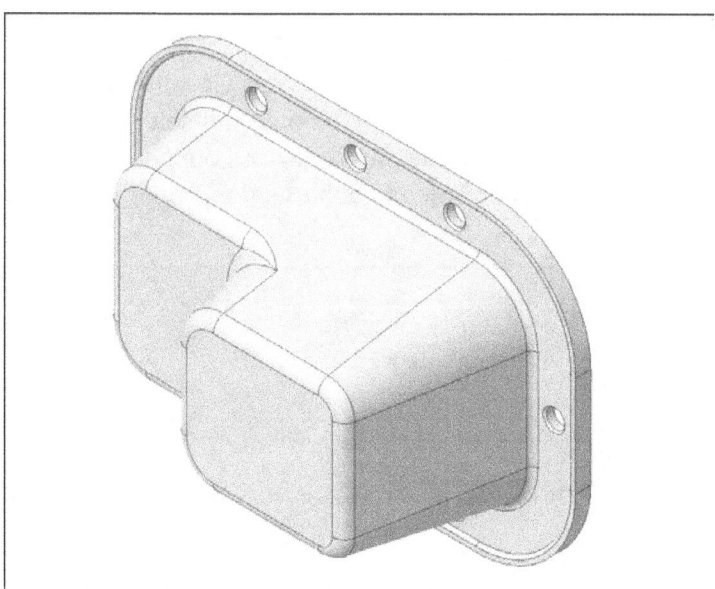

Figure 17.1 Oil Sink

OBJECTIVES

- Master the use of the **Shell Tool**
- **Reorder** features
- **Insert** a feature at a specific point in the design order
- Create a **Hole Pattern** using a **Table**
- **Render** the part.
- Create a **3D PDF**
- **Detail** the part

REFERENCES AND RESOURCES

For **Resources** go to www.cad-resources.com > click on the PTC Creo Parametric 5.0 Book cover

- Lesson 17 Lecture at **YouTube Creo Parametric Lecture Videos**

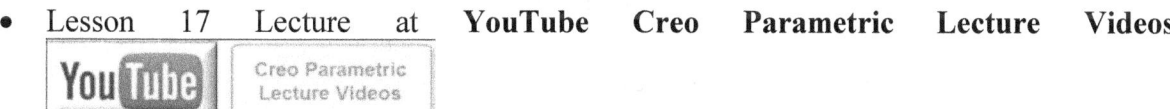

- Lesson 17 3D PDF models embedded in a PDF

SHELL, REORDER, AND INSERT MODE

The **Shell Tool** removes a surface or surfaces from the solid and then hollows out the inside of the solid, leaving a shell of a specified wall thickness, as in the Oil Sink (Fig. 17.1). When Creo Parametric makes the shell, all the features that were added to the solid before you chose the Shell Tool are hollowed out. Therefore, the *order of feature creation* is very important when you use the Shell Tool. You can alter the feature creation order by using the **Reorder** option. Another method of placing a feature at a specific place in the feature/design creation order is to use the **Insert Mode** option.

Creating Shells

The Shell Tool [Figs. 17.2(a-c)] enables you to remove a surface or surfaces from the solid, then hollows out the inside of the solid, leaving a shell of a specified wall thickness. If you flip the thickness side by entering a negative value, dragging a handle, or using the ⬛ **Change thickness direction** icon, the shell thickness is added to the outside of the part. If you do not select a surface to remove, a "closed" shell is created, with the whole inside of the part hollowed out and no access to the inside. In this case, you can add the necessary cuts or holes to achieve the proper geometry at a later time.

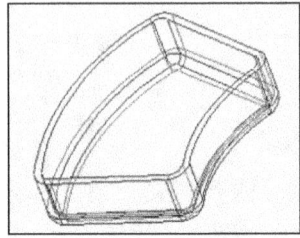 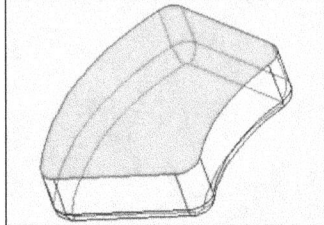 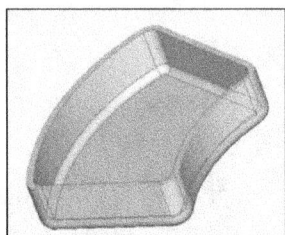

Figures 17.2(a-c) Shell

When defining a shell, you can also select surfaces where you want to assign a different thickness. You can specify independent thickness values for each such surface. However, you cannot enter negative thickness values, or flip the thickness side, for these surfaces. The thickness side is determined by the default thickness of the shell. When Creo Parametric 5.0 makes the shell, all the features that were added to the solid before you started the Shell Tool are hollowed out. Therefore, the order of feature creation is very important when you use the Shell Tool. To access the Shell Tool, click ⬛ icon in the **Model** tab Engineering Group. The Thickness box lets you change the value for the default shell thickness. You can type the new value, or select a recently used value from the drop-down list.

In the graphics window, you can use the shortcut menu (press **RMB**) to access the following options:

- **Removed Surfaces** Activates the collector of surfaces. You can select any number of surfaces
- **Non Default Thickness** Activates the collector of surfaces with a different thickness
- **Exclude Surfaces** Activates the collector of excluded surfaces
- **Clear** Remove all references from the collector that is currently active
- **Flip** Change the shell side direction

The Shell Dashboard displays the following slide-up/down panels (tabs):

- **References** Contains the collector of references used in the Shell feature
- **Options** Contains the collector of Excluded surfaces
- **Properties** Contains the feature name and an icon to access feature information

The **References** slide-up/down panel (tab) contains the following elements:

- The **Removed surfaces** collector lets you select the surfaces to be removed. If you do not select any surfaces, a "closed" shell is created.
- The **Non-default thickness** collector lets you select surfaces where you want to assign a different thickness. For each surface included in this collector, you can specify an individual thickness value.

The **Properties** panel (tab) contains the Name text box `Name SHELL_ID_200  ⓘ`, where you can type a custom name for the shell feature, to replace the default name. It also contains the ⬛ icon that you can click to display information about this feature in the Browser.

Reordering Features

You can move features forward or backward in the feature creation (regeneration) order list, thus changing the order in which features are regenerated [Figs. 17.3(a-b)]. You can reorder features in the Model Tree by dragging one or more features to a new location in the feature list. If you try to move a child feature to a higher position than its parent feature, the parent feature moves with the child feature in context, so that the parent/child relationship is maintained.

You can reorder multiple features in one operation, as long as these features appear in *consecutive* order. Feature reorder *cannot* occur under the following conditions:

- **Parents** Cannot be moved so that their regeneration occurs after the regeneration of their children
- **Children** Cannot be moved so that their regeneration occurs before the regeneration of their parents

You can select the features to be reordered by choosing an option:

- **Select** Select features to reorder by picking on the screen and/or from the Model Tree
- **Layer** Select all features from a layer by selecting the layer
- **Range** Specify the range of features by entering the regeneration numbers of the starting and ending features

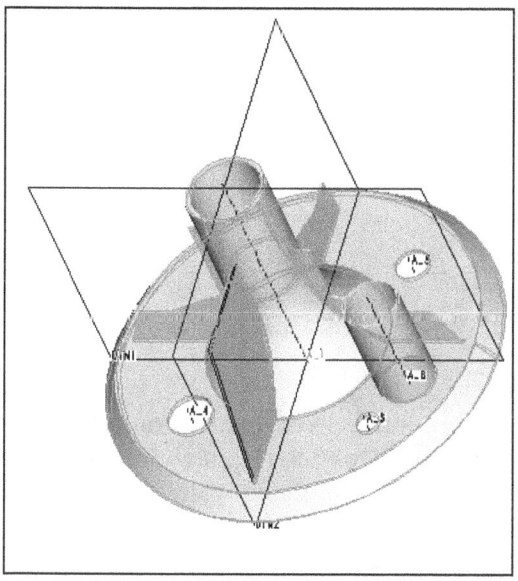

Figure 17.3(a) Reorder (Note the gusset closest to you)

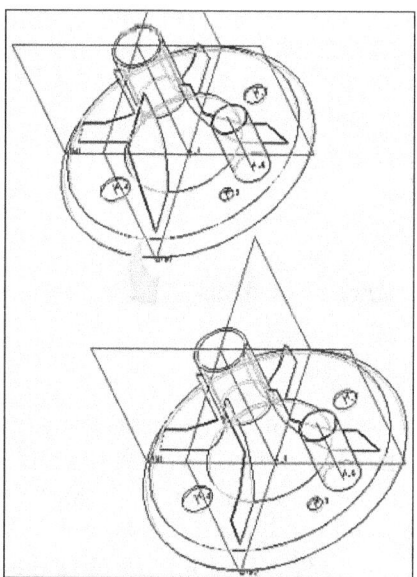

Figure 17.3(b) Reordered feature

Inserting Features

Normally, Creo Parametric adds a new feature after the last existing feature in the part, including suppressed features. Insert Mode allows you to add new features at any point in the feature sequence, except before the base feature or after the last feature. You can also insert features using the Model Tree. There is an arrow-shaped icon on the Model Tree that indicates where features will be inserted upon creation. By default, it is always at the end of the Model Tree. You may drag the location of the *arrow* higher or lower in the tree to insert features at a different point. When the *arrow* is dropped at a new location, the model is rolled backward or forward in response to the insertion *arrow* being moved higher or lower in the tree.

Lesson 17 STEPS

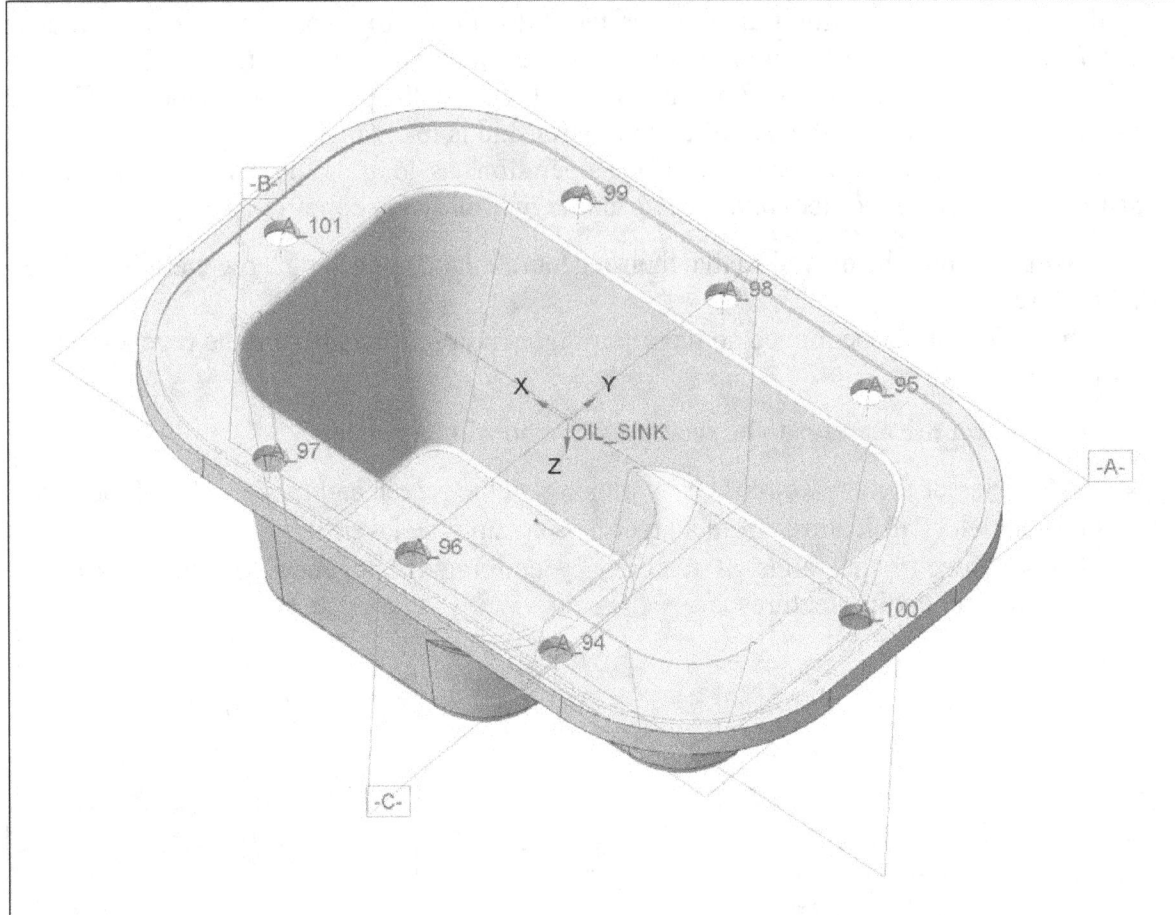

Figure 17.4 Oil Sink

Oil Sink

The Oil Sink (Fig. 17.4) requires the use of the **Shell Tool**. The shelling of a part should be done after the desired extrusions and most rounds have been modeled. This lesson part will have you create an extrusion, a cut, and a set of rounds. Some of the required rounds will be left off the part model on purpose.

Creo Parametric's **Insert Mode** option enables you to insert a set of features at an earlier stage in the design of the part. In other words, you can create a feature after or before a selected existing feature even if the whole model has been completed. You can also *move the order in which a feature was created* and therefore have subsequent features affect the reordered feature. A round created after a shell operation can be reordered to appear before the shell, to have the shell be affected by the round.

In this lesson, you will also insert a round or two before the existing shell feature using Insert Mode. The rounds will be shelled after the **Resume** option is picked, because the rounds now appear before the shell feature. The details shown in Figures 17.5(a) through (h) provide the design dimensions.

.625
.3125
R.125
DETAIL B
SCALE 1.500
.500
.125
82°
ø.875
ø.75
DETAIL A
SCALE 1.500
.3750 THICK
SEE DETAIL B
4.00
7.00
R.50
10.0°
.75
.375
SEE DETAIL A
8.00
R4.00
10.00
8X .750
R.50
SECTION B-B
18.00
12.000

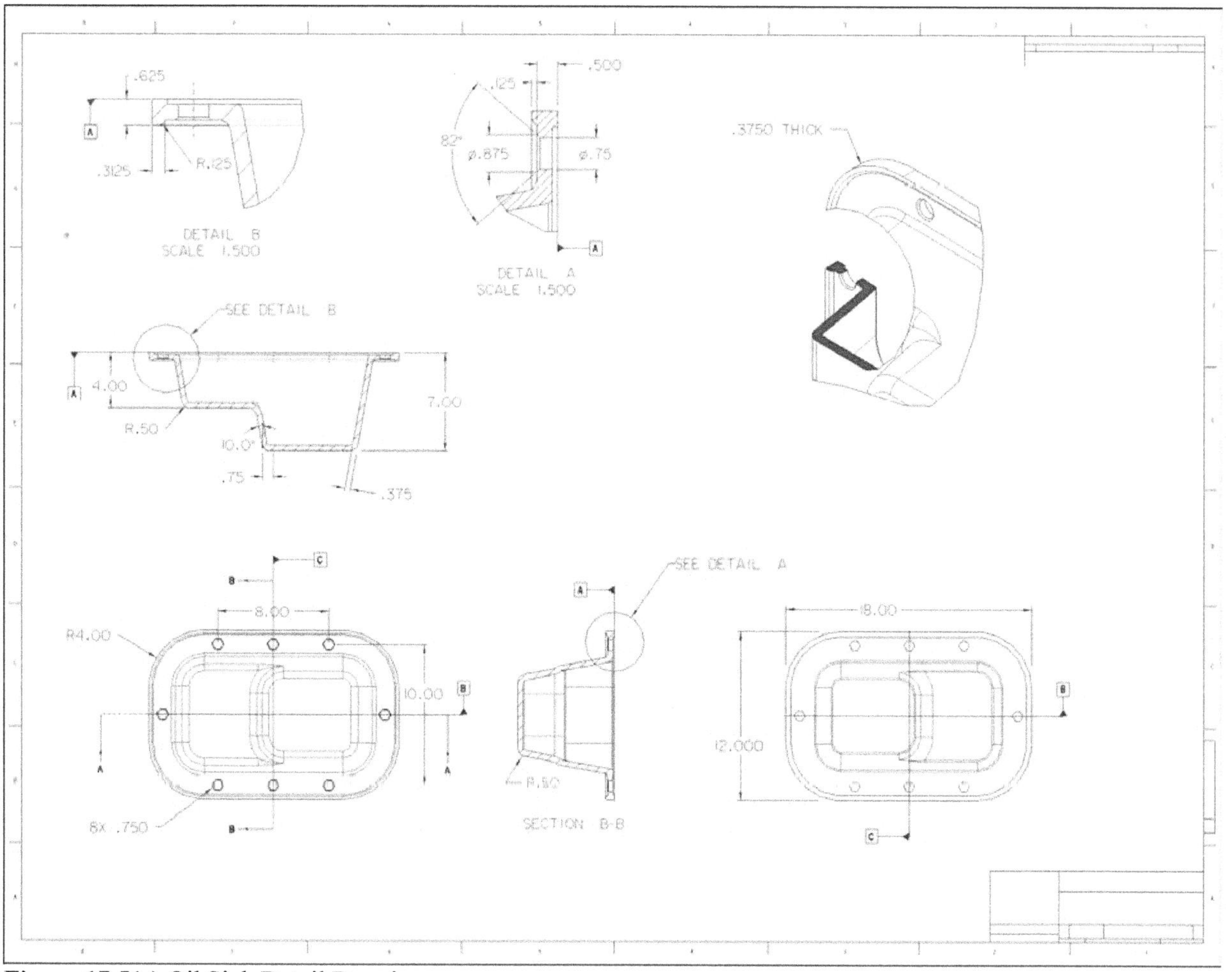

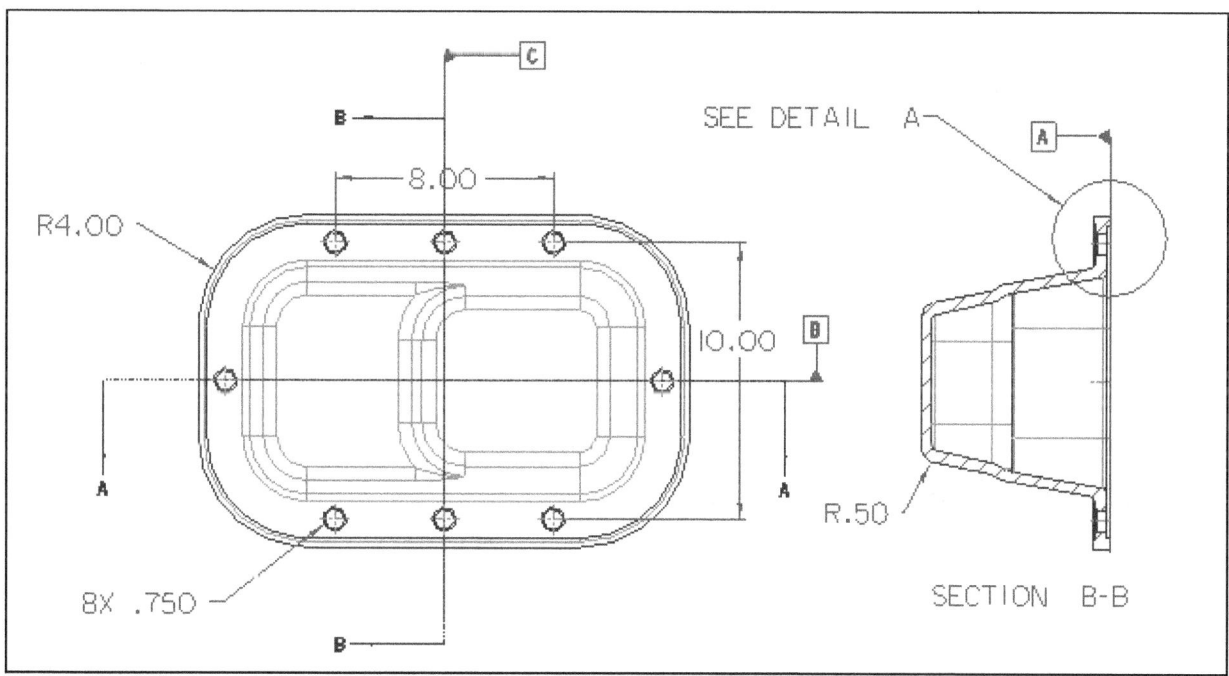

Figure 17.5(a) Oil Sink Detail Drawing

C
B
SEE DETAIL A
A
8.00
R4.00
10.00
B
A
A
R.50
8X .750
SECTION B-B
B

Figure 17.5(b) Oil Sink Front and Right Side Views

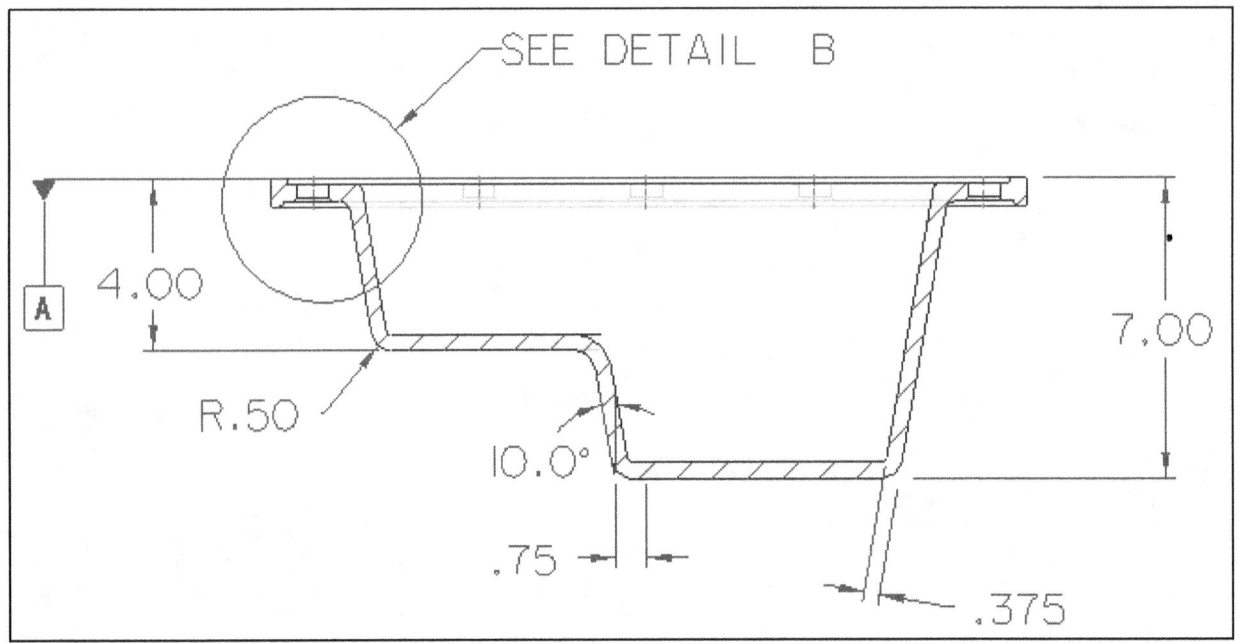

Figure 17.5(c) Oil Sink Section A-A

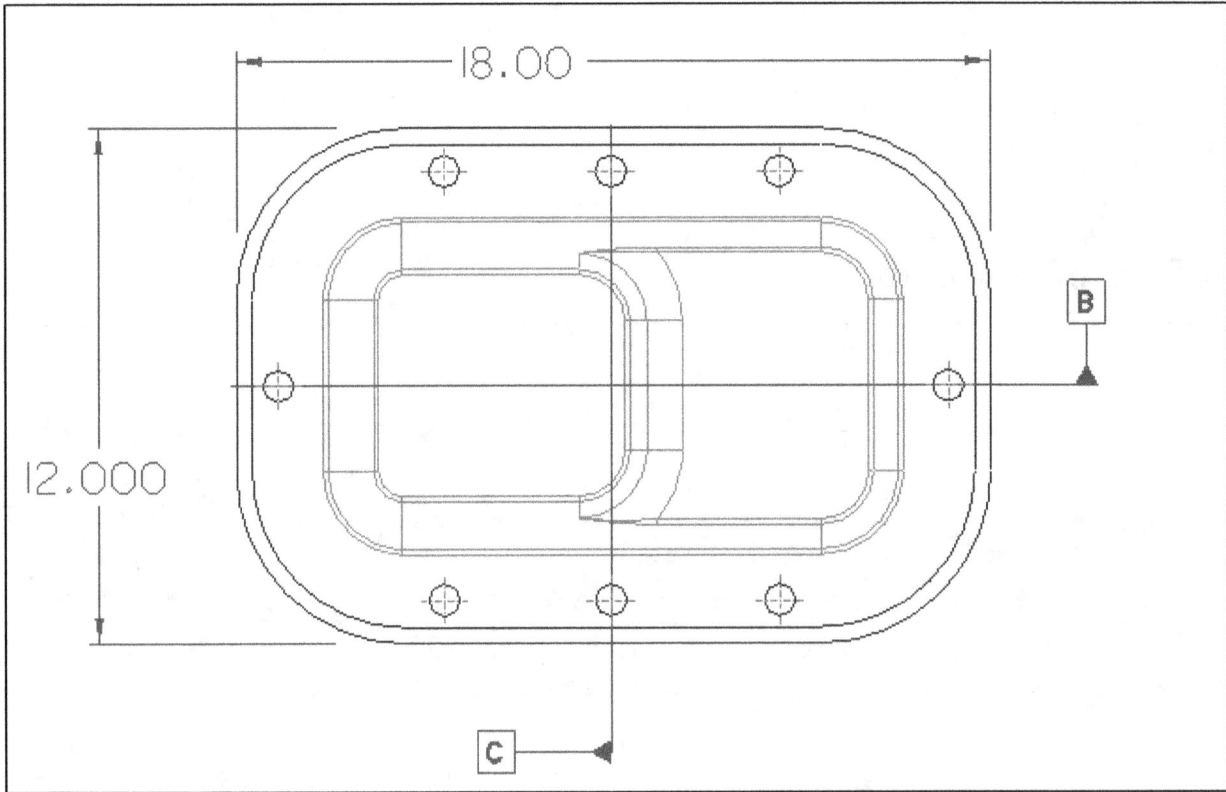

Figure 17.5(d) Oil Sink Back View

726

.3750 THICK

Figure 17.5(e) Oil Sink Cutaway View

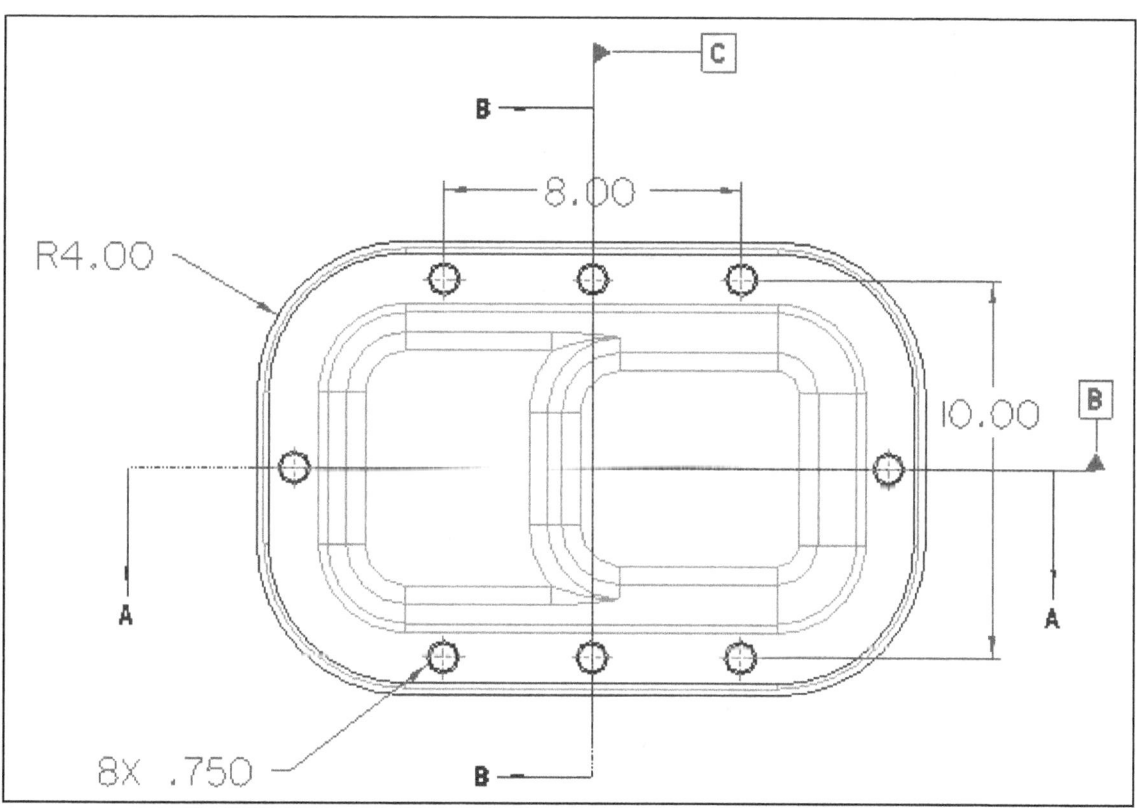

C
B
8.00
R4.00
10.00
B
A
A
8X .750
B

Figure 17.5(f) Oil Sink Front View Dimensions

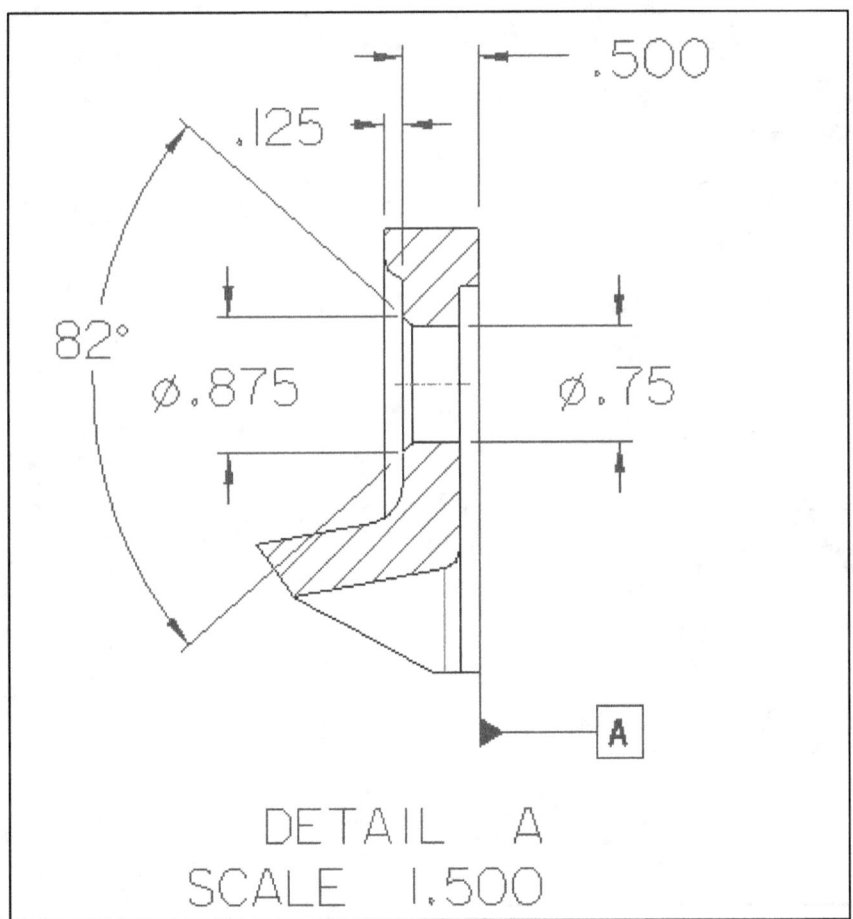

.500

.125

82°

Ø.875

Ø.75

DETAIL A
SCALE 1.500

Figure 17.5(g) Oil Sink DETAIL A

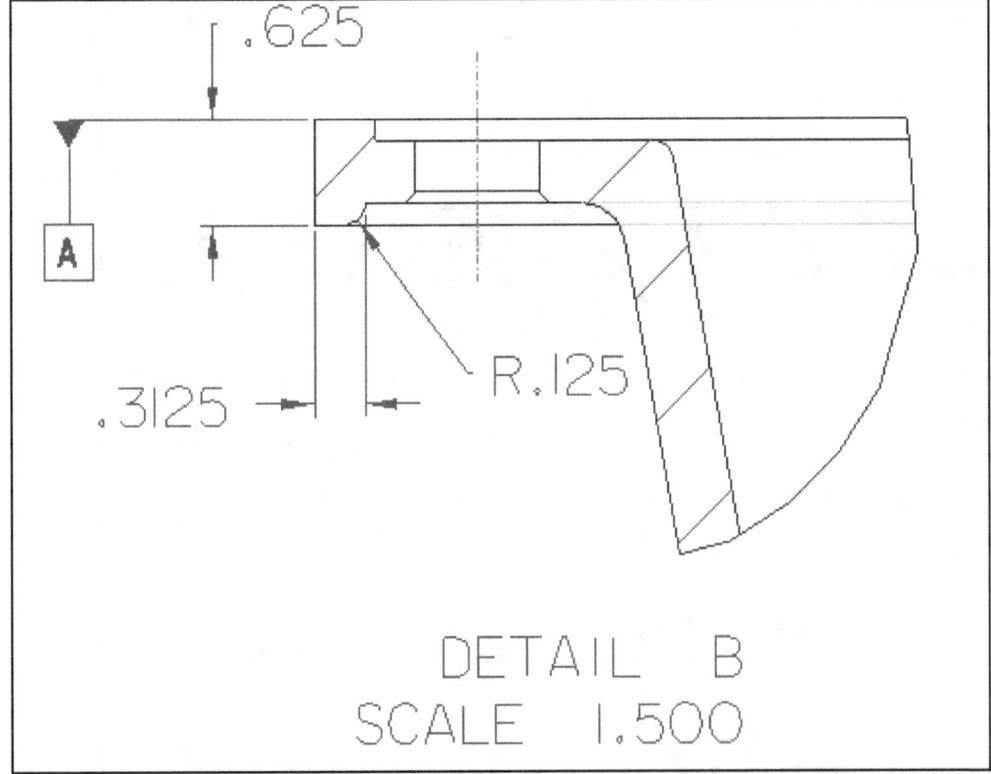

.625

A

.3125

R.125

DETAIL B
SCALE 1.500

Figure 17.5(h) Oil Sink DETAIL B

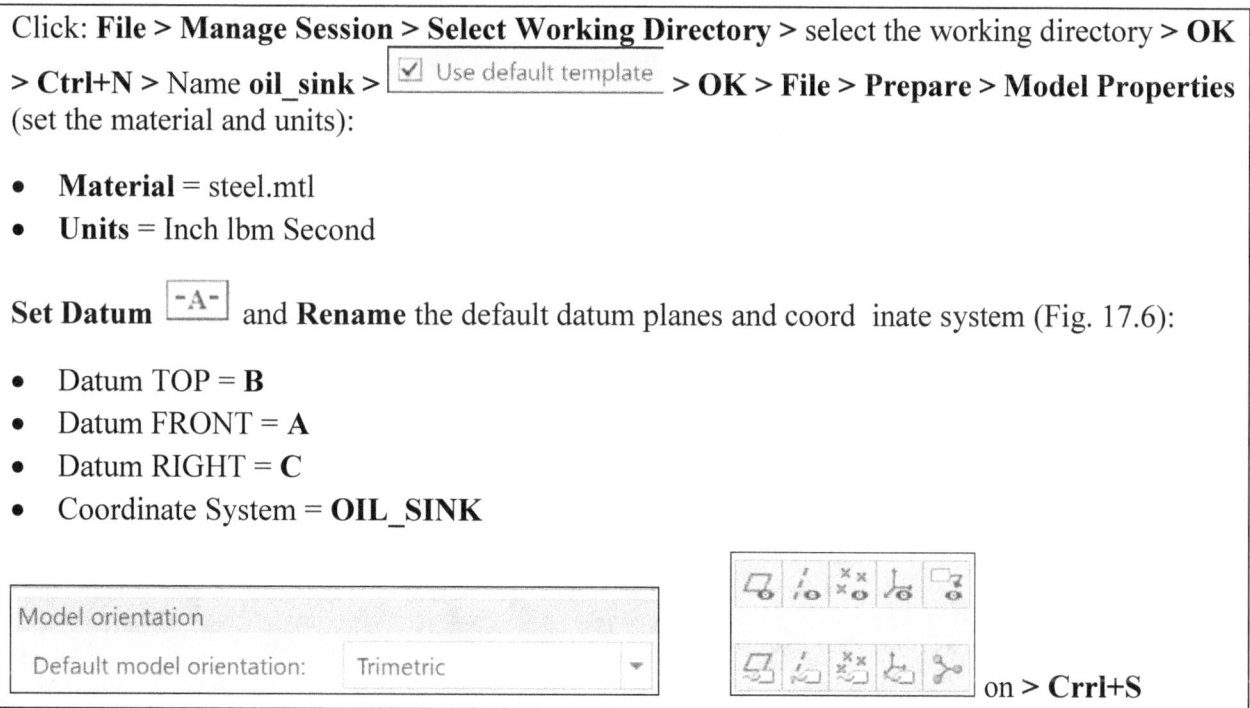

Click: **File > Manage Session > Select Working Directory** > select the working directory > **OK**
> **Ctrl+N** > Name **oil_sink** > ☑ Use default template > **OK** > **File > Prepare > Model Properties**
(set the material and units):

- **Material** = steel.mtl
- **Units** = Inch lbm Second

Set Datum -A- and **Rename** the default datum planes and coord inate system (Fig. 17.6):

- Datum TOP = **B**
- Datum FRONT = **A**
- Datum RIGHT = **C**
- Coordinate System = **OIL_SINK**

Model orientation

Default model orientation: Trimetric   ▼   on > **Crrl+S**

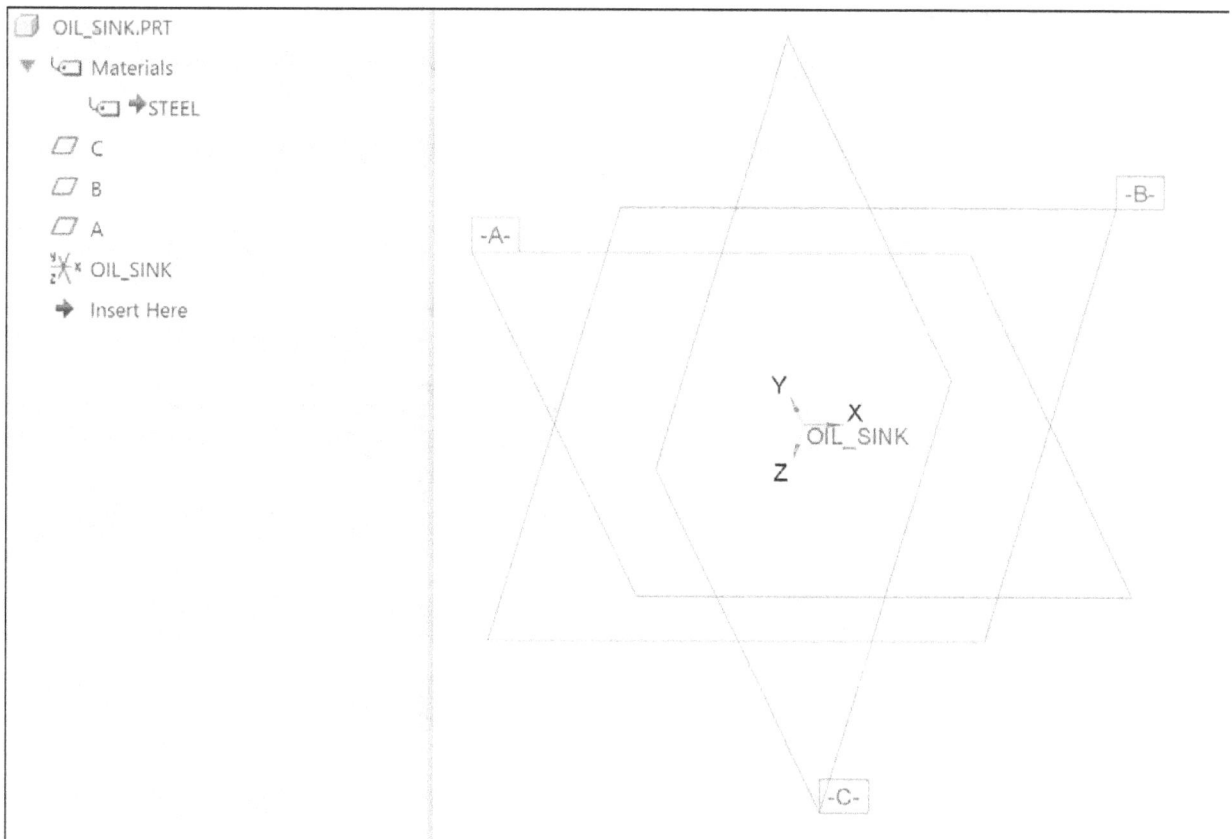

Figure 17.6 Set Datums and Renamed Coordinate System

729

Make the first extrusion .50 (thickness) **X 12.00** (height) **X 18.00** (length), with **R5.00** rounds (add the fillets to the sketch). Sketch on datum plane **A** and center the first extrusion horizontally on datum **B** and vertically on datum **C** [Figs. 17.7(a-b)].

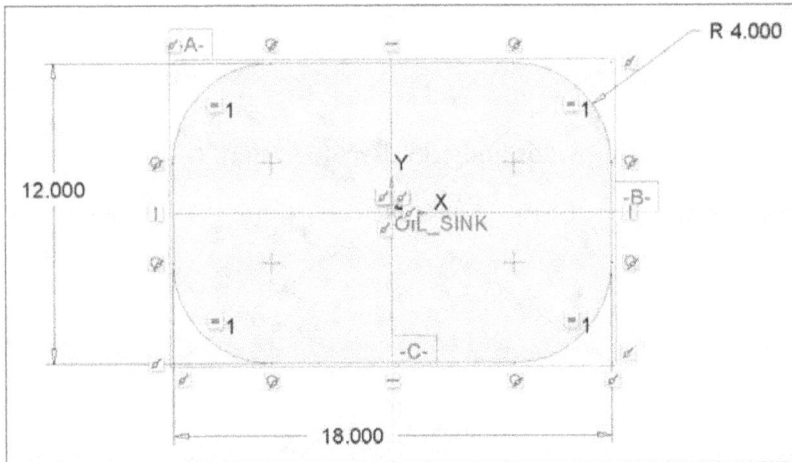

Figure 17.7(a) Dimensions for the First Extrusion

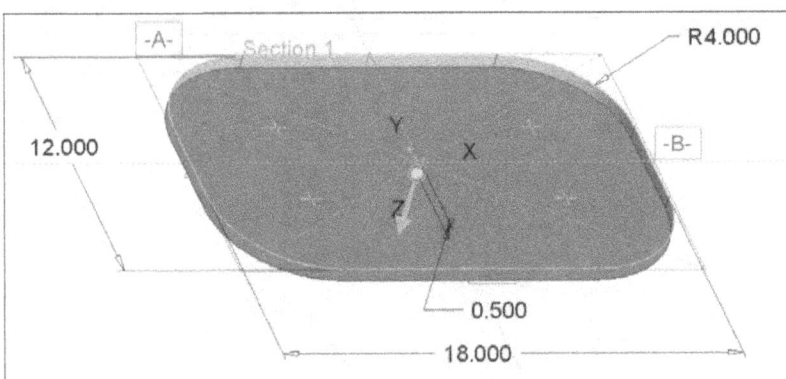

Figure 17.7(b) Standard Orientation

Make the second extrusion offset from the edge of the first extrusion **-3.00**, with a height of **7.00** [Figs. 17.8(a-b)]. Sketch on the top surface of the first extrusion; then, create the cut [Figs. 17.9(a-b)]. Save.

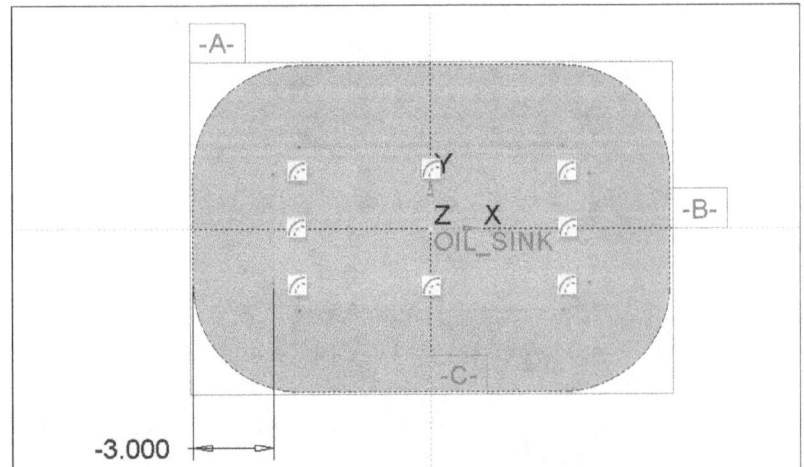

Figure 17.8(a) Second Extrusion is Offset from the Edge of the First Extrusion

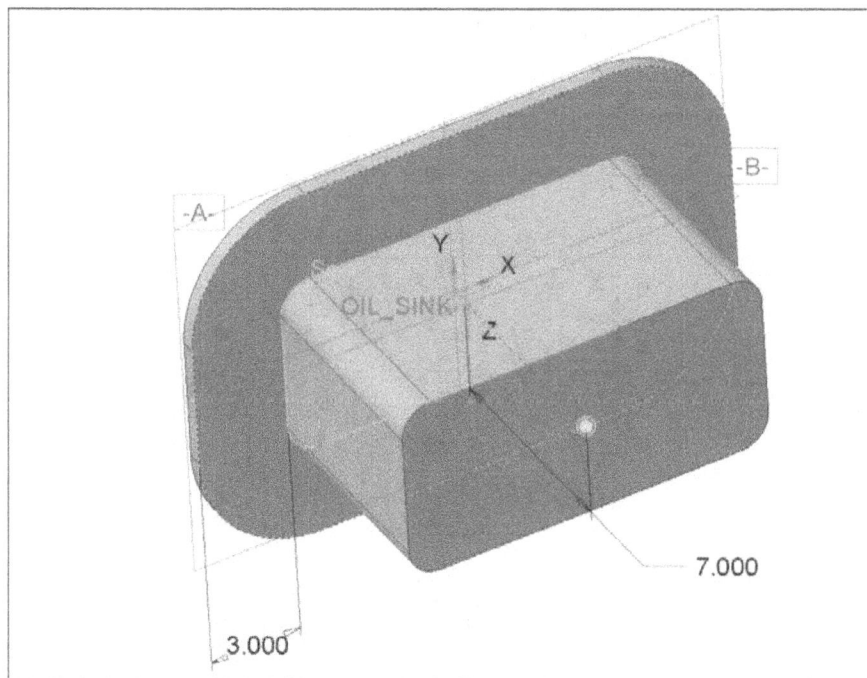

Figure 17.8(b) Second Extrusion

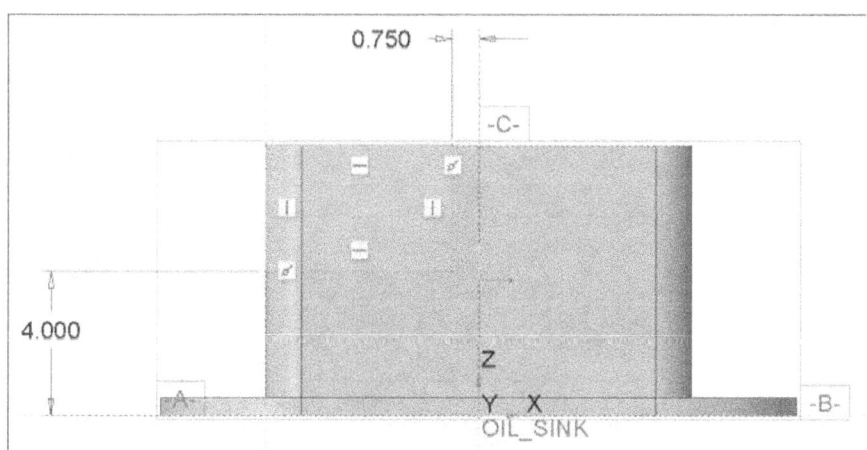

Figure 17.9(a) Dimensions for the Cut

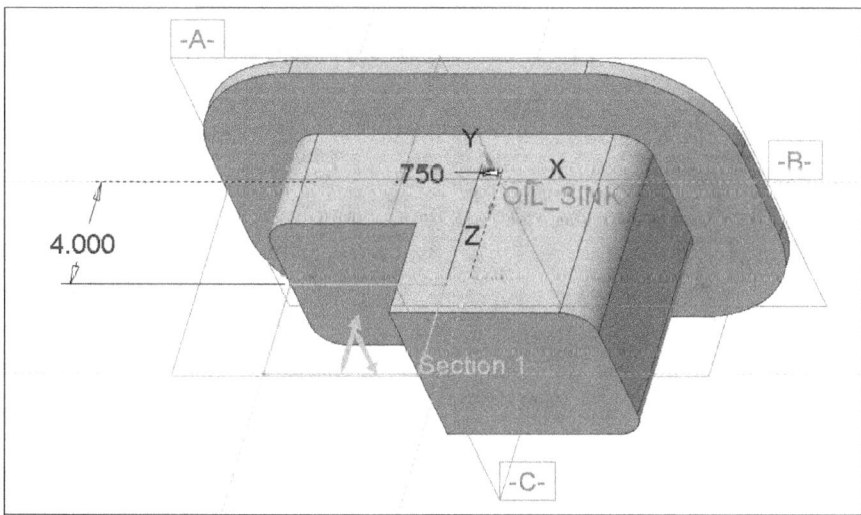

Figure 17.9(b) Standard Orientation of the Cut

Add the **R1.50** rounds [Figs. 17.10(a-b)]. Draft all vertical surfaces of the second extrusion **10** degrees. Select one vertical surface to establish the drafted surfaces. Use the top surface as the Draft hinge [Figs. 17.11(a-b)]. Change the model color as desired. Save.

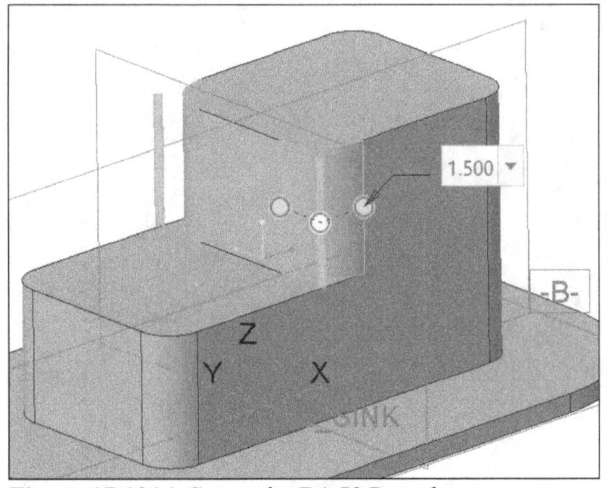

Figure 17.10(a) Create the **R1.50** Rounds

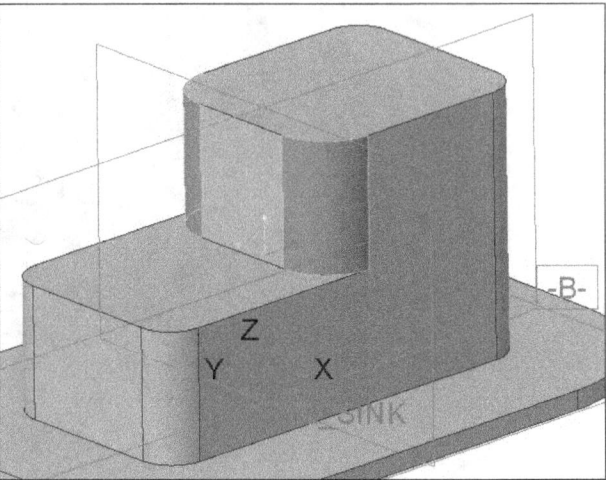

Figure 17.10(b) Completed **R1.50** Rounds

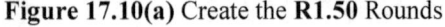

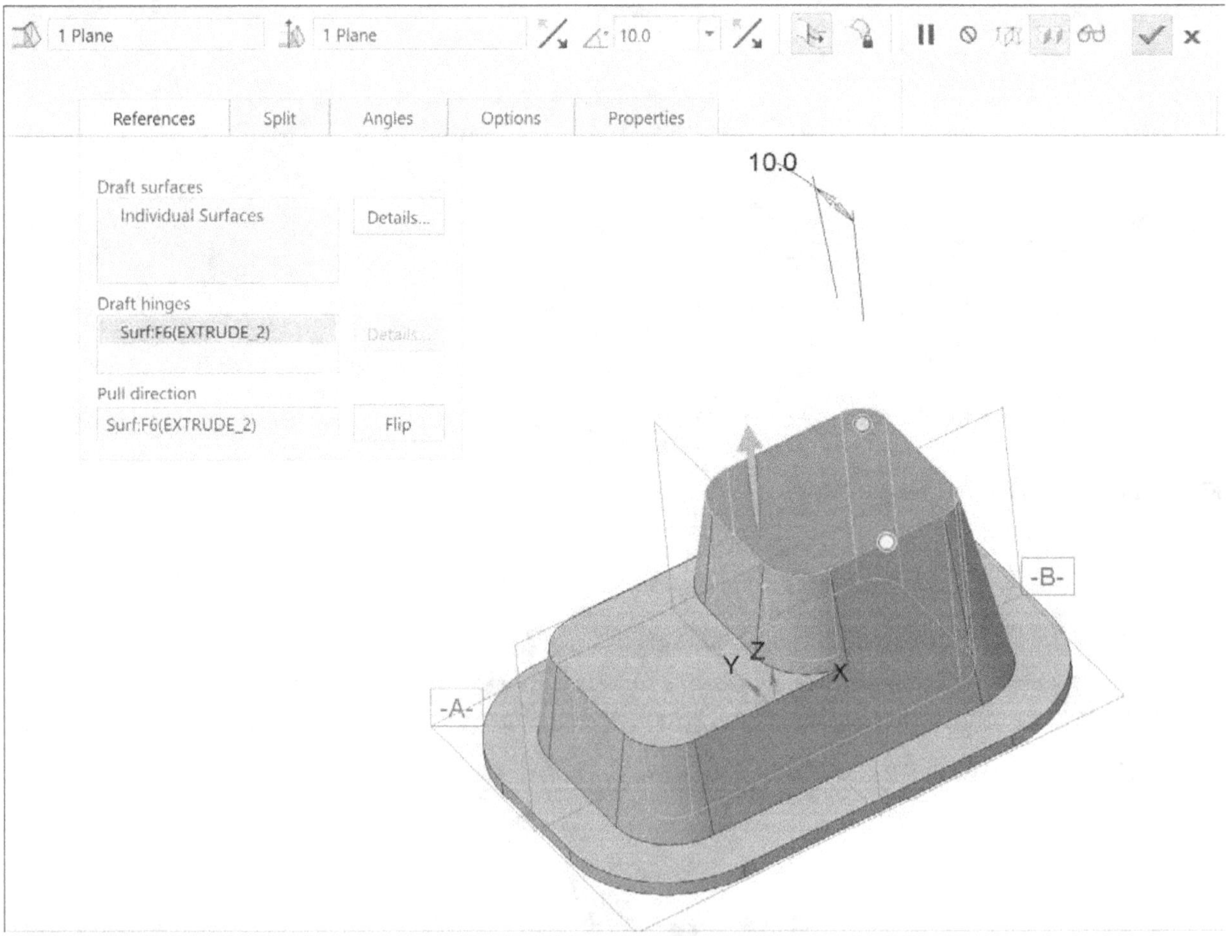

Figure 17.11(a) Draft References

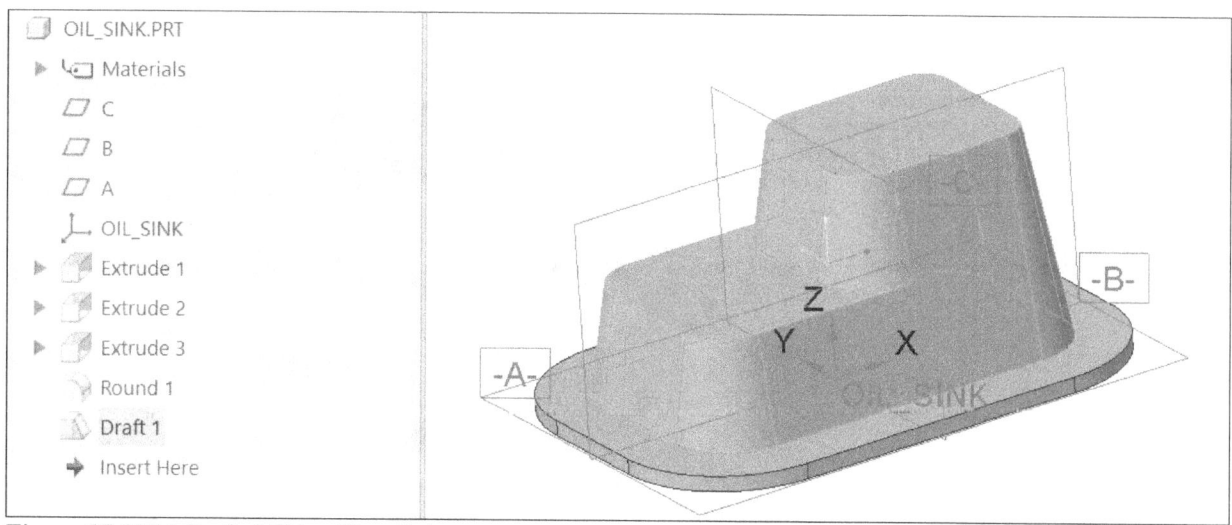

OIL_SINK.PRT
Materials
C
B
A
OIL_SINK
Extrude 1
Extrude 2
Extrude 3
Round 1
Draft 1
Insert Here
-A-
-B-
Z
Y
X
OIL_SINK

Figure 17.11(b) Drafted Sides. New Color.

Click: Shell > Thickness **.375 > Enter >** spin the model **> References** tab > Removed surfaces-- select the bottom surface of the part [Fig. 17.12(a)] **> MMB > LMB > Ctrl+R > Ctrl+S** [Fig. 17.12(b)]

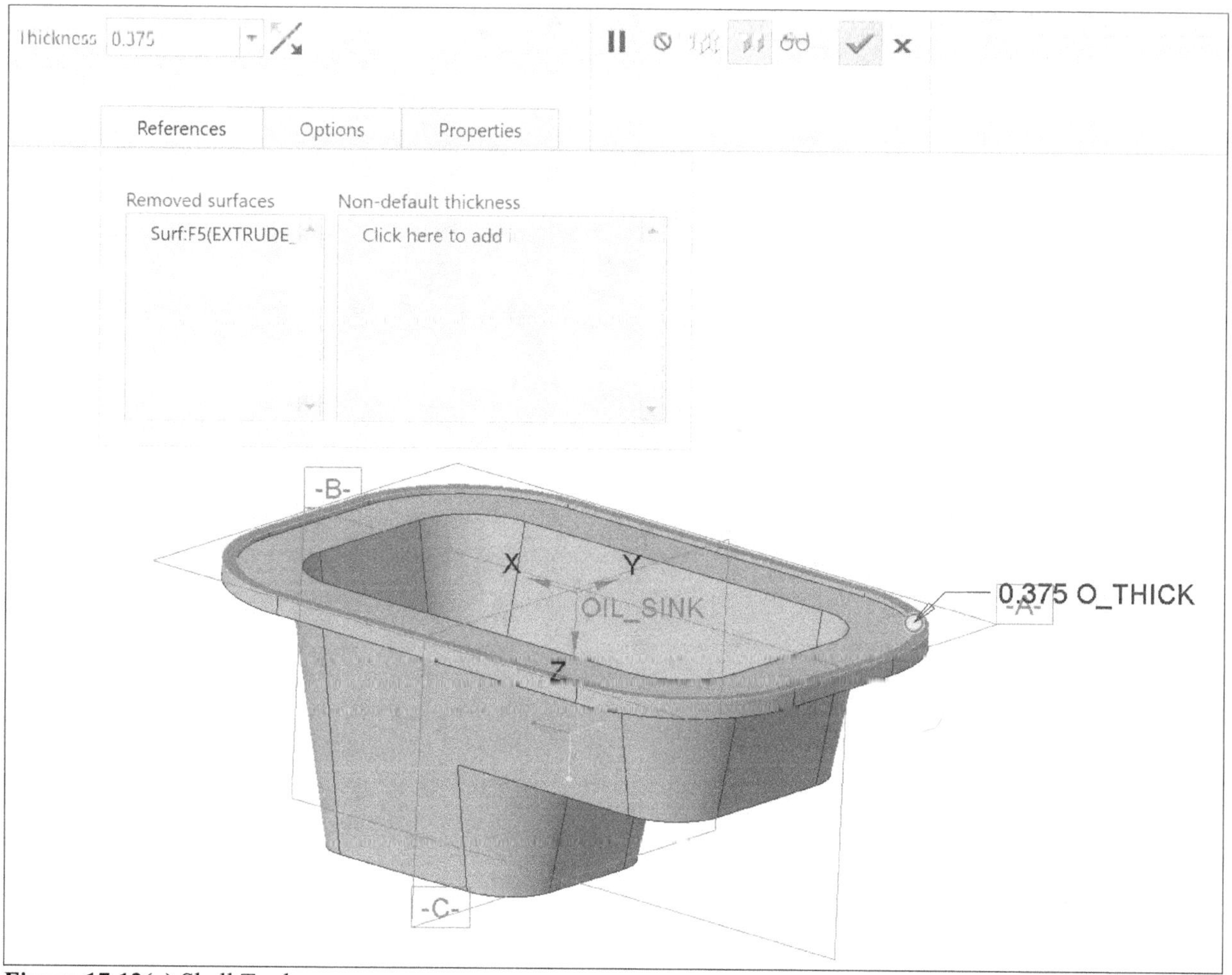

Thickness 0.375
References
Options
Properties
Removed surfaces
Surf:F5(EXTRUDE_
Non-default thickness
Click here to add
-B-
X
Y
OIL_SINK
Z
0.375 O_THICK
-A-
-C-

Figure 17.12(a) Shell Tool

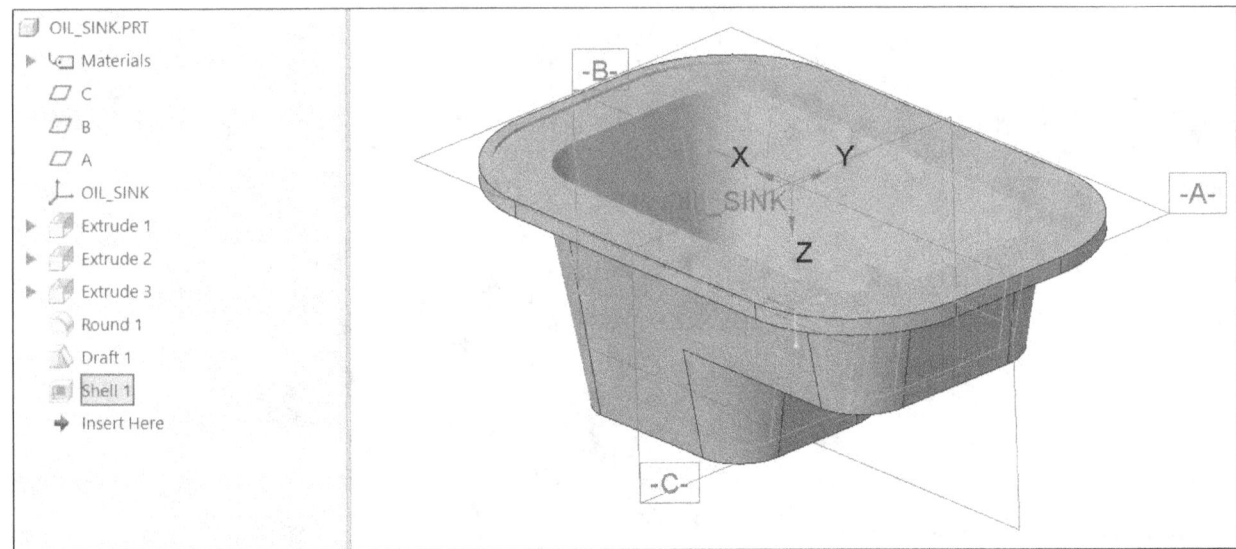

Figure 17.12(b) Shelled Part

The next feature you need to create is a *"lip"* around the part using an extrusion. Sketch on the top surface of the first extrusion > sketch two closed loops using Fig. 17.13(a)]. Use the edge of the first extrusion for the first loop and then create an offset edge (**-.3125**) for the second loop [Fig. 17.13(b)]. > The depth of the lip extrusion is **.125** [Figs. 17.13(c-d)].

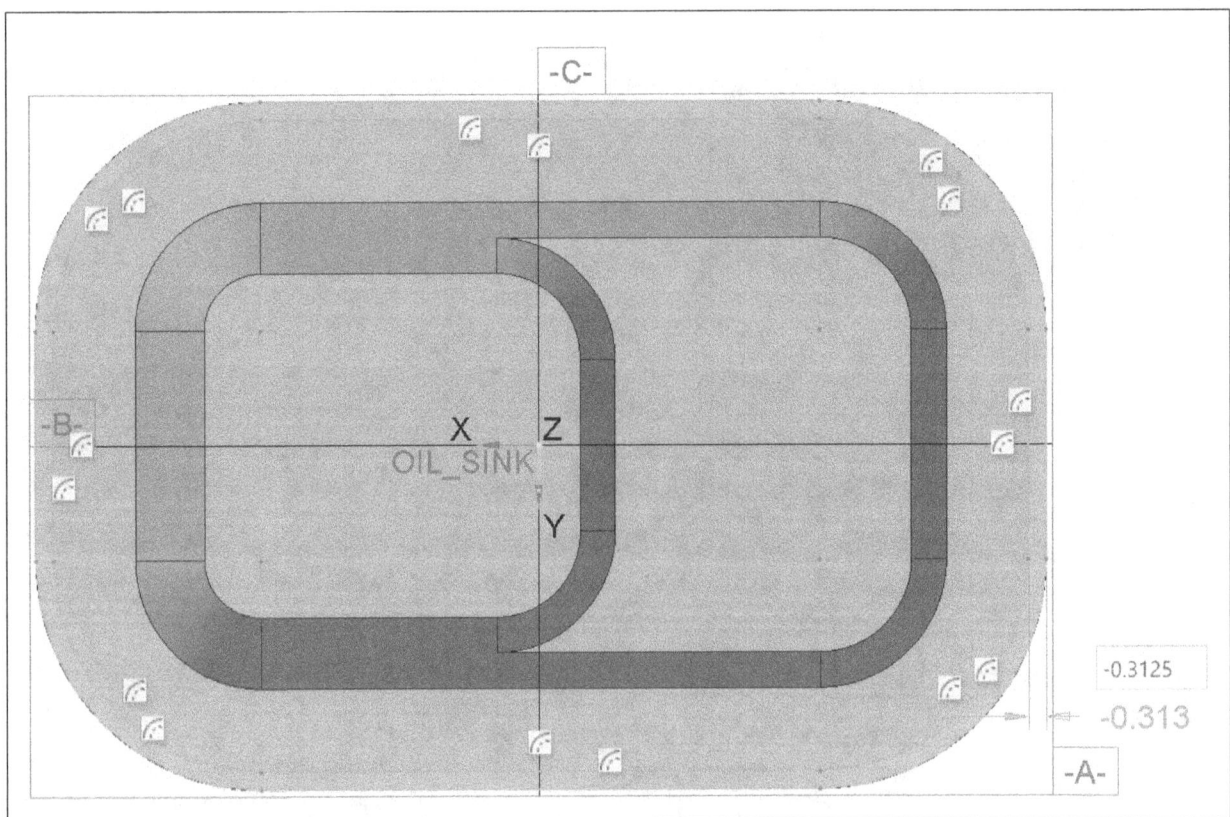

Figure 17.13(a) Sketch Two Closed Loops

734

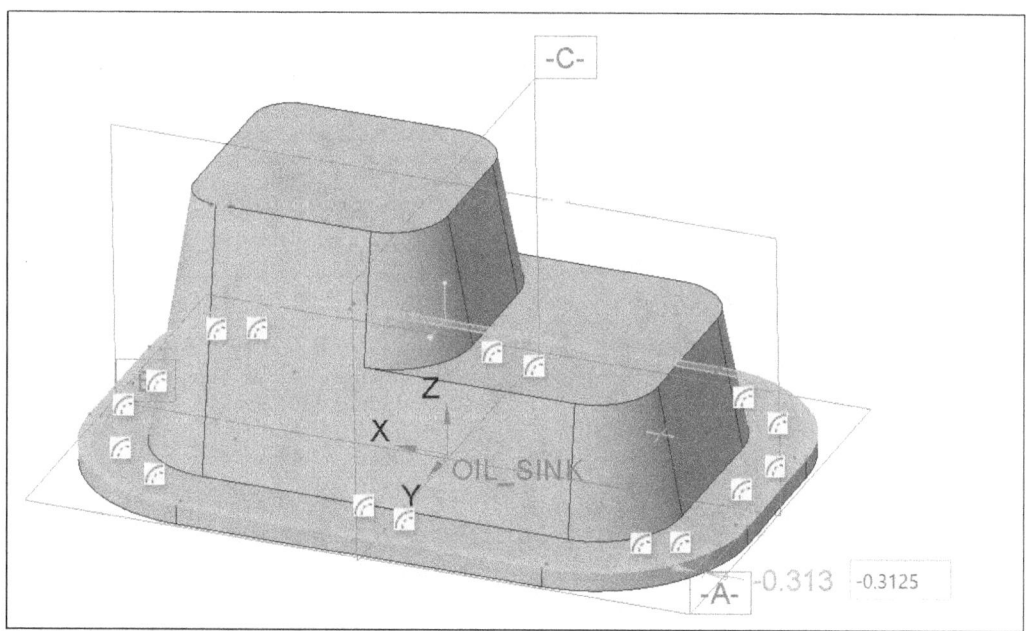

Figure 17.13(b) 3D View of the Sketch

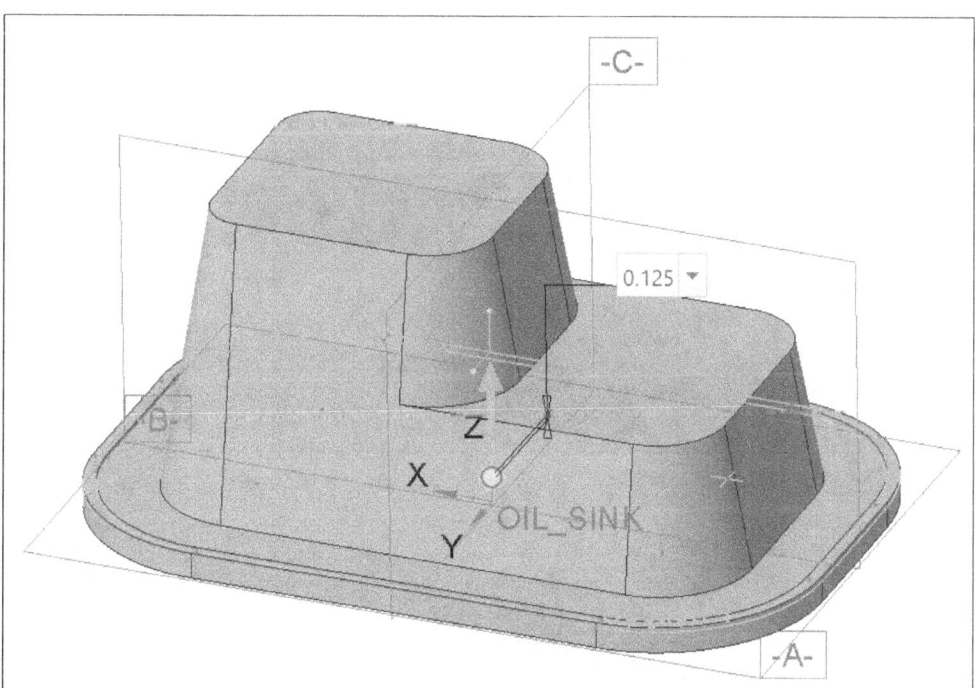

Figure 17.13(c) Depth **.125**

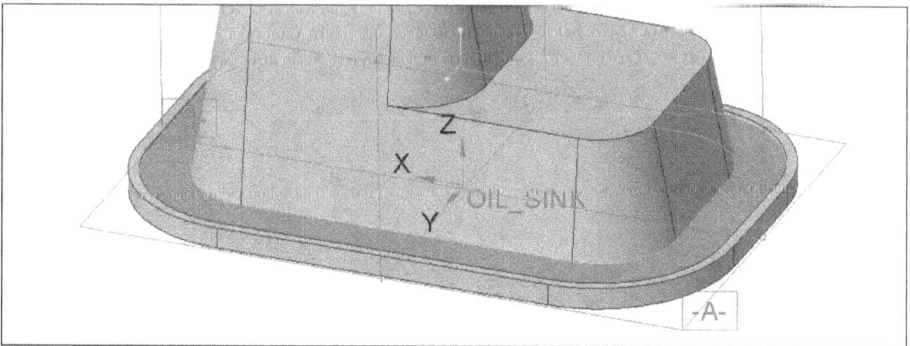

Figure 17.13(d) Completed "Lip" Extrusion

735

Add the rounds with one feature (three Sets): **R.125** round on the inside of the *"lip"* [Fig. 17.14(a)] > **R.125** round to the inside edge [Fig. 17.14(b)] > **R.250** round between the first two extrusions [Fig. 17.14(c)]

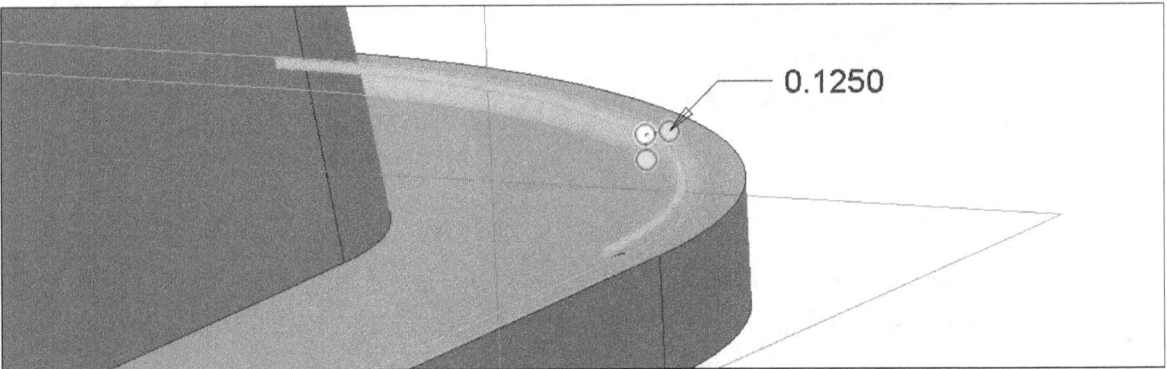

Figure 17.14(a) Set 1 Round **R.125**

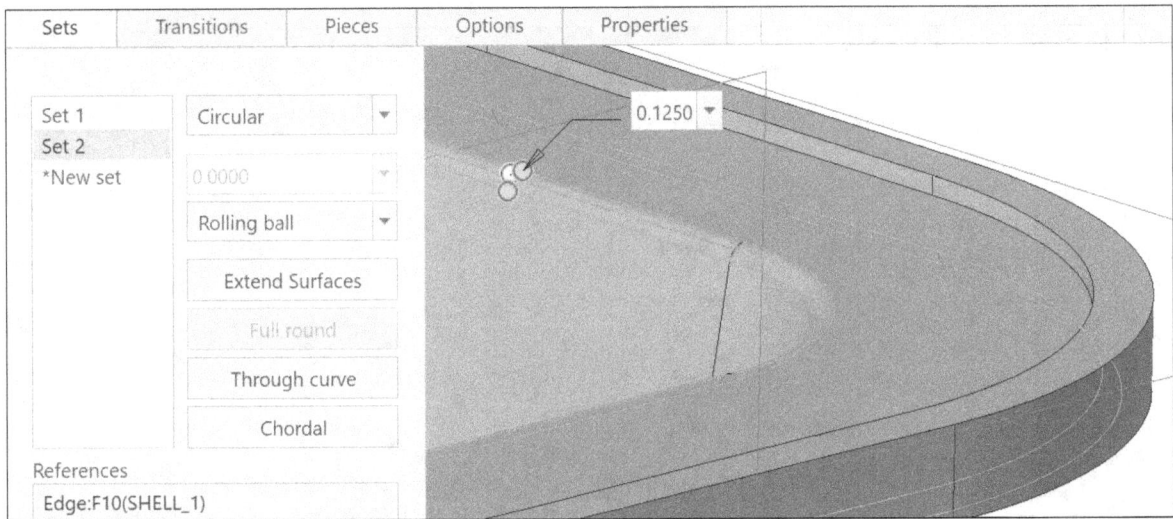

Figure 17.14(b) Set 2 Round **R.125**

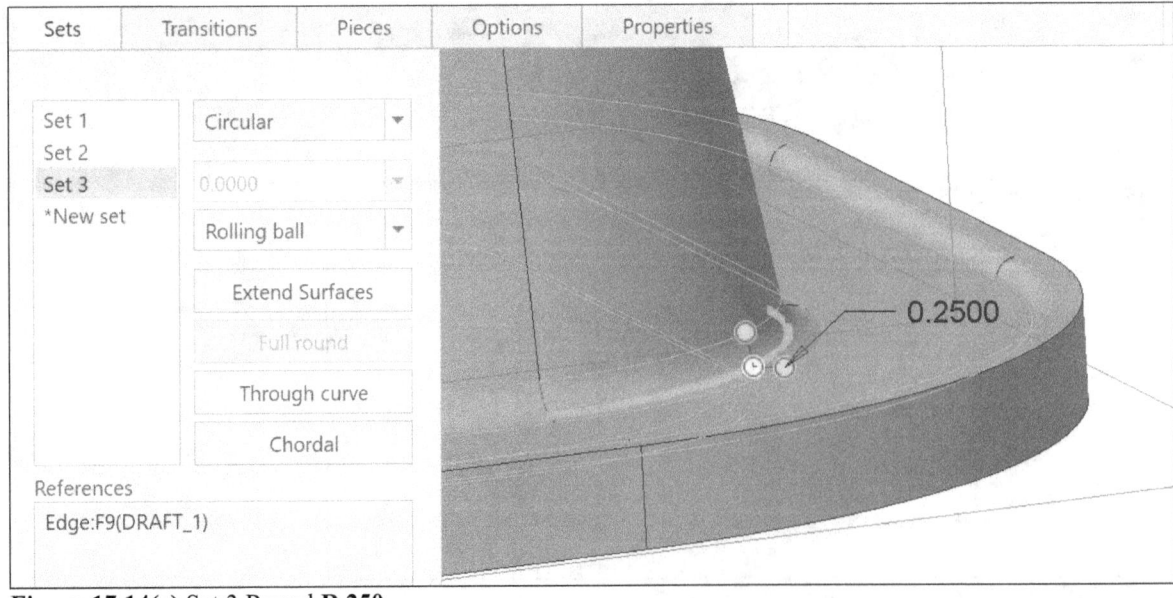

Figure 17.14(c) Set 3 Round **R.250**

The countersunk holes will be added next, click: [Hole] > spin the part > [⊟⊟] **Drill to intersect with all surfaces** > change the diameter to **.750** > **Enter** > **Placement** tab > select a location on the surface for the hole placement [Figs. 17.15(a-c)]

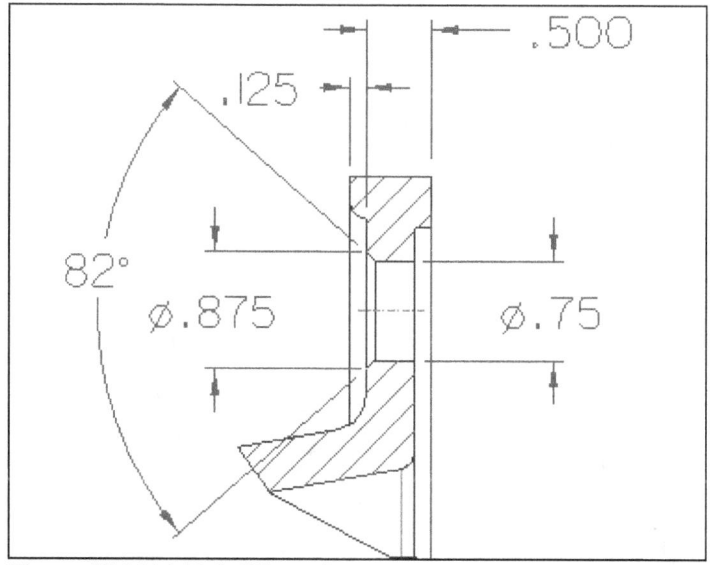

Figure 17.15(a) Hole Dimensions

Figure 17.15(b) X-Section of Hole

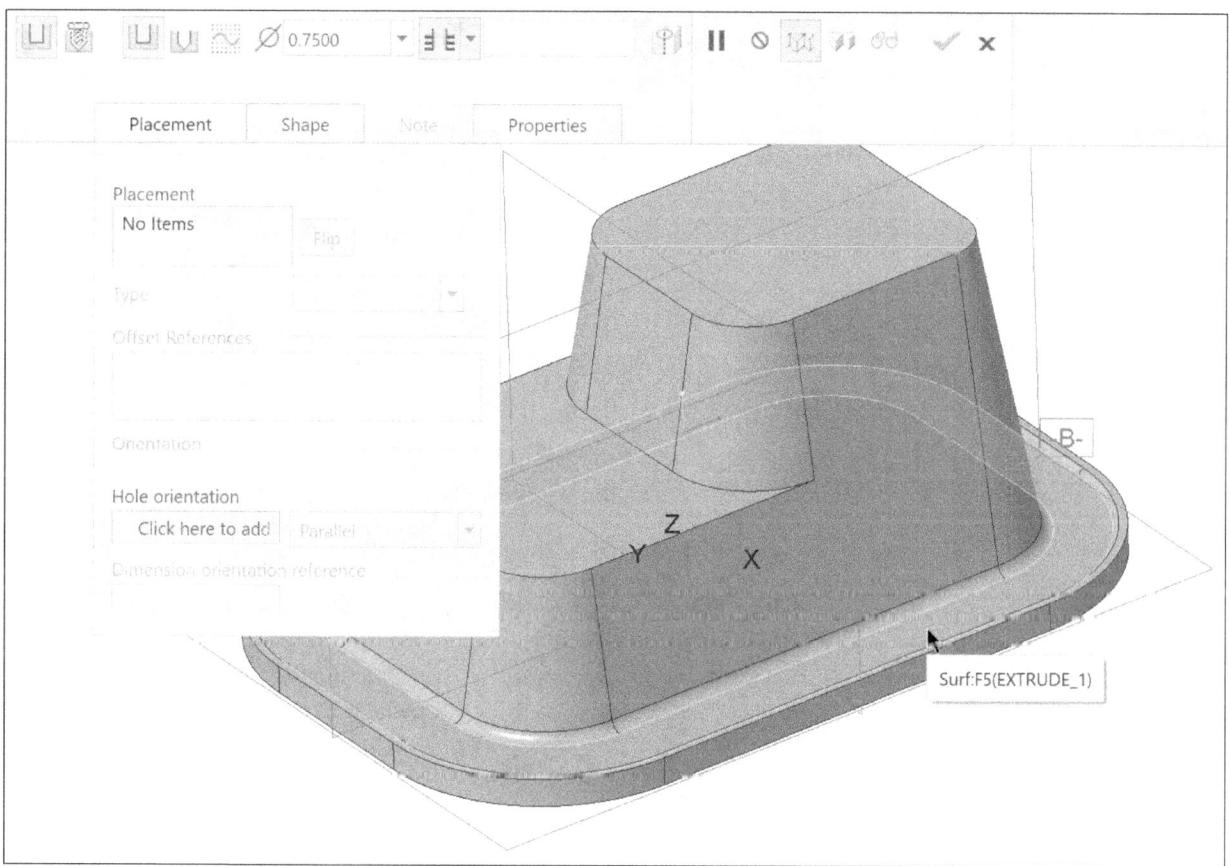

Figure 17.15(c) Hole Placement View Orientation

Place the pointer on a drag handle > press **LMB** > move the pointer to datum **C** (highlights) > similarly, move the other drag handle to datum **B** [Figs. 17.15(d-e)]

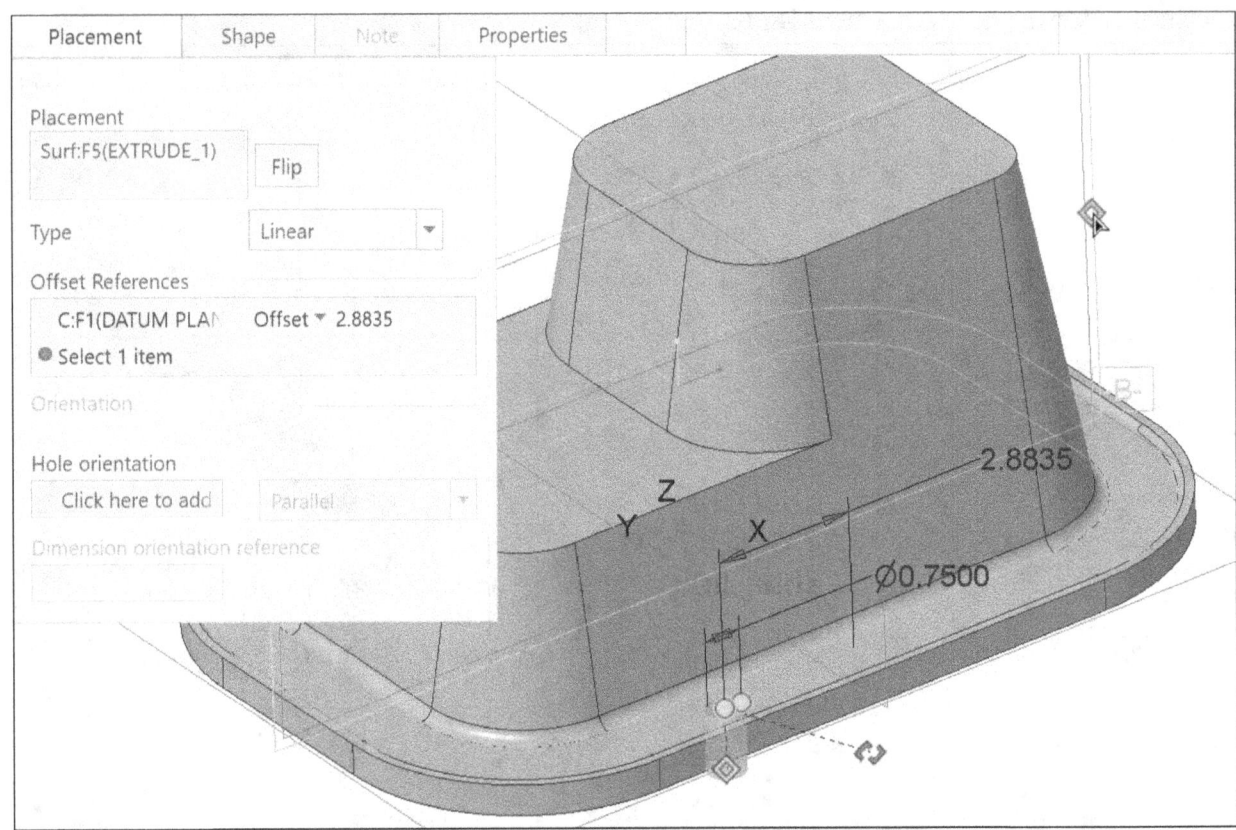

Figure 17.15(d) Hole Offset Reference to Datum **C** Established. Drag Remaining Position Handle to Datum **B**

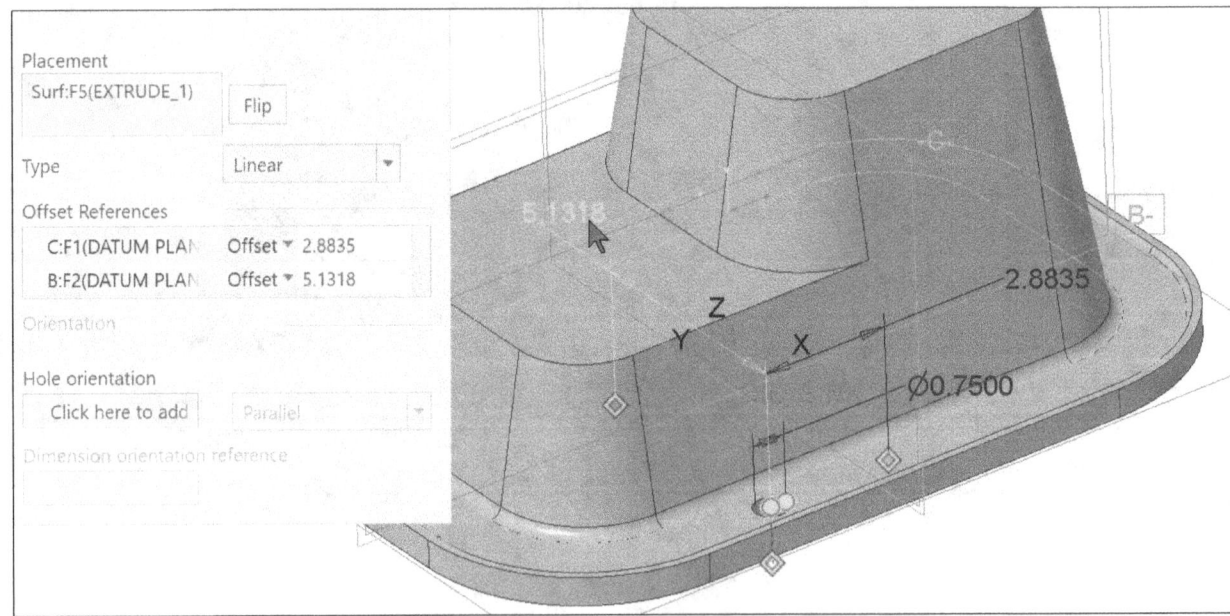

Figure 17.15(e) Initial Offset References Dimensions

Modify the values to be **4.00** from datum **C** and **5.00** from datum **B** [Fig. 17.15(f)] > **Shape** tab [Fig. 17.15(g)]

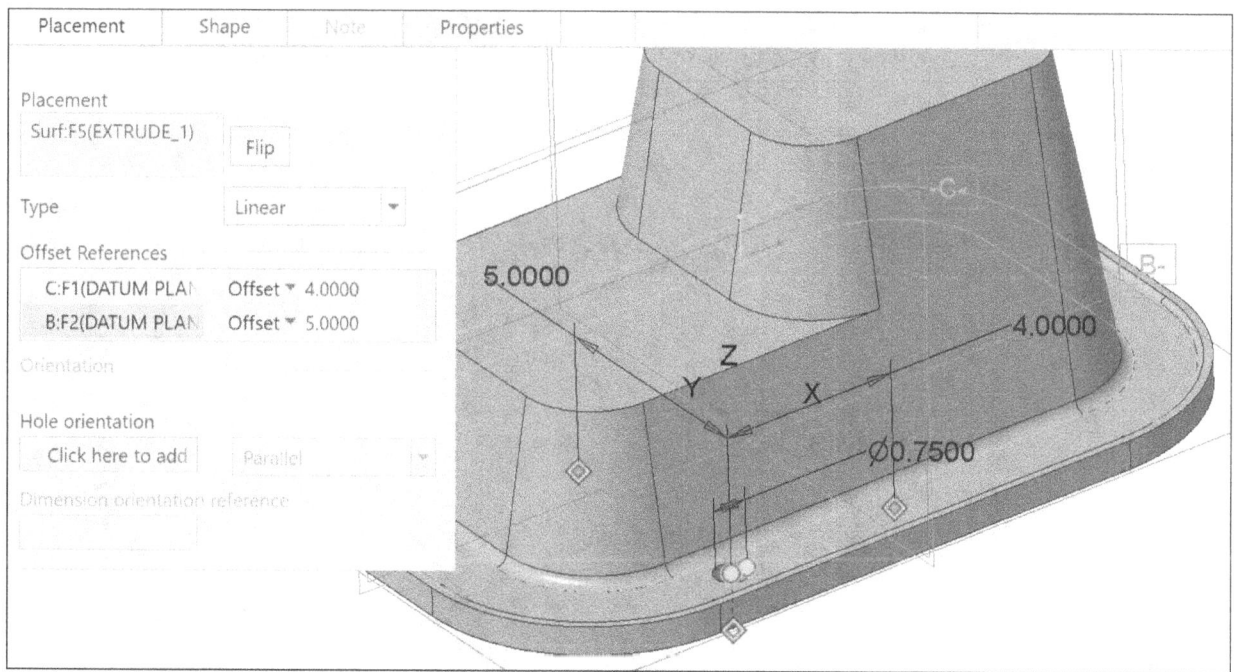

Figure 17.15(f) Offset References Dimensions (design dimensions)

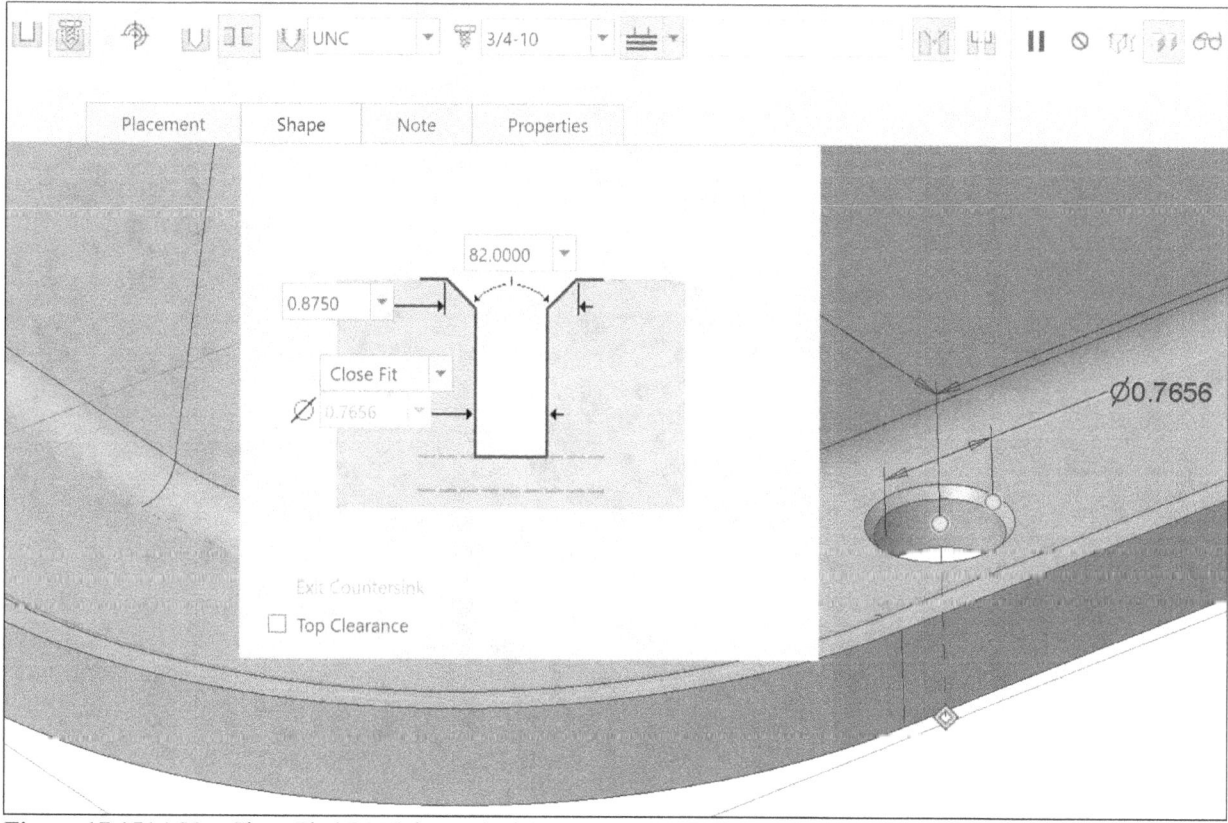

Figure 17.15(g) Use Close Fit (**.7656**) instead of the drawings .750.

Click: 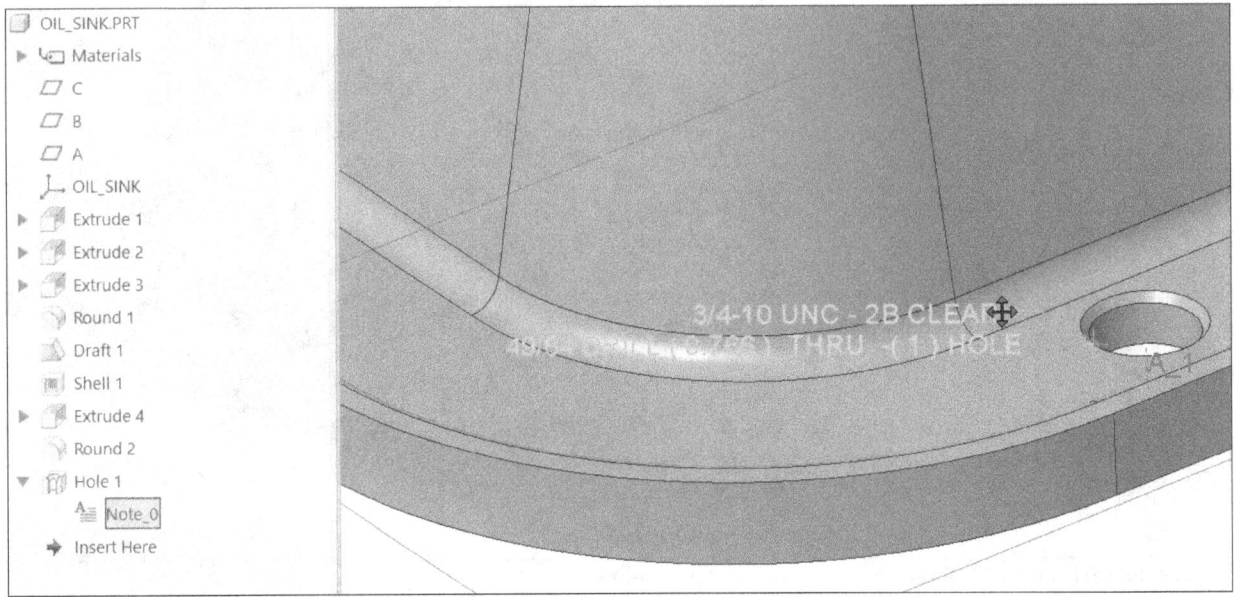 > **Ctrl+S** > check the Navigator settings, click: **Settings** > **Tree Filters** > toggle *on* all Display options > **OK** > next to the Hole feature in the Model Tree, pick ▶ (to expand) > pick on the Note [Fig. 17.15(h)] > select the Hole feature > **RMB** > ⊞ **Pattern** [Fig. 17.16(a)]

Figure 17.15(h) Completed Countersunk Hole *(your Hole id and Note id may be different)*

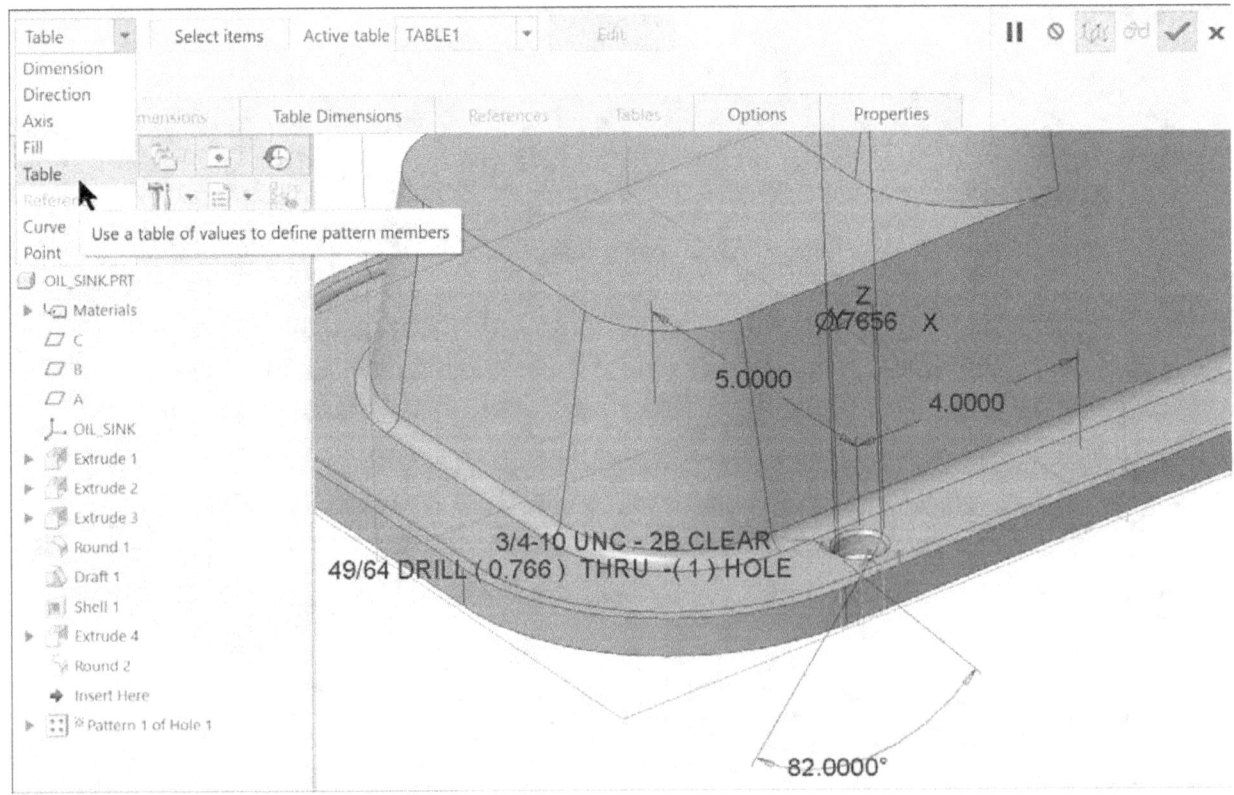

Figure 17.16(a) Pattern Members Defined by Table

Click: **Table Dimensions** tab > with the **Ctrl** key pressed, select the **4.00** dimension and then the **5.00** dimension [Fig. 17.16(b)] > **Edit** [Fig. 17.16(c)] > add the information [Figs. 17.16(d-f)]

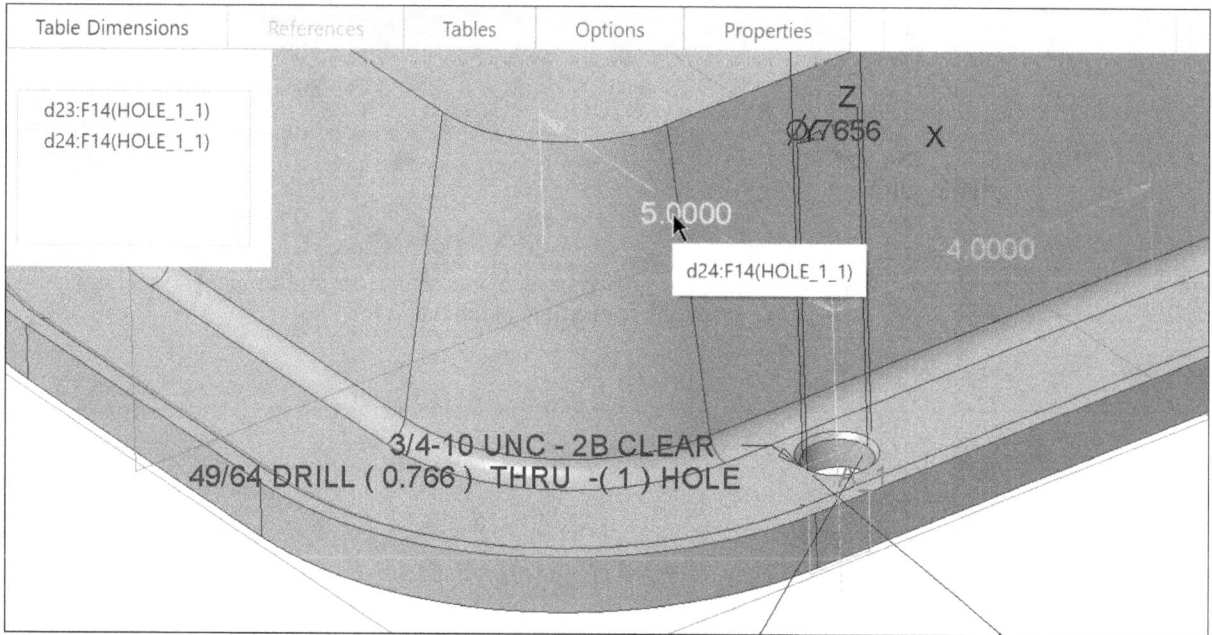

Figure 17.16(b) Table Dimensions Tab with the **5.00** and the **5.00** Dimensions Added to the Table *(your display may be different)*

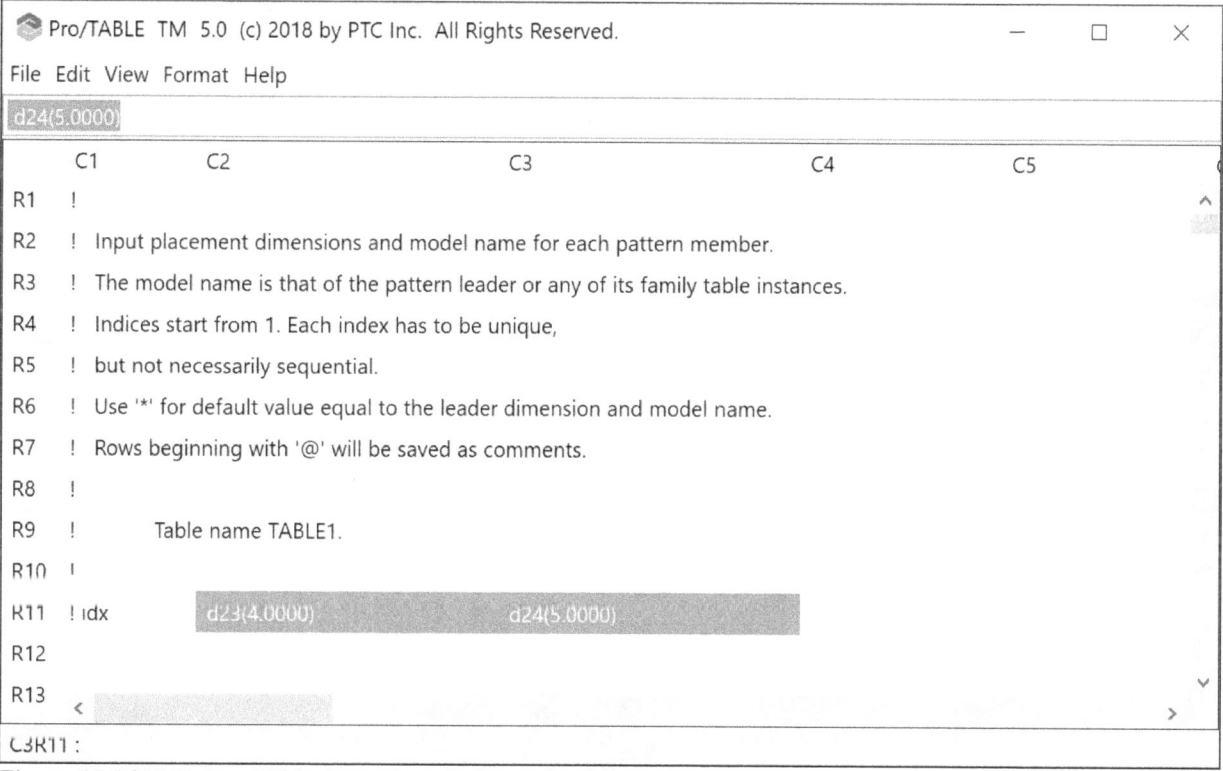

Figure 17.16(c) Pattern Table *(your d symbols may be different)*

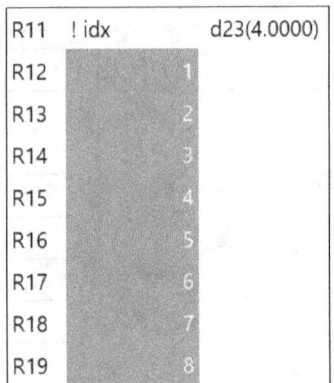

R11	! idx	d23(4.0000)
R12		1
R13		2
R14		3
R15		4
R16		5
R17		6
R18		7
R19		8

Figure 17.16(d) Add numbers 1-7

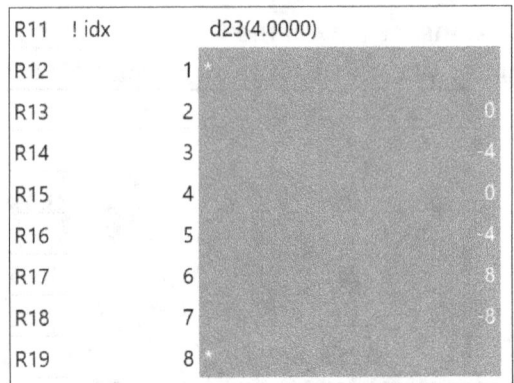

R11	! idx	d23(4.0000)	
R12		1	*
R13		2	0
R14		3	-4
R15		4	0
R16		5	-4
R17		6	8
R18		7	-8
R19		8	*

Figure 17.16(e) Add Values in the Second Column
(* means identical to parent value)

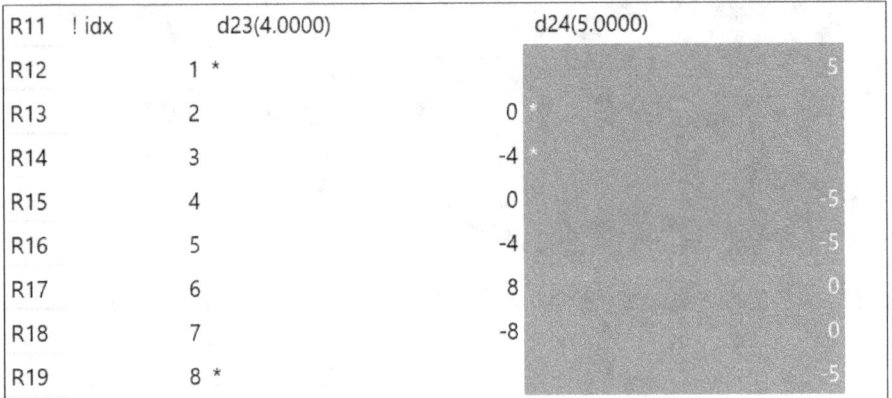

R11	! idx	d23(4.0000)		d24(5.0000)	
R12	1 *			5
R13	2	0	*	
R14	3	-4	*	
R15	4	0		-5
R16	5	-4		-5
R17	6	8		0
R18	7	-8		0
R19	8 *			-5

Figure 17.16(f) Add Values in the Third Column

From the Pro/TABLE window, click: **File > Exit** [Fig. 17.16(g)]

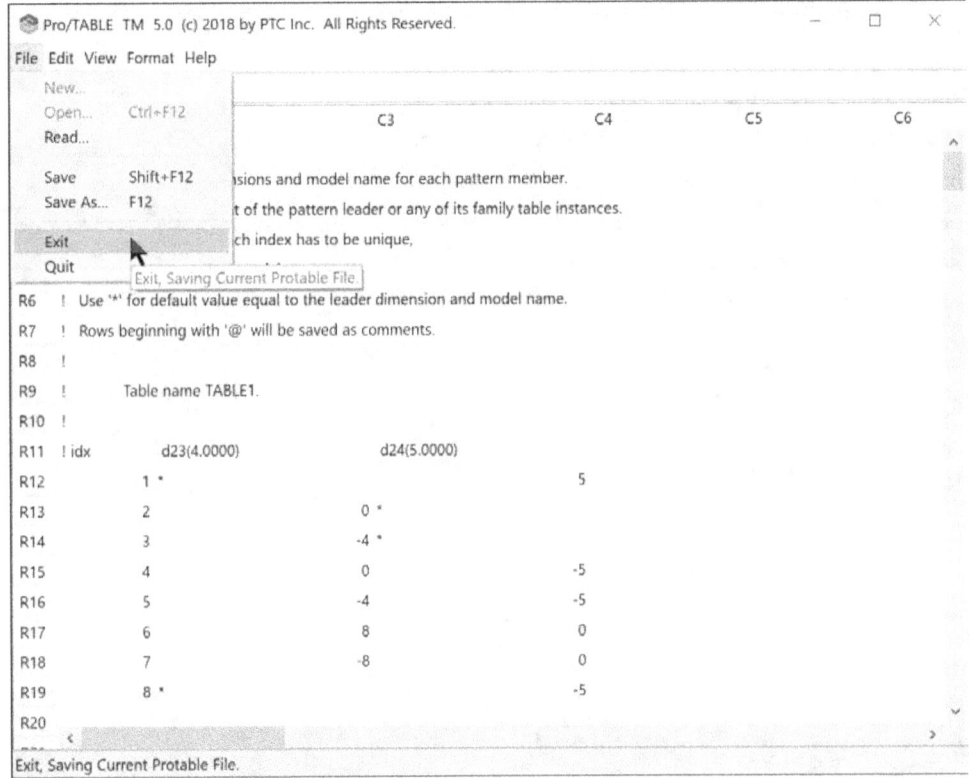

Figure 17.16(g) Completed Table

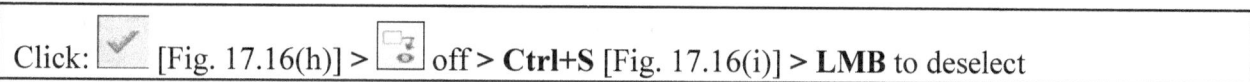

Click: ✓ [Fig. 17.16(h)] > 🔲 off > **Ctrl+S** [Fig. 17.16(i)] > **LMB** to deselect

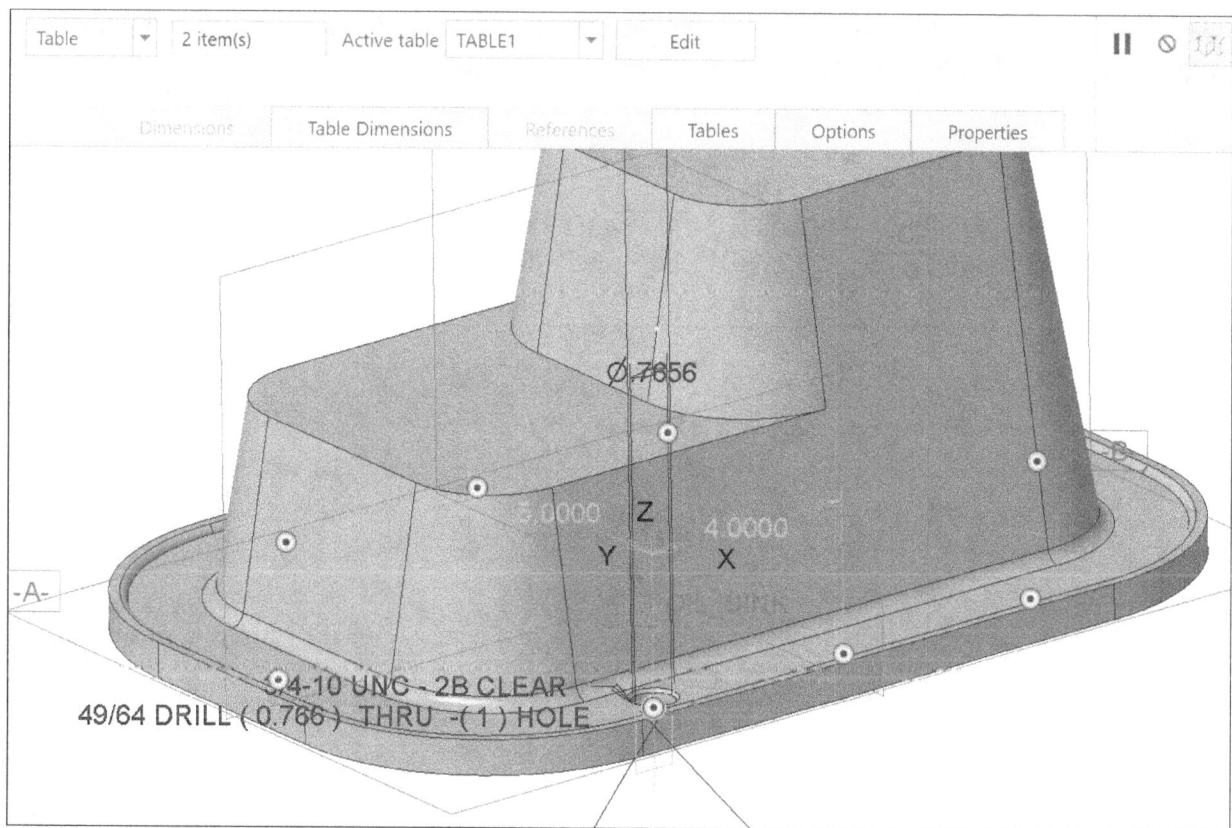

Figure 17.16(h) Previewed Pattern

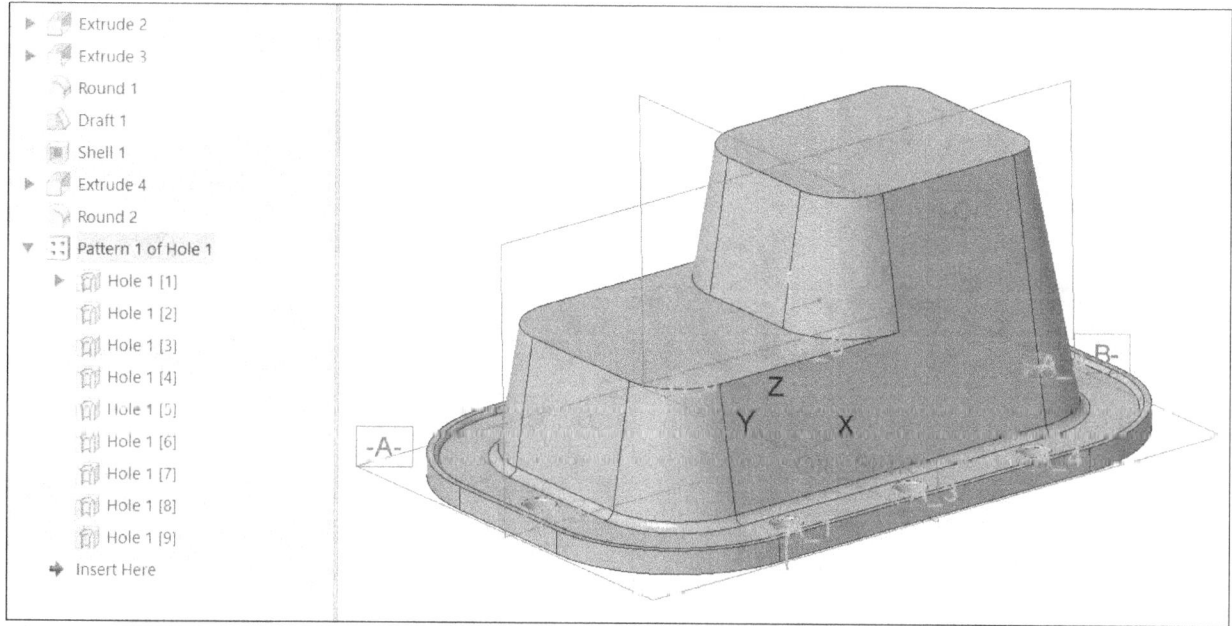

Figure 17.16(i) Completed Pattern

The next series of *features will be created purposely at the wrong stage* in this project. You will now create the **R.50** round. Because the design intent is to have a constant thickness for the part, the round should have been created before the shell. The reorder capability will be used to change the position of this round in the design sequence. Using the Model Tree, you can pick and drag the round to a new location in the feature list.

Select the top edge of the part > [icon] **Round** [Fig. 17.17(a)] > move a drag handle to **.50** or double-click on the dimension and *type* **.50** [Fig. 17.17(b)] > **Enter** > [icon] [Fig. 17.17(c)] > spin the part > select the inner surface [Fig. 17.17(d)] (the rounds did not propagate to the internal edges) > **LMB** to deselect > **Ctrl+S**

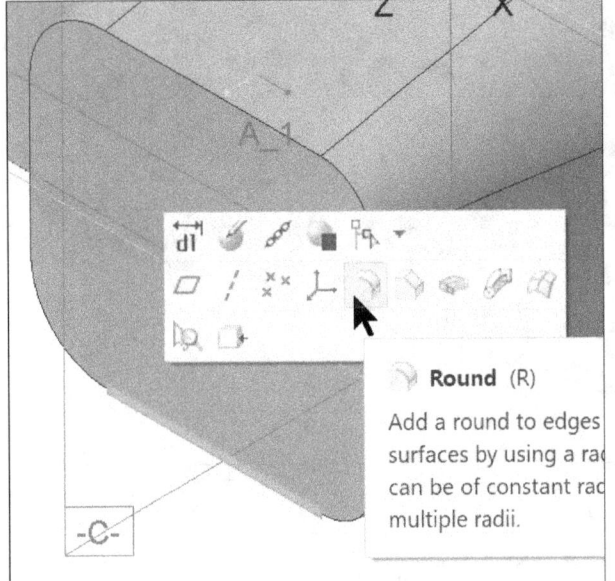

Figure 17.17(a) Add a **R.50** Round

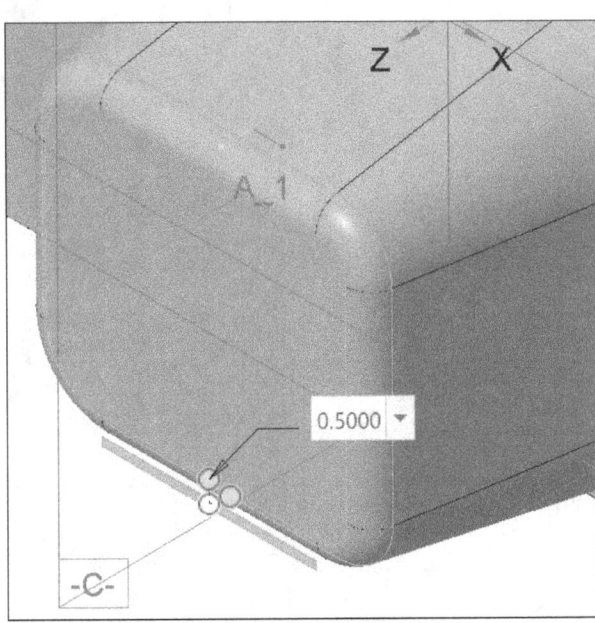

Figure 17.17(b) Move a Drag Handle to .50 or *type* .50

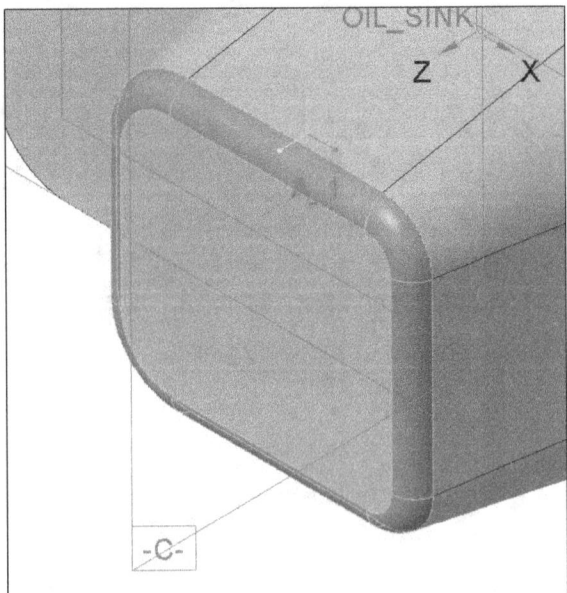

Figure 17.17(c) Completed Round

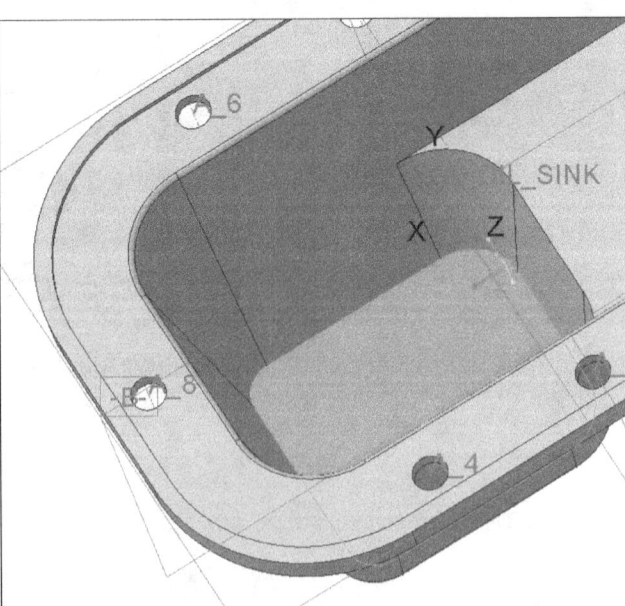

Figure 17.17(d) Interior is not Rounded

Reorder the round to appear before the Shell. Click: on the last **Round** in the Model Tree [Fig. 17.18(a)] > press **LMB** and drag **Round 3** to a position before/above the **Shell 1** feature [Fig. 17.18(b)] > release the **LMB** to drop the Round feature [Fig. 17.18(c)] > **Ctrl+S**

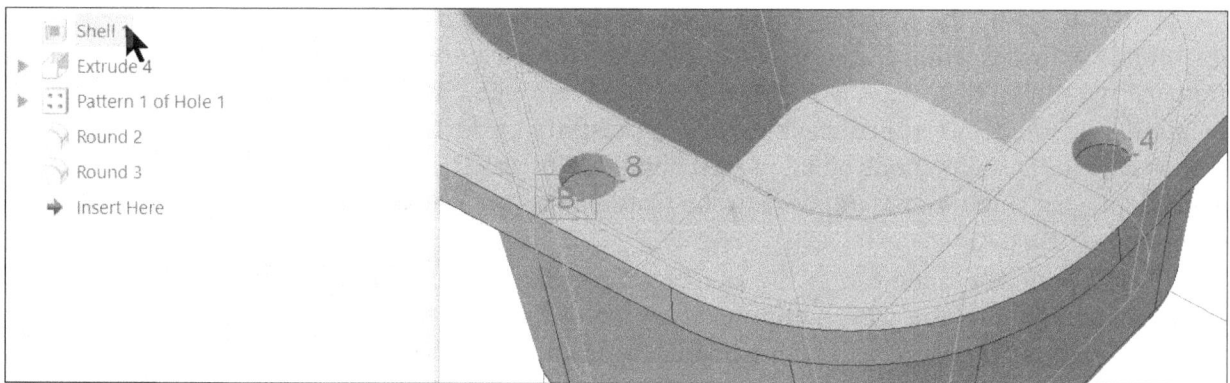

Figure 17.18(a) Click on the Shell in the Model Tree *(your Model Tree may look different)*

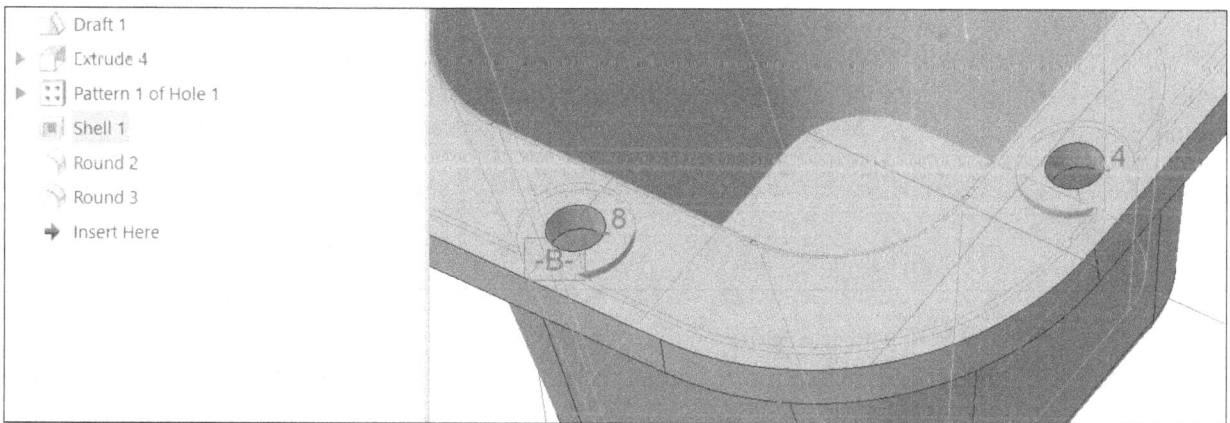

Figure 17.18(b) Move Cursor above the Shell Feature and then the Pattern and Round 3

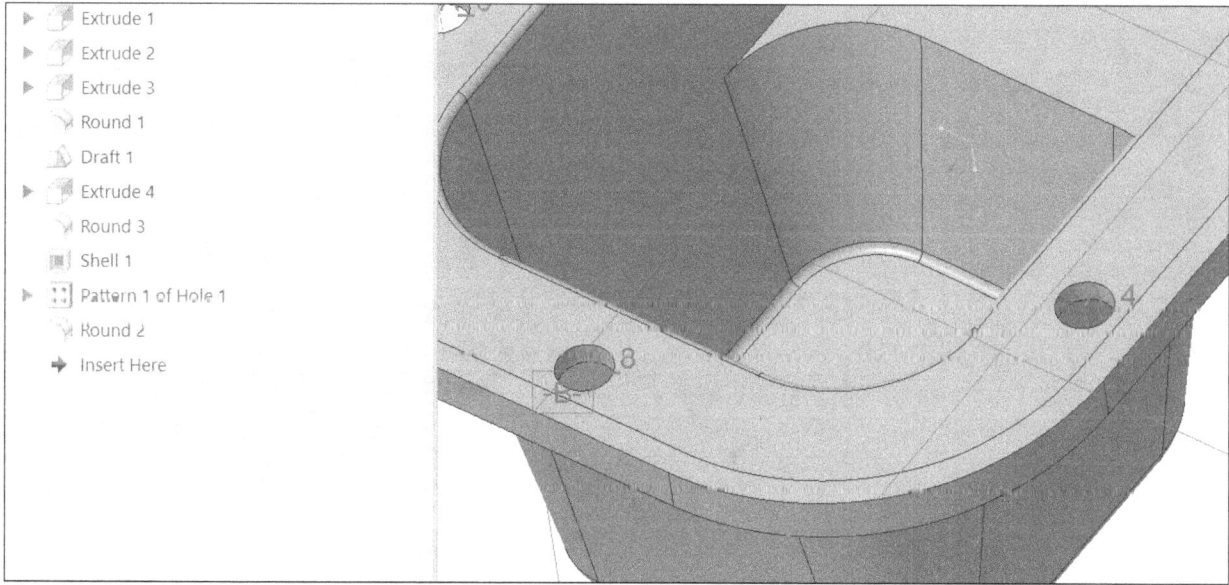

Figure 17.18(c) Reordered Round Shows on the Inside of the Part (**.50** minus the shell thickness)

You can also insert new features using the Model Tree. The arrow-shaped icon 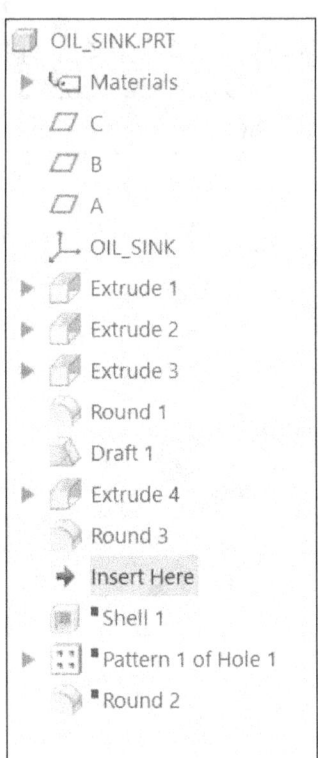 in the Model Tree indicates where features will be inserted upon creation and is by default at the end (or bottom) of the Model Tree.

By dragging the location of the insert node higher, so that its position is before existing features, you can insert a new feature at that stage of the model history. When the *insert node* is dropped at a new location, the model is rolled backward (suppressed) or forward in response to the insertion node being moved higher or lower. The Model Tree displays a small square (■) next to the features that are not active (suppressed).

The previous round was created at the wrong stage in the design sequence and then reordered. To eliminate the reordering of a feature, the remaining **R.50** rounds will be created using Insert Mode with the Model Tree.

Insert Mode allows you to insert a feature at a previous stage of the design sequence. This is like going back into the past and doing something you wish you had done before--not possible with life, but with Creo Parametric less of a problem. Add the additional **R.50** rounds.

(Your Model Tree may look different.)

Spin the model > **Ctrl+R** > in the Model Tree, place the pointer on [Insert Here] [Fig. 17.19(a)] > press and hold down the **LMB** > move (drag) the pointer to a position before/above the **Shell 1** feature > release the **LMB** to drop [Insert Here] [Fig. 17.19(b)]

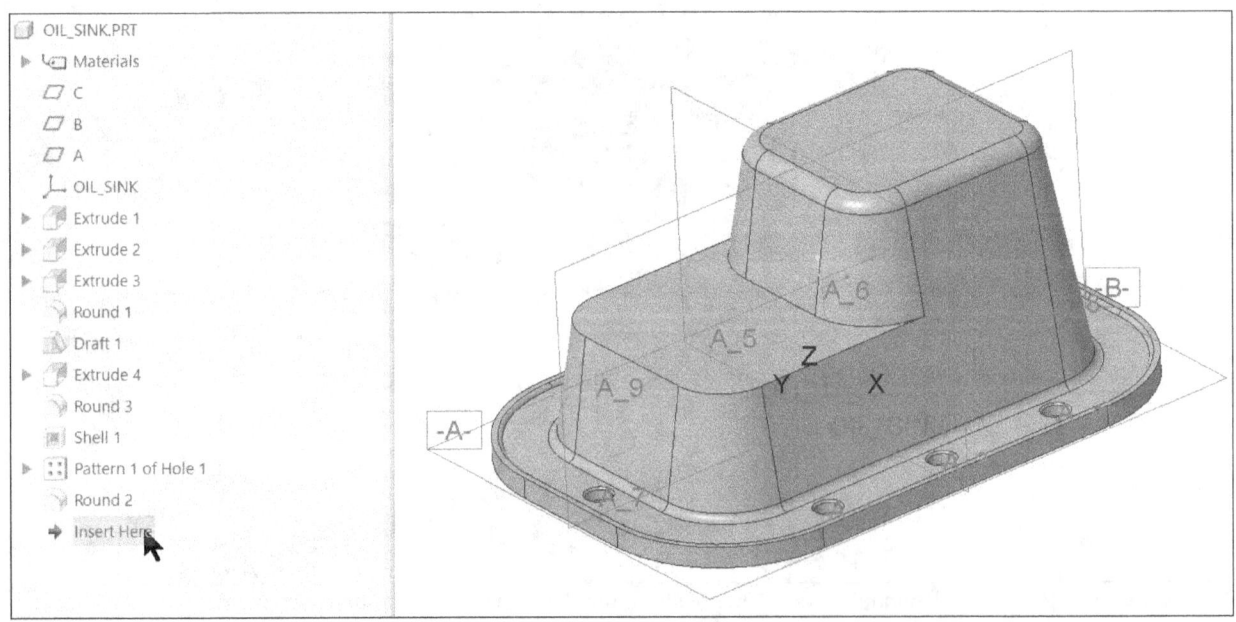

Figure 17.19(a) Place Pointer on [Insert Here] *(your Model Tree may Look Different)*

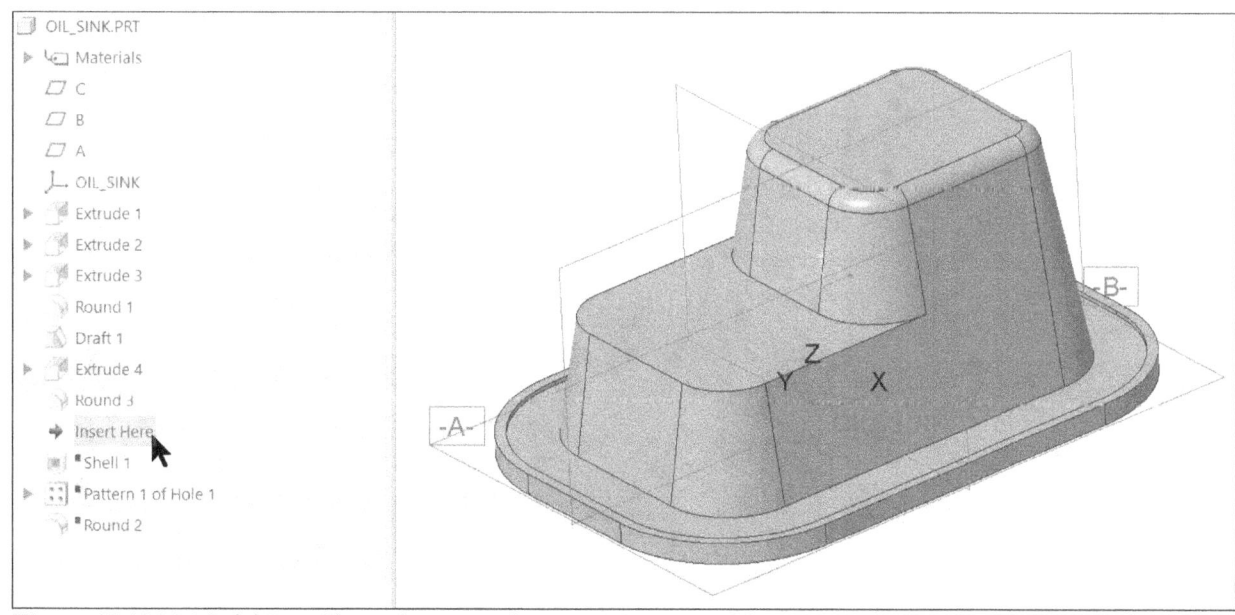

Figure 17.19(b) Model Tree Shows Suppressed Features (*your Model Tree may Look Different*)

Create two separate round features. Click: [Round] > select the upper edge > Radius **.50** > **Enter** [Fig. 17.20(a)] > MMB > **Ctrl+C > Ctrl+V** > select the front edge [Fig. 17.20(b)] > **MMB** > press the **Ctrl** key > select the previously created round features from the Model Tree > release the **Ctrl** key [Fig. 17.20(c)]

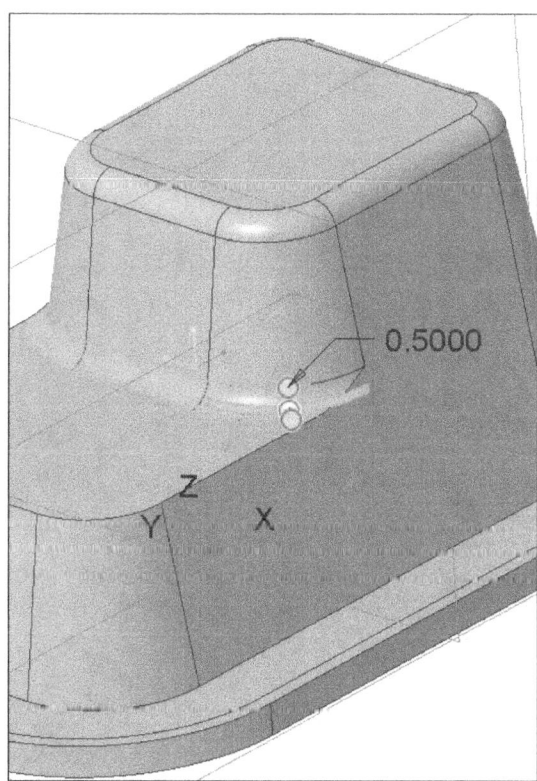

Figure 17.20(a) First Round

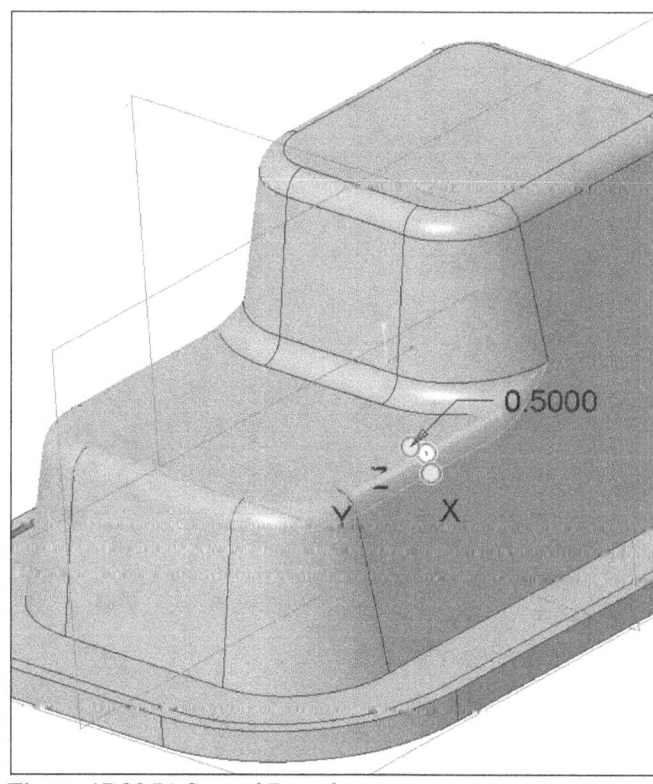

Figure 17.20(b) Second Round

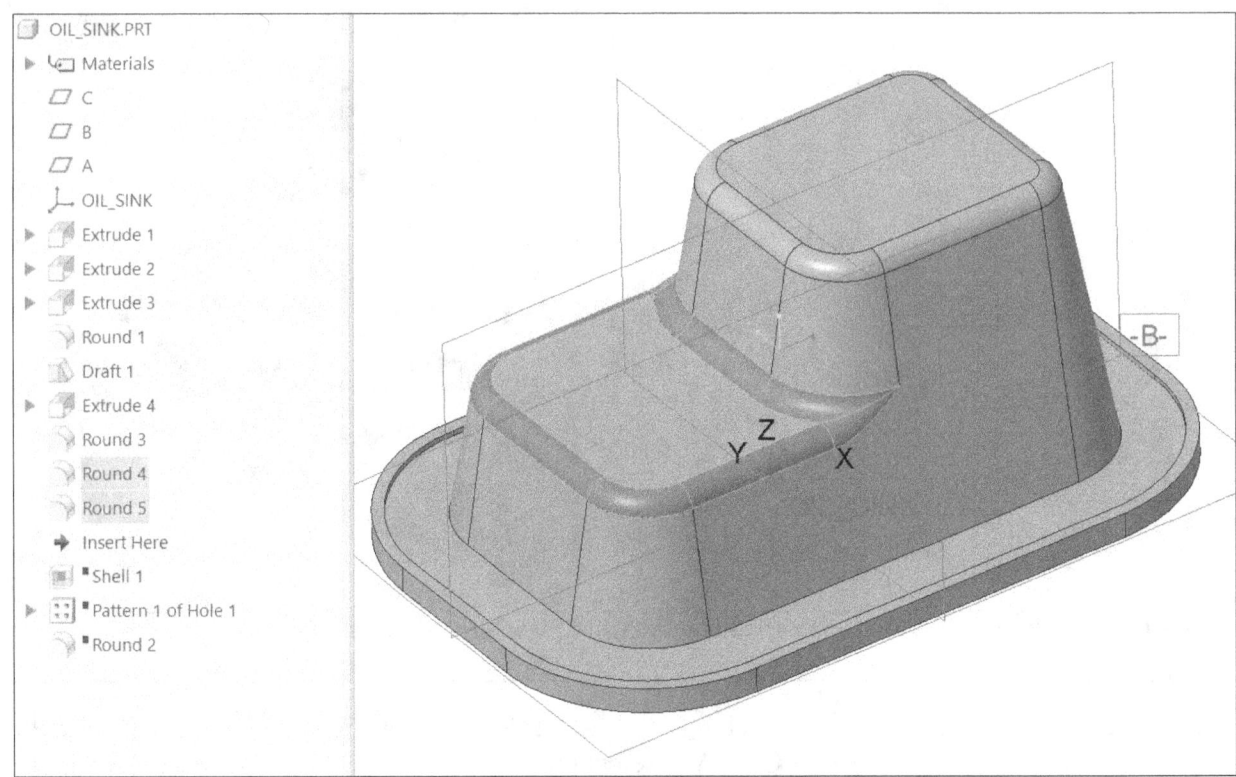

Figure 17.20(c) New Rounds Added

Rotate the model > in the Model Tree, place the pointer on ⮕ Insert Here > hold down the **LMB** [Fig. 17.21(a)]

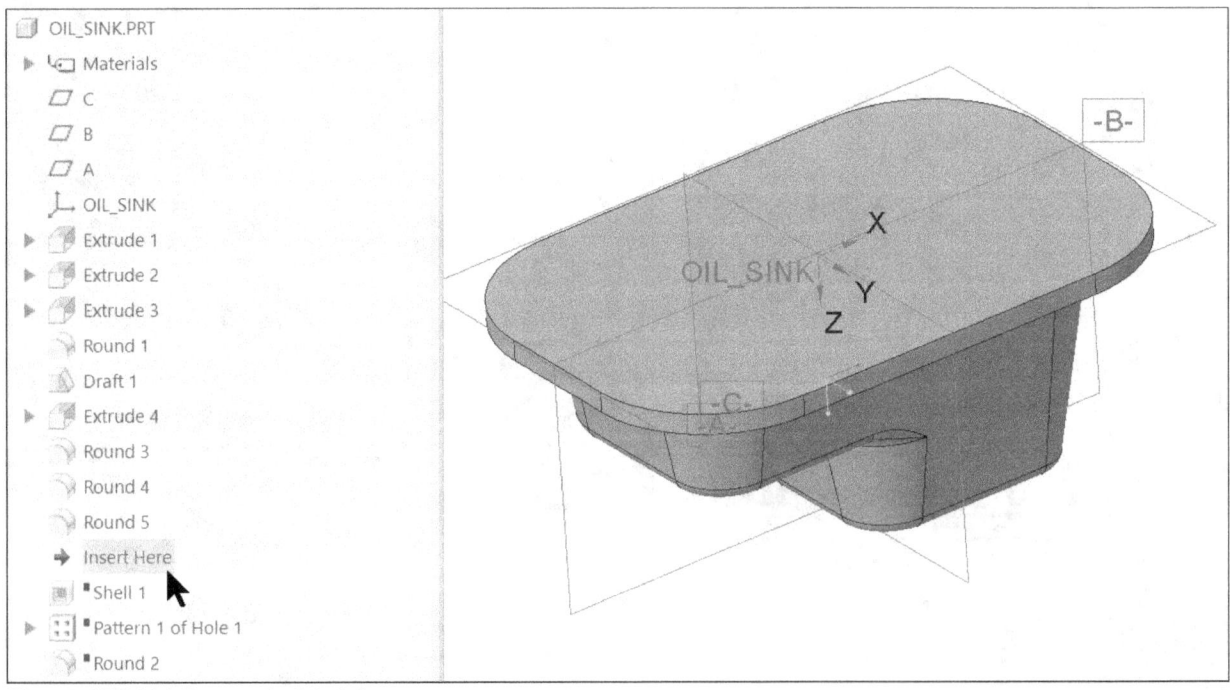

Figure 17.21(a) Rotate the Model

Drag the pointer to the bottom of the Model Tree list > release the **LMB** to drop ⇥ Insert Here [Fig. 17.21(b)] > select the propagated internal round surfaces in the model [Fig. 17.21(c)] > set the model display as **Isometric > Ctrl+D > Ctrl+S > File > Manage File > Delete Old Versions > Enter**

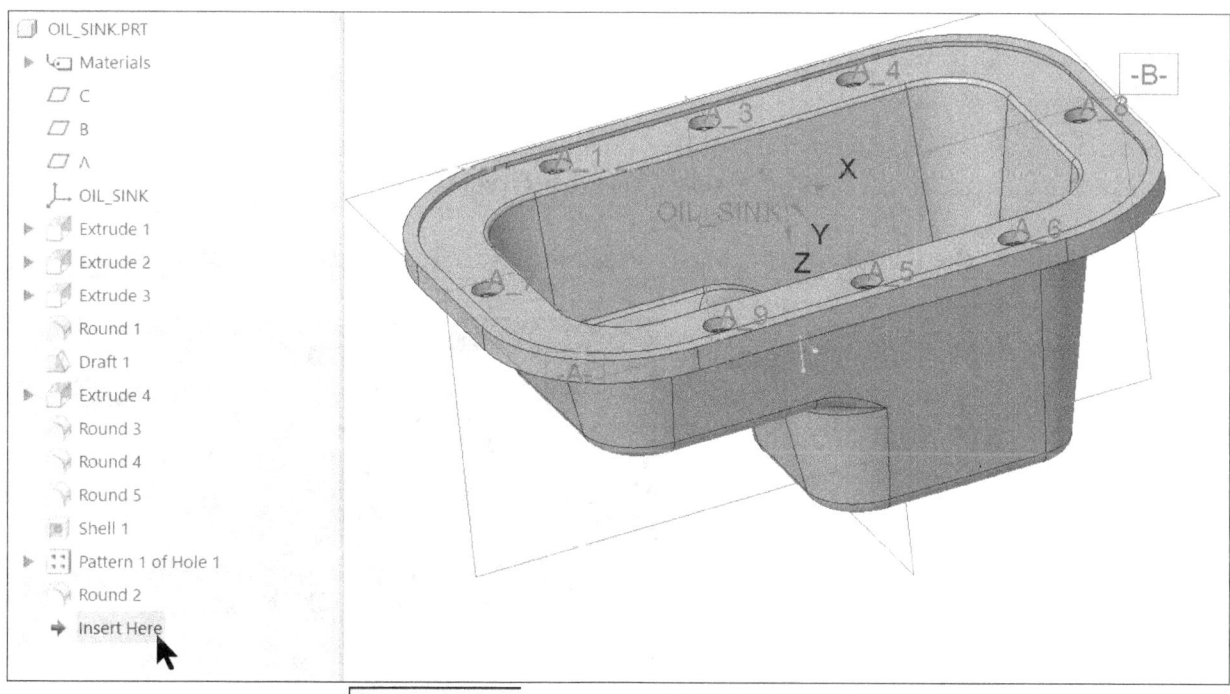

Figure 17.21(b) Drag and Drop ⇥ Insert Here , All Features are Resumed *(your Model Tree will look different)*

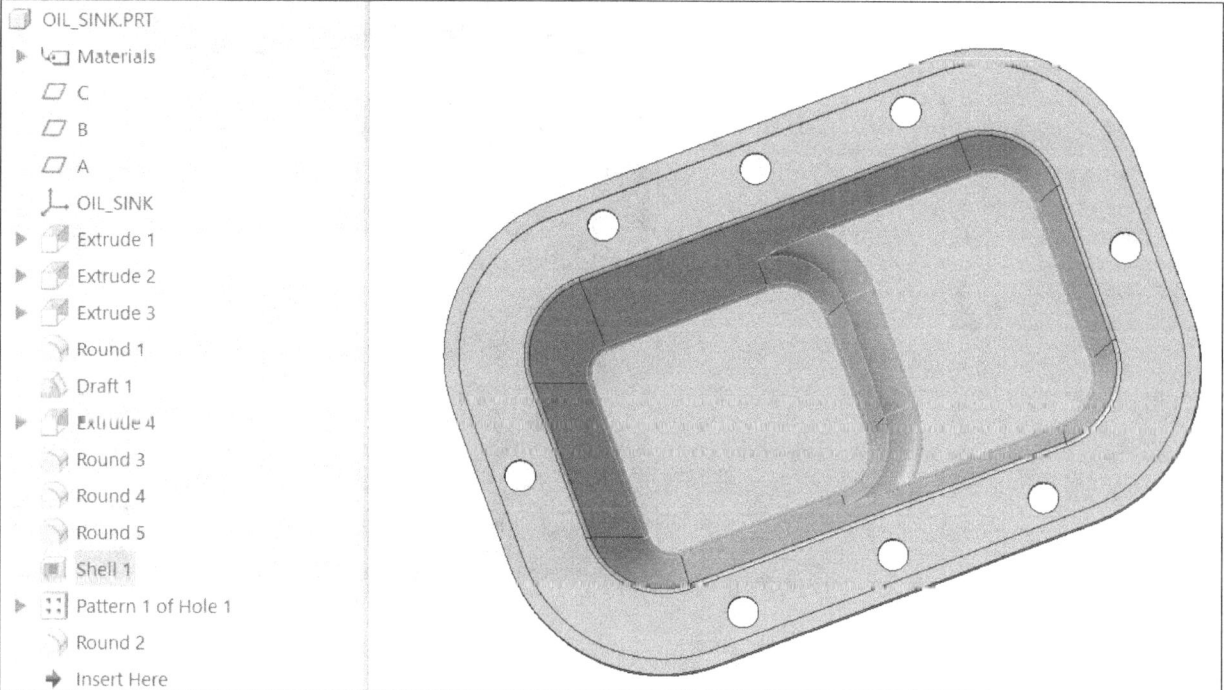

Figure 17.21(c) Propagated Internal Rounds

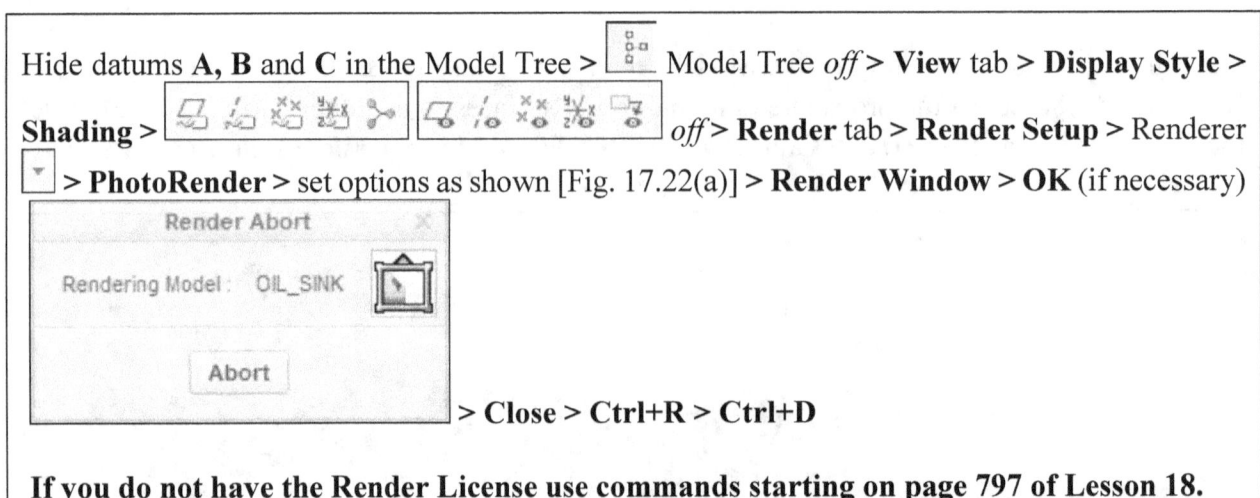

Hide datums **A, B** and **C** in the Model Tree > [icon] Model Tree *off* > **View** tab > **Display Style** > **Shading** > [icons] [icons] *off* > **Render** tab > **Render Setup** > Renderer [icon] > **PhotoRender** > set options as shown [Fig. 17.22(a)] > **Render Window** > **OK** (if necessary)

Render Abort

Rendering Model : OIL_SINK [icon]

Abort

> **Close** > **Ctrl+R** > **Ctrl+D**

If you do not have the Render License use commands starting on page 797 of Lesson 18.

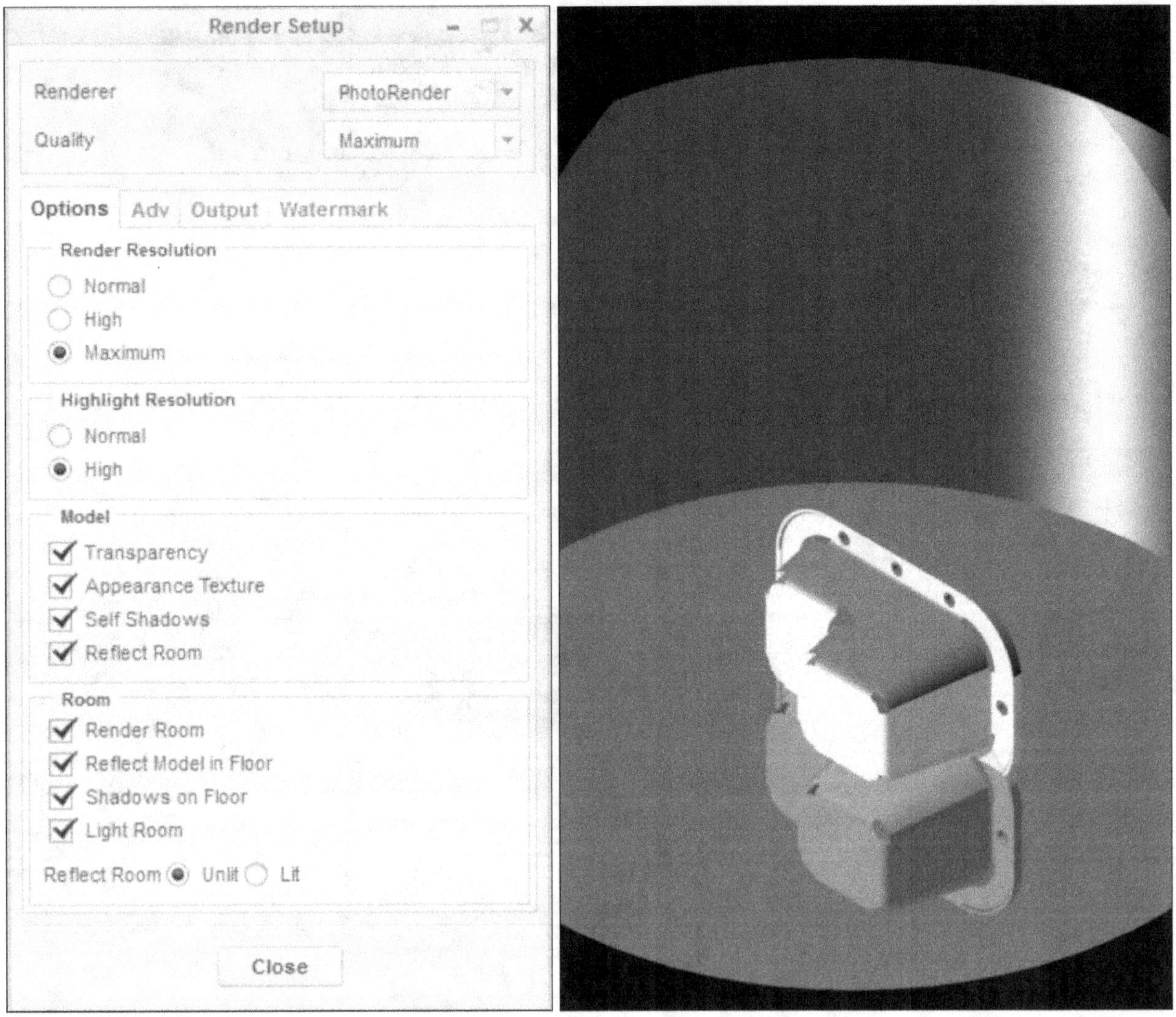

Figure 17.22(a) Render

Click: **Scene > Lights** tab > 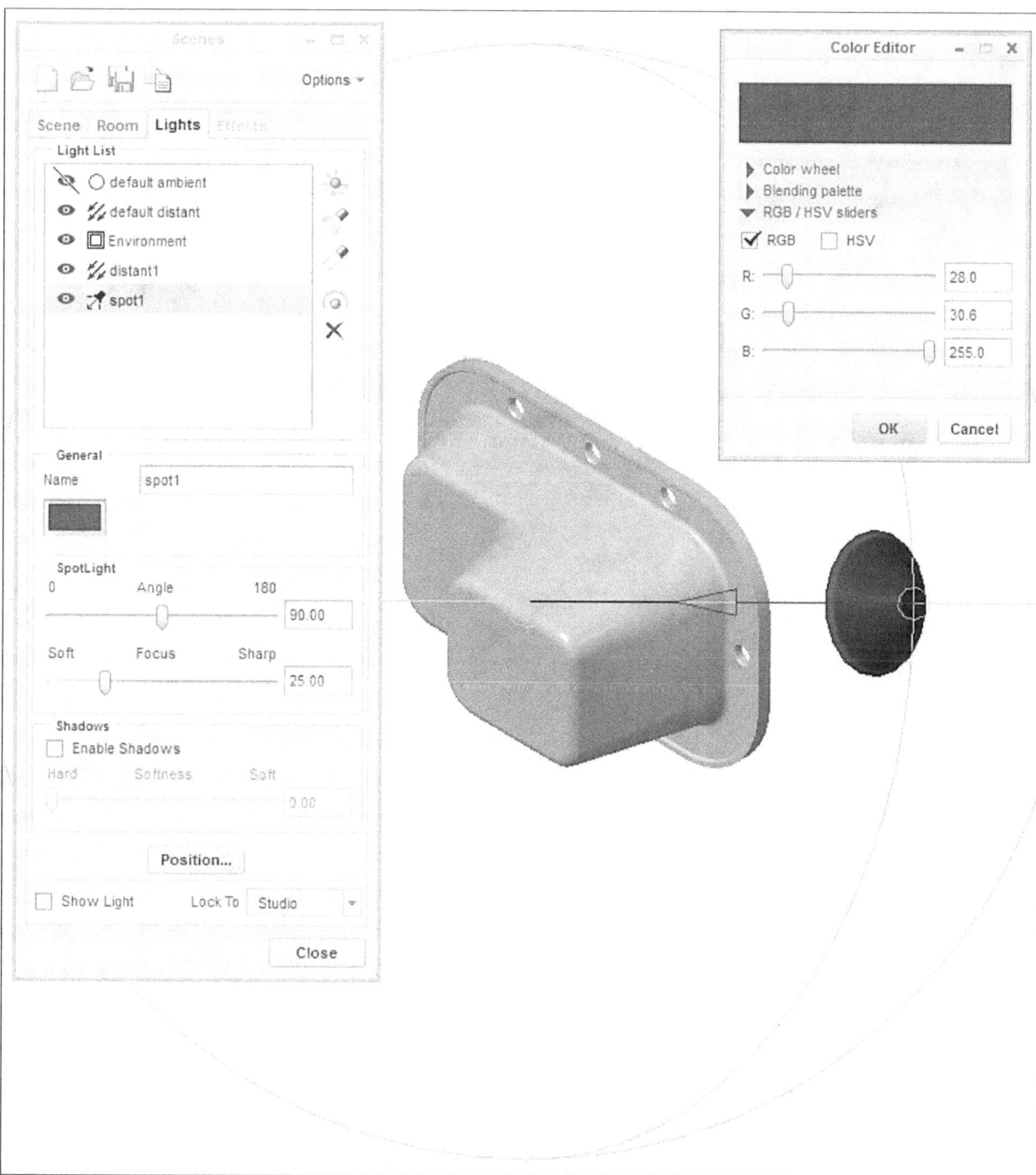 **Add new spotlight** > **Color for lighting** > adjust the slide bars in the Color Editor Dialog Box to the RGB values you desire [Fig. 17.22(b)] > **OK** (from the Color Editor Dialog Box) > *Enable Shadows* > *Show Light*

Figure 17.22(b) New Spot Light *(your display may appear differently)*

Click: [image] **Add new distance light** > [image] **Color for lighting** > adjust the slide bars in the Color Editor Dialog Box to the RGB values you desire > **OK** (from the Color Editor Dialog Box) > move the light to a new position > ✓ Enable Shadows > ✓ Show Light [Fig. 17.22(c)]

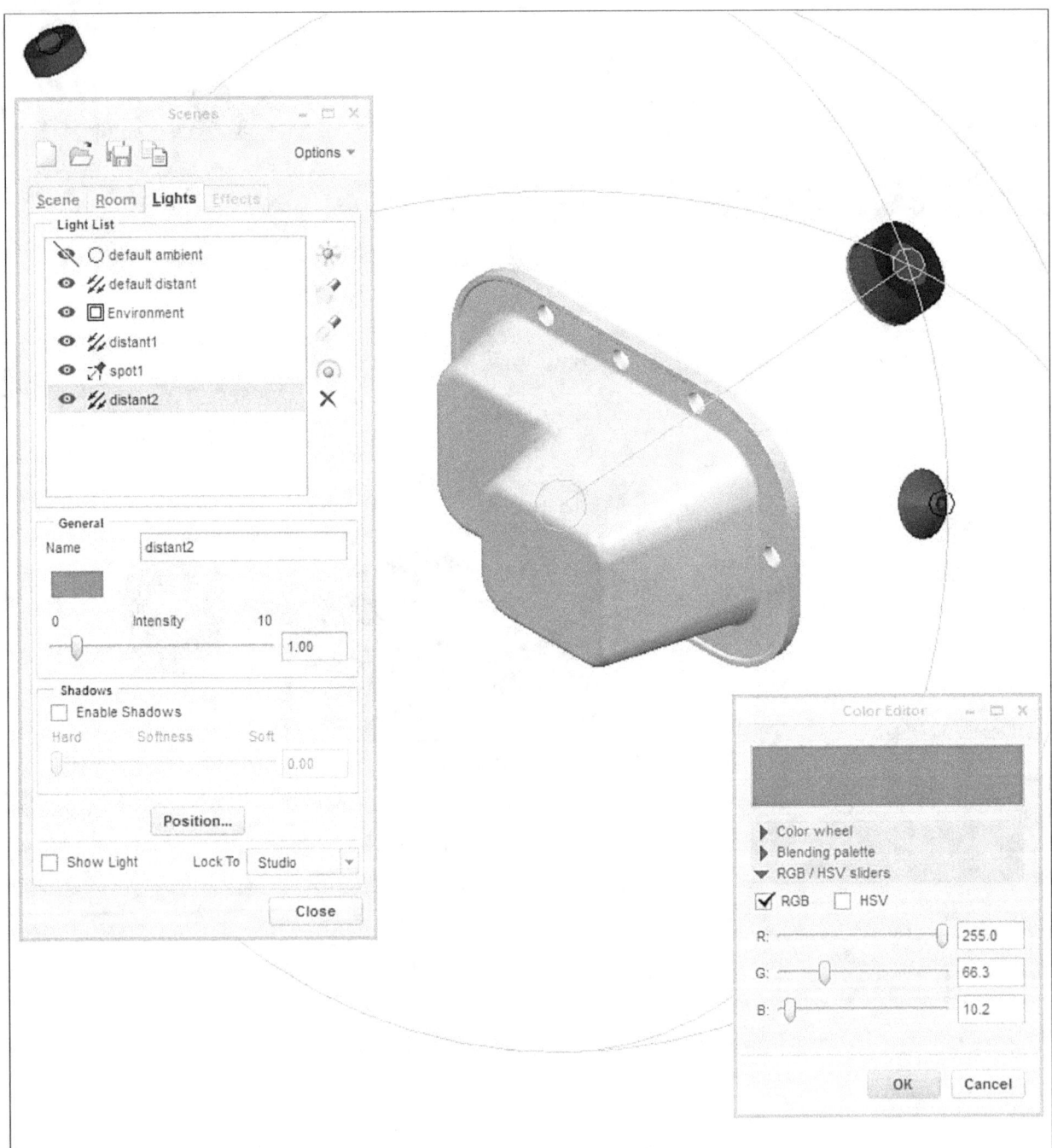

Figure 17.22(c) New Distant Light *(your display may appear differently)*

Click: ⊙ **Add new skylight** > ▨ **Color for lighting** > adjust the slide bars in the Color Editor Dialog Box to the RGB values you desire > **OK** (from the Color Editor Dialog Box) > move the skylight to a new position > ☑ Enable Shadows > ☑ Show Light [Fig. 17.22(d)] > **Close** the Scenes Dialog Box

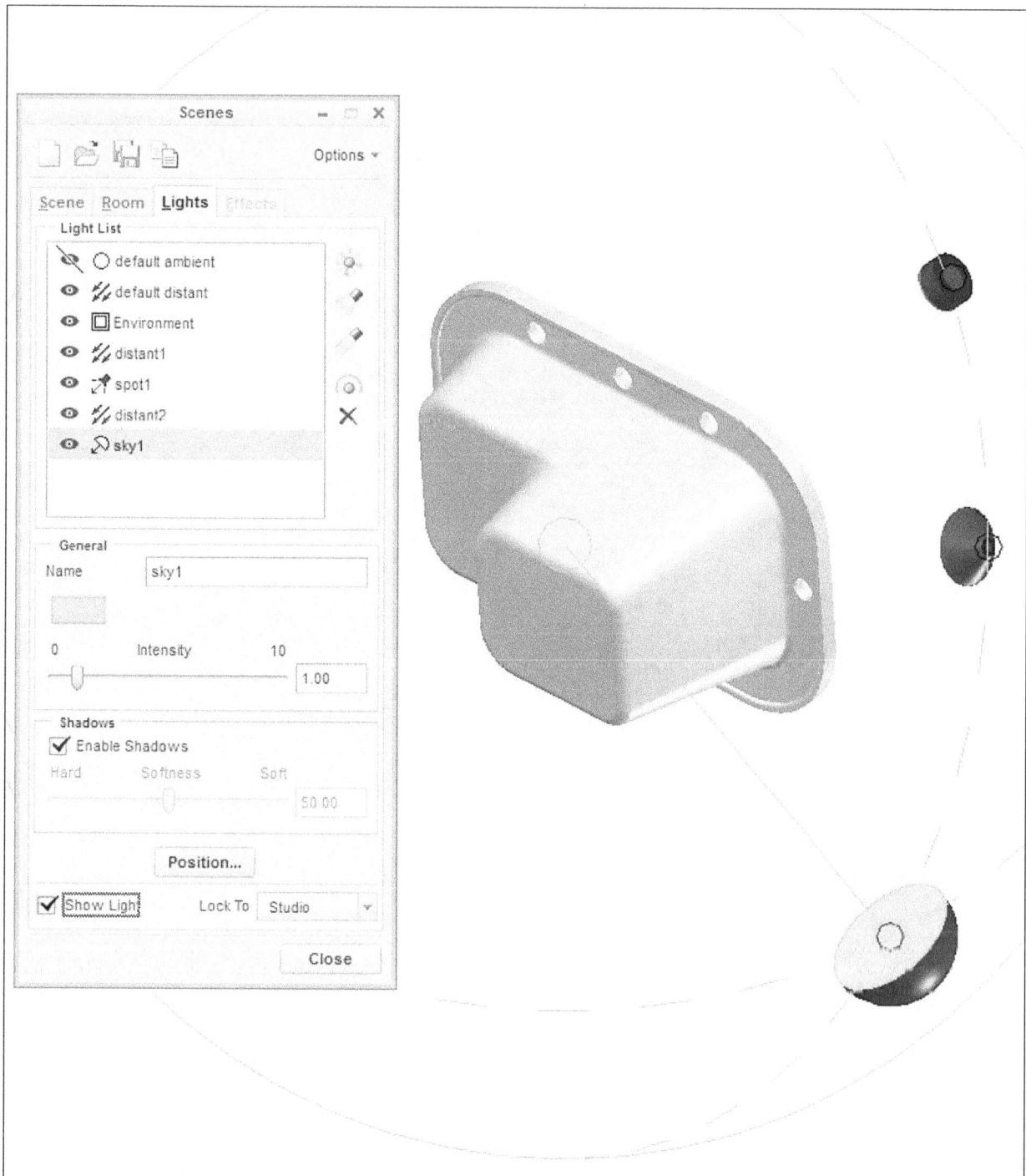

Figure 17.22(d) New Sky Light *(your display may appear differently)*

Figure 17.22(e) Shading With Reflections *(the quality of your graphics card and graphics settings may prevent this display)*

Click: **File > Save As > Save a Copy >** Type **PDF U3D (*.pdf)** [Fig. 17.23(a)] **> OK** [Fig. 17.23(b)] **> OK**

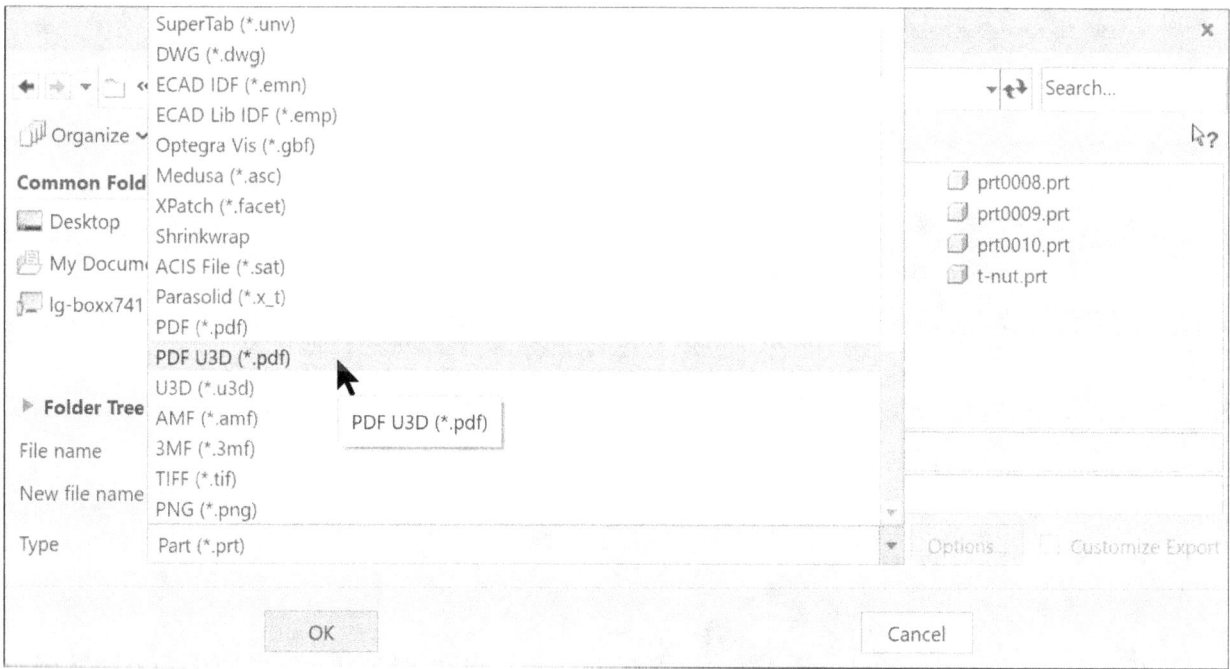

Figure 17.23(a) PDF U3D (*.pdf)

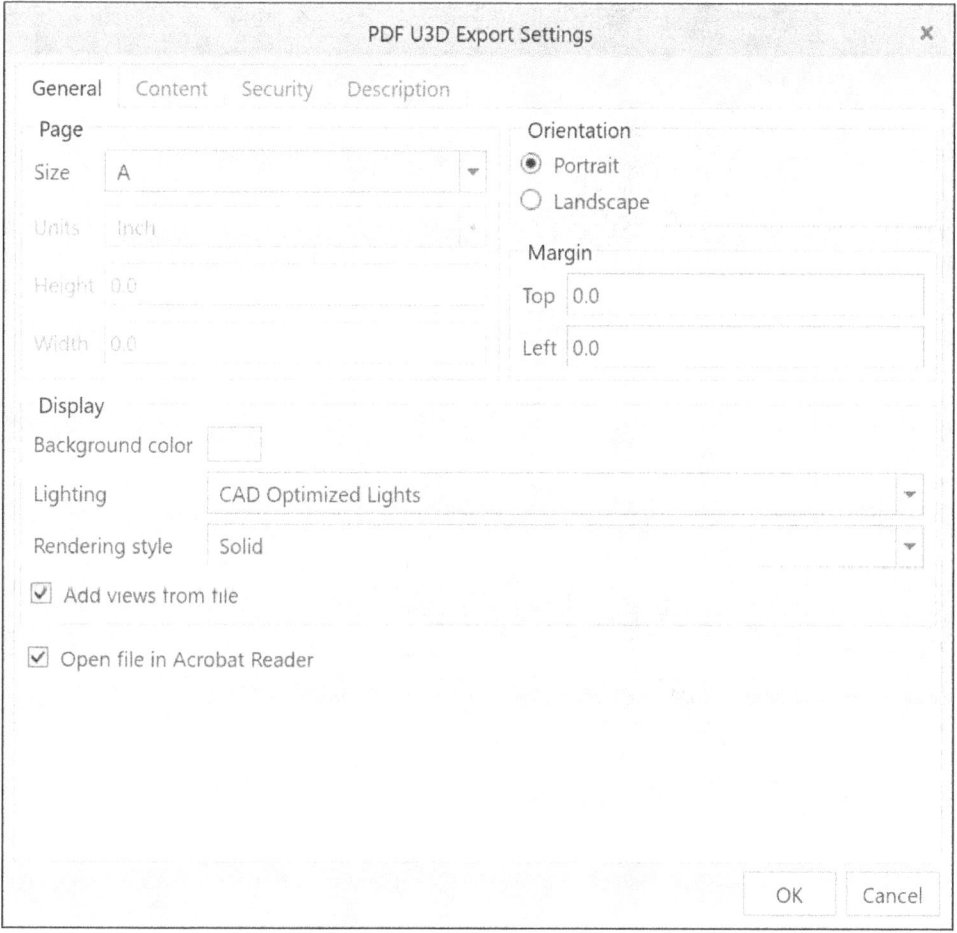

Figure 17.23(b) PDF U3D Export Settings

Click to activate [Fig. 17.23(c)] > Trust this document always > double-click on the model in the PDF

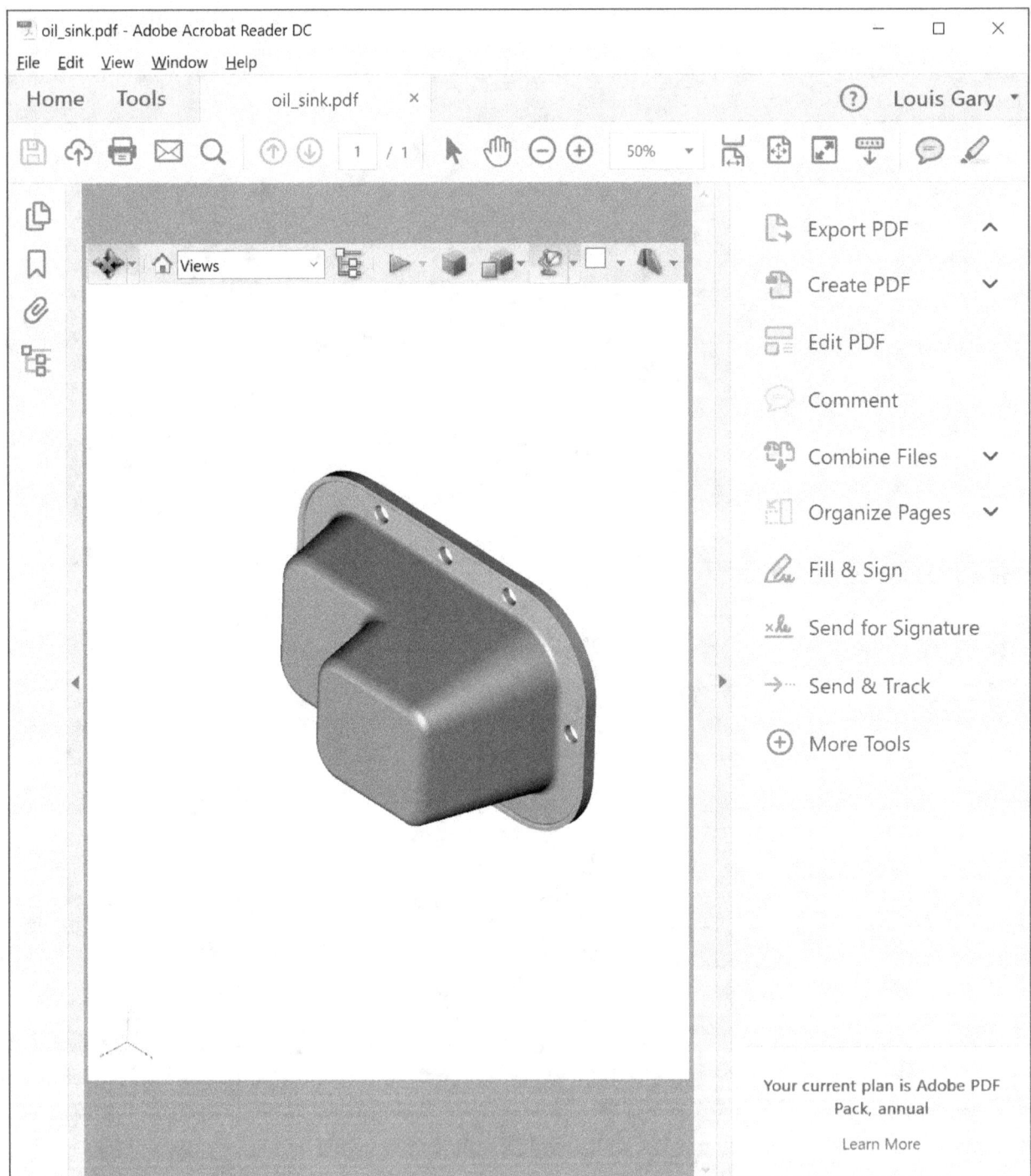

Figure 17.23(c) Part Displayed in PDF Reader *(make sure are using the latest available PDF Reader)*

Click: 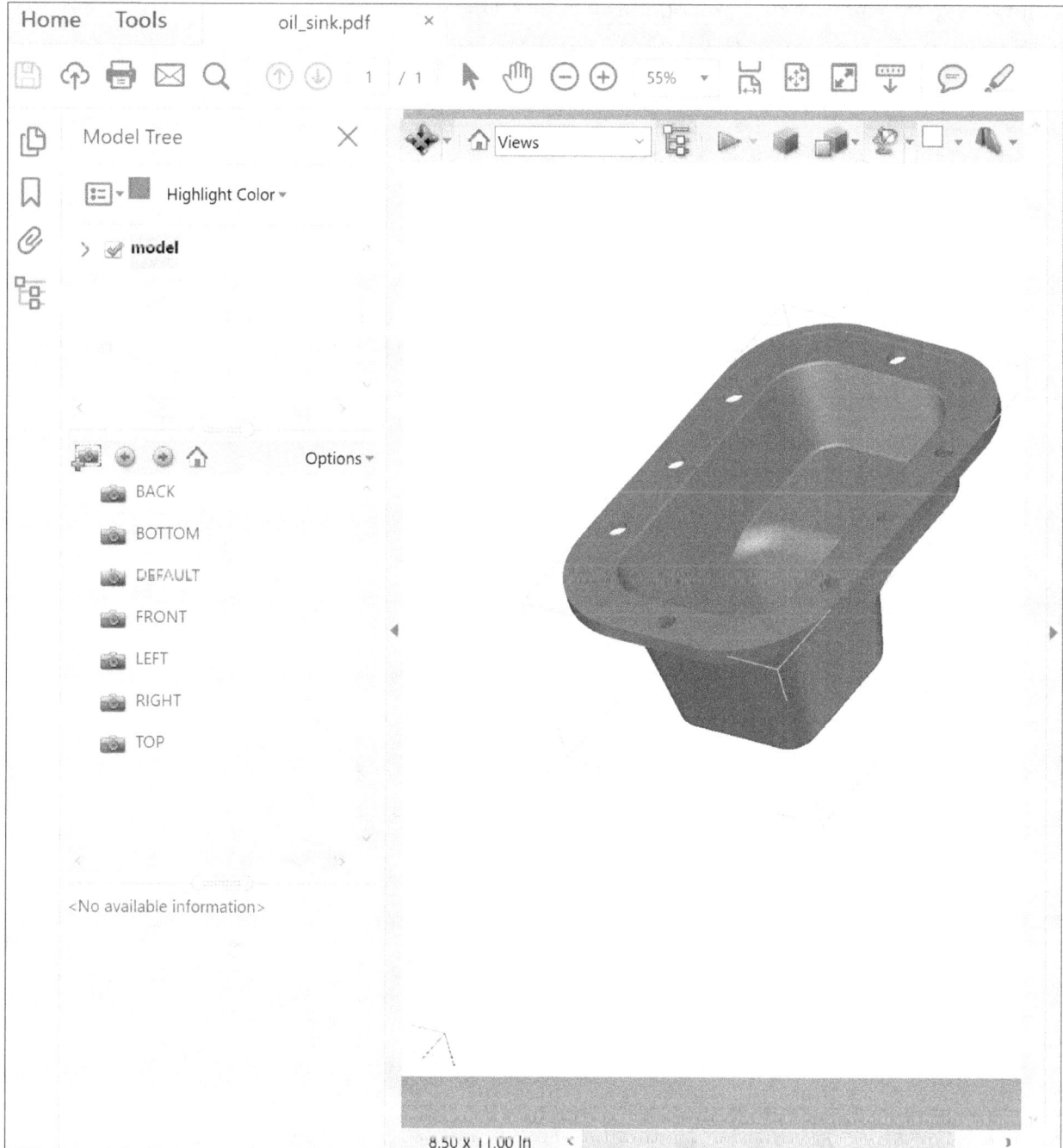 **Toggle Model Tree** [Fig. 17.23(d)]

Figure 17.23(d) Model Tree Displayed

Click: 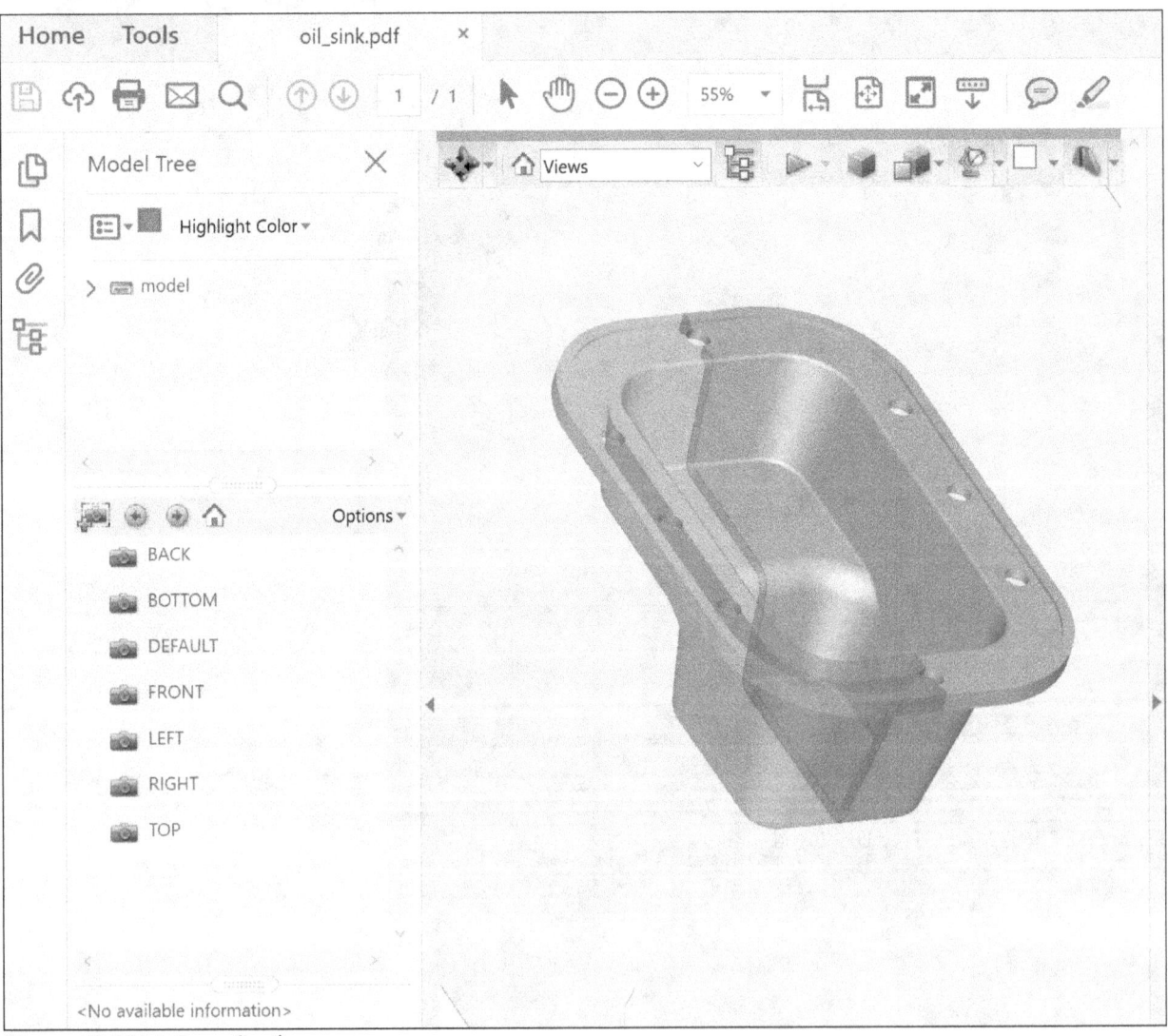 **Use Perspective Projection** > ⬛ *off* > ☐ ▾ **Background Color** (select a different color) > experiment with different options > **Cross Section Properties**

Alignment	Display Settings	
○ X-Axis	☑ Show Intersections	⬛
◉ Y-Axis	☑ Show Cutting Plane	☐
○ Z-Axis	☐ Ignore Selected Parts	
Align to Face	☑ Add Section Caps	
Align to 3 Points	☑ Show Transparent	

> **Save Section View** > ⬛ **Close** [Fig. 17.23(e)] > **File** > **Properties** > fill in the information as desired > **OK** > **File** > **Send File** > **Attach to Email** > email/upload as required

Figure 17.23(e) Cross Section

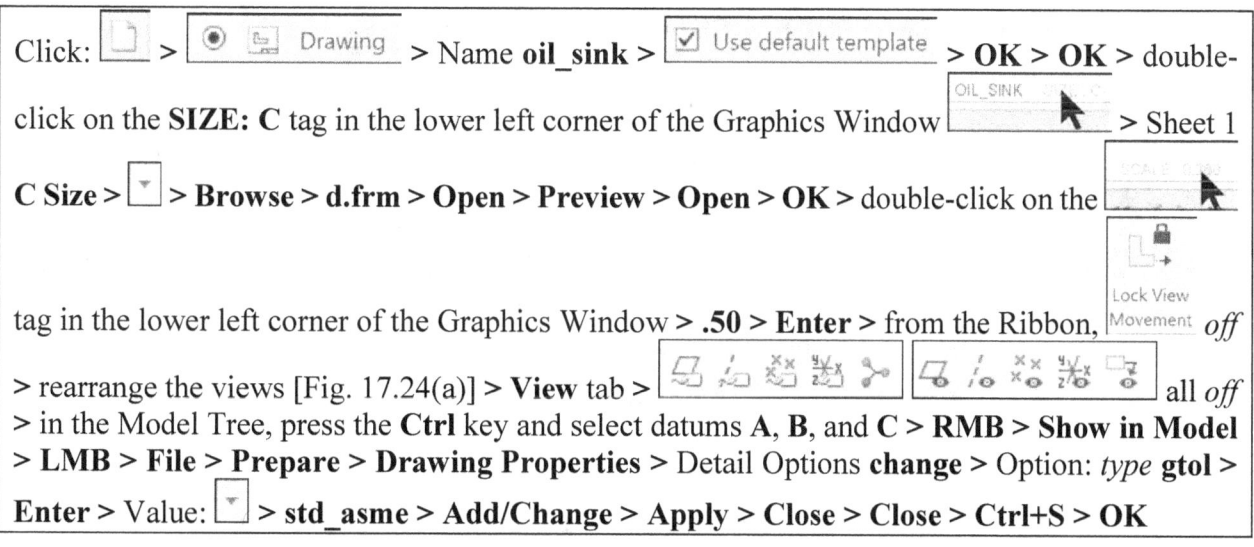

Click: ⬚ > ◉ 🖳 Drawing > Name **oil_sink** > ☑ Use default template > **OK > OK** > double-click on the **SIZE: C** tag in the lower left corner of the Graphics Window 〔OIL_SINK〕 > Sheet 1

C Size > ▾ > **Browse > d.frm > Open > Preview > Open > OK** > double-click on the 〔 〕

tag in the lower left corner of the Graphics Window > **.50 > Enter** > from the Ribbon, Lock View Movement *off*
> rearrange the views [Fig. 17.24(a)] > **View** tab > 〔 〕 〔 〕 all *off*
> in the Model Tree, press the **Ctrl** key and select datums **A**, **B**, and **C** > **RMB > Show in Model**
> **LMB > File > Prepare > Drawing Properties** > Detail Options **change** > Option: *type* **gtol** >
Enter > Value: ▾ > **std_asme > Add/Change > Apply > Close > Close > Ctrl+S > OK**

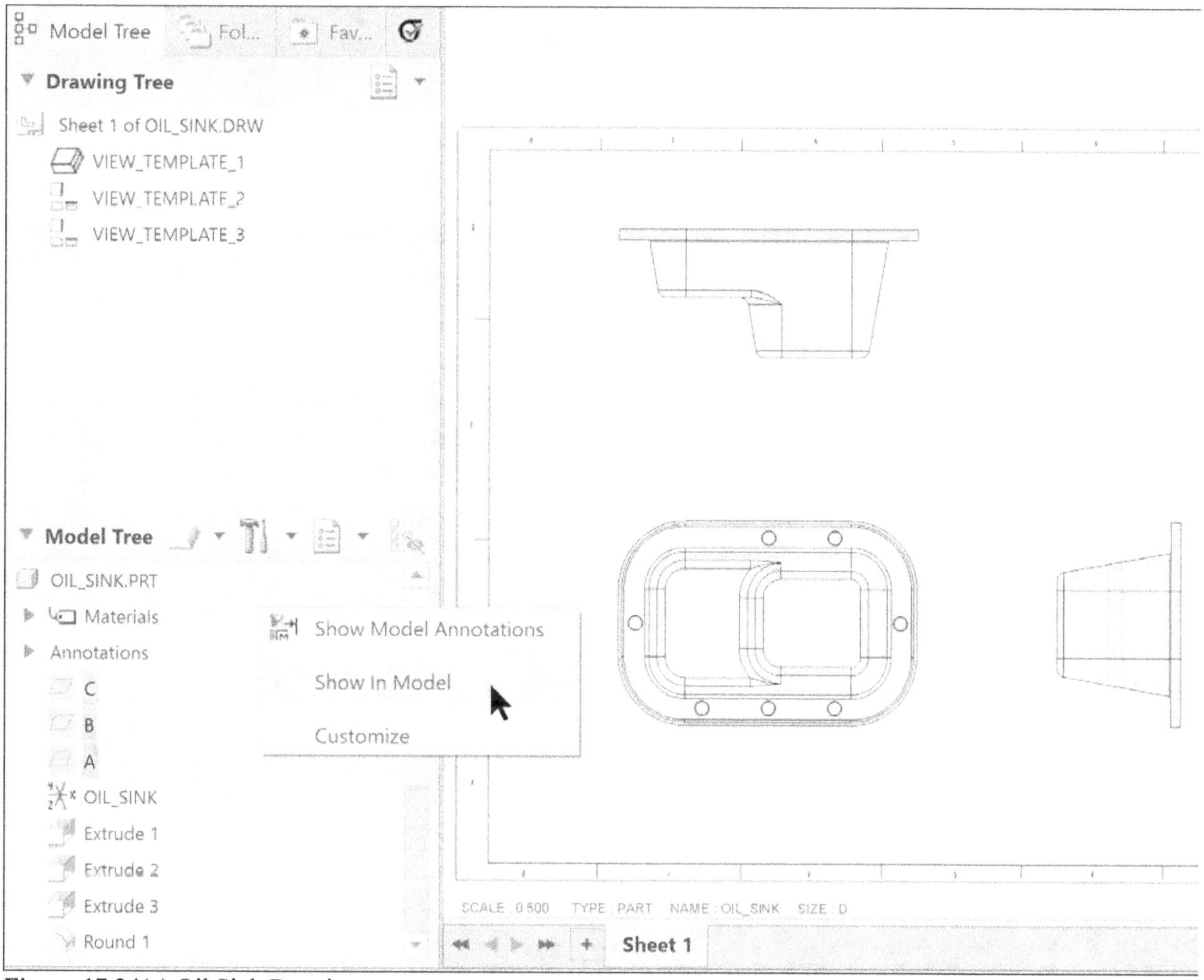

Figure 17.24(a) Oil Sink Drawing

759

Click: **Annotate** tab > from the Ribbon, **Show Model Annotations** > select the front view > hold down the **Ctrl** key > select the top view > release the **Ctrl** key > [⊢⊣] tab > [✓] (select all) > **Apply** > [⬛] tab > [✓] (select all) > **Apply** > [✕] (close the dialog box) > in the Drawing Tree, expand Annotations [Fig. 17.24(b)] > **Layout** tab > select the right view > press **RMB** > **Delete** > rearrange the remaining two views to fit the sheet > Change the sheet size, add views and sections to completely describe the part. Erase, delete, and reposition the axes and annotations as per ASME standards [Figs. 17.24(c-d)]. Use additional sheets as needed.

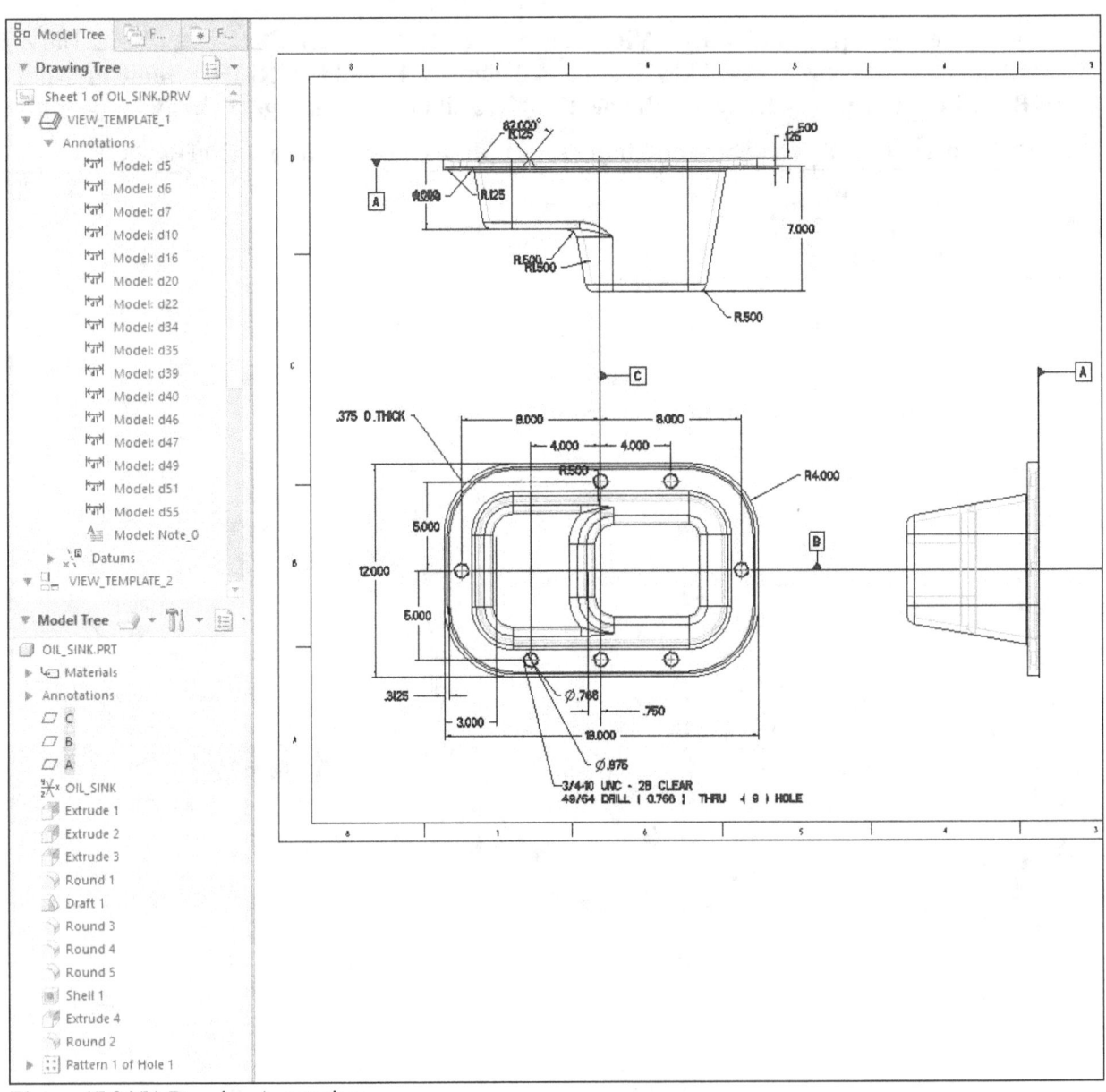

Figure 17.24(b) Drawing Annotations

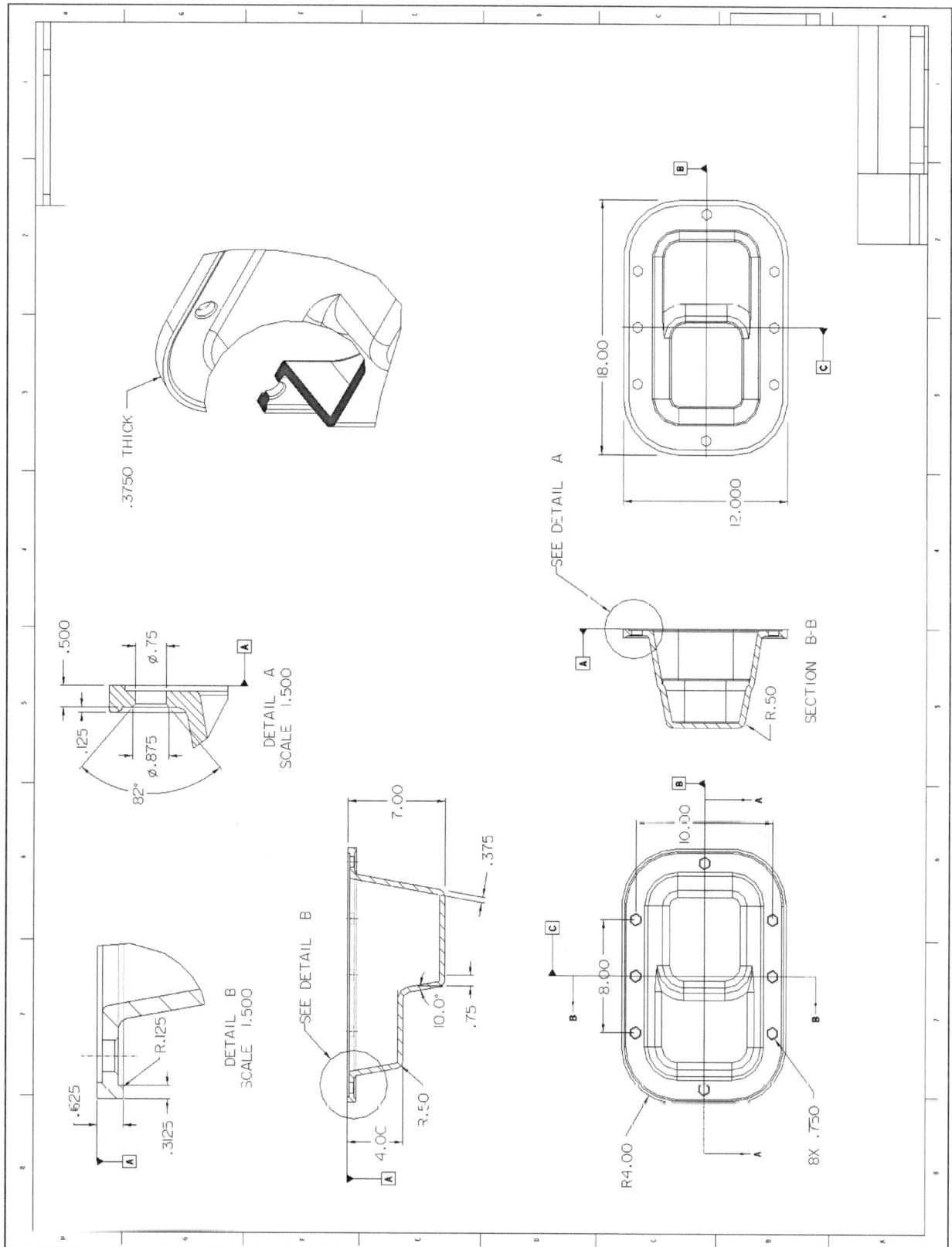

Figure 17.24(c) Oil Sink Detail Drawing, Sheet 1

761

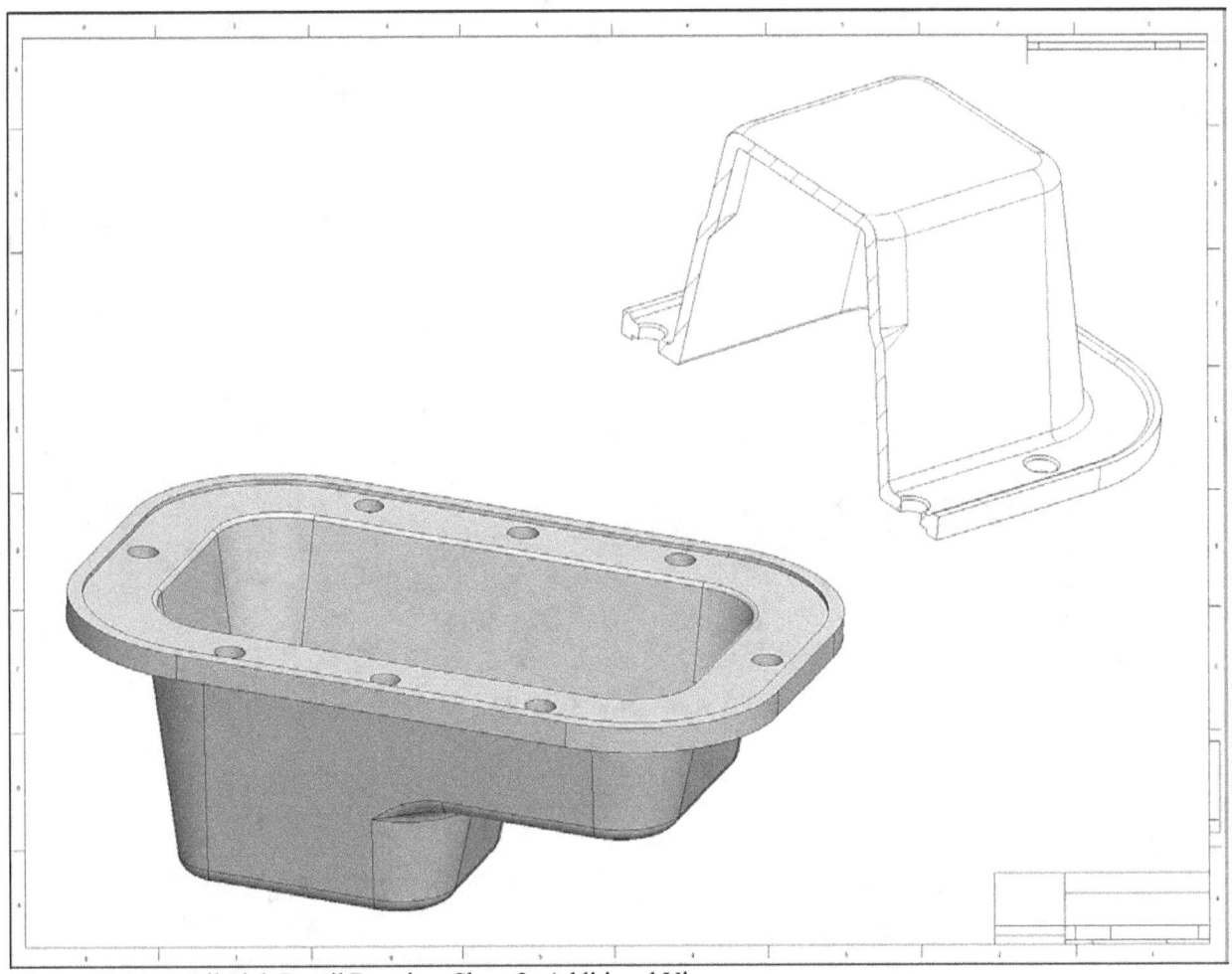

Figure 17.24(d) Oil Sink Detail Drawing, Sheet 2- Additional Views

Press: **Ctrl+S > File > Manage File > Delete Old Versions > Enter > File > Save As >** Type ⬚ **> Zip File (*.zip) > OK > upload** the zip file to your course interface or attach to an email and send to your instructor and/or yourself **> File > Close > File > Close > File > Exit > Yes**

Download additional projects from ***www.cad-resources.com*** > *click on the image of your book cover.*

Lesson 18 Drafts, Suppress, and Text Extrusions

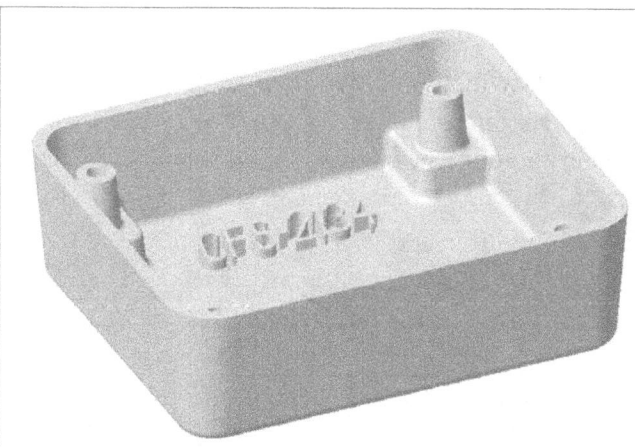

Figure 18.1 Enclosure

OBJECTIVES

- Create **Draft** features
- **Shell** a part
- **Suppress** features to decrease regeneration time
- **Resume** a set of suppressed features
- Create **Text** features on parts
- **Render** the part using **room scenes**

REFERENCES AND RESOURCES

For **Resources** go to www.cad-resources.com > click on the PTC Creo Parametric 5.0 Book cover

- Lesson 18 Lecture at **YouTube Creo Parametric Lecture Videos**
- Lesson 18 3D PDF models embedded in a PDF

DRAFTS, SUPPRESS, AND TEXT EXTRUSIONS

The **Draft** feature adds a draft angle between surfaces. A wide range of parts incorporate drafts into their design. Casting, injection mold, and die parts normally have drafted surfaces. The ENCLOSURE in Figure 18.1 is a plastic injection-molded part. Suppressing features by using the **Suppress** command temporarily removes them from regeneration. Suppressed features can be "unsuppressed" (**Resume**) at any time. It is sometimes convenient to suppress text extrusions and rounds to speed up regeneration of the model. Suppressing removes the item from regeneration and requires you to resume the item later. **Hide** is another option. Creo Parametric allows you to hide and unhide some types of model entities. When you hide an item, Creo Parametric removes the item from the graphics window. The hidden item remains in the Model Tree list, and its icon dims to reveal its hidden status. When you unhide an item, its icon returns to normal display (undimmed) and the item is redisplayed in the graphics window. The hidden status of items is saved with the model. Unlike the suppression of items, hidden items are regenerated. **Text** can be included in a sketch for extruded extrusions and cuts, trimming surfaces, and cosmetic features. To decrease regeneration time of the model, text can be suppressed after it has been created. Text can also be drafted.

763

Drafts

The **Draft Tool** adds a draft angle between two individual surfaces or to a series of selected planar surfaces. During draft creation, remember the following:

- You can draft only the surfaces that are formed by tabulated cylinders or planes.
- The draft direction must be normal to the neutral plane if a draft surface is cylindrical.
- You cannot draft surfaces with fillets around the edge boundary. However, you can draft the surfaces first, and then fillet the edges.

The following table lists the terminology used in drafts.

TERM	DEFINITION
Draft surfaces	Model surfaces selected for drafting.
Draft Hinges	Draft surfaces are pivoted about the intersection of the neutral plane with the draft surfaces.
Pull direction	Direction that is used to measure the draft angle. It is defined as normal to the reference plane.
Draft angle	Angle between the draft direction and the resulting drafted surfaces. If the draft surfaces are split, you can define two independent angles for each portion of the draft.
Direction of rotation	Direction that defines how draft surfaces are rotated with respect to the neutral plane or neutral curve.
Split areas	Areas of the draft surfaces to which you can apply different draft angles. Split object is also a choice.

Suppressing and Resuming Features

Suppressing a feature is similar to removing the feature from regeneration temporarily. You can "unsuppress" (**Resume**) suppressed features at any time. Features on a part can be suppressed to simplify the part model and decrease regeneration time. For example, while you work on one end of a shaft, it may be desirable to suppress features on the other end of the shaft. Similarly, while working on a complex assembly, you can suppress some of the features and components for which the detail is not essential to the current assembly process.

Unlike other features, the base feature cannot be suppressed. If you are not satisfied with your base feature, you can redefine the section of the feature, or you can delete it and start over again. Select feature(s) to suppress by: selecting it, picking on it from the Model Tree, specifying a *range*, entering its *feature number* or *identifier*, or using *layers*.

You can use **Suppress** and **Resume** to simplify the part before inserting features such as text extrusions. In addition, you may wish to suppress the text extrusion if there is other work to be done on the part. Text extrusions take time to regenerate, and increase the file size considerably.

Text Extrusions

When you are modeling, **Text** can be included in a sketch for extruded extrusions and cuts, trimming surfaces, and cosmetic features. The characters that are in an extruded feature use the font **font3d** as the default. Other fonts are available.

Lesson 18 STEPS

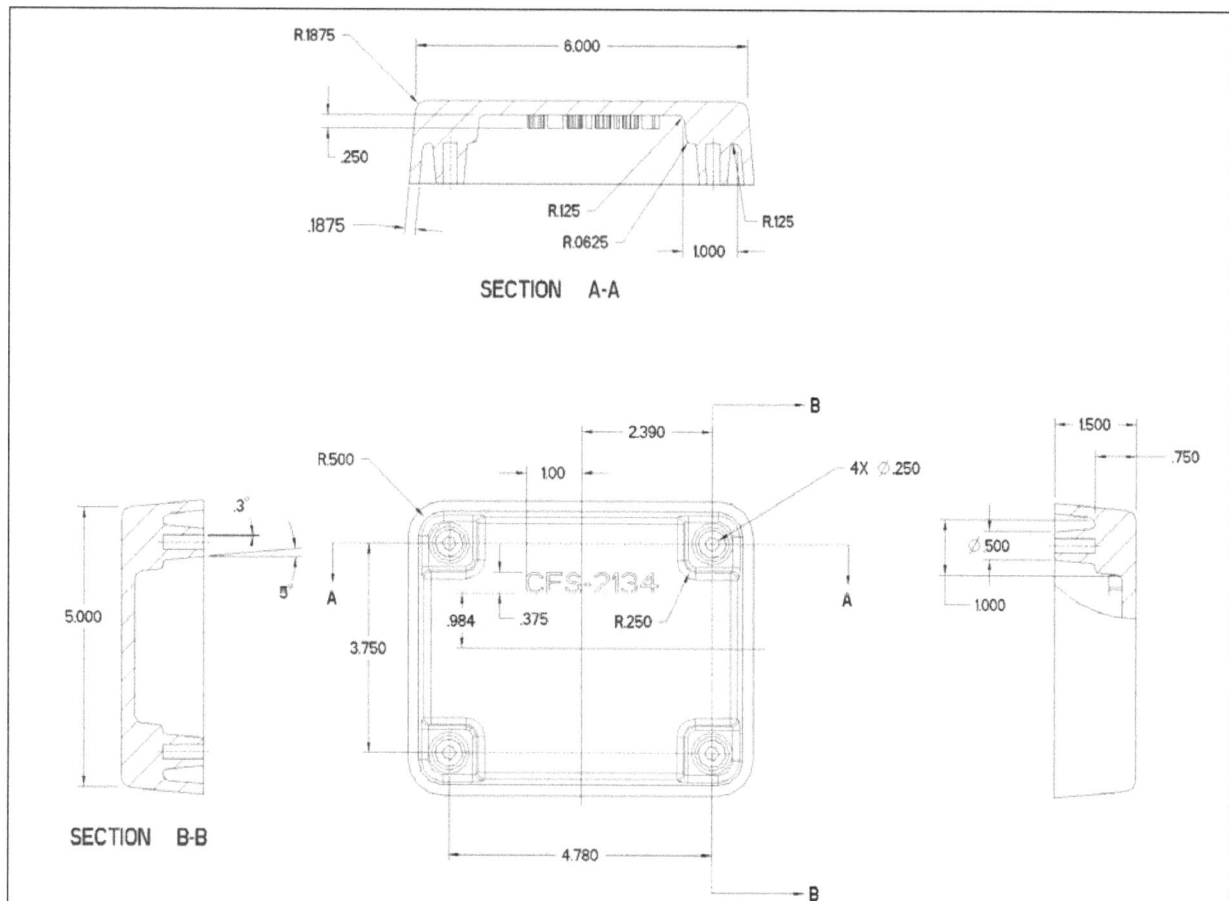

Figure 18.2(a) Enclosure

Enclosure

The Enclosure is a plastic injection-molded part. A variety of drafts will be used in the design of this part. A *raised text extrusion* will be modeled on the inside of the Enclosure, as shown in Figure 18.1. The dimensions for the part are provided in Figures 18.2(a) through 18.2(e).

Click: **File > Manage Session > Select Working Directory** > select the working directory > **OK > Ctrl+N** > Name **enclosure > OK > File > Options > Configuration Editor > Import/Export > Import configuration file** > select your previously created and saved file (**Creo_texbook.pro**) > **Open > Find** > 1. Type keyword: *default_dec_places* > **Find Now** > 3. Set value: **3 > Enter > Close > OK > No > File > Prepare > Model Properties** > Units **change** > Units Manager **Inch lbm Second (Creo Parametric Default) > Close** > Material **change** > double-click on **fe20.mtl > OK > Close** > change the default coordinate system name in the Model Tree-- **PRT_CSYS_DEF** > *type* **CSYS_ENCLOSURE > Enter > Ctrl+S > Enter > File > Options > Customize Ribbon > Import > Import customization file** > select your previously saved **.ui** file from Lesson Two (creo_parametric_customization.ui) [only available if you created a **.ui** file and saved it as was previously instructed] > **Open > Import > OK > LMB** to deselect

765

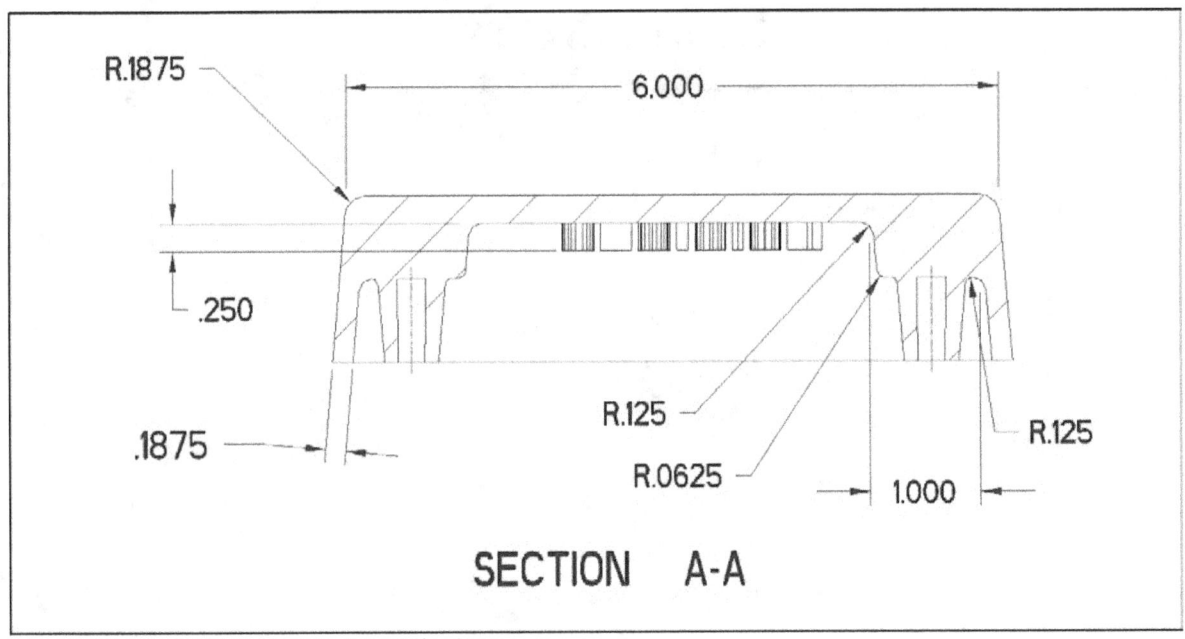

R.1875

6.000

.250

.1875

R.125

R.0625

R.125

1.000

SECTION   A-A

Figure 18.2(b) SECTION A-A (Top View)

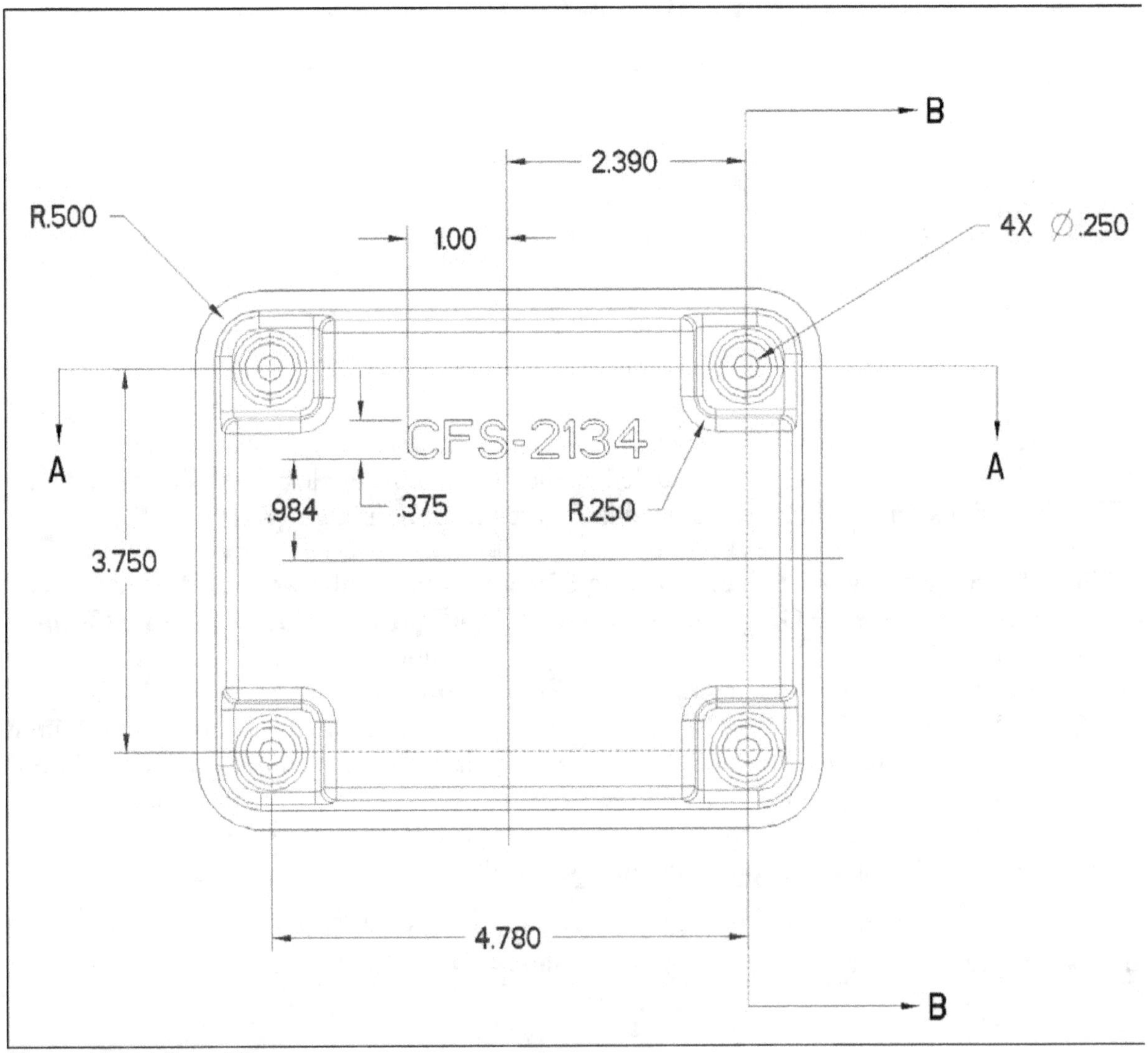

B

2.390

R.500

1.00

4X ⌀.250

CFS-2134

A

.984

.375

R.250

A

3.750

4.780

B

Figure 18.2(c) Front View

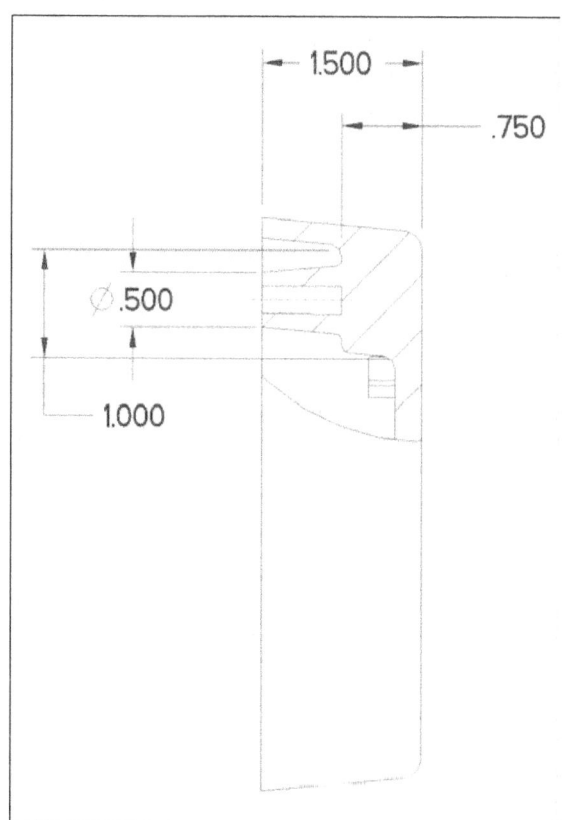

Figure 18.2(d) Right Side View

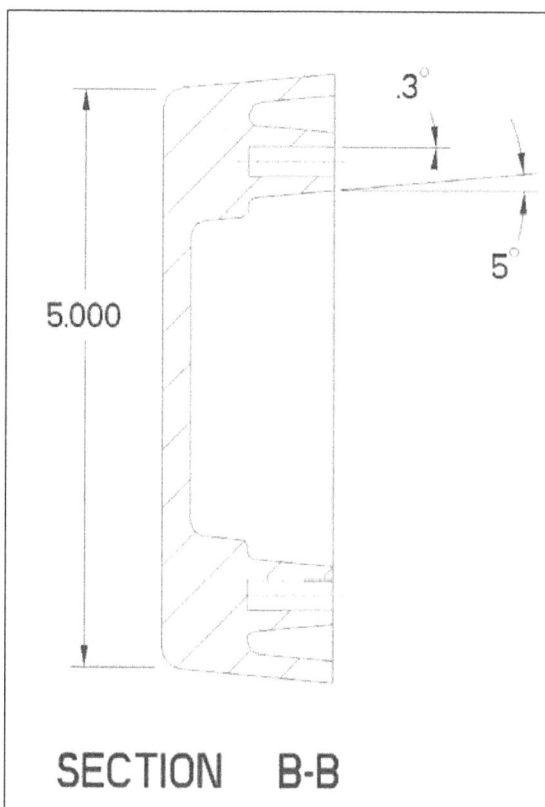

SECTION   B-B

Figure 18.2(e) SECTION B-B (Left Side View)

Make the first extrusion **6.00** (width) **X 5.00** (height) **X 1.50** (depth), with **R.50** rounds. Add the fillets in the sketch instead of rounds after the first extrusion is complete. Sketch on datum **FRONT**. Center the first extrusion horizontally on datum **TOP** and vertically on datum **RIGHT**. Add constraints as needed to control your sketch geometry. Incorporate the draft angle into the first extrusion instead of adding a separate draft feature [Figs. 18.3(a-c)].

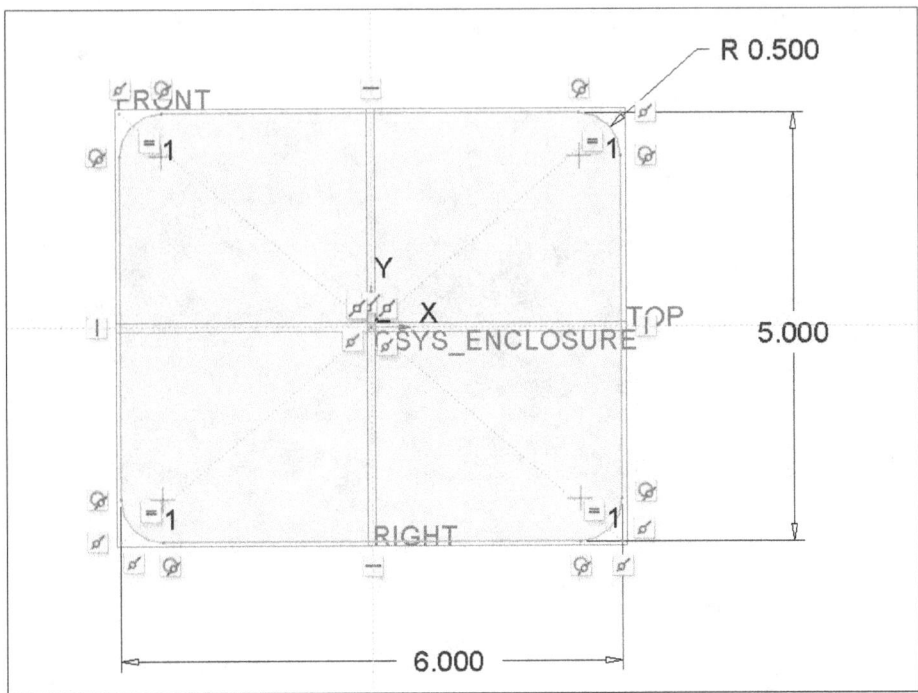

Figure 18.3(a) Sketch on the Front Datum

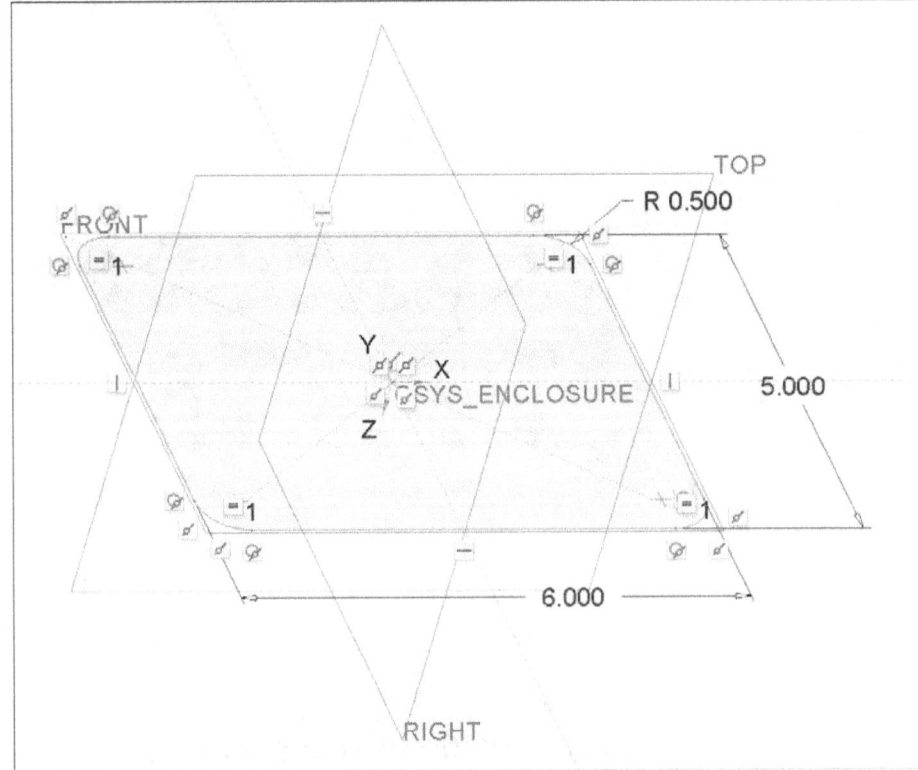

Figure 18.3(b) Standard Orientation of the Sketch

Click: [✓] to compete the sketch > **Options** tab > [✓ Add taper] > **5** > **Enter** [Fig. 18.3(c)] > [✓]
> **Ctrl+S** > **LMB** > **View** tab > **Appearance Gallery** > set a new color for the part

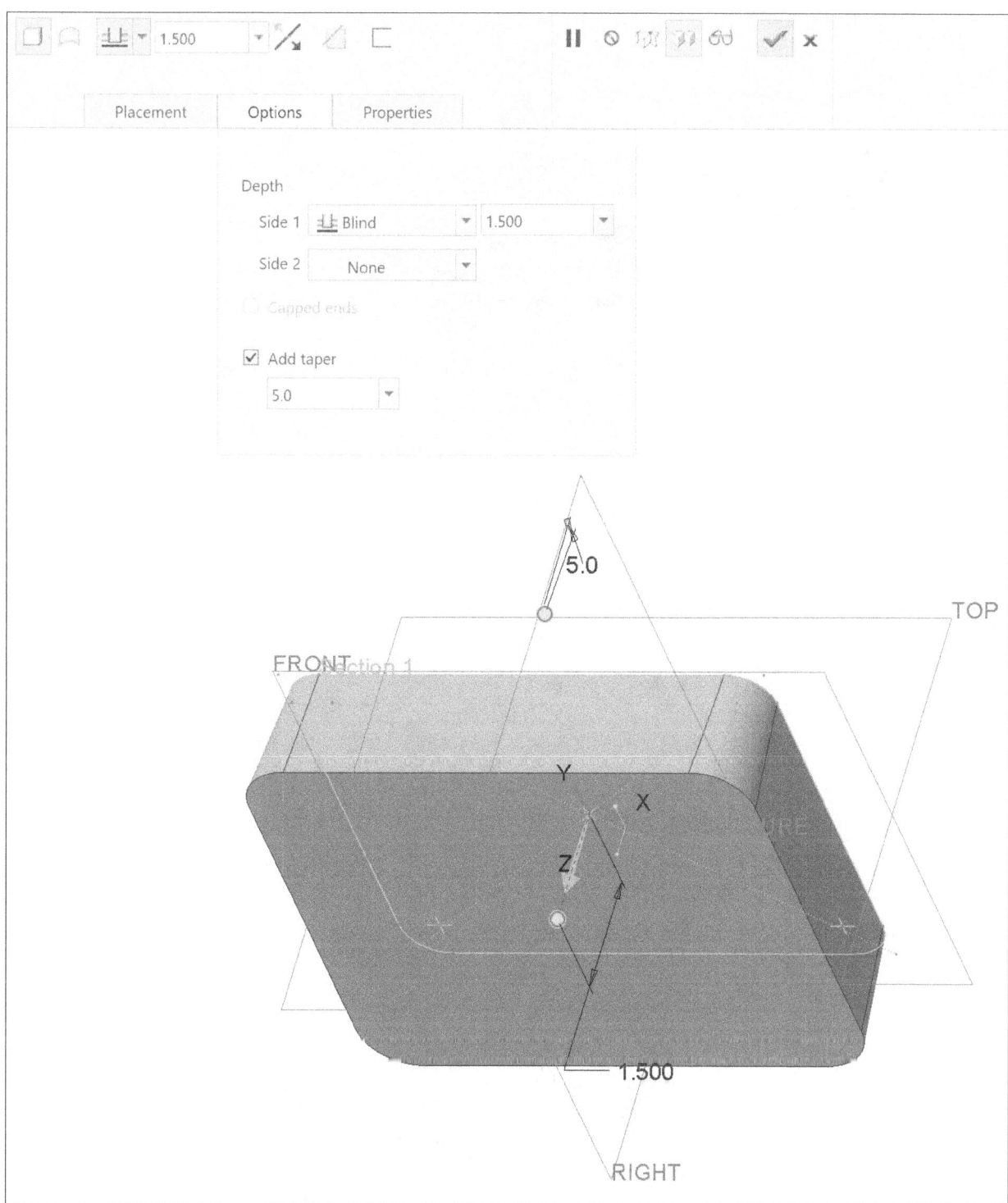

Figure 18.3(c) Adding a Taper

Click: **Model** tab > 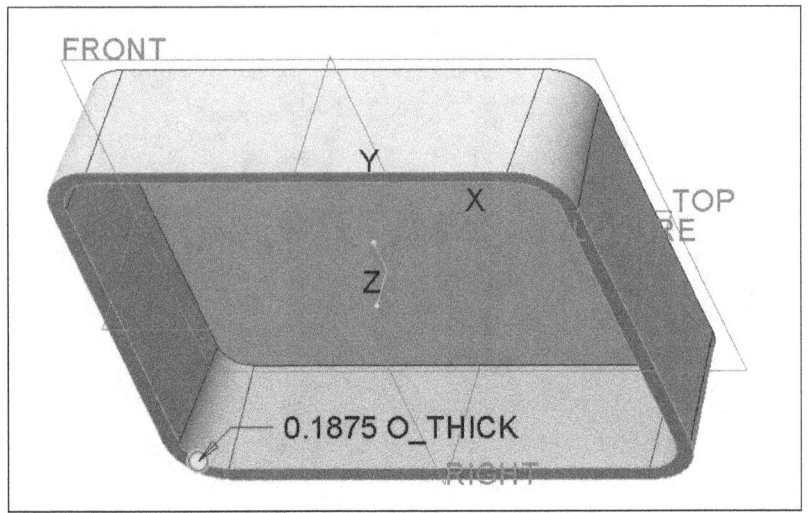 Shell > Thickness **.1875** > **Enter** select the face to be removed [Fig. 18.4(a)] > **References** tab [Fig. 18.4(b)]

Figure 18.4(a) Select the Surface to Remove

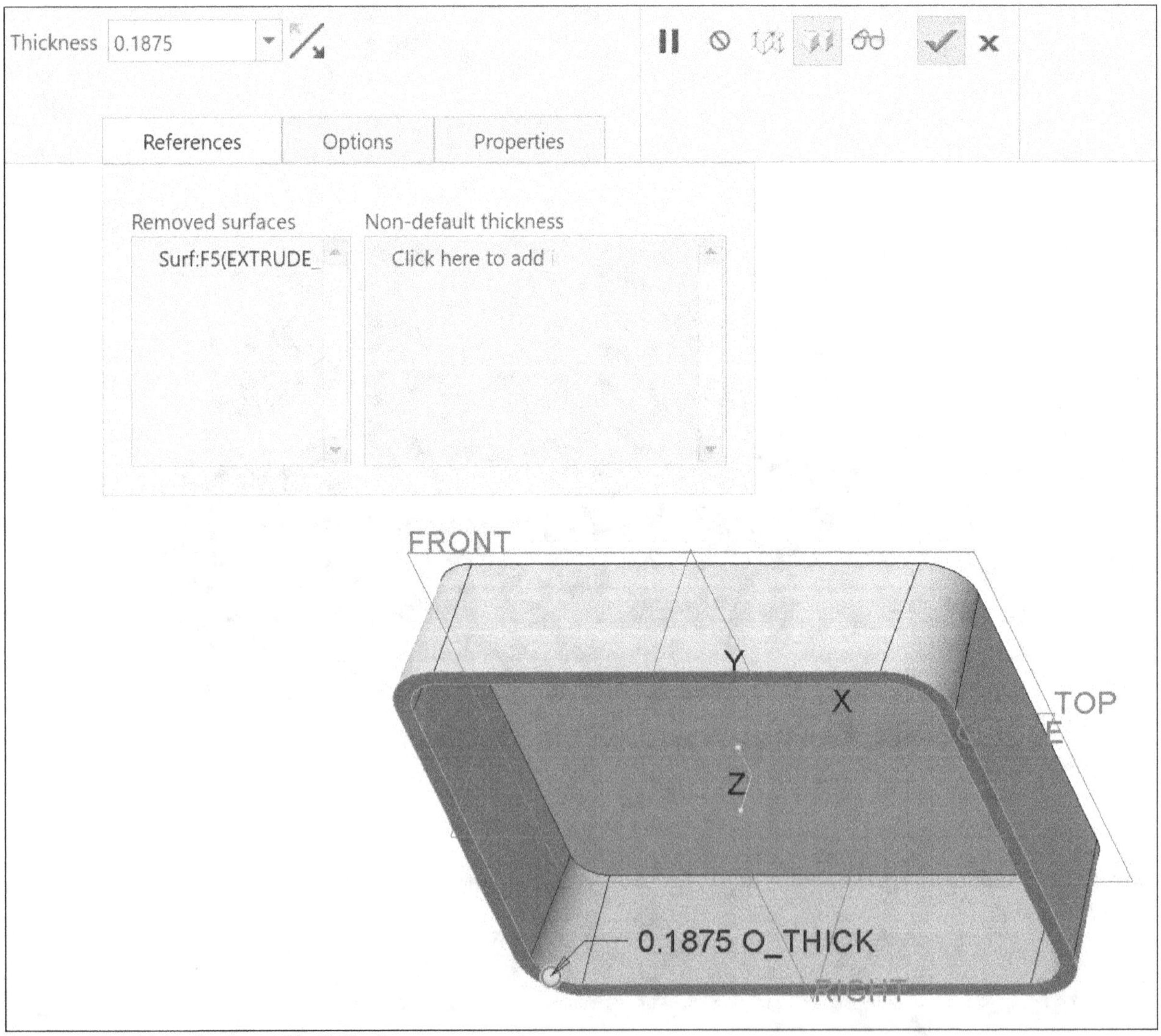

Thickness 0.1875

References   Options   Properties

Removed surfaces   Non-default thickness

Surf:F5(EXTRUDE_   Click here to add i

Figure 18.4(b) References Shell Tool

Change the thickness of the top of the enclosure to be **.25**, the walls will remain **.1875**. In the Graphics Window, press: **RMB >** [⊙ Non Default Thickness] [Fig. 18.4(c)] **>** select the face (*highlights*) [Fig. 18.4(d)] **>** *type* **.250** in the Dimension field

Removed surfaces	Non-default thickness
Surf:F5(EXTRUD... ▲	Surf:F5(EXTRUD... 0.250 ▲

> Enter [Fig. 18.4(e)] **>** [✓] **> LMB** to deselect **> Ctrl+D > Ctrl+S**

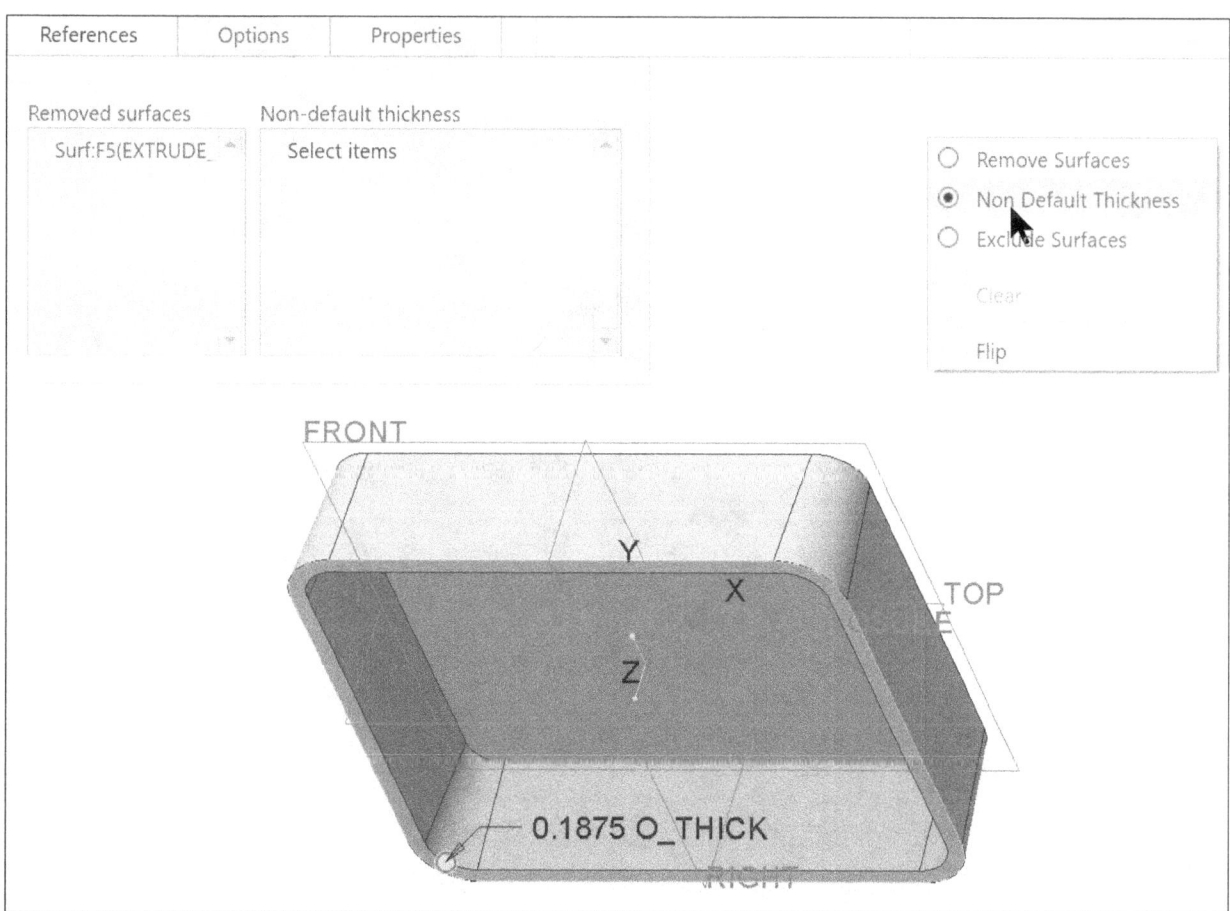

Figure 18.4(c) Non Default Thickness

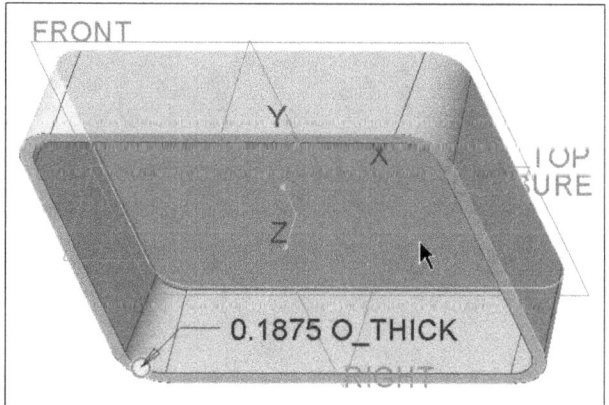

Figure 18.4(d) Select the Non Default Surface

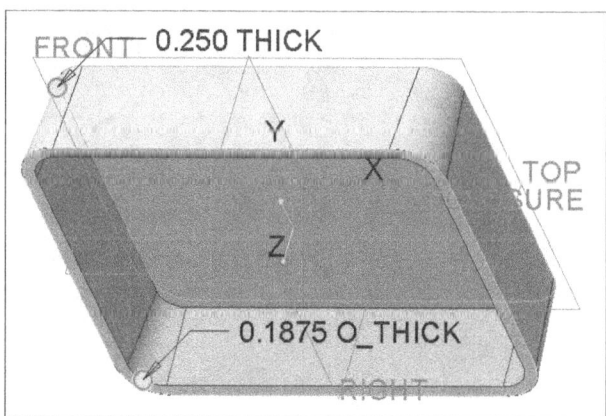

Figure 18.4(e) Shelled Extrusion

Create the raised pedestal-like extrusion. Click: [☐ Plane] > select datum **FRONT** > drag the depth handle forward by **.75** [Fig. 18.5(a)] > **OK** (**DTM1** will be used to control the height of the pedestal.) > in the Graphics Window, **LMB** to deselect > [Extrude] > **Placement** tab > **Define** > select the inside surface of the shell as the sketching plane > **Sketch** > press **RMB** > **References** > delete the datum references > select only the *edges (if needed toggle with RMB until an edge is highlighted)* [Fig. 18.5(b)] > **Solve** > **Close**

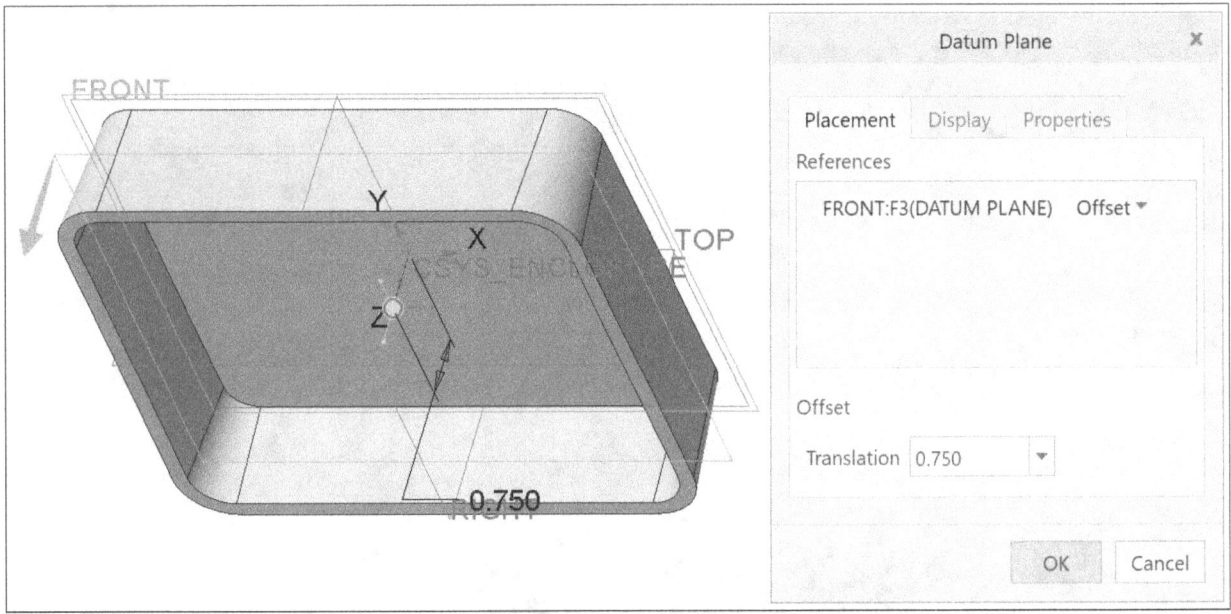

Figure 18.5(a) Offset Datum DTM1

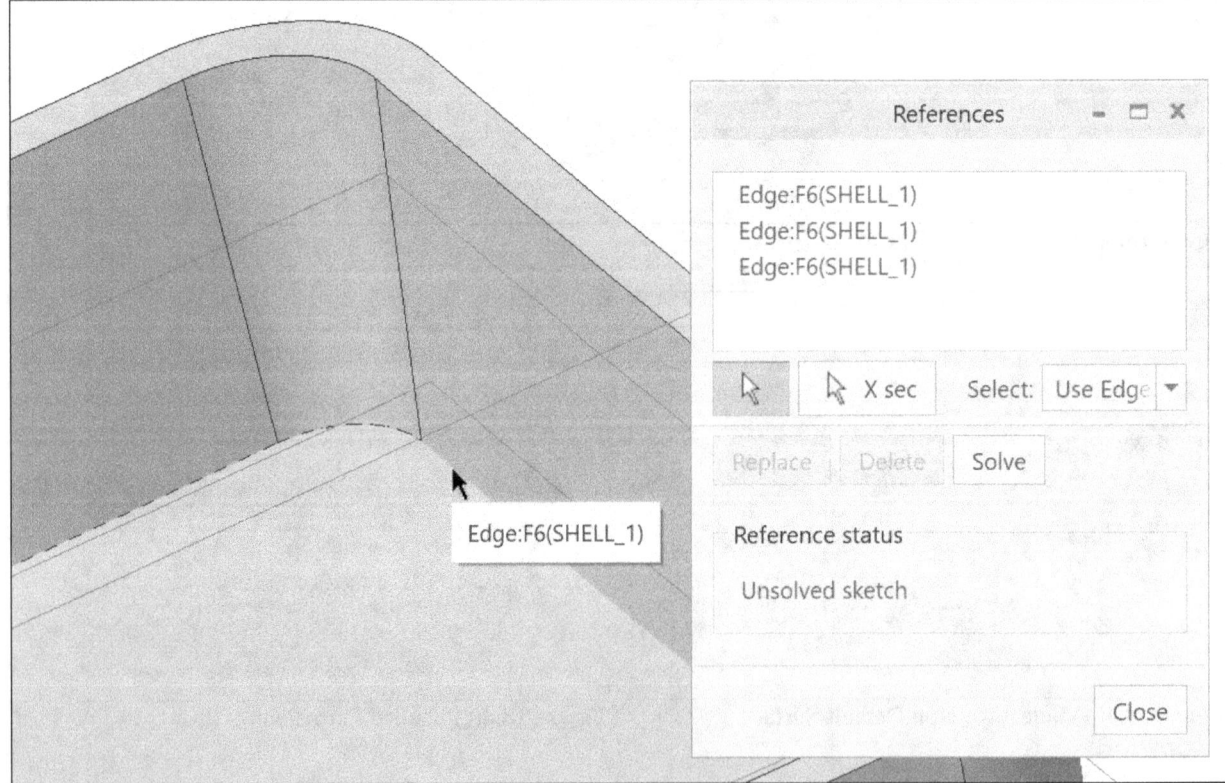

Figure 18.5(b) Extrusion References

772

Click: ⬚ **Project** > select an existing *internal_shelled_edge* to start the section and then the arc and other edge > **Close** > add two lines and a fillet [Fig. 18.5(c)] > trim the edge lines to create a closed section > add, move, and modify dimensions [Fig. 18.5(d)] > ✓ [Fig. 18.5(e)]

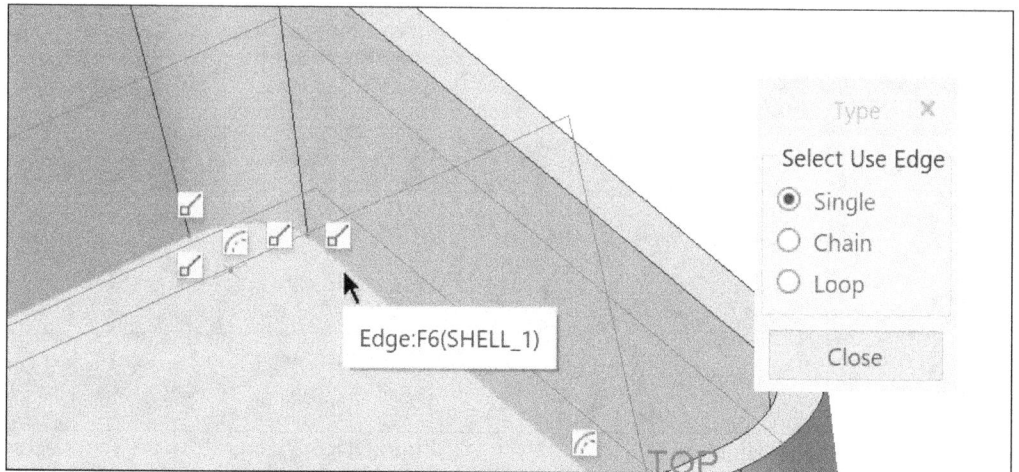

Figure 18.5(c) Create the First Entity using: ⬚ **Create entities by projecting curves or edges onto the sketch plane**

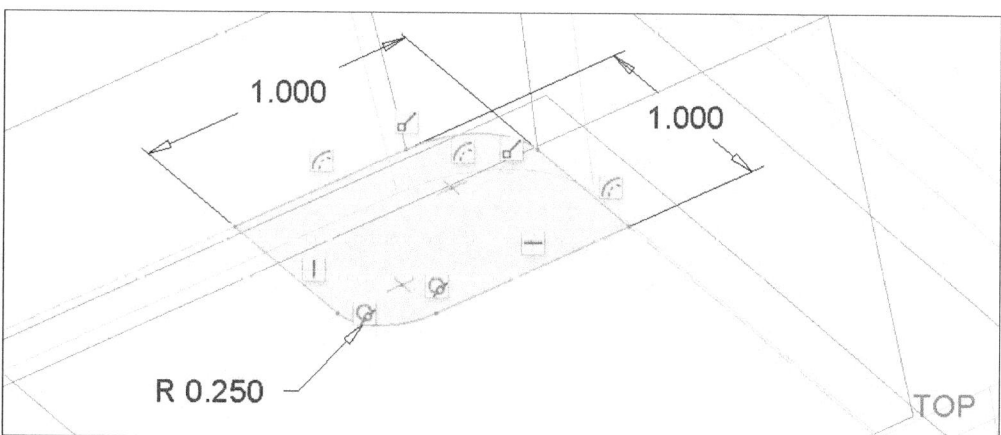

Figure 18.5(d) Section Sketch

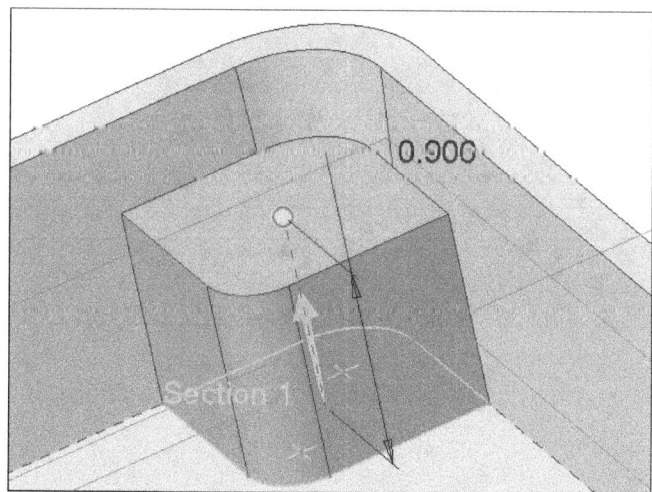 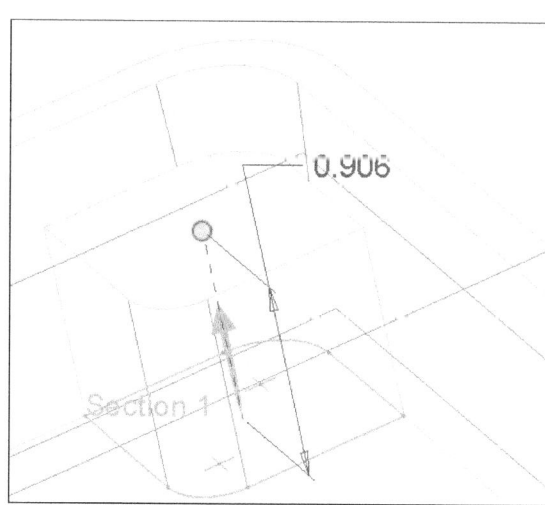

Figure 18.5(e) Initial Previewed Extrusion Depth

773

Place the pointer over the depth drag handle [Fig. 18.5(f)] > press **RMB** > **To Selected** [Fig. 18.5(g)] > select **DTM1** [Fig. 18.5(h)] > ☑ > **RMB** > 🔲 **Edit** [Fig. 18.5(i)] > **Ctrl+S** > **LMB**

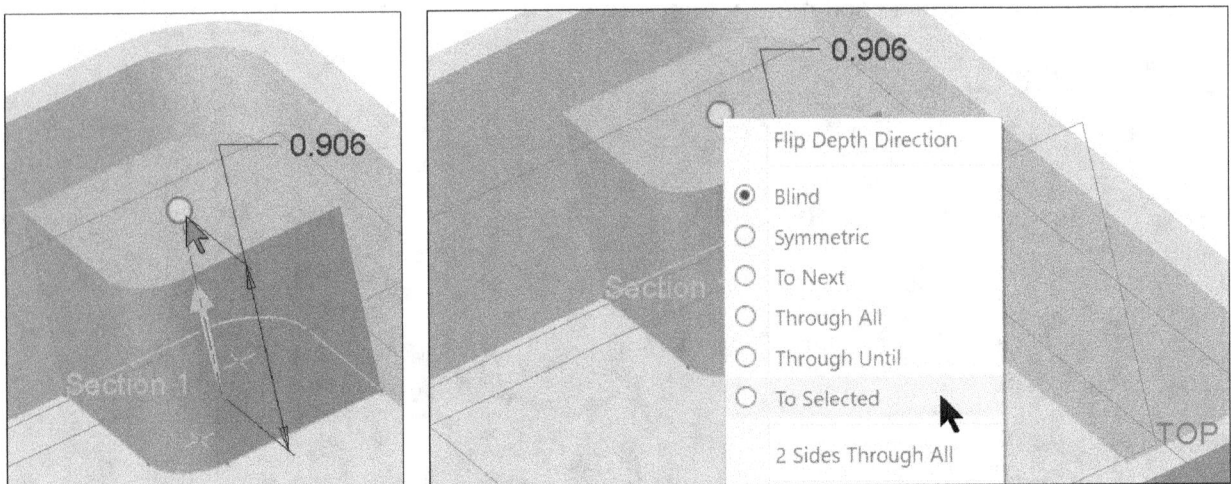

Figure 18.5(f) Place Pointer Over the Drag Handle   **Figure 18.5(g)** Press RMB > To Selected

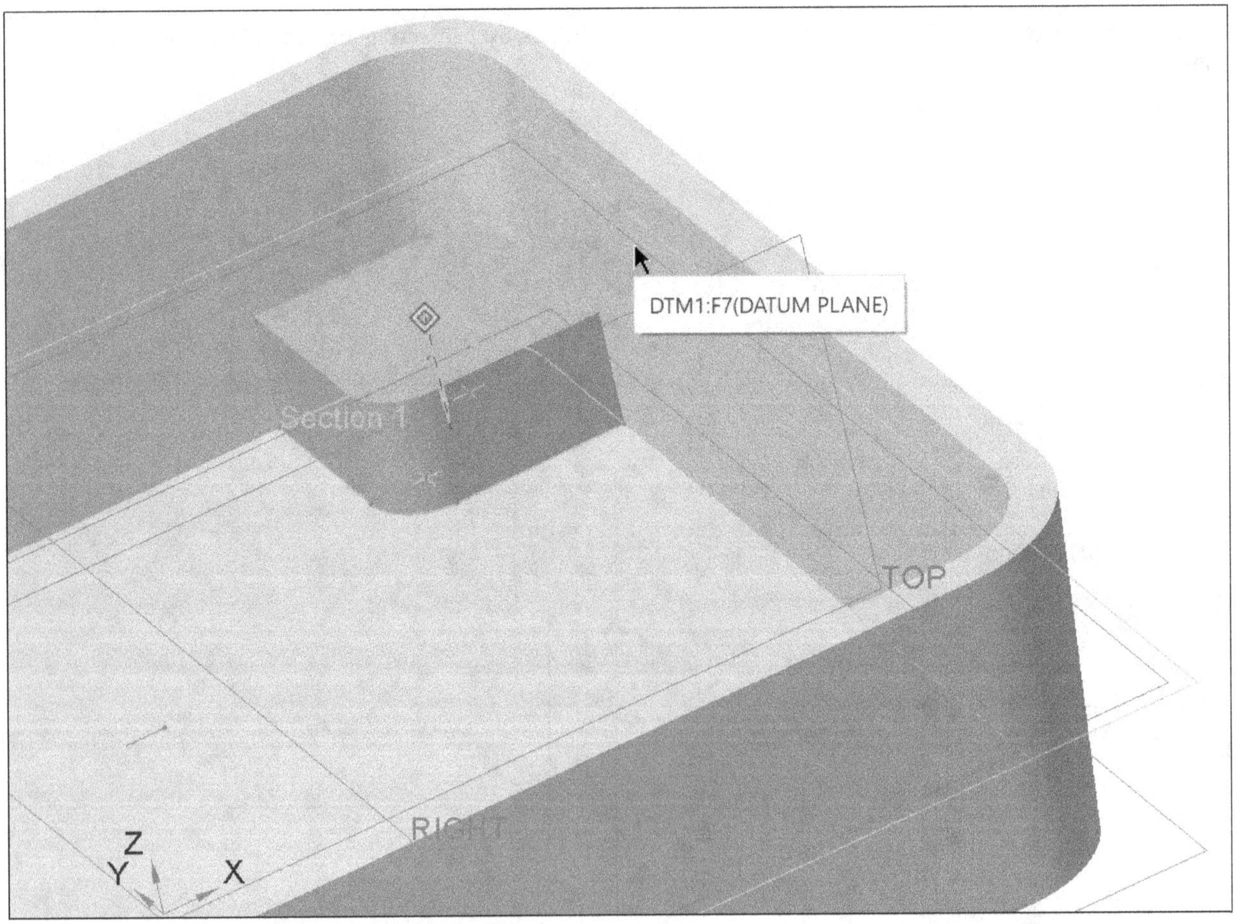

Figure 18.5(h) Select DTM1 to Establish the Depth

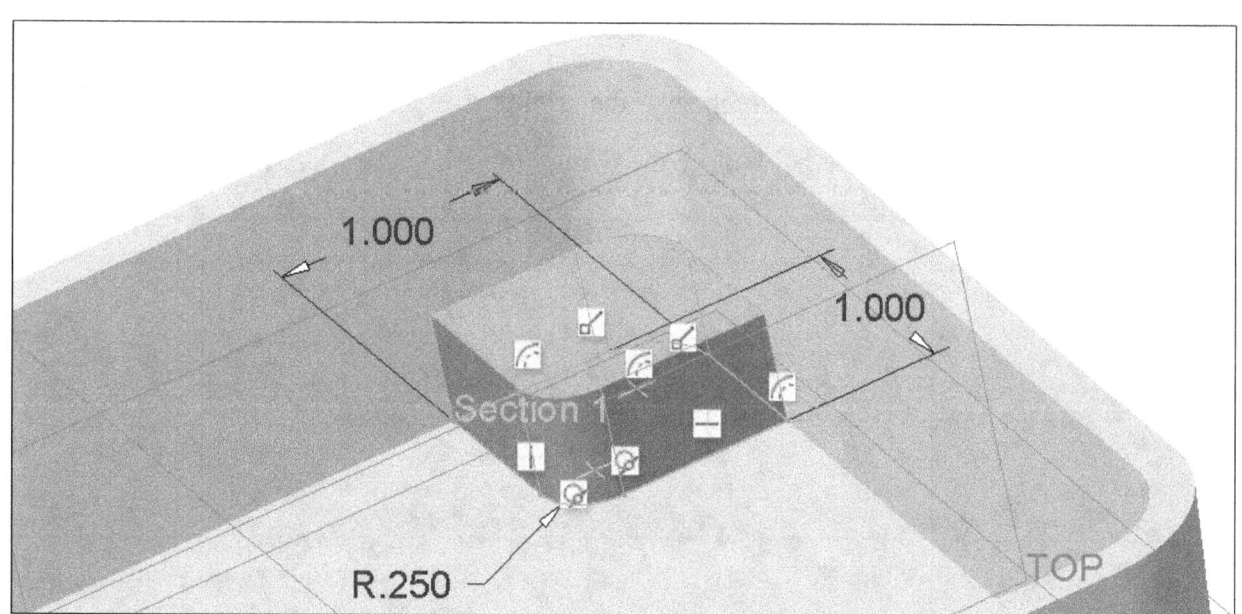

Figure 18.5(i) Completed Pedestal Extrusion

Click: ⟨ Draft ⟩ > **References** tab > select a vertical surface of the pedestal > press **RMB** > **Draft Hinges** > select datum **FRONT** > *type* **5** in the Angle field > **Enter** > ⟨ ⟩ **Reverse angle** (Fig. 18.6) > ⟨✓⟩ > ⟨💾⟩

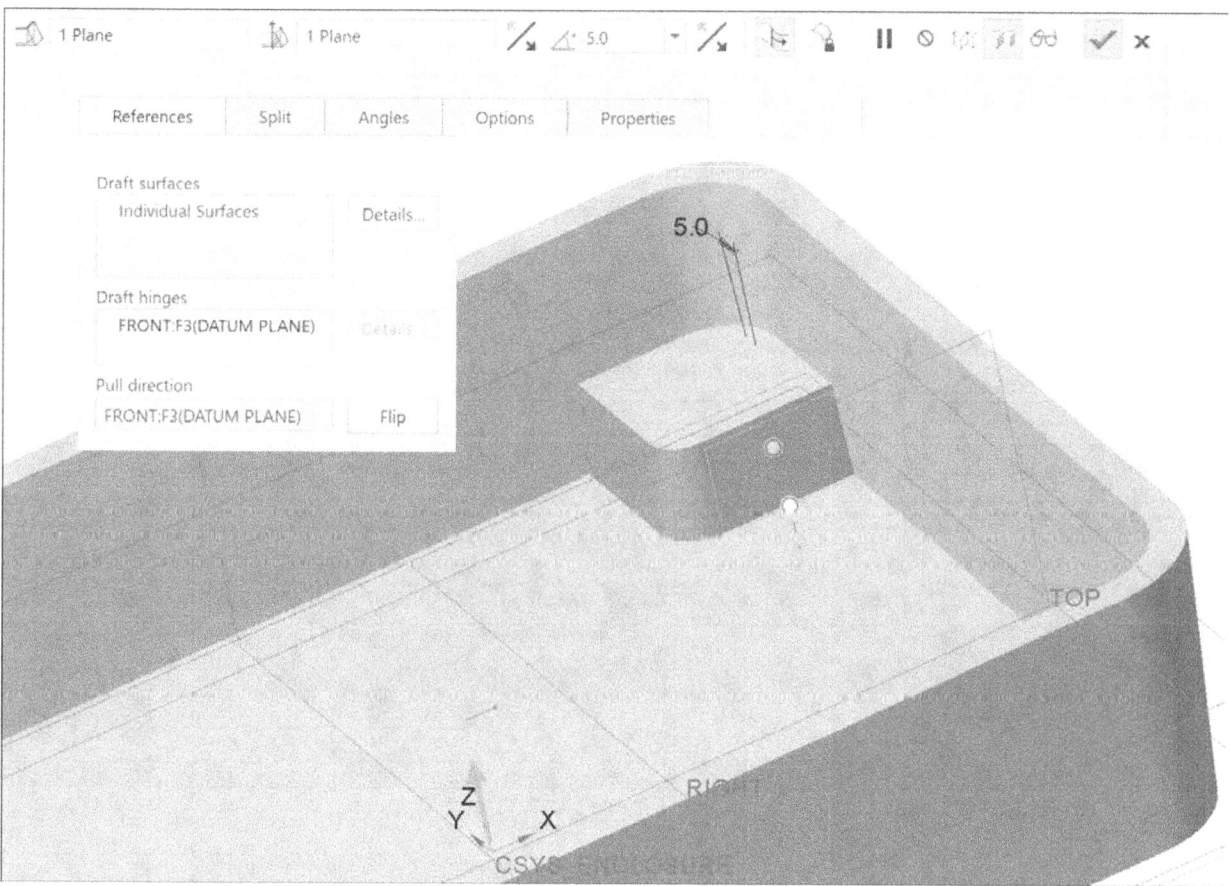

Figure 18.6 Draft Three Lateral Surfaces of the Pedestal

Model the circular extrusion. Click: [Extrude] > select the top surface of the pedestal as the sketching plane > keep the default references > the section consists of one circle [Fig. 18.7(a)] > [✓] > rotate the model [Fig. 18.7(b)] > place the pointer over the depth drag handle > press **RMB** > **To Selected** [Fig. 18.7(c)]

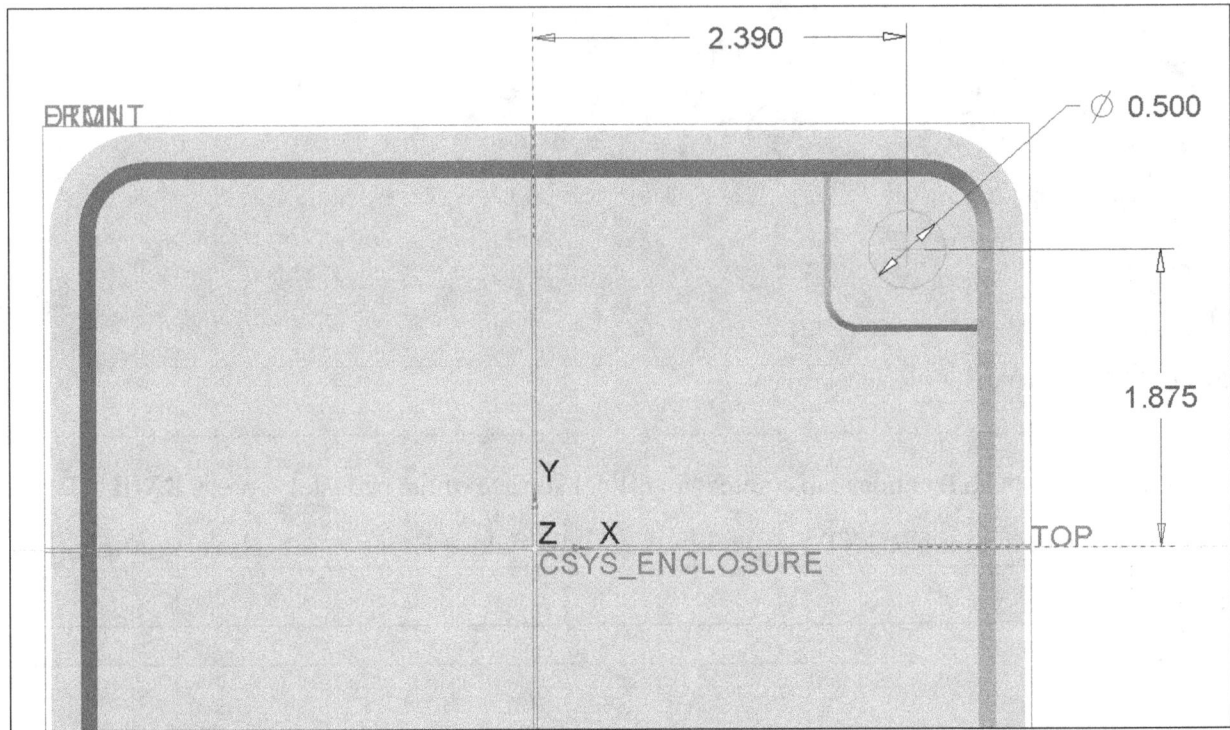

Figure 18.7(a) Section Sketch for the Circular Extrusion

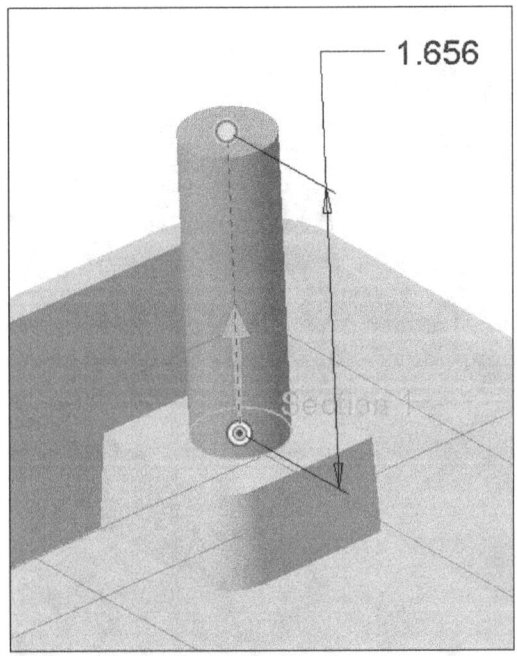

Figure 18.7(b) Extrusion Depth

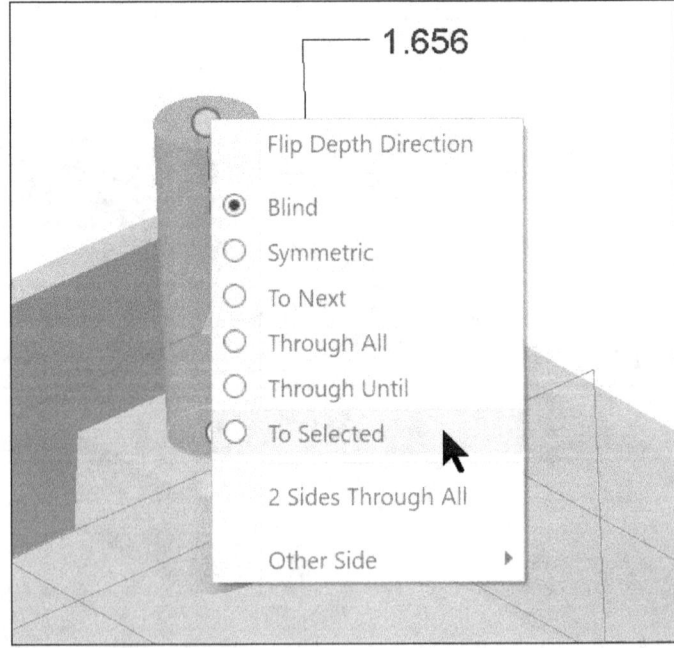

Figure 18.7(c) Depth Options

Select the surface [Fig. 18.7(d)] > 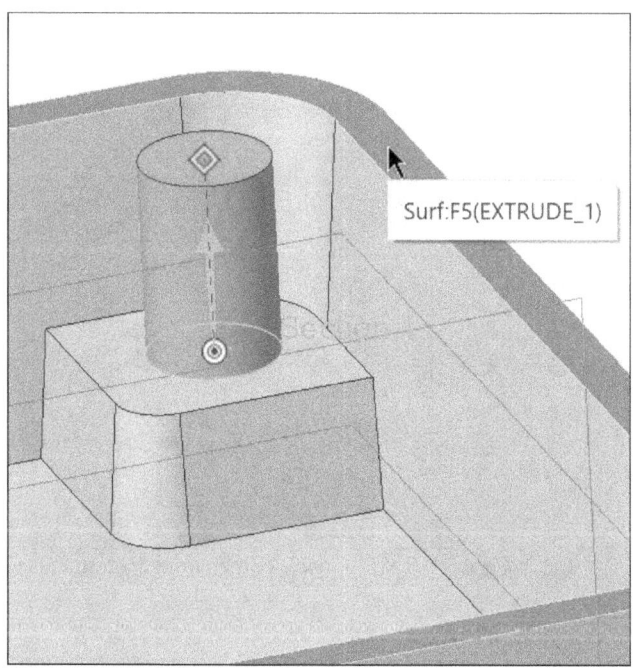 > double-click on the feature [Fig. 18.7(e)] > **LMB** > **Ctrl+S**

Figure 18.7(d) Select Surface to Extrude To   **Figure 18.7(e)** Circular Extrusion

Select the pedestal extrusion in the Graphics Window > 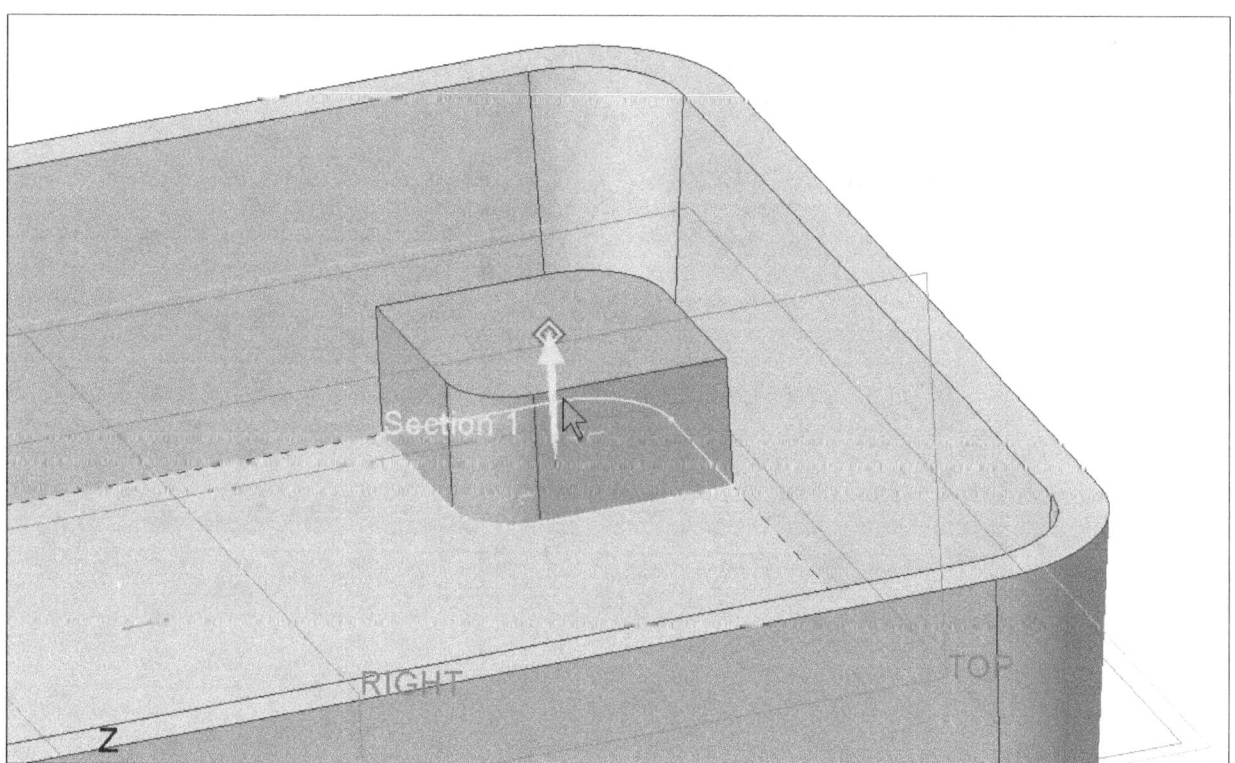 **Edit Definition** [Fig. 18.8(a)]

Figure 18.8(a) Redefining the Pedestal

Press: **RMB** > 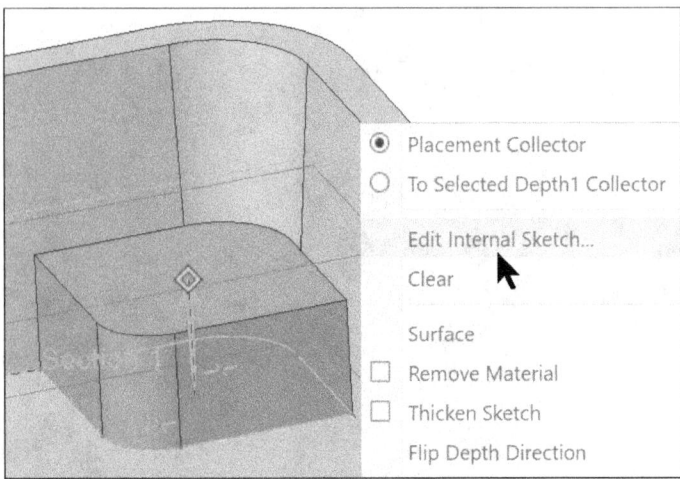 **Edit Internal Sketch** [Fig. 18.8(b)] > check the defining dimensions references [Fig. 18.8(c)] (make sure one of the 1.00 dimensions does not go to the center of the arc) > ☑ > ☑ > [Fig. 18.8(d)] > **LMB** > **Ctrl+S**

Figure 18.8(b) Original Dimensioning Scheme

Figure 18.8(c) Check Dimensions References

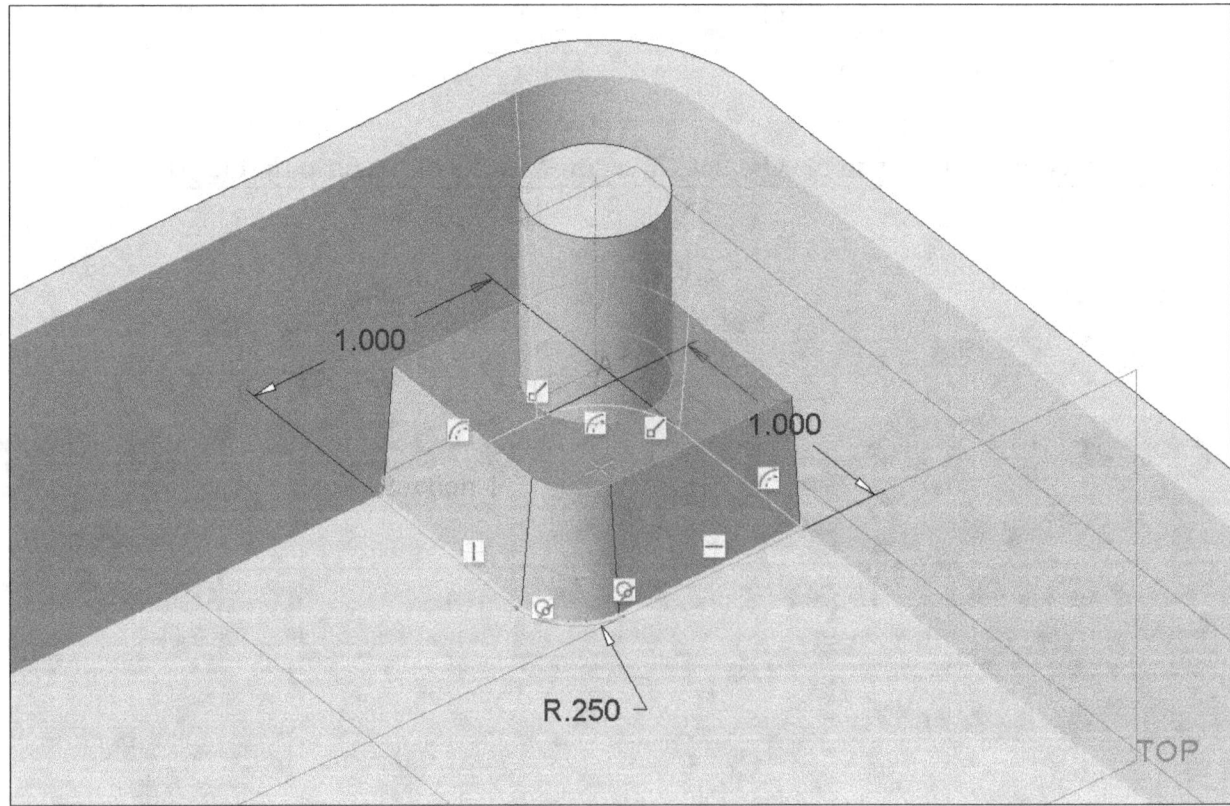

Figure 18.8(d) Pedestal

Click: 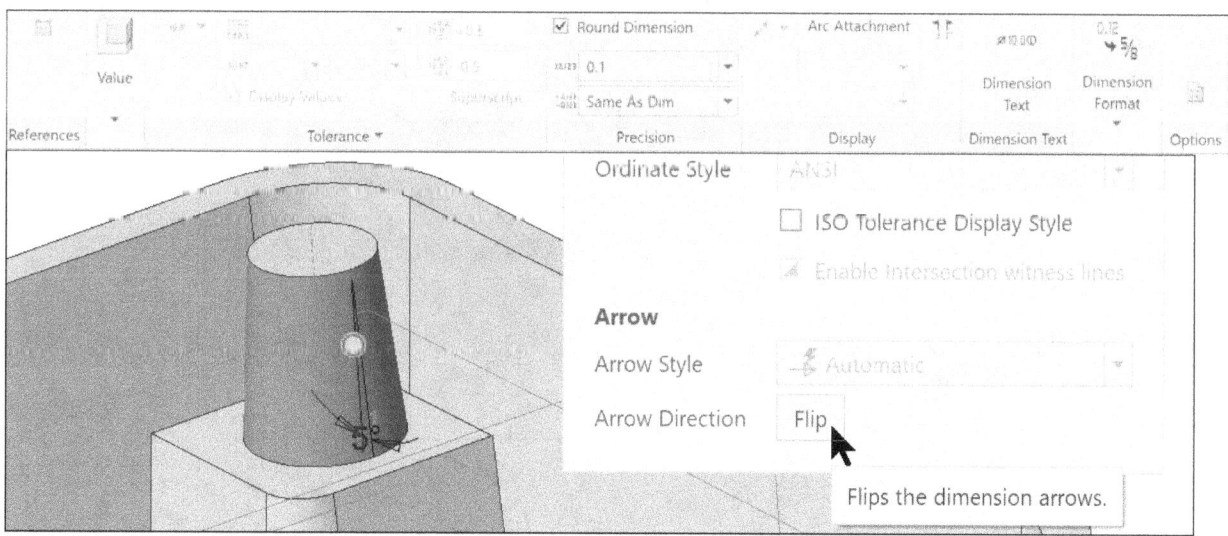 Draft (draft the circular extrusion at **5°**, using its top surface as the draft hinge [Fig. 18.9(a)]) > ⟋ **Reverse angle** > ✓ > **RMB** > ⟨dl⟩ **Edit** > double-click on the **5** degree dimension > **Display** > | Arrow Direction   Flip | > | :0.123 | 0 | ▼ | Decimal Places [Fig. 18.9(b)] > **LMB** > **LMB**

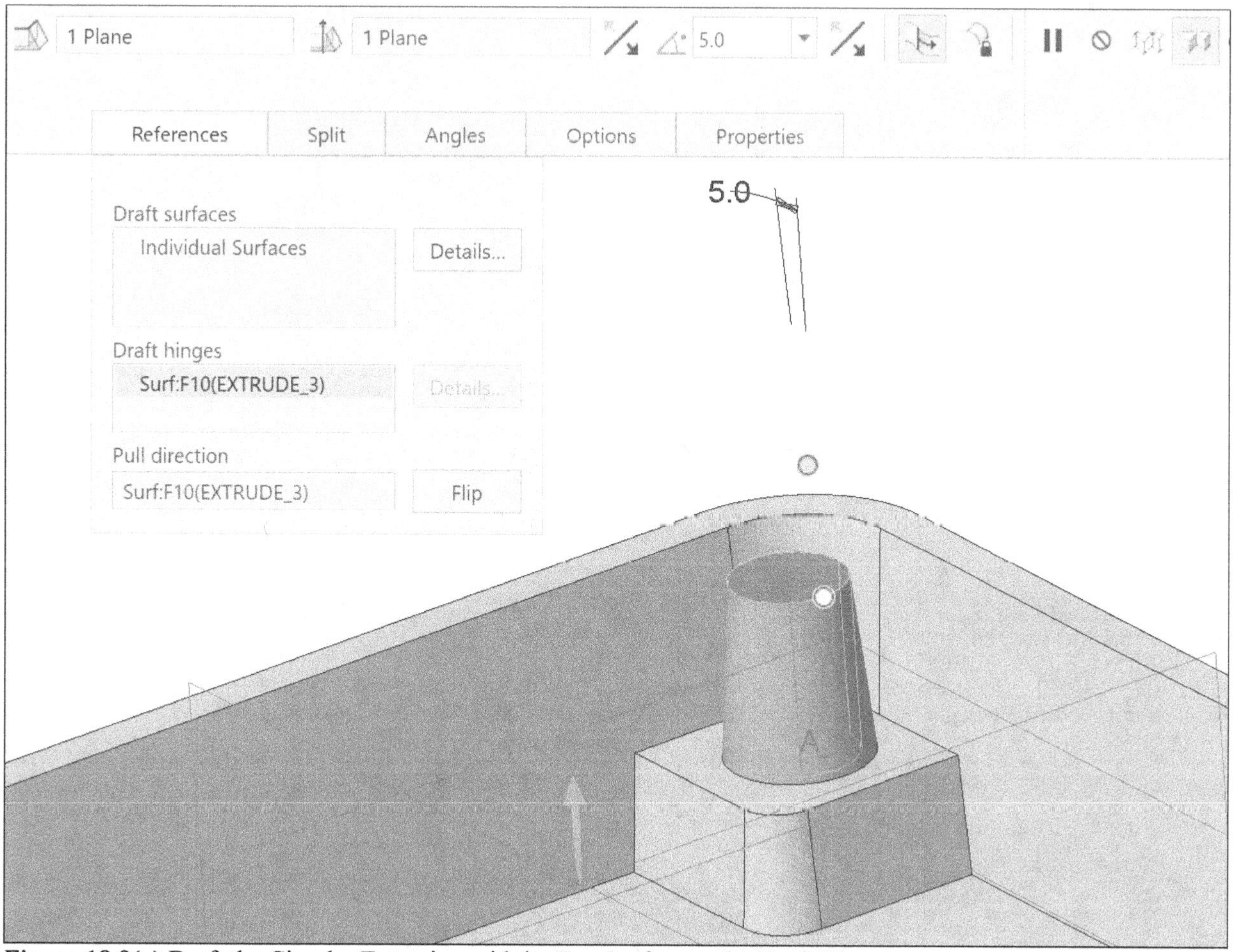

| 1 Plane | | 1 Plane | | | ⟋ | ∠° 5.0 | ▼ | ⟋ | | | ‖ | ⊘ | | |

| References | Split | Angles | Options | Properties |

Draft surfaces
   Individual Surfaces   Details...

5.0

Draft hinges
   Surf:F10(EXTRUDE_3)   Details...

Pull direction
   Surf:F10(EXTRUDE_3)   Flip

Figure 18.9(a) Draft the Circular Extrusion with its Top Surface as the Draft Hinge

---

				☑ Round Dimension		Arc Attachment	⌀10.0⌀	0.1⌀ ↦ 5/8
	Value			⌖ 0.1 ▼			Dimension Text	Dimension Format
		Display Values	Superscript	Same As Dim ▼				
References		Tolerance ▼		Precision	Display		Dimension Text	Options

Ordinate Style   ANSI   ▼

☐ ISO Tolerance Display Style

☐ Enable Intersection witness lines

Arrow

Arrow Style   Automatic   ▼

Arrow Direction   Flip

5°

Flips the dimension arrows.

Figure 18.9(b) Edit the Dimensions Properties

779

Create a **.250** diameter coaxial hole on the upper surface of the circular extrusion [Fig. 18.10(a)]. Use "To Selected" to establish the hole's depth to the top surface of the pedestal [Fig. 18.10(b)].

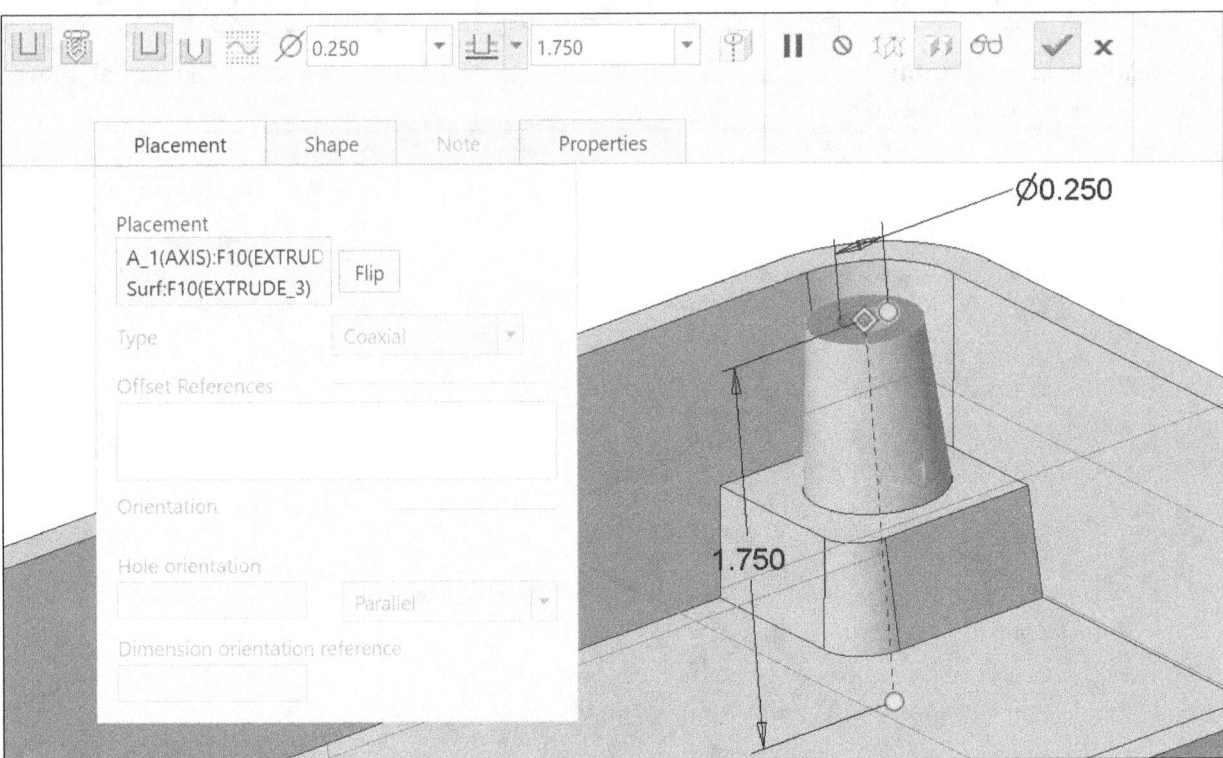

Figure 18.10(a) Coaxial Hole

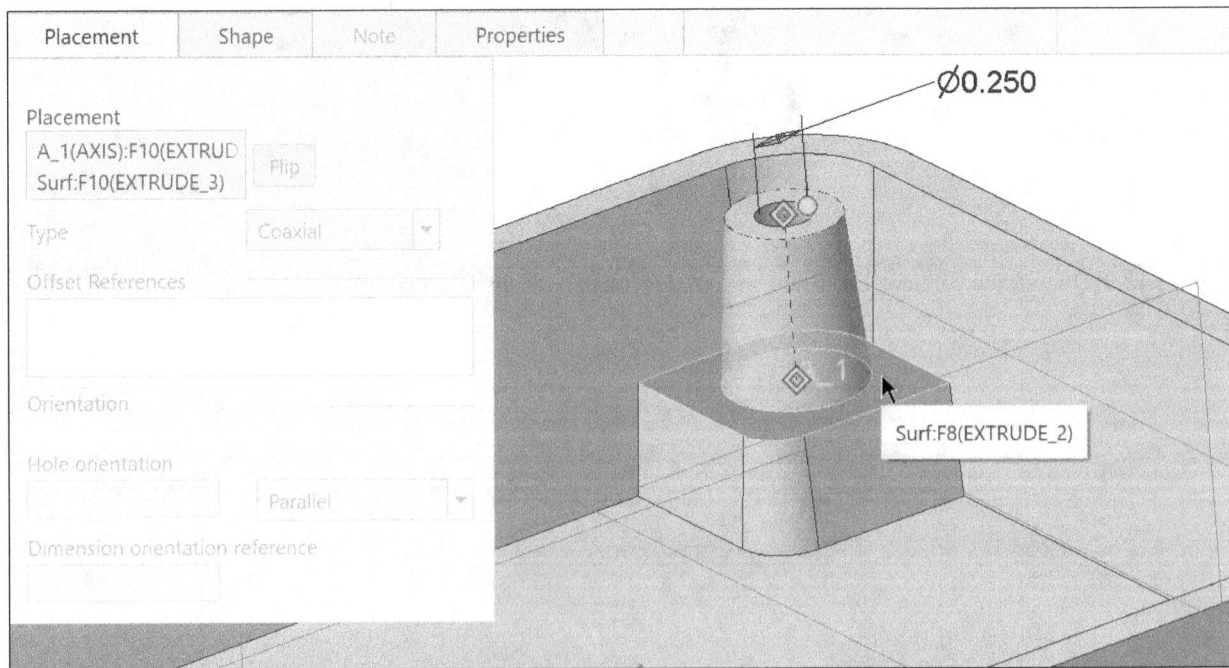

Figure 18.10(b) Hole Depth to Selected Surface

Next, add an internal draft of **.3°** to the coaxial hole (top surface of the cylinder as the draft hinge) (Fig. 18.11).

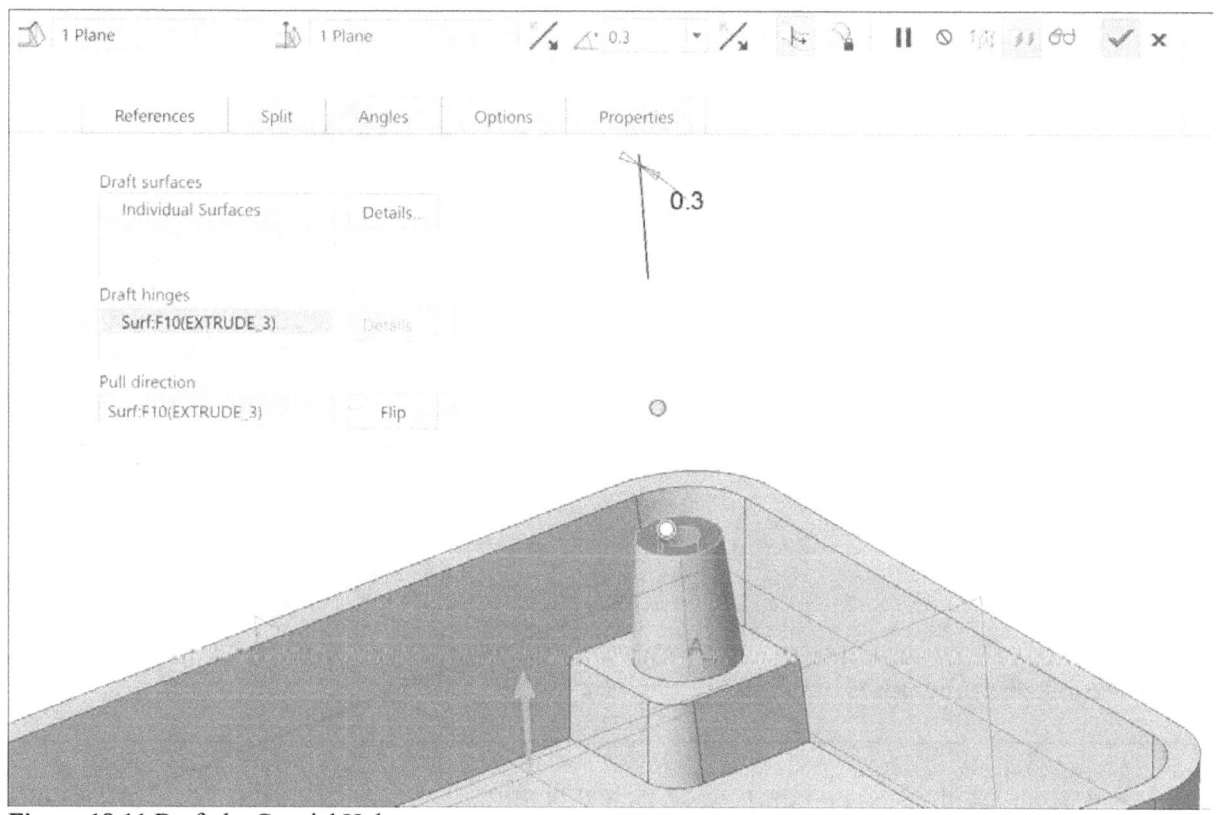

Figure 18.11 Draft the Coaxial Hole

Create one round feature for the **.0625** [Fig. 18.12(a)] and the **.125** rounds [Fig. 18.12(b)].

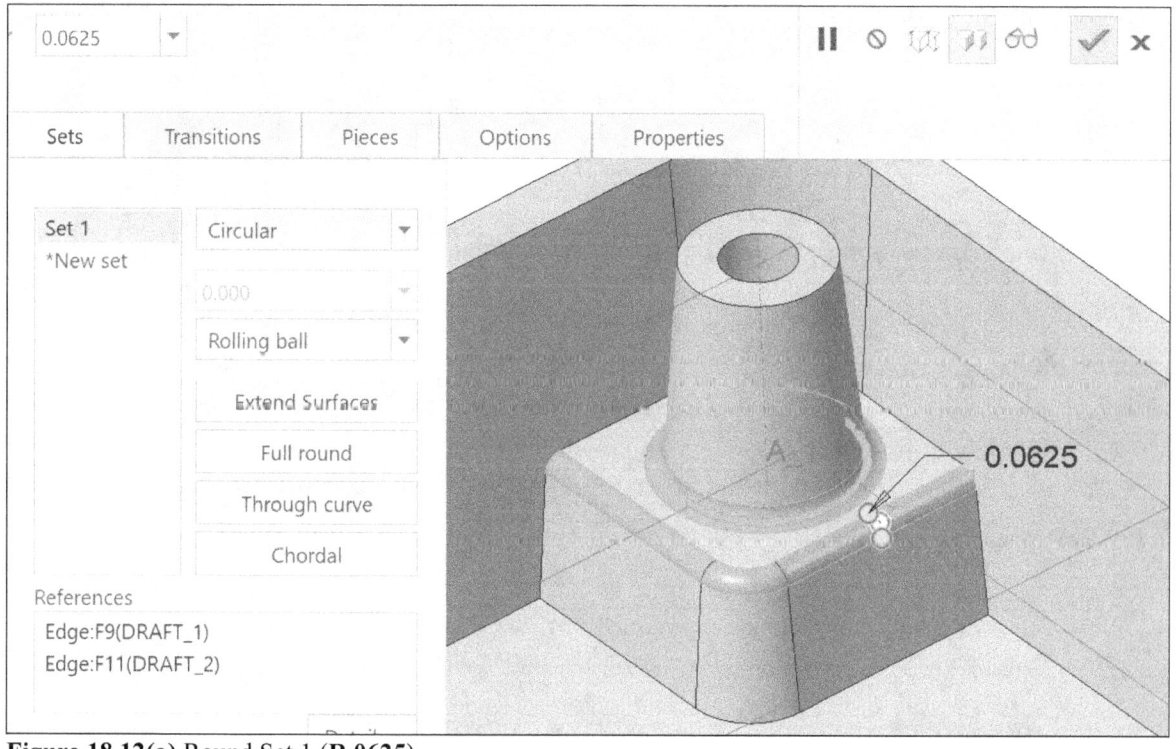

Figure 18.12(a) Round Set 1 (**R.0625**)

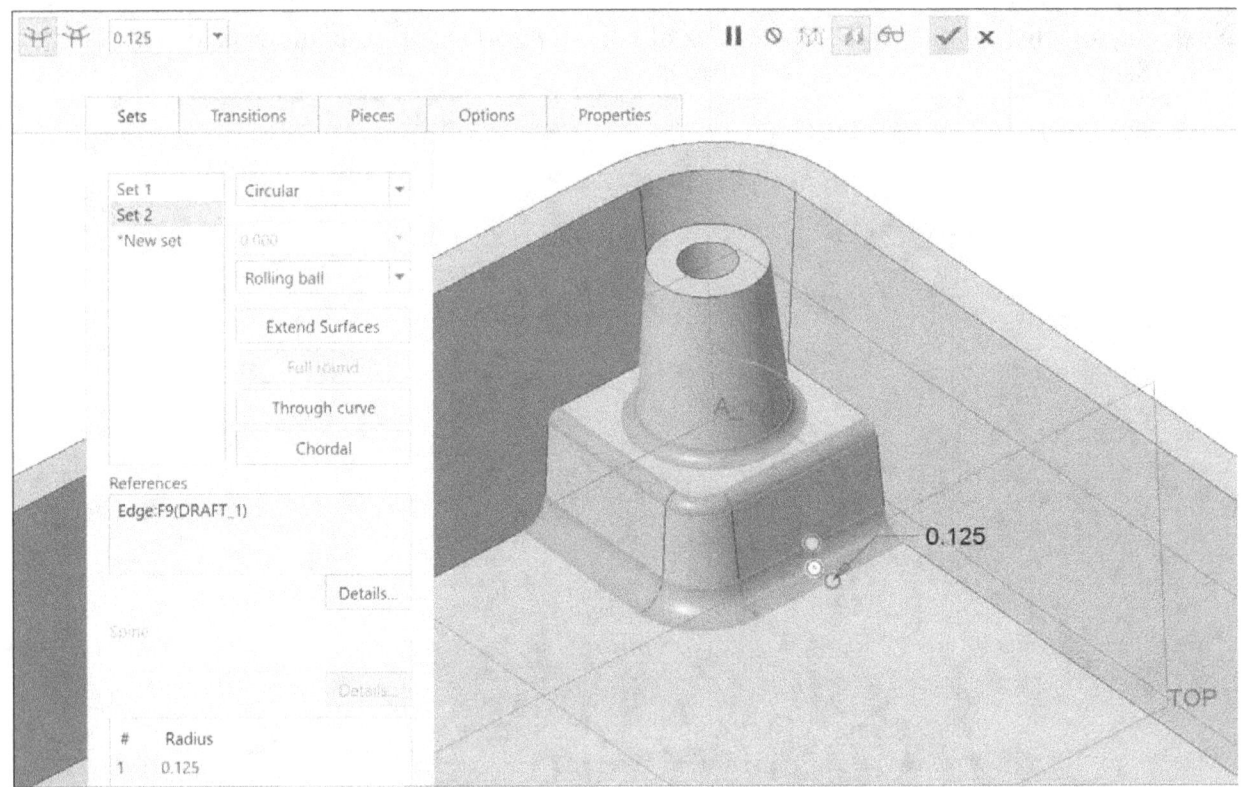

Figure 18.12(b) Round Set 2 (**R.125**) pedestal

Group the extrusions, the hole and the rounds. Select features > [⚙ Group] [Fig. 18.13(a)] > click twice on the group name > *type* **PEDESTAL** > **Enter** (▶ ⚙ Group PEDESTAL) [Fig. 18.13(b)] > **LMB > Ctrl+S**

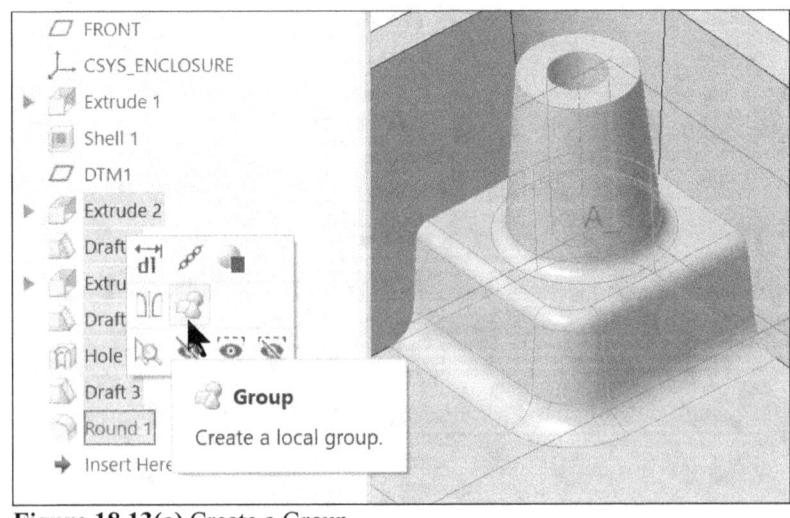

Figure 18.13(a) Create a Group

Figure 18.13(b) Group PEDESTAL

From the Model Tree, select: Group **PEDESTAL** > � **Mirror** > select datum **RIGHT** [Fig. 18.14(a)] > ✓ > with the **Ctrl** key pressed, select Group **PEDESTAL** and **Mirror 1** from the Model Tree > ⎯ **Mirror** > select datum **TOP** [Fig. 18.14(b)] > ✓ > **Ctrl+S** > **File** > **Manage File** > **Delete Old Versions** > **Enter**

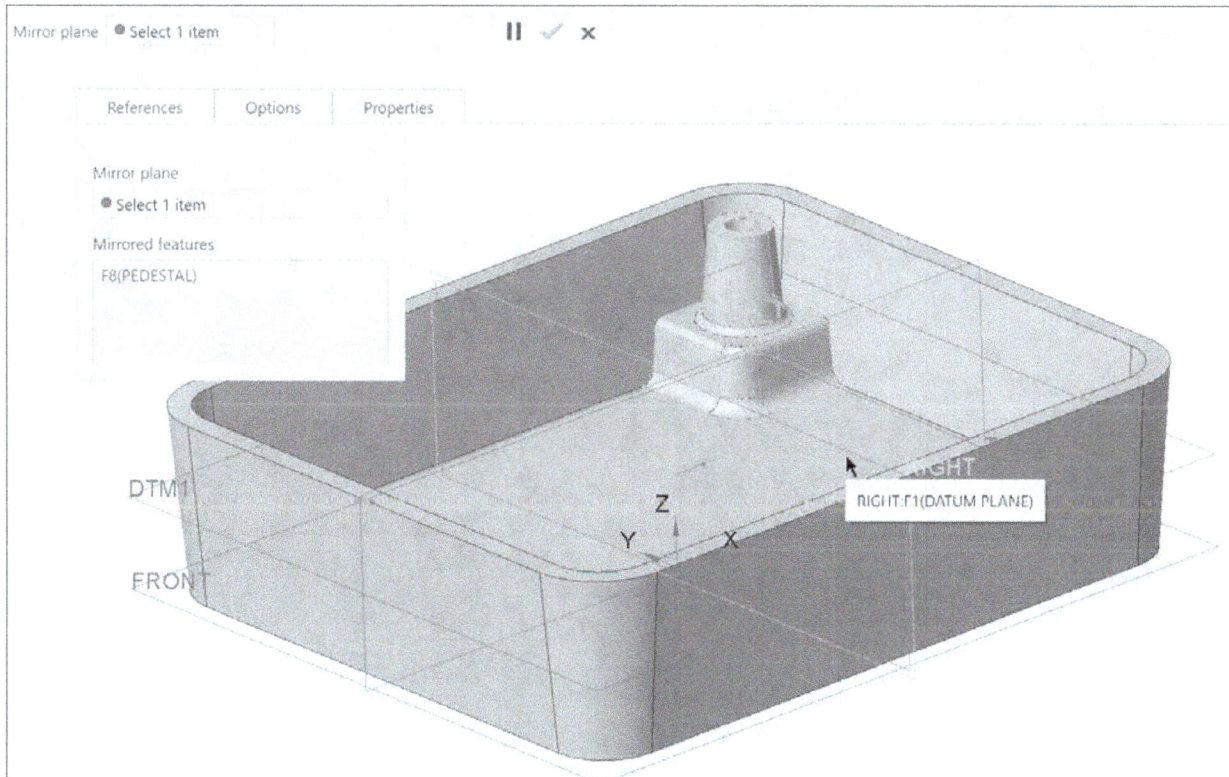

Figure 18.14(a) Group Copied and Mirrored about Datum RIGHT

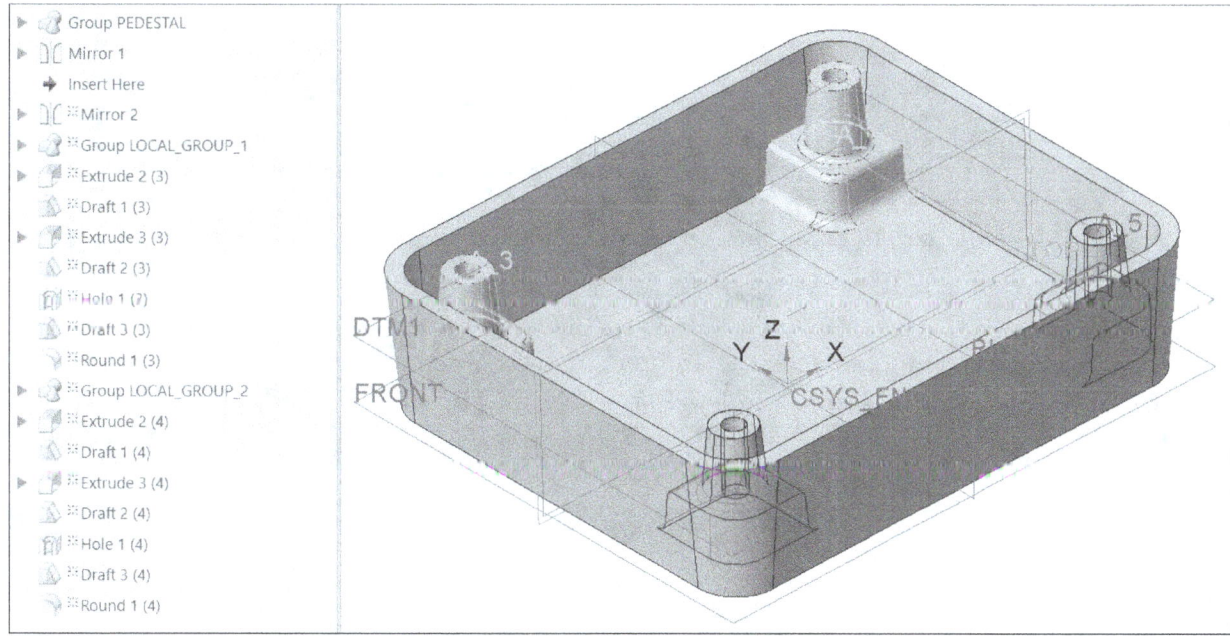

Figure 18.14(b) Previewed Groups Copied and Mirrored about Datum TOP

783

Create the internal round, click: 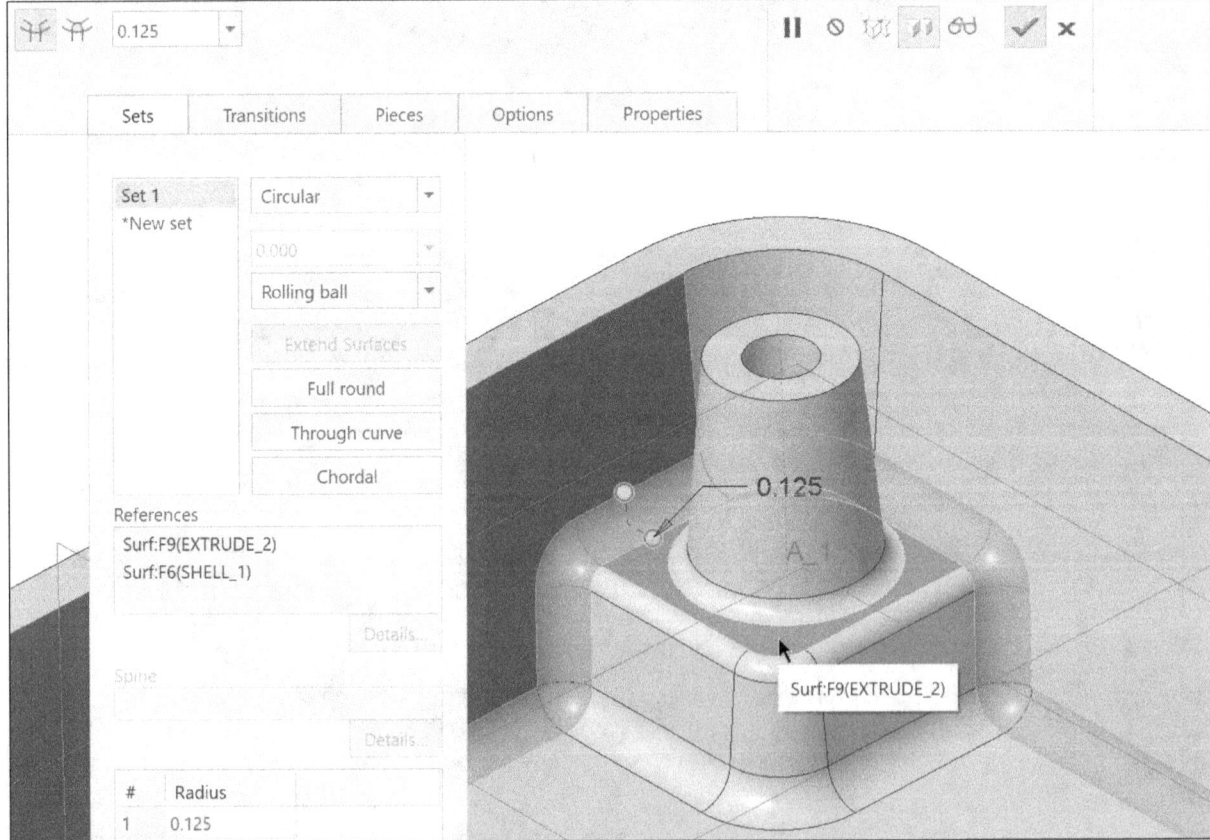 Round > **.125** > select the inside of the shelled wall as the first reference [Fig. 18.15(a)] > **Sets** tab > press and hold the **Ctrl** key > select the top surface of the pedestal as the second reference [Fig. 18.15(b)] > release the **Ctrl** key > ✓ > **LMB** to deselect > **Ctrl+S**

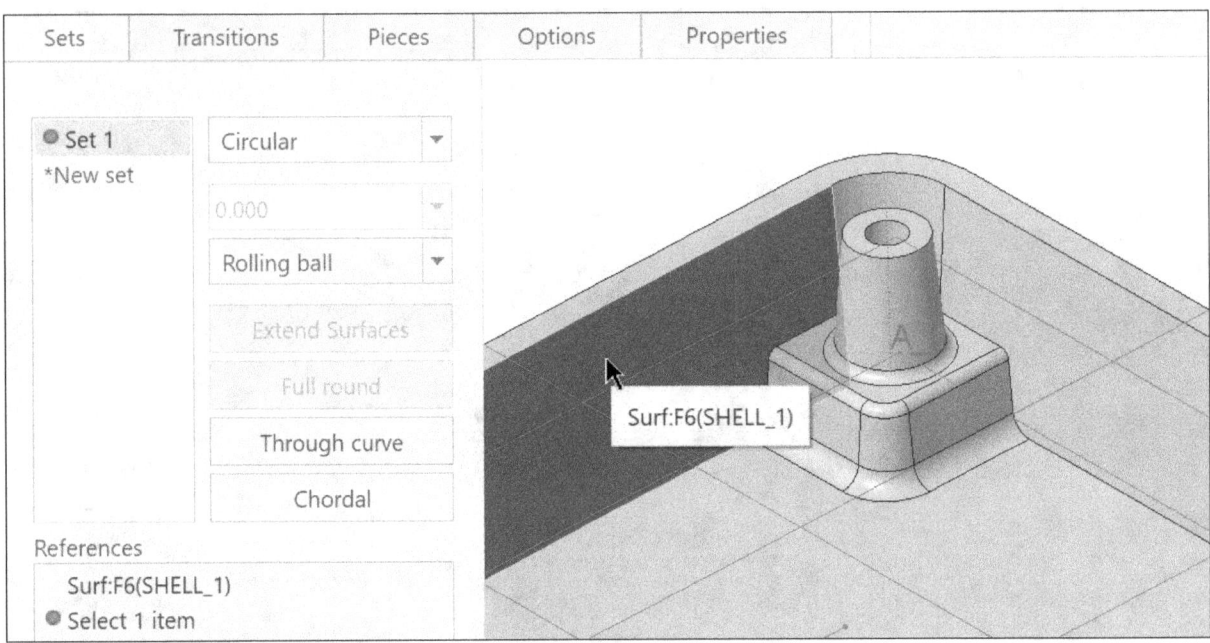

Figure 18.15(a) Select First Reference

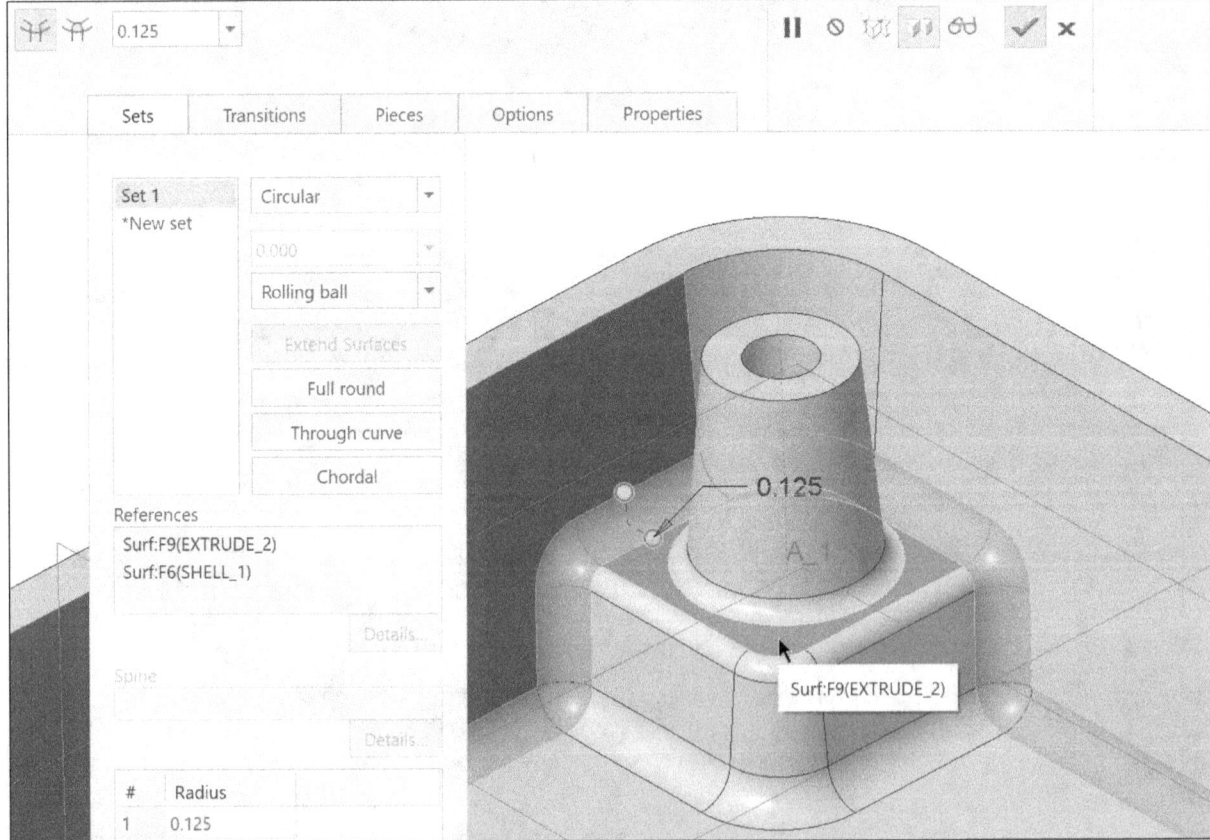

Figure 18.15(b) Select Second Reference

Click: ⟦ Plane ⟧ > References: select axis **A_1** [Fig. 18.16(a)] > press and hold the **Ctrl** key >
References: select axis **A_3** *(your id's may be different)* [Fig. 18.16(b)] > release the **Ctrl** key >
OK (creates DTM2)

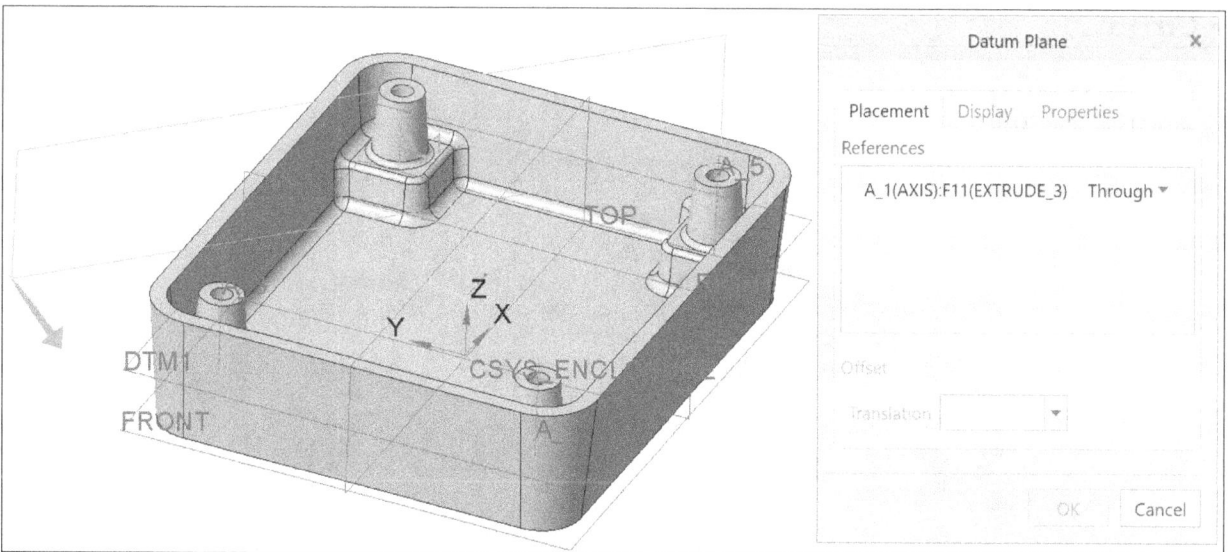

Figure 18.16(a) Select Axis A_1 *(your id's may be different)*

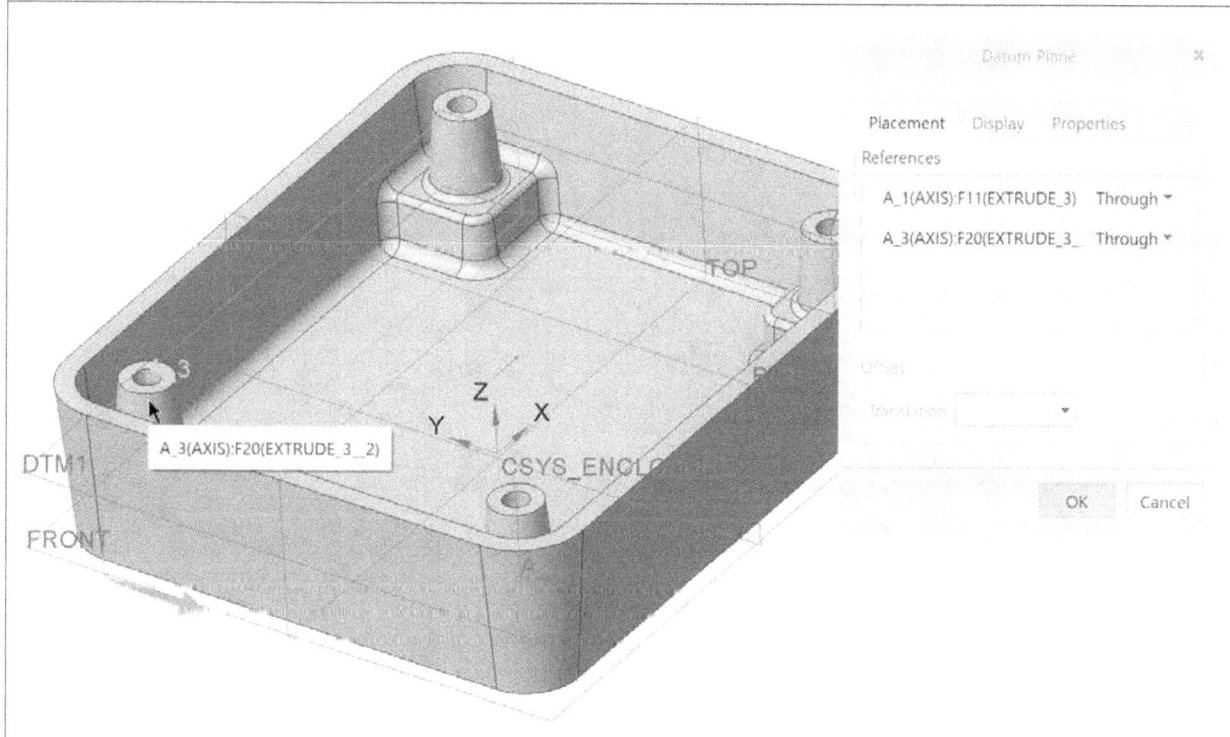

Figure 18.16(b) Select Axis A_7 *(your id's may be different)*

Create a cross section through the part. With DTM2 selected, lick: 📷 **View Manager > Sections** tab > **New > Planar** > *type name* **A > Enter** > set options as shown [Fig. 18.17(a)] > ☑ > **RMB** > ☐ Show Section [Fig. 18.17(b)] > select **No Cross Section** > press **RMB > Activate > Close** > **Ctrl+S**

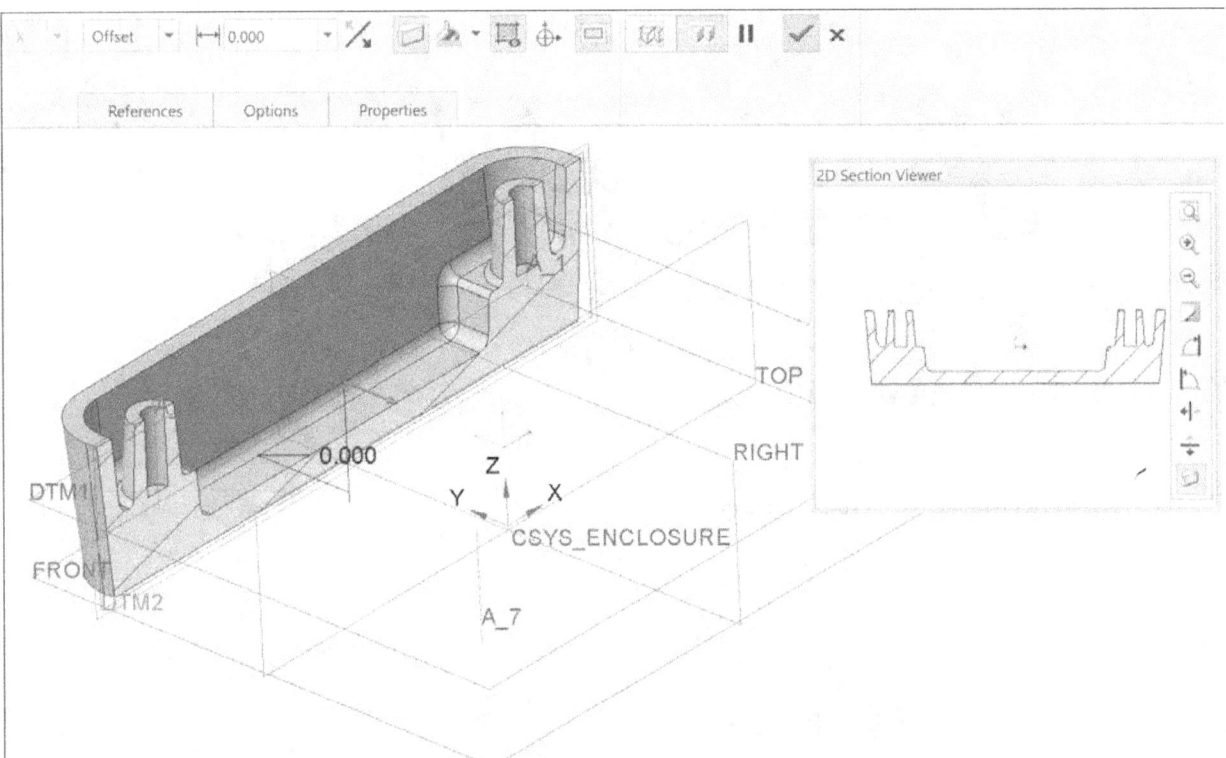

Figure 18.17(a) Active Section A with Hatching

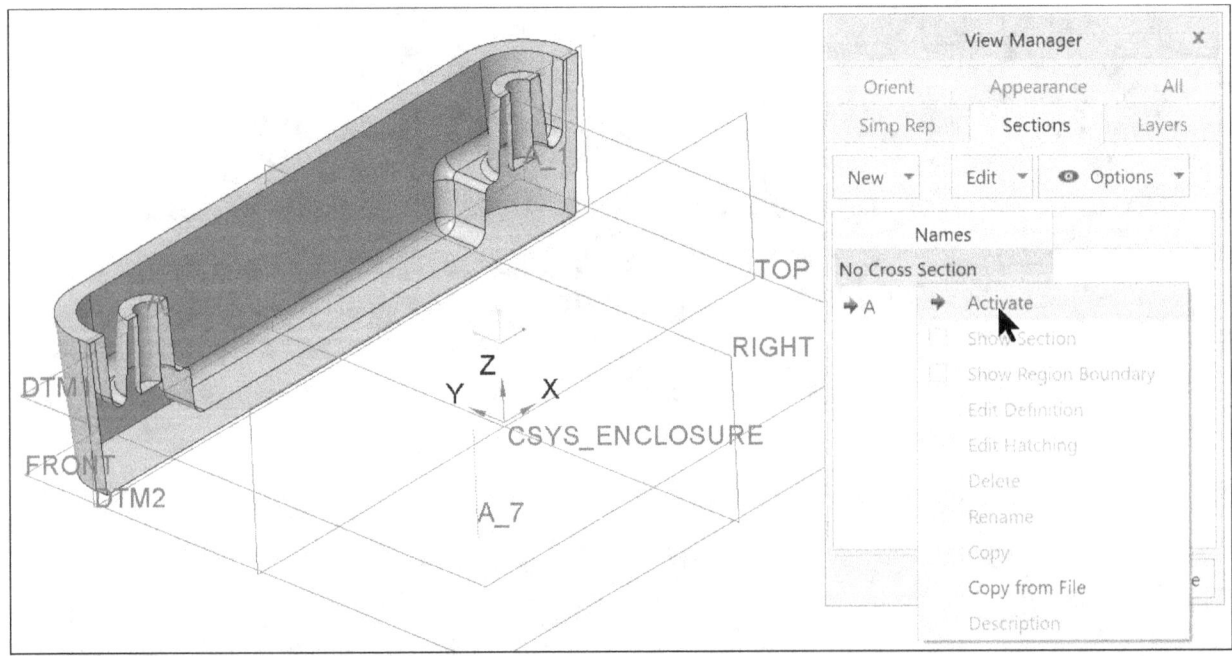

Figure 18.17(b) Activate No Cross Section

Expand the Model Tree to include the feature number and status. Click: [icon] **Settings > Tree Columns > Feat # >** [>> button] **> Feature Status >** [>> button] **> Apply > OK** (Fig. 18.18) > [icon] **> Tree Filters > toggle** *on* all options **> Apply > OK**

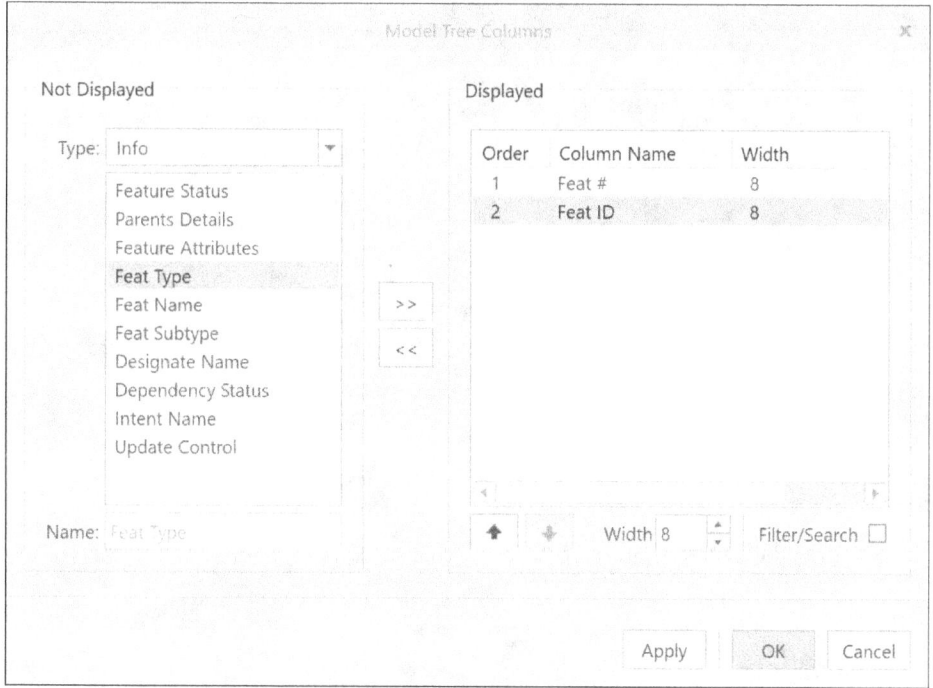

Figure 18.18 Model Tree Items Dialog Box

Click on **Group PEDESTAL** in the Model Tree **>** press and hold the **Shift** key **>** click **DTM2** in the Model Tree (or the last feature in the Model Tree before *Insert Here*) [Fig. 18.19(a)] **>** release the **Shift** key **>** press [icon] **Suppress** [Fig. 18.19(b)] **> OK** [Fig. 18.19(c)] **> LMB > Ctrl+S**

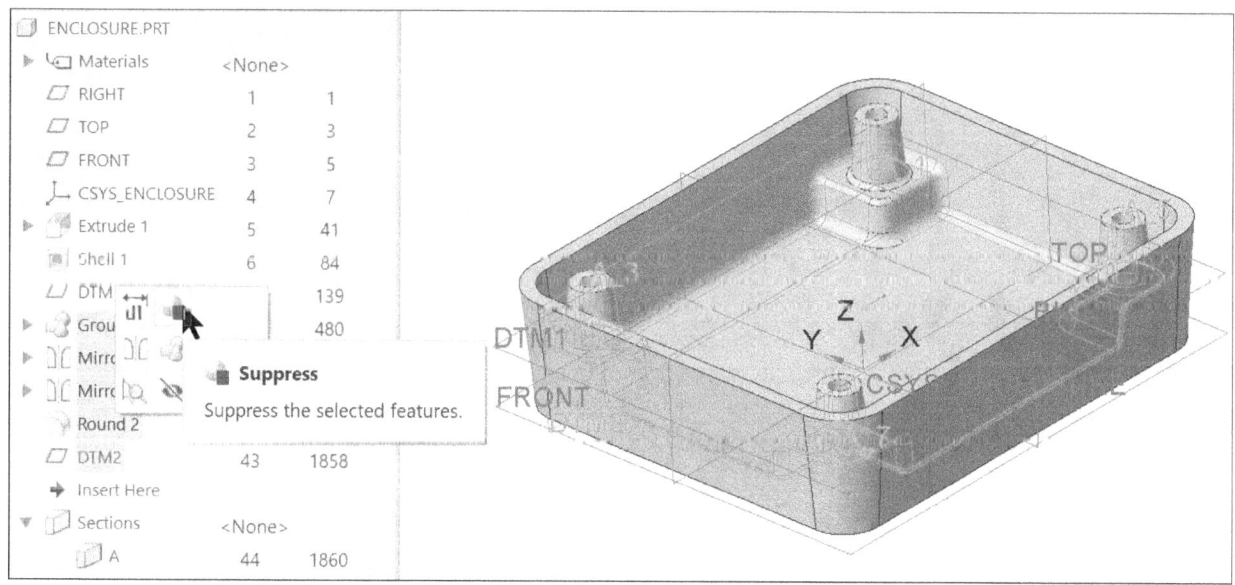

Figure 18.19(a) Select the Features in the Model Tree to be Suppressed *(Your Model Tree will Look Different.)*

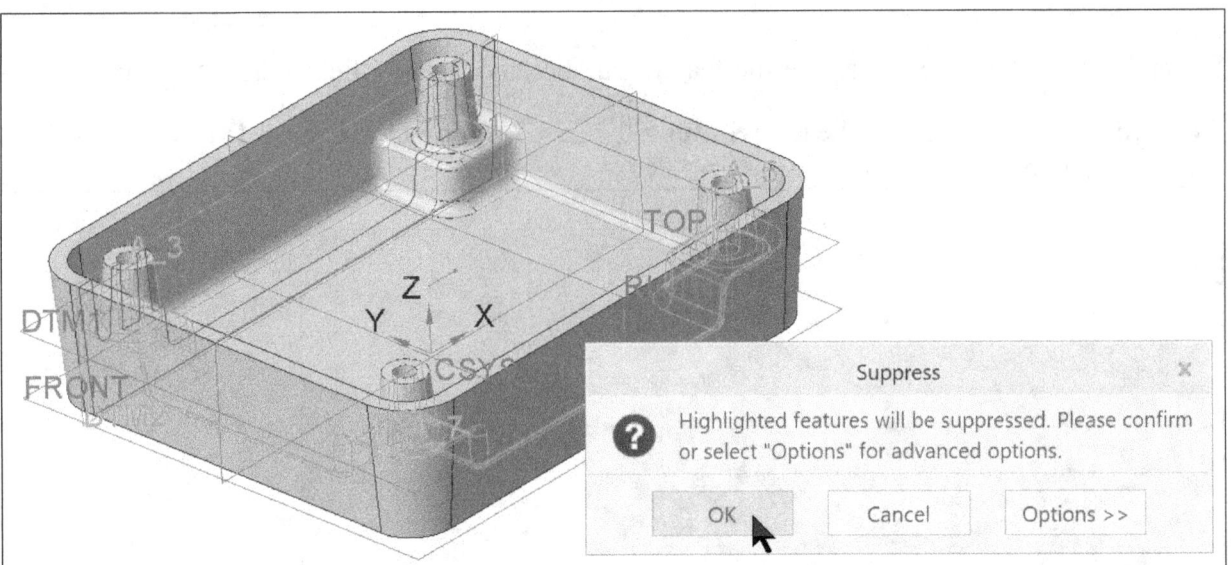

Figure 18.19(b) Highlighted Features

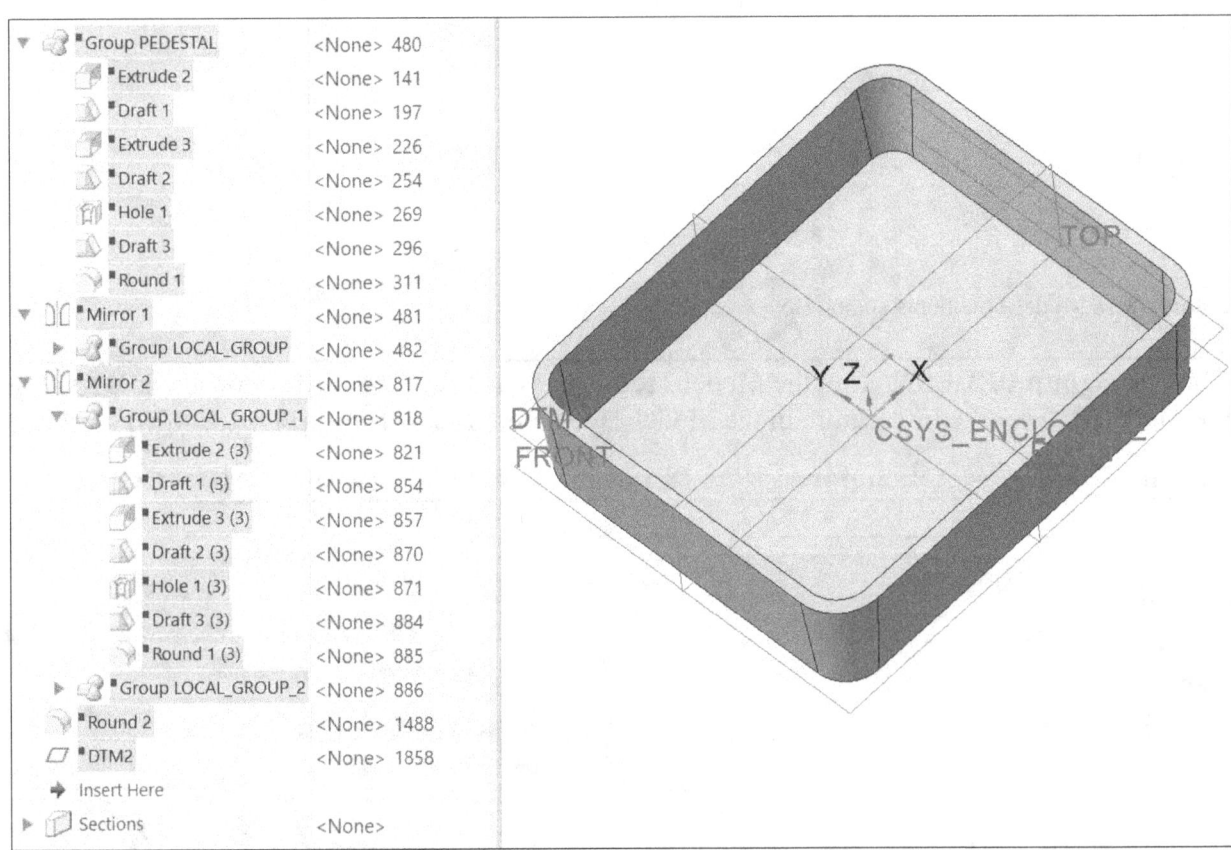

Figure 18.19(c) Make sure all ⬚ Tree Filters are checked to display ☑ Suppressed objects)

The regeneration time for your model will now be shorter. Next, add the text extrusion.

Press: **Ctrl+D** > [Extrude] > **Placement** tab > **Define** > select the *inside surface of the enclosure* for the sketching plane > **Sketch** [Fig. 18.20(a)] > [ ] > [A Text] > select *start point* of a line to determine text starting position [Fig. 18.20(b)] > select *second point* of a line to determine text height and orientation [Fig. 18.20(c)]

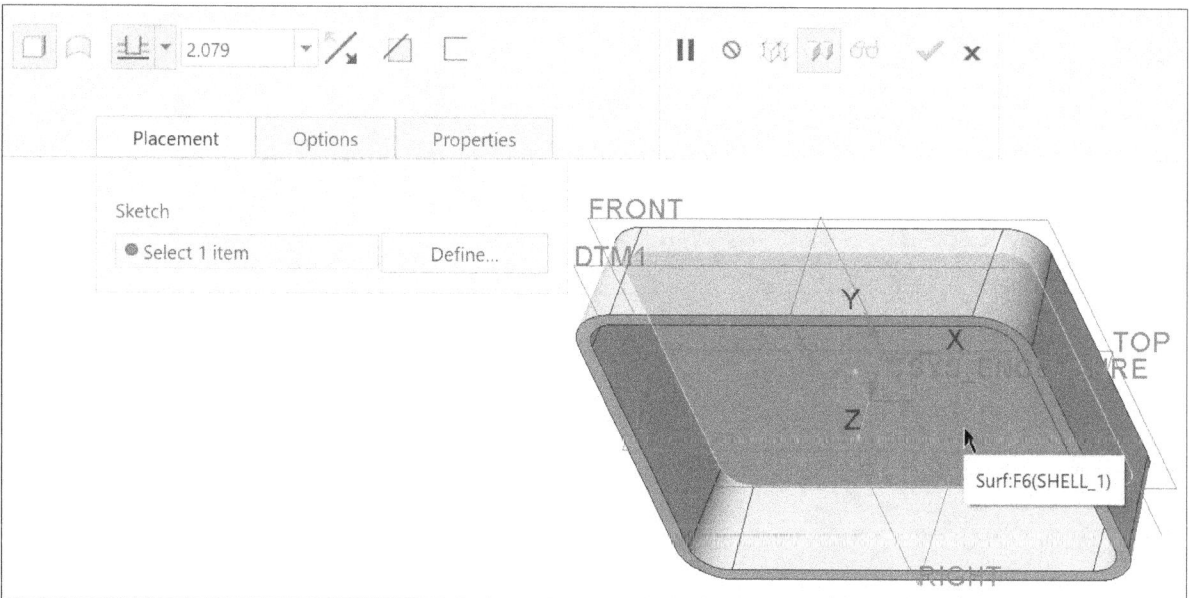

Figure 18.20(a) Sketching Plane, Inside Surface

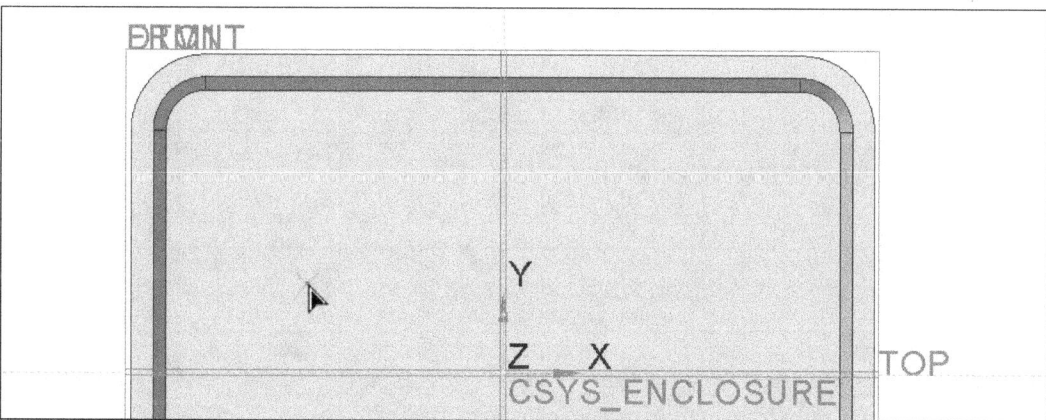

Figure 18.20(b) Pick First Point to Determine the Starting Point of the Lettering

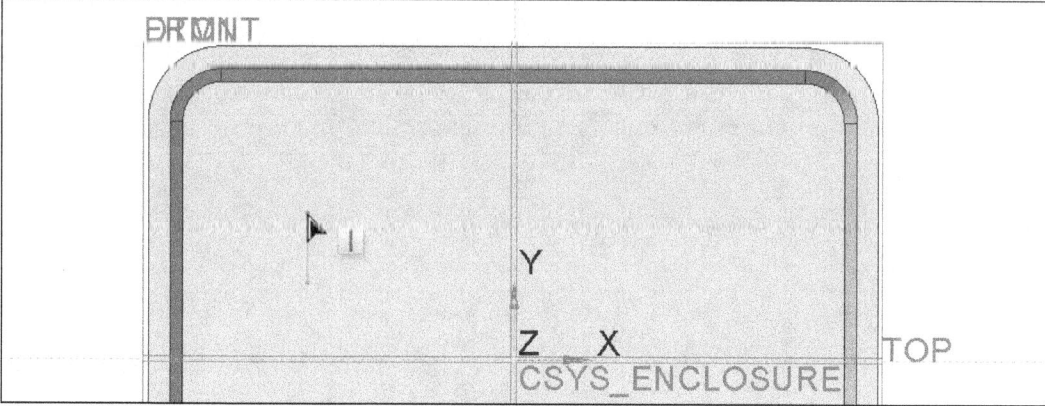

Figure 18.20(c) Pick Second Point to Determine the Height of the Lettering

Click: 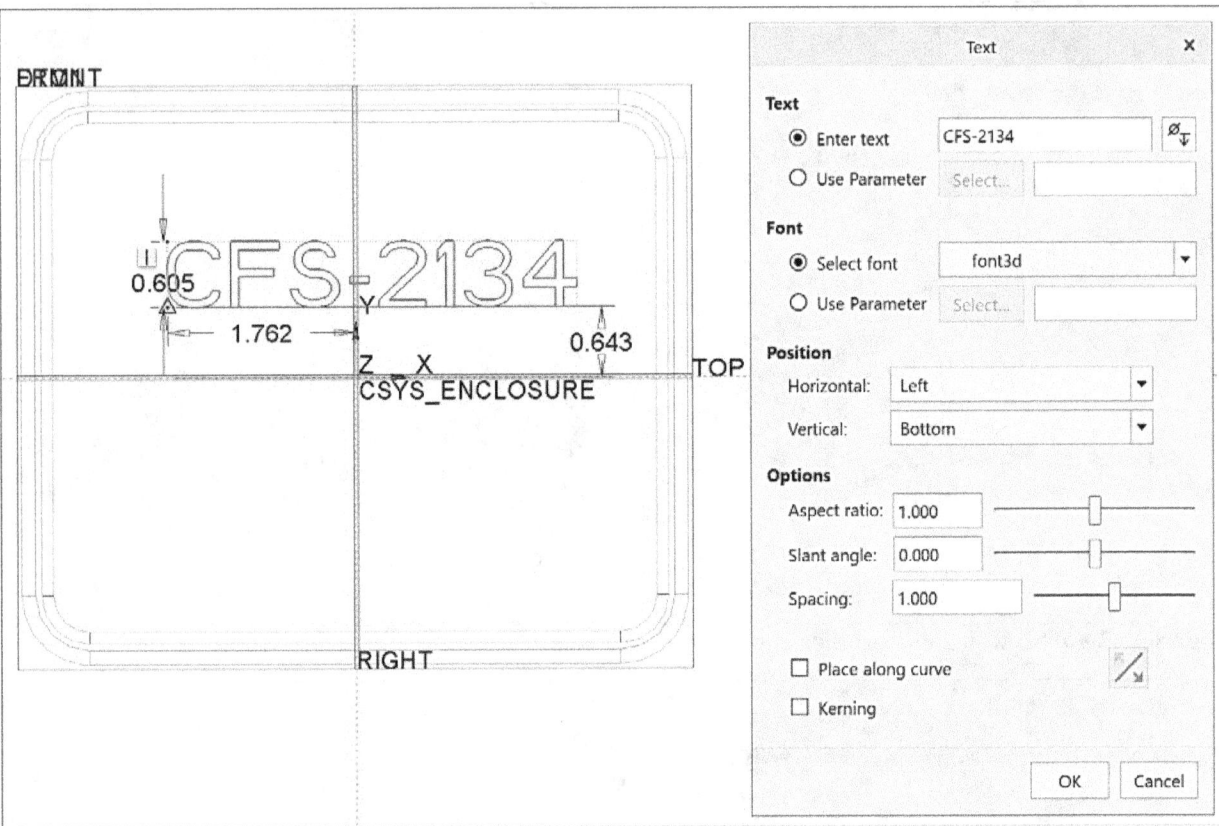 > ☐ (Select All) all *off* > **LMB** in the Graphics Window > *type* **CFS-2134** in Text line field [Fig. 18.20(d)] > **OK** > **MMB** > window-in (select) the sketch to capture all dimensions > **Modify** > modify the dimensions [Fig. 18.20(e)] > **OK** > **LMB** > ✓

Figure 18.20(d) Type the Text "**CFS-2134**" (case sensitive)

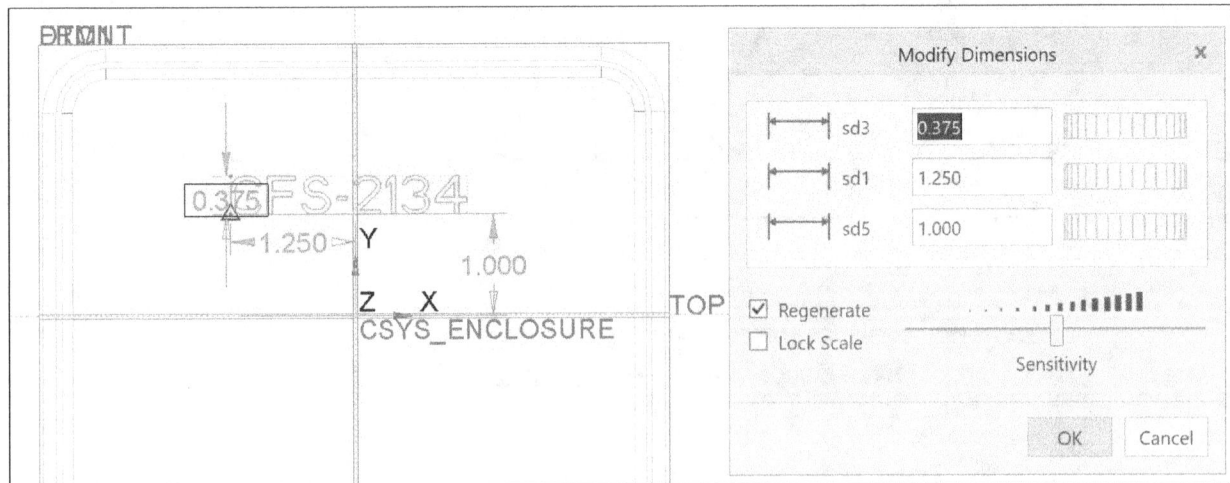

Figure 18.20(e) Modified Dimensions

Press: **MMB** to spin the model [Fig. 18.20(f)] > double-click on the height dimension and modify it to **.0625** > **Enter** [Fig. 18.20(g)] > ✓ [Fig. 18.20(h)] > **Ctrl+S** > **LMB**

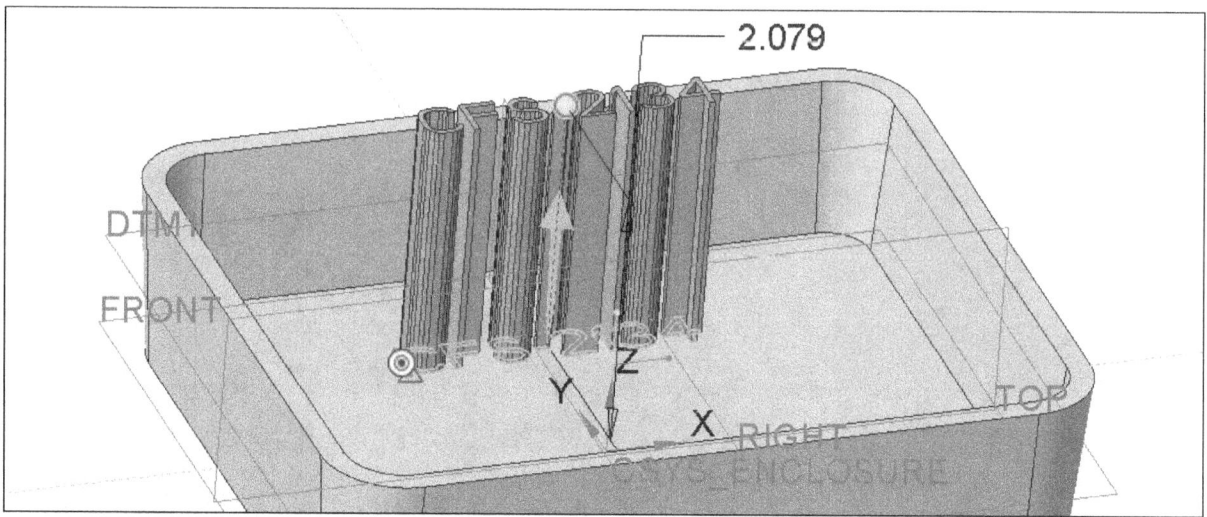

Figure 18.20(f) Dynamic Preview

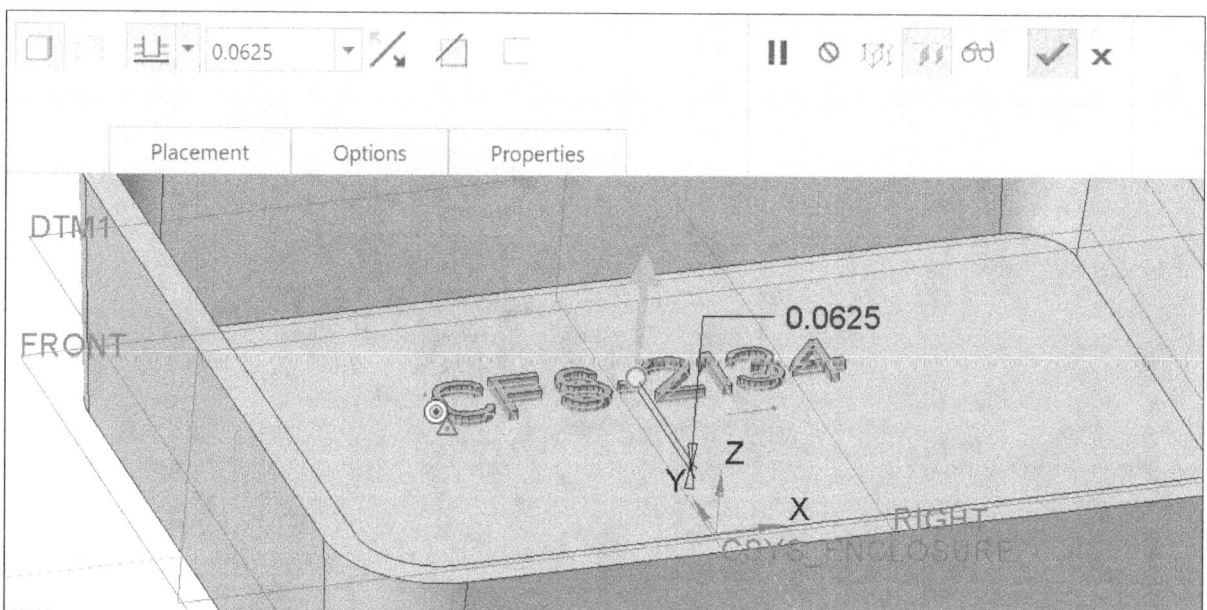

Figure 18.20(g) Modified Depth

Figure 18.20(h) Completed Text Extrusion *(Your extrude id may be different) (Your Model Tree may Look Different.)*

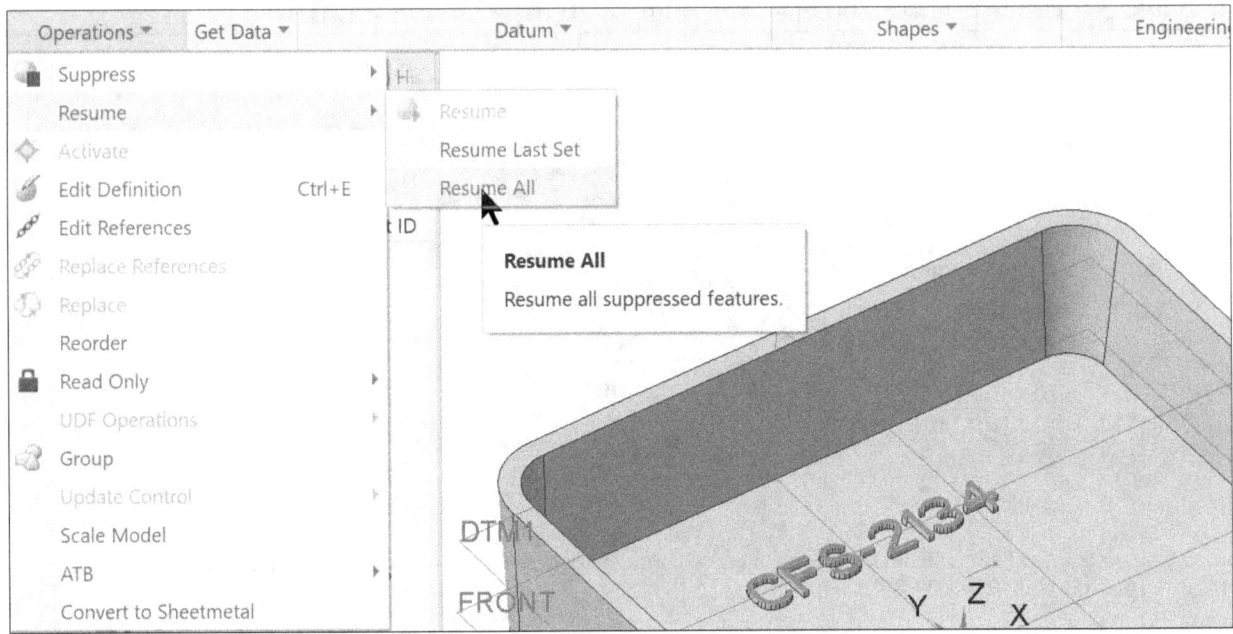

Figure 18.21(a) Resume

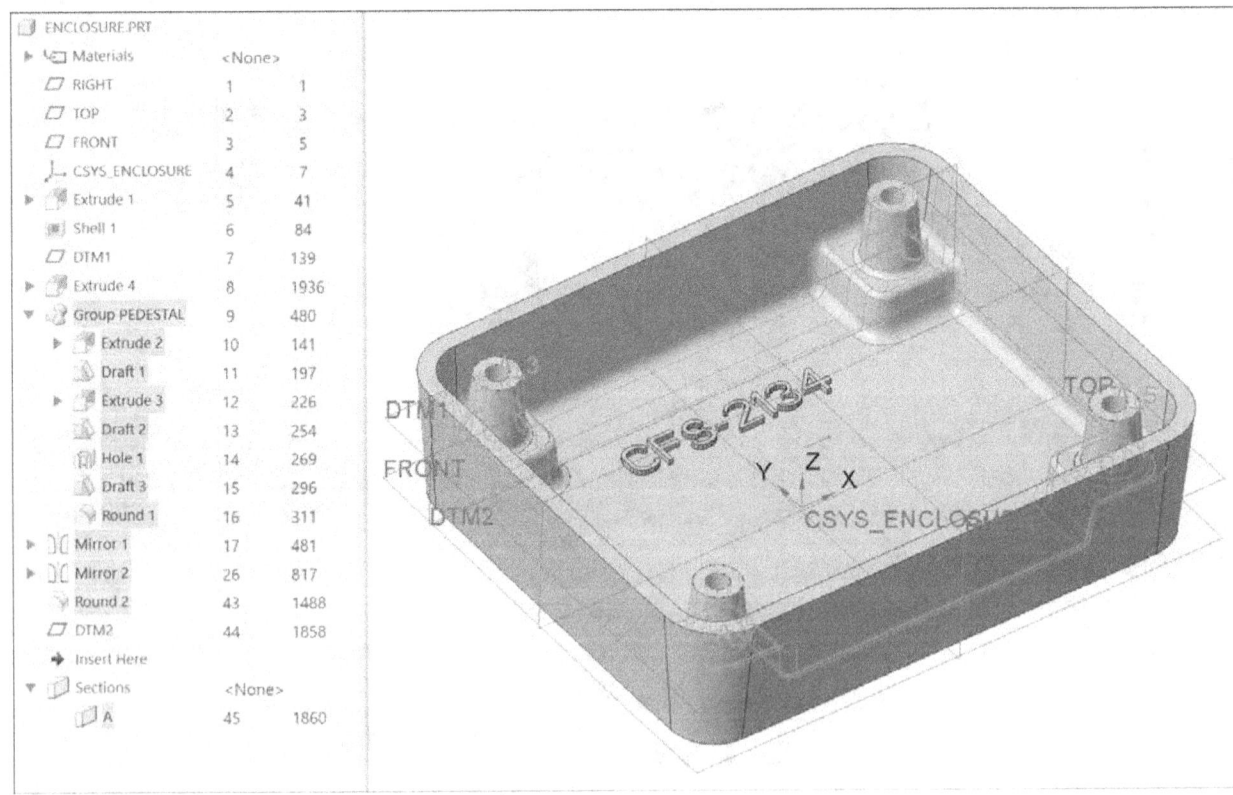

Figure 18.21(b) Suppressed Features Resumed

Press: **MMB** to spin the part > 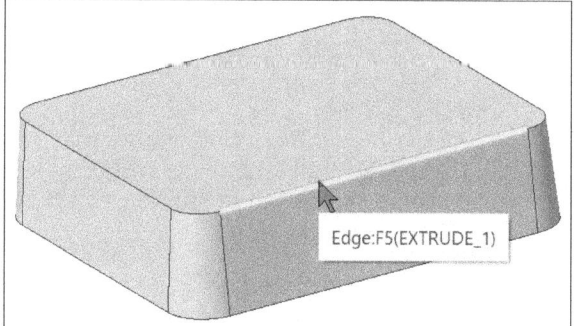 > shade   ✕ 🔍 >
Temporary **Shade** > select the parts' edge [Fig. 18.22(a)] > press and hold down the **Shift** key >
select the top surface > [Fig. 18.22(b)] > pull the drag handle [Fig. 18.22(c)] > modify the
round to **.1875** [Fig. 18.22d)] > **Enter** > ✓ > **LMB** to deselect > **View tab** > **Model Display** >
Temporary Shade [Fig. 18.22(e)]

Figure 18.22(a) Select the Edge

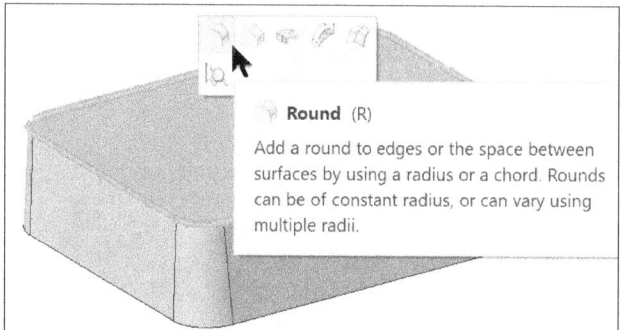

Figure 18.22(b) Shift + Select the Surface

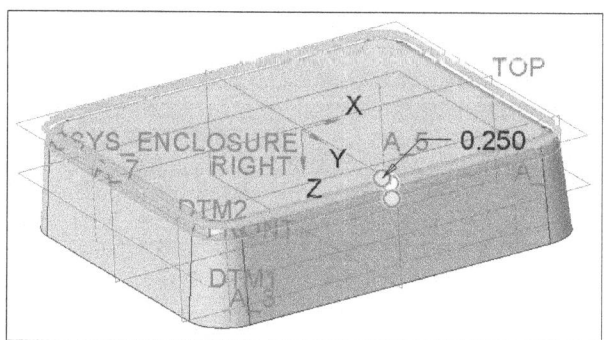

Figure 18.22(c) Move Drag Handle

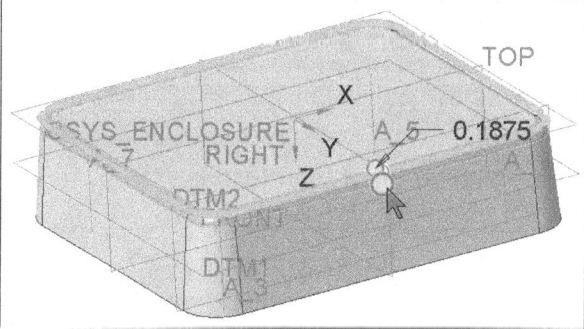

Figure 18.22(d) Modify to **.1875**

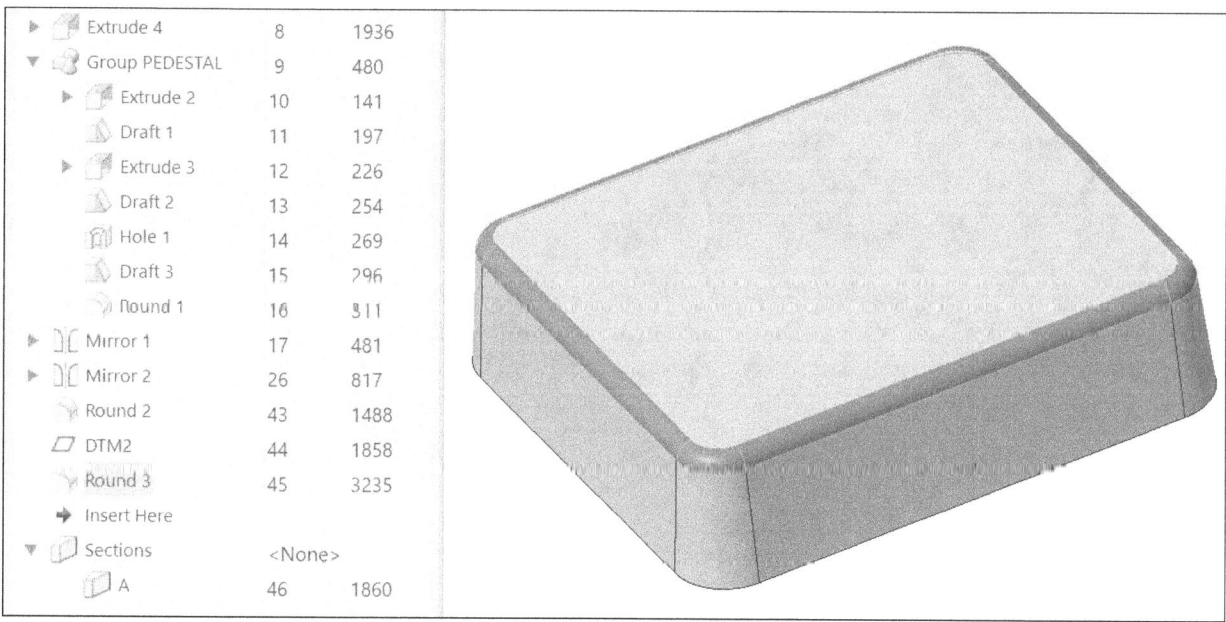

Figure 18.22(e) Completed Round

Press: **Ctrl+D** > select the text extrusion in the Model Tree > **View** tab > Orientation ▼ > Orient Mode > **RMB** > ● Velocity [Fig. 18.23(a)] > hold down the **MMB** and move the cursor about the screen to orbit the model [Fig. 18.23(b)] > release the **MMB** > press **RMB** > **Exit Orient Mode**

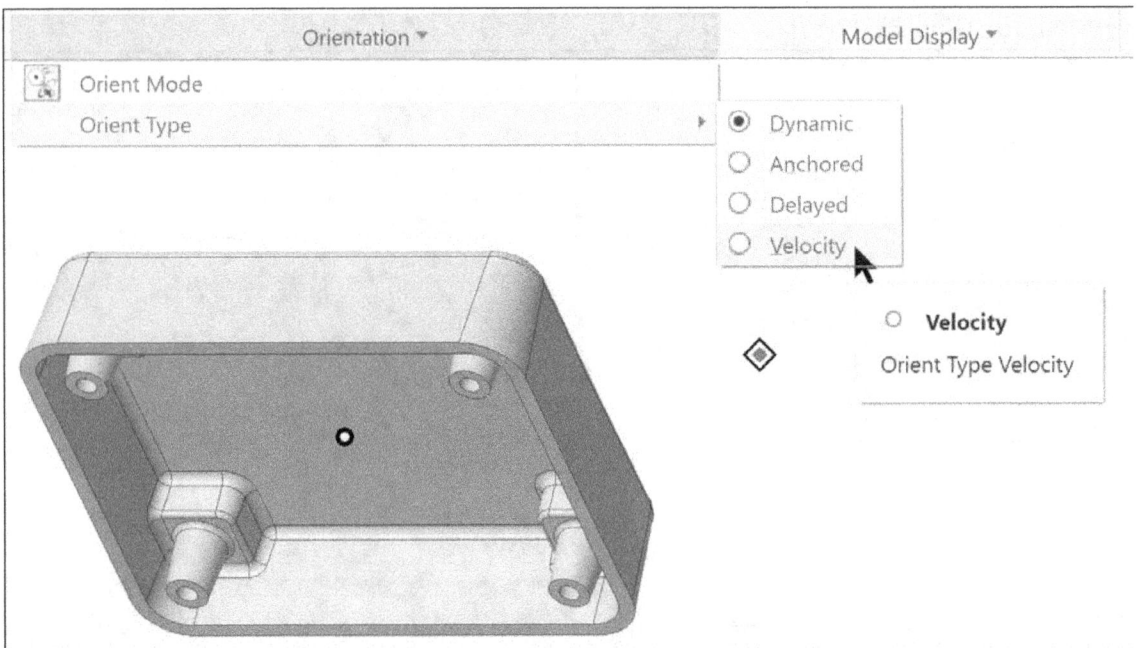

Figure 18.23(a) Orient Mode Velocity

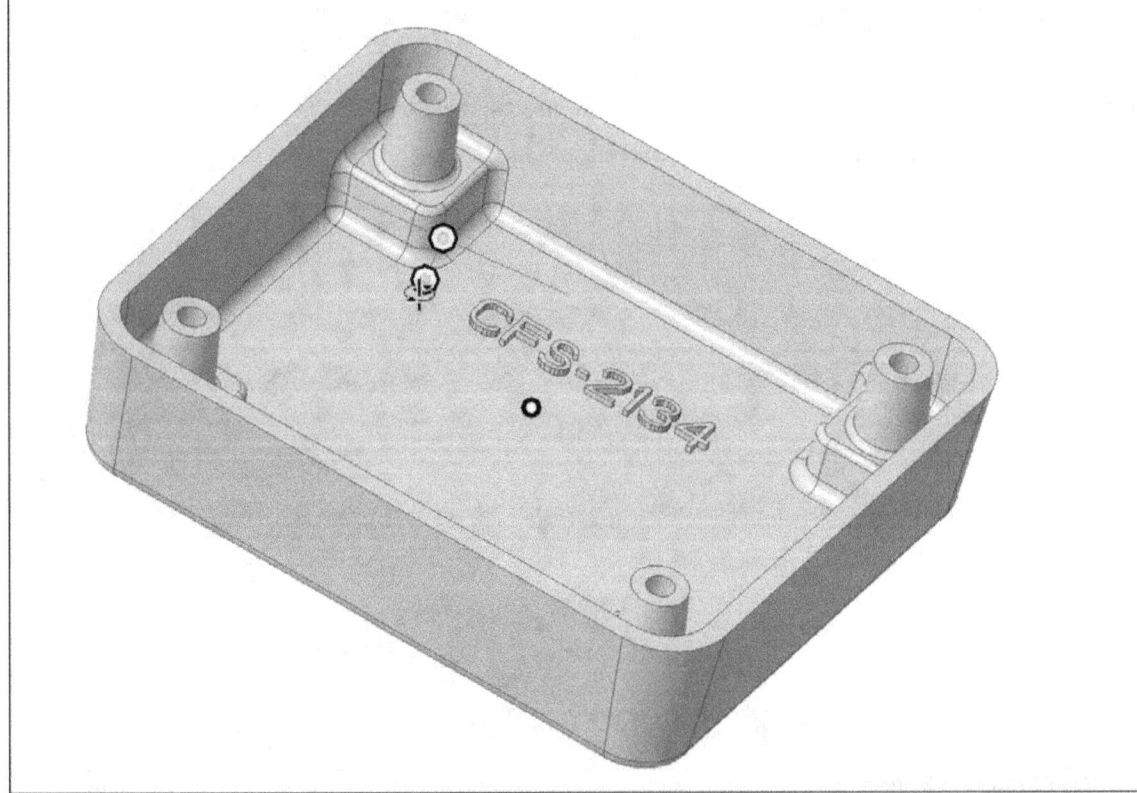

Figure 18.23(b) Spinning the Model in Real Time

With the text extrusion selected, press: **RMB >** 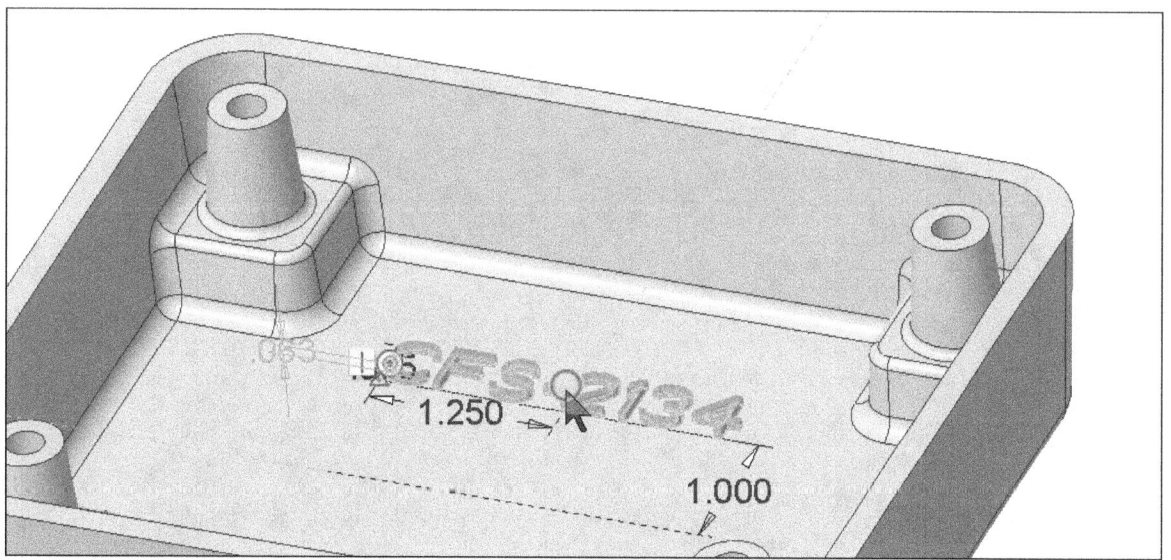 **Edit > Model** tab [Fig. 18.24(a)] > drag the (depth) handle to **.250** [Fig. 18.24(b)] > **Ctrl+G** Regenerate > **Ctrl+S** > **File > Manage File > Delete Old Versions > Enter** [Fig. 18.24(c)] > **Ctrl+D > LMB** to deselect

Figure 18.24(a) Edit the Text Extrusion

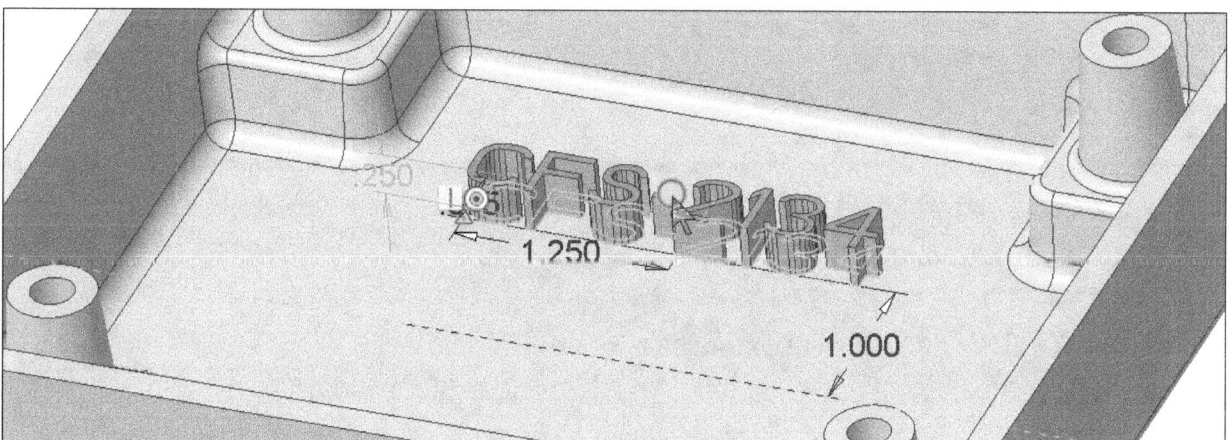

Figure 18.24(b) Drag the (White) Square Drag Handle to **.250**

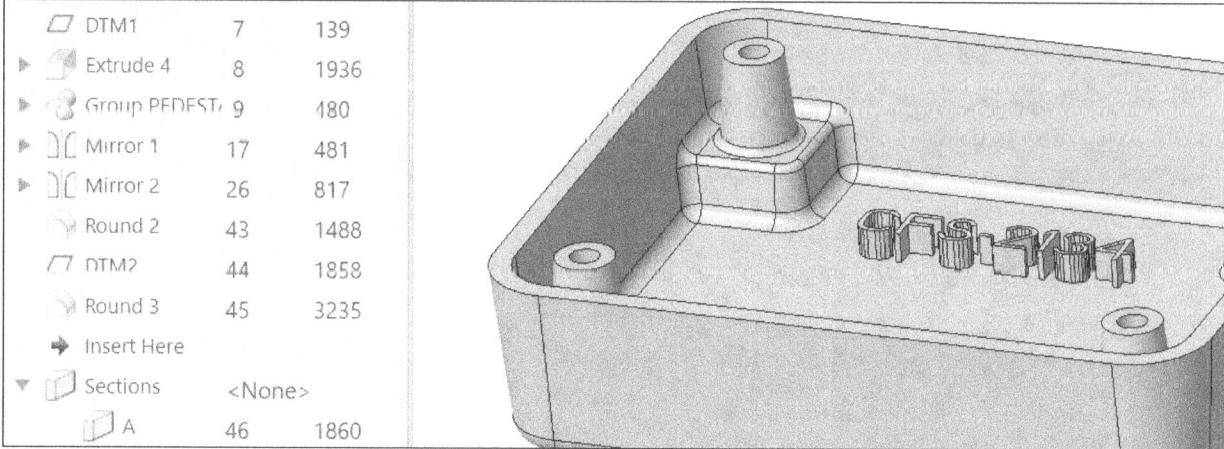

☐ DTM1	7	139	
▶ Extrude 4	8	1936	
▶ Group PEDEST/	9	180	
▶ Mirror 1	17	481	
▶ Mirror 2	26	817	
Round 2	43	1488	
☐ DTM2	44	1858	
Round 3	45	3235	
➡ Insert Here			
▼ ☐ Sections	<None>		
☐ A	46	1860	

Figure 18.24(c) Completed Enclosure

Click: **File > Prepare > ModelCHECK Geometry Check** [Fig. 18.25(a)] **> OK >** ⬚ [Fig. 18.25(b)] **> Ctrl+S >** 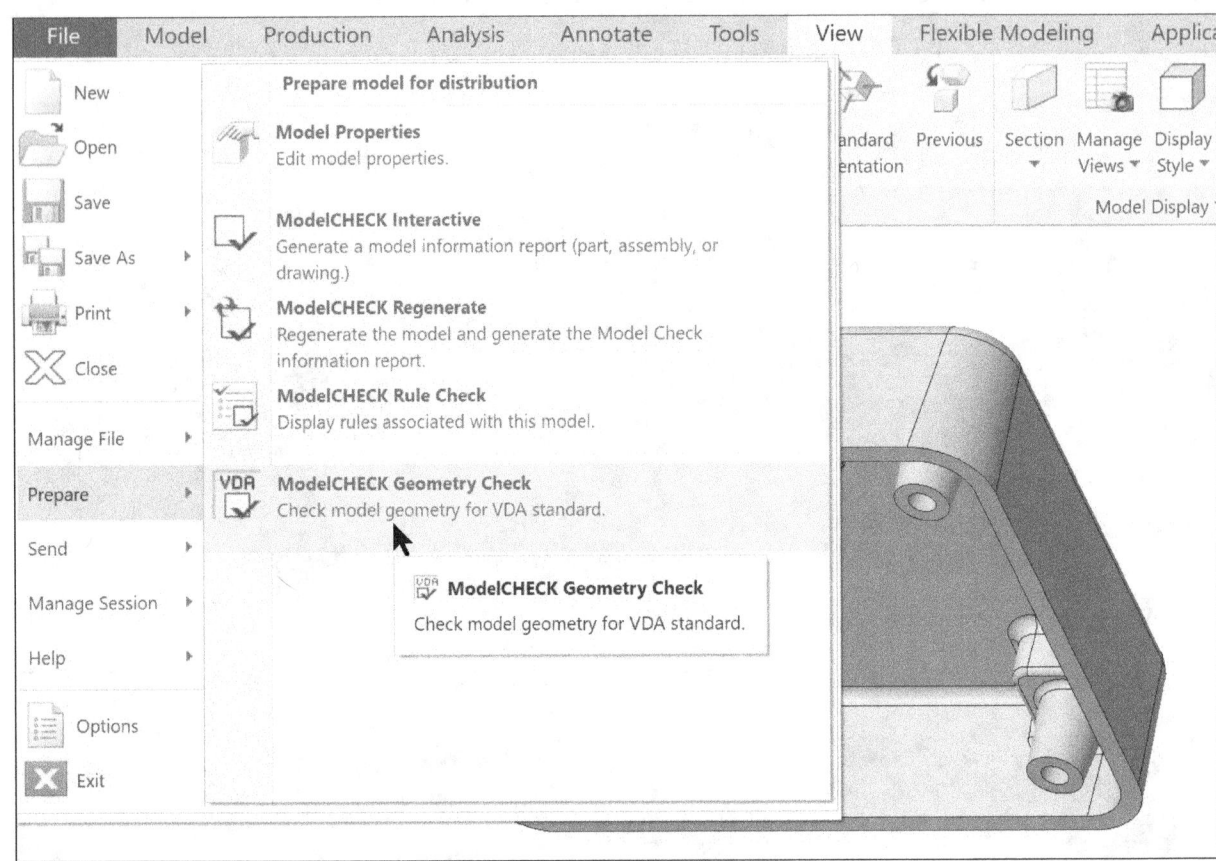 close the Browser **> Ctrl+D**

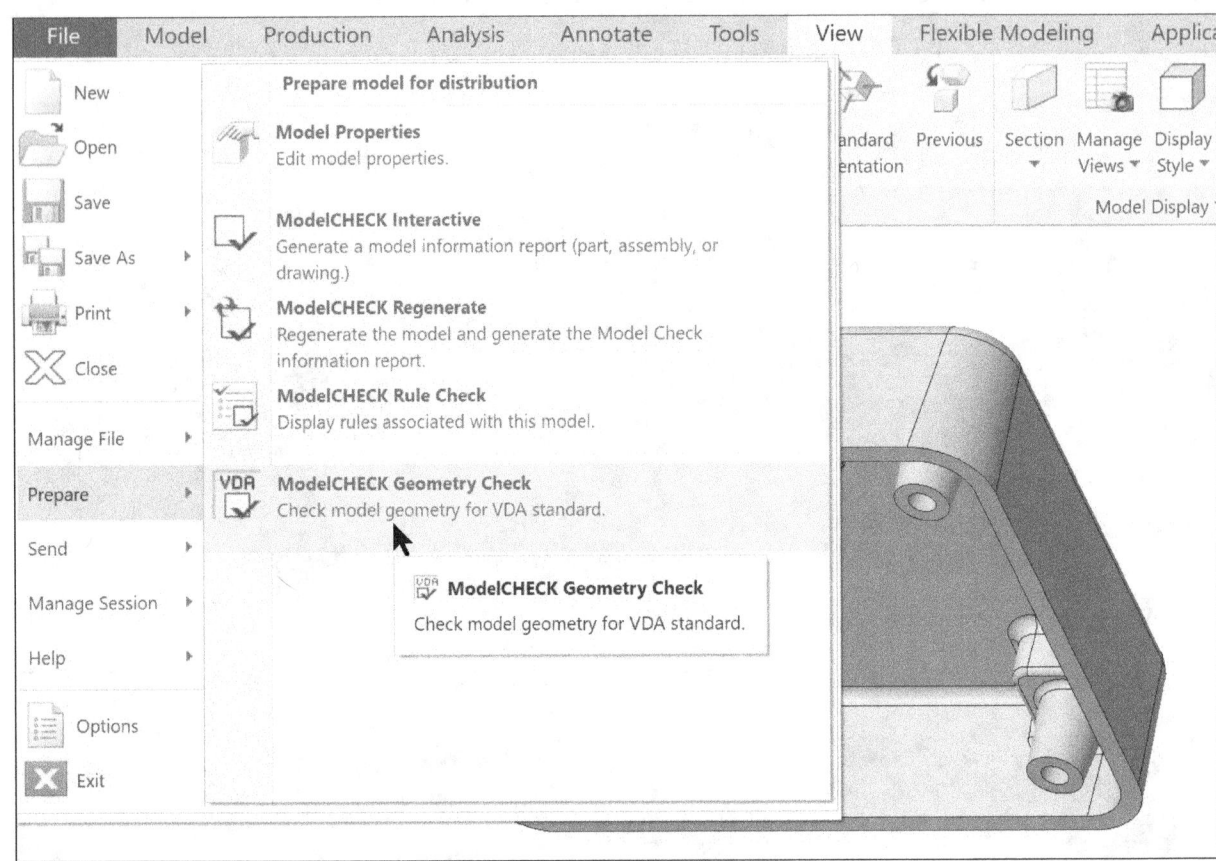

Figure 18.25(a) ModelCHECK Geometry Check

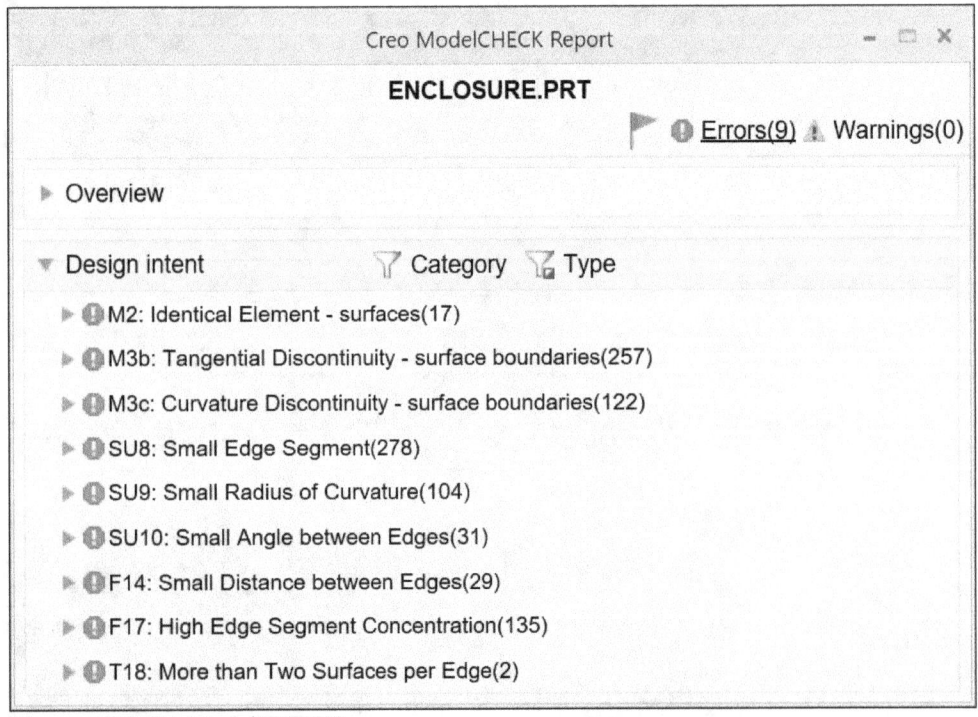

Figure 18.25(b) ModelCHECK

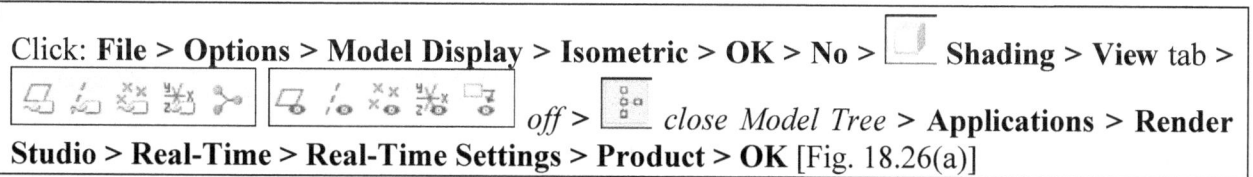

Click: **File > Options > Model Display > Isometric > OK > No >** ▭ **Shading > View** tab > ▭▭▭▭▭ ▭▭▭▭▭ *off >* ▭ *close Model Tree >* **Applications > Render Studio > Real-Time > Real-Time Settings > Product > OK** [Fig. 18.26(a)]

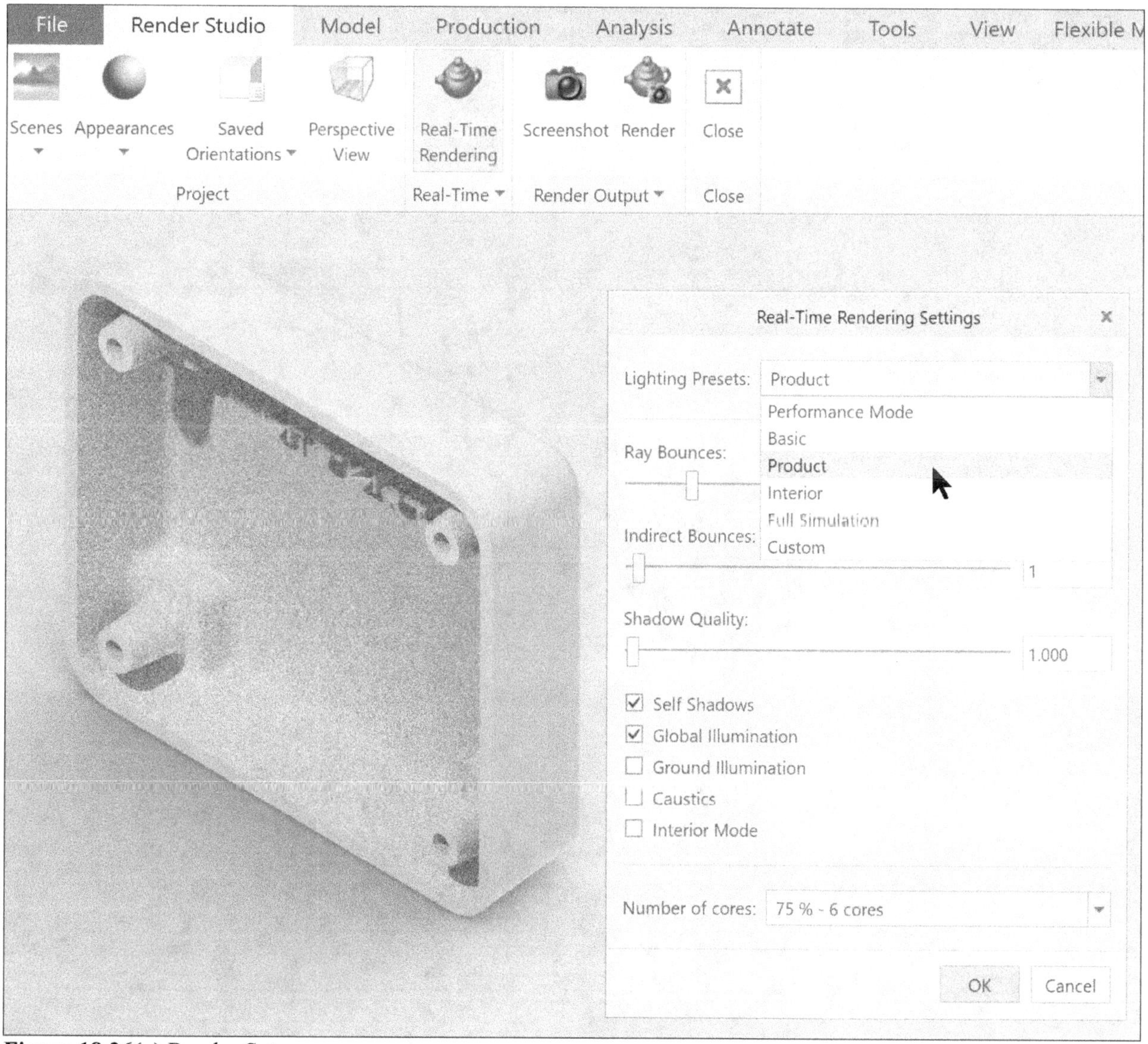

Figure 18.26(a) Render Setup

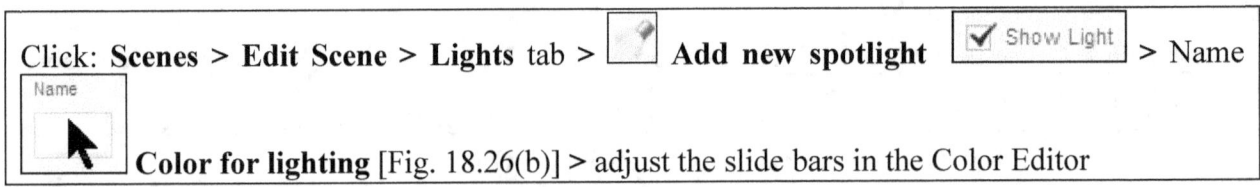

Click: **Scenes > Edit Scene > Lights** tab > ☐ **Add new spotlight** ☑ Show Light > Name

Color for lighting [Fig. 18.26(b)] > adjust the slide bars in the Color Editor

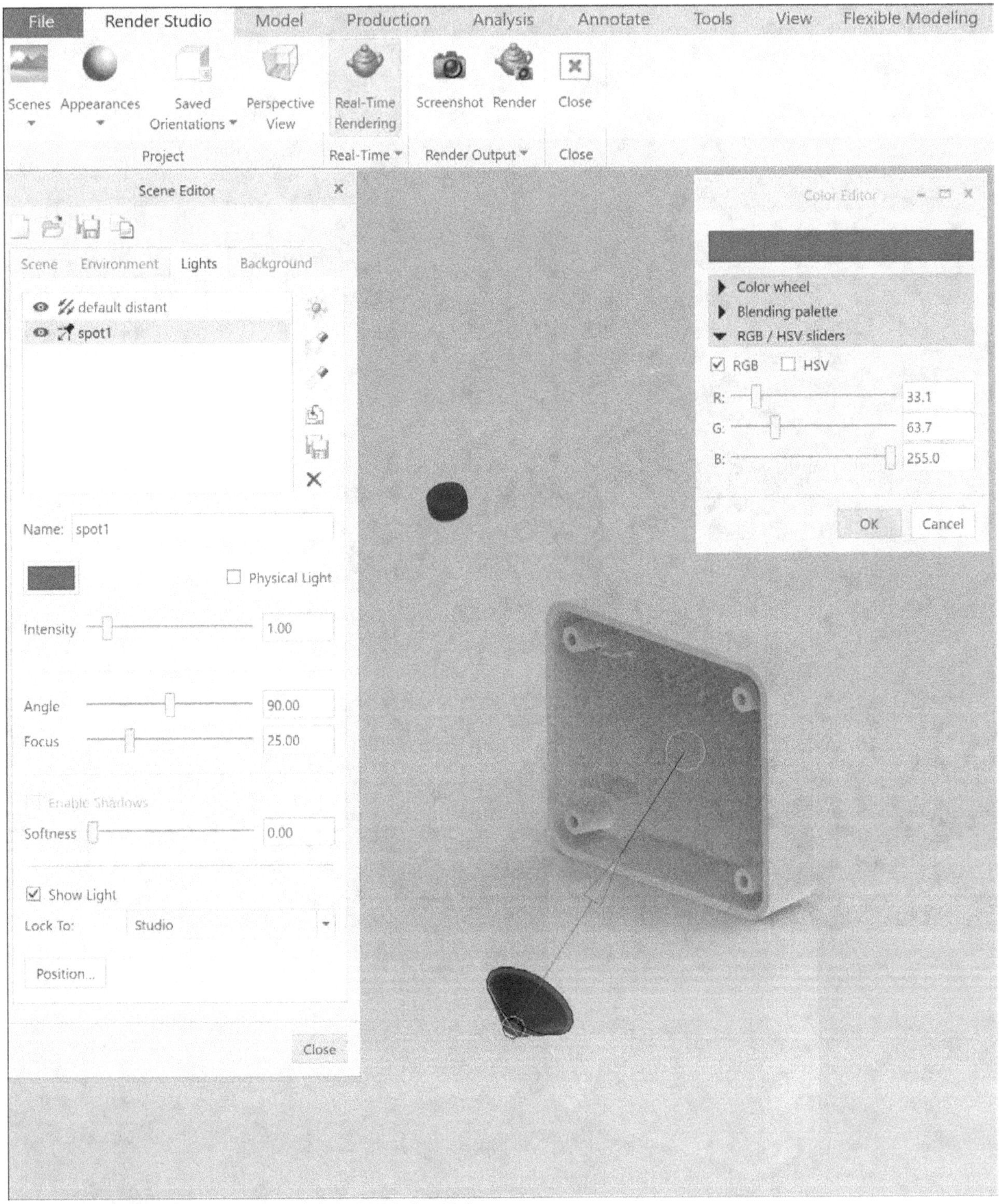

Figure 18.26(b) Adjust the Spot Light RGB Color Values

Click: **OK** (from the Color Editor dialog box) > place the pointer on the circle just behind the spot light (highlights) > **LMB** [Fig. 18.26(c)] > move again [Fig. 18.26(d)]

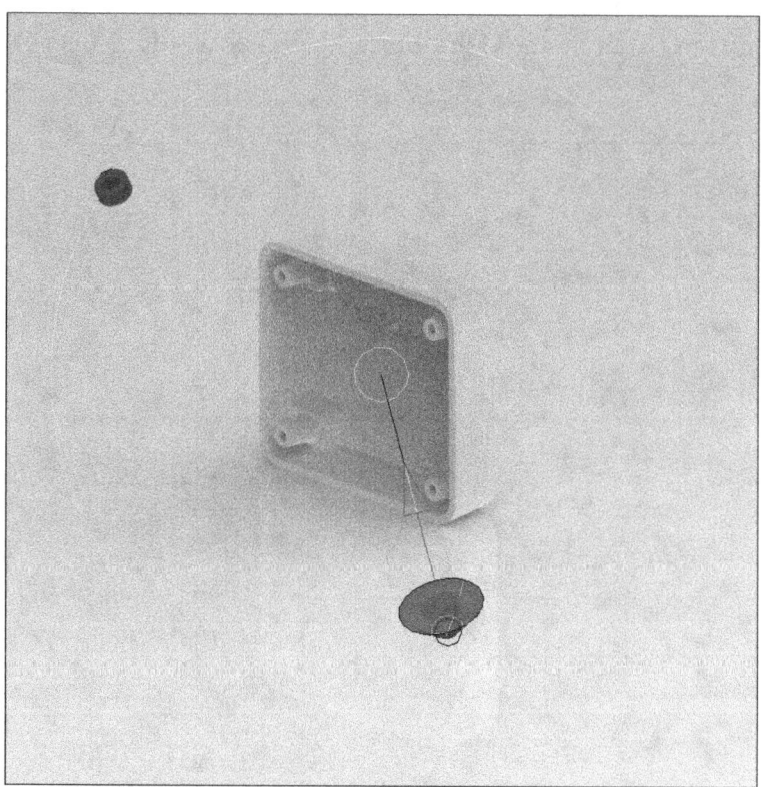

Figure 18.26(c) Change Spot Light Position

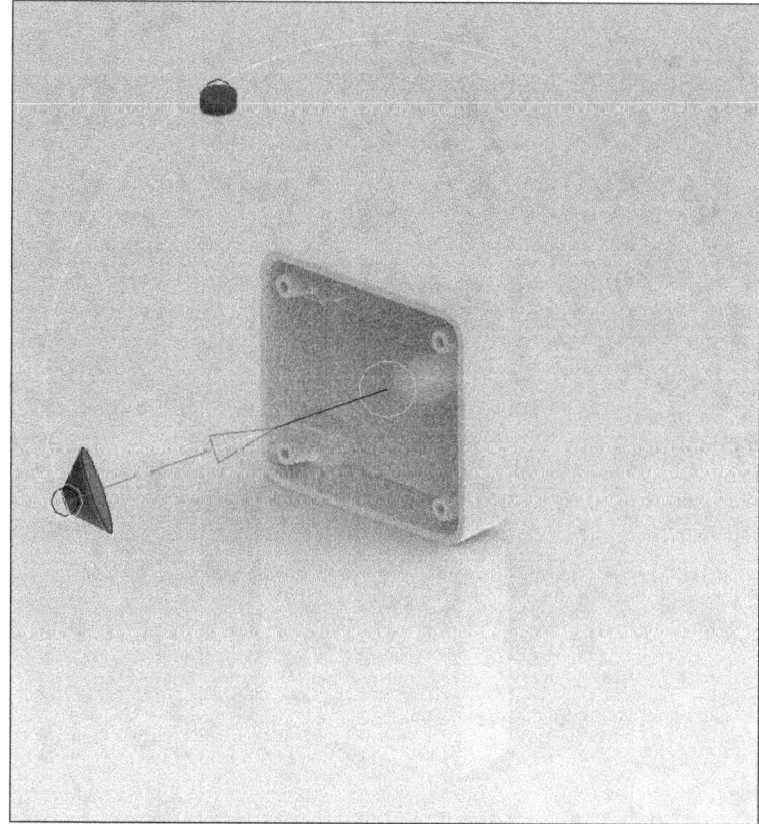

Figure 18.26(d) Change Focus

Click: 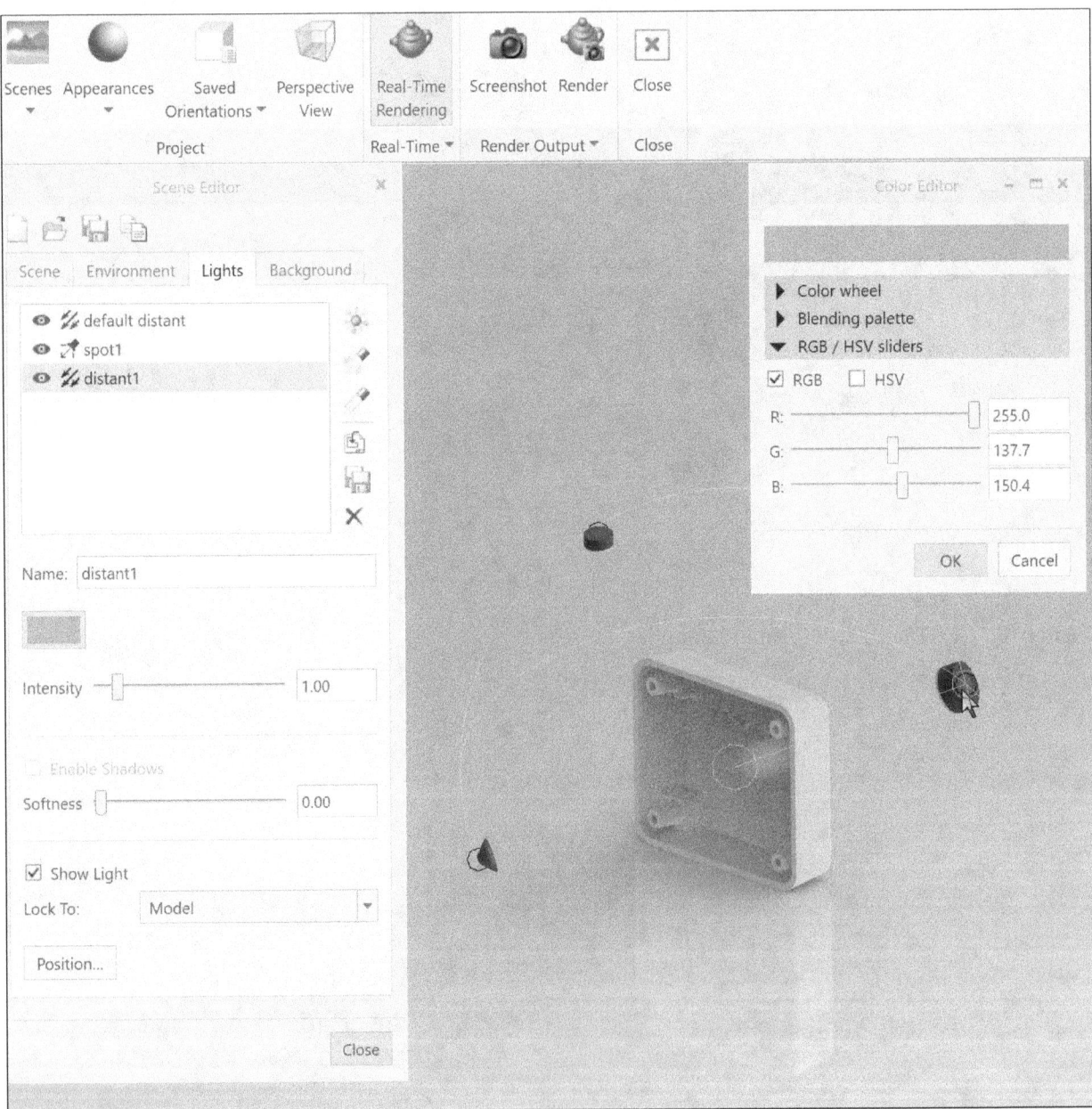 **Add new distance light >** ☑ Show Light **>** Lock To: Model **> Name** ☐ **Color for lighting >** adjust the slide bars in the Color Editor to the RGB values you desire > select the new distance light and change its position > **OK** (from the Color Editor dialog box) [Fig. 18.26(e)]

Figure 18.26(e) New Distance Light

Click: **Scenes > workshop_scene** [Fig. 18.26(f)] **> Edit Scene > Save scene with model** [Fig. 18.26(g)] **> repeat the process and change the scene** [Fig. 18.26(h)]

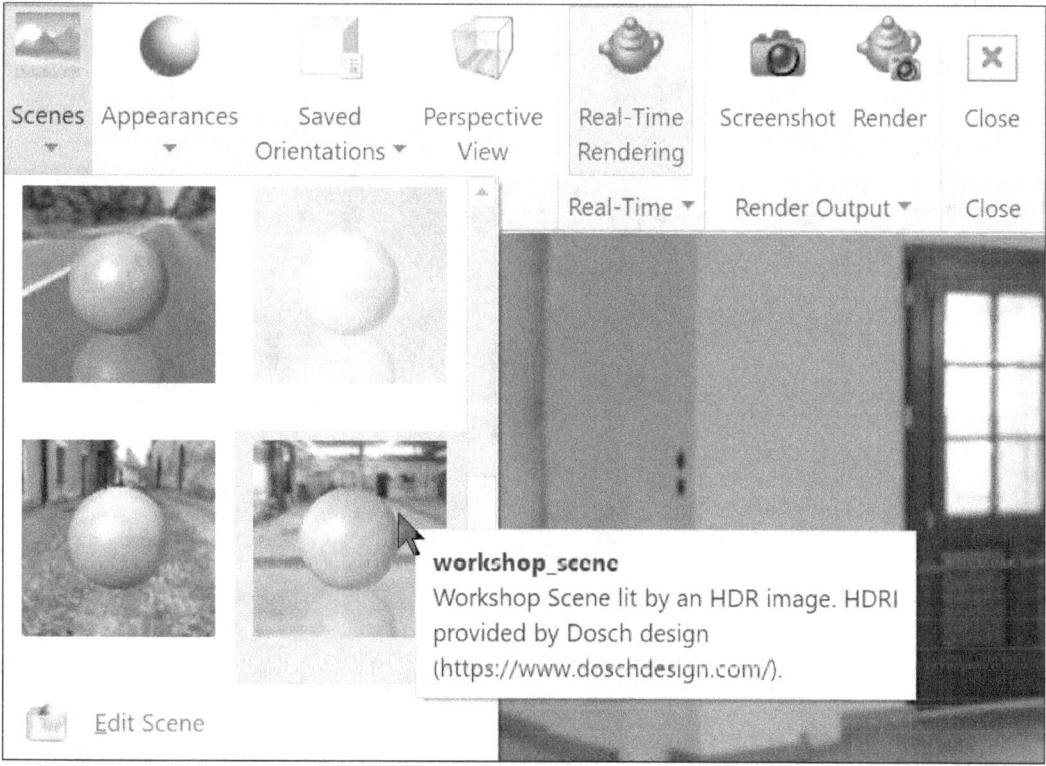

Figure 18.26(f) New Wall Selection

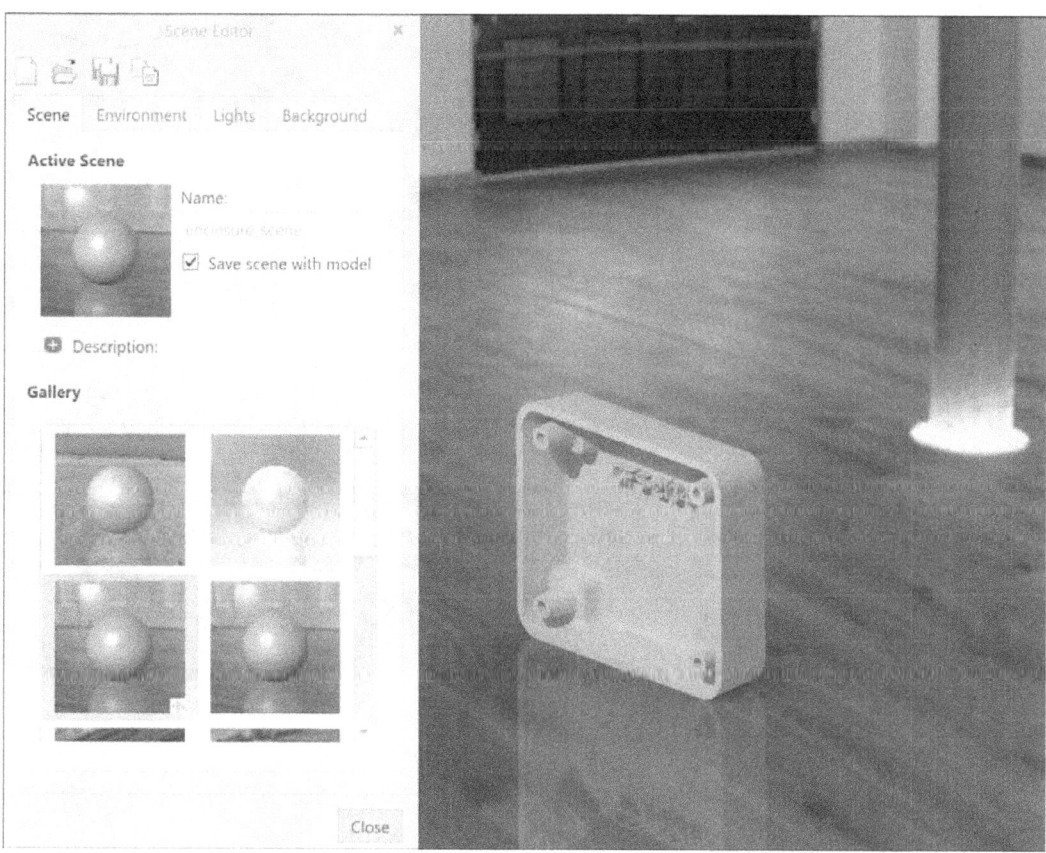

Figure 18.26(g) Room Appearance Editor

Figure 18.26(h) enclosure_scene

Click: **Background > Image** > click on image > navigate to an image from your folders [Fig. 18.26(j)] > **Close > Real-Time Rendering** off > **Ctrl+S > File > Close**

Figure 18.26(i) Render Window

Click: [image] > [image] Drawing > Name **enclosure** > **OK** > Template **d_drawing** > **OK** > **Layout** tab > [image] Sheet Setup > Sheet 1 Format [D Size] > [image] > **Browse** > **d.frm** > **Open** > **OK** > **Ctrl+S** > **OK** > [image] > [image] (Select All) all *off* > **LMB** in the Graphics Window > **Ctrl+R** > select the front view > press **RMB** > **Insert Projection View** > select on the left of the front view > press **RMB** > **Lock View Movement** (uncheck) > move the views as required > select **left_6** *(your view id may be different)* view from the Drawing Tree > press **RMB** > **Properties** (Fig. 18.27) > **View Display** > **Follow Environment** > **Hidden** > **OK** > **LMB** to deselect > complete the detail [see Fig. 18.2(a)] > **Ctrl+S** > **File** > **Save As** > **Save a Copy** > Type [image] > **Zip File (*.zip)** > **OK** > **upload** the zip file to your course interface or attach to an email and send to your instructor and/or yourself > **File** > **Close**

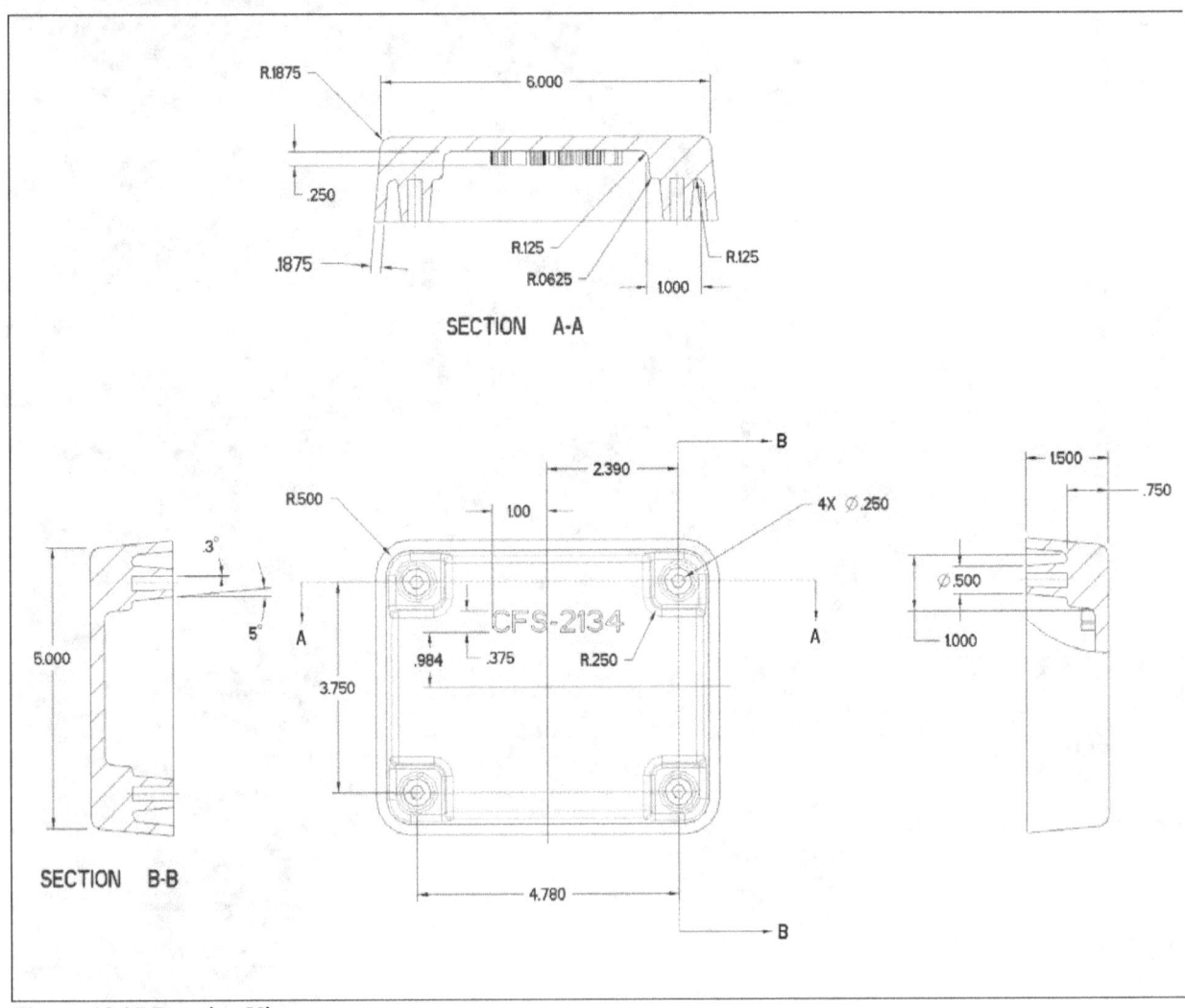

Figure 18.27 Drawing Views

If your system has the Flexible Modeling application, complete the following commands.

Click: **Flexible Modeling** tab > spin the model > 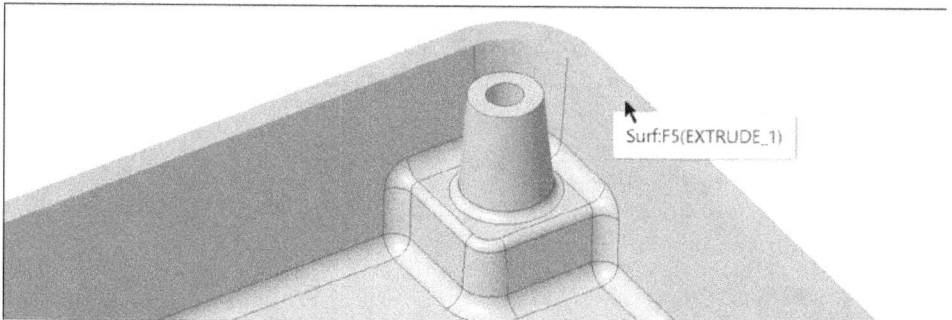 **Move using Dragger** > select the top face of the part [Figs. 18.28(a-b)] > place the pointer on the vertical axis (translation) arrow > press and hold down the **LMB** > move the pointer to drag the axis vertically [Fig. 18.28(c)] > release the **LMB** at the desired position >

Surf:F5(EXTRUDE_1)

Figure 18.28(a) Select the Surface

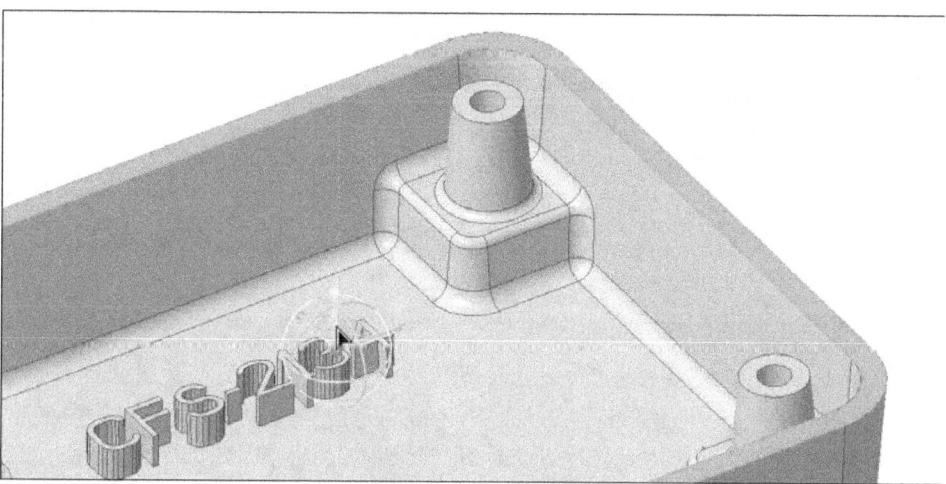

Figure 18.28(b) 3D Dragger Displays

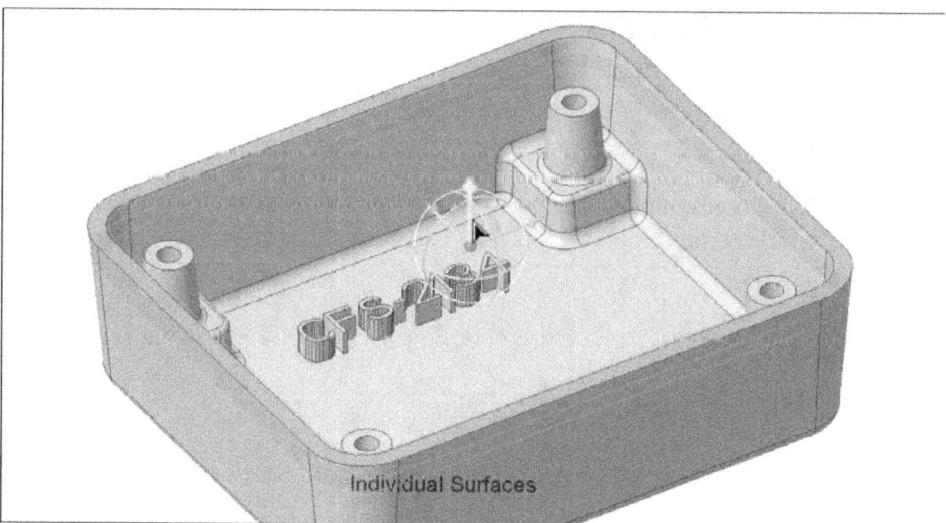

Individual Surfaces

Figure 18.28(c) Drag the Vertical Axis

Click: 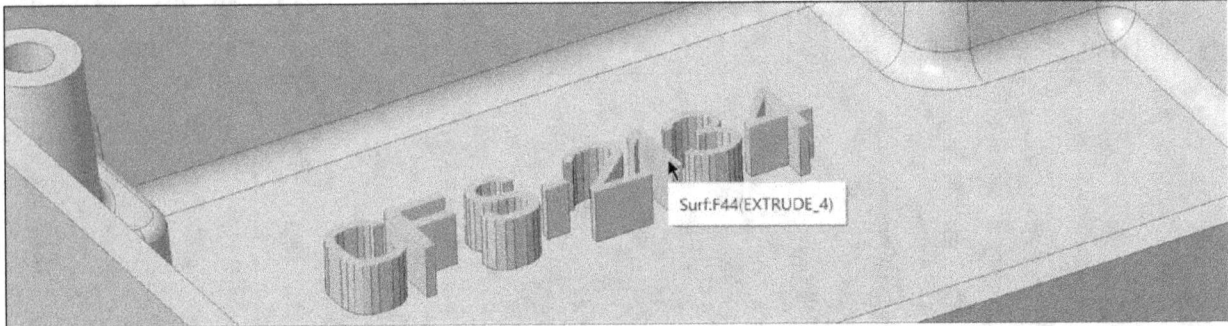 > select the top face of the lettering [Fig. 18.29(a)], 3D Dragger displays [Fig. 18.29(b)] > place the pointer on a rotation ring (arc) [Fig. 18.29(c)] > using your **LMB** > rotate the ring (arc) > ✓

Figure 18.29(a) Select the Surface

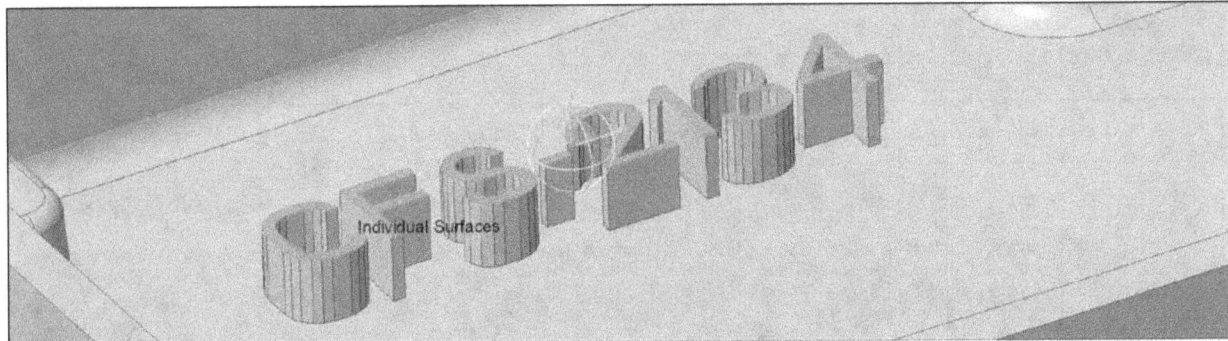

Figure 18.29(b) 3D Dragger Displays

Figure 18.29(c) Drag the Rotation Ring (Arc)

806

Click: **Analysis** tab >  Mass Properties > **Saved** > **Preview** [Fig. 18.30(a)] > **OK** > **Ctrl+D** > **File** > **Save As** > **Save a Copy** > New Name **enclosure_one_off** > **OK** > **File** > **Save As** > **Save a Copy** > Type ▼ > **Zip File (*.zip)** > New Name **enclosure_one_off.prt.zip** > **OK** > **upload** the zip file to your course interface or attach to an email and send to your instructor and/or yourself

Figure 18.30(a) Mass Properties

Click: 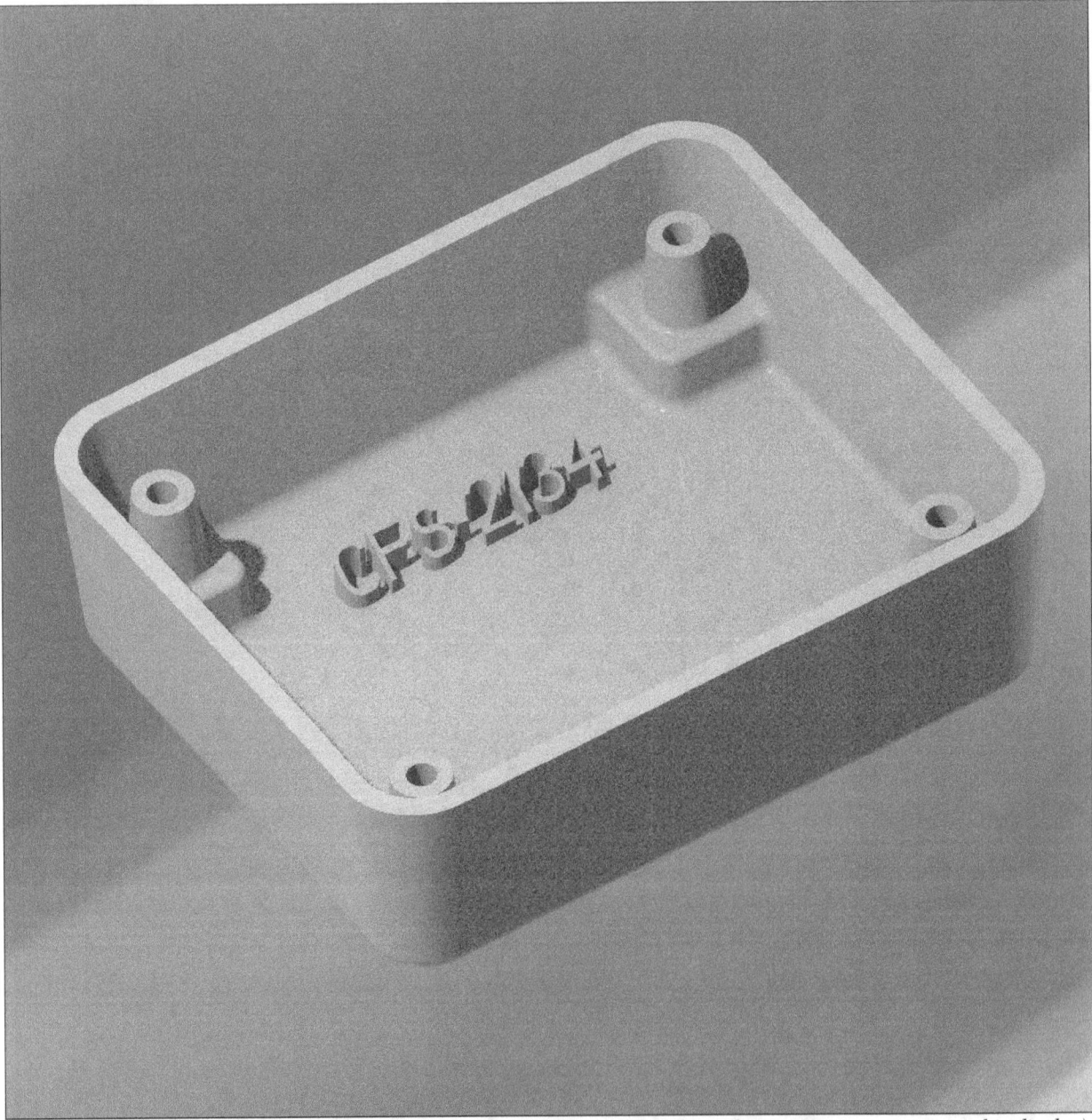 **Shading With Reflections > Applications** or **Render Studio** tab **> Scenes >**

> Render > Yes > Real-Time Rendering [Fig. 18.30(b)] **> Cancel > Close >**
Ctrl+D > Ctrl+S > File > Close

Figure 18.30(b) Rendered Design *(the quality of your graphics card and graphics settings may prevent this display)*

Video Lessons 19-22, and **Lesson Projects** can be downloaded from: http://www.cad-resources.com

Lesson 19 Gear Assembly (Video Lesson)

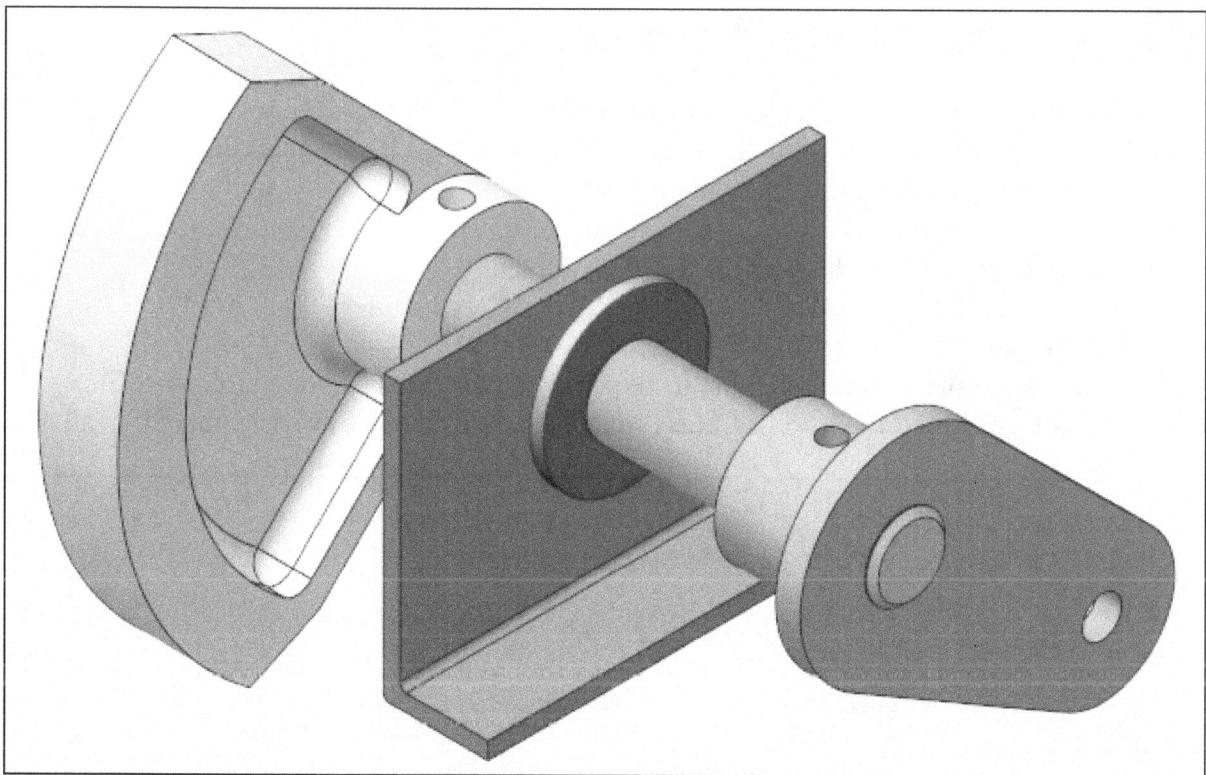

Figure 19.1 Gear Assembly

OBJECTIVES

- Use **Assembly** commands
- Establish and edit **Units**
- Apply **Analysis** to determine **Global Interference** and **Global Clearance**
- **Measure** features to analyze components fit
- Use **Edit Definition** features to resolve design problems

RESOURCES

- Lesson 19 Lecture at **YouTube Creo Parametric Lecture Videos**

- Lesson 19 3D PDF models embedded in a PDF

- Choose **commercial** or **academic** Gear Assembly for parts

Steps

Follow the Steps to complete the lesson:

- **Download** the Lesson Lecture onto your hard drive
- **Play** directly from your computer using *Windows Media Player* (**.wmv** file)
- **Create** the assembly using the available parts
- **Check** your Units
- Use **Measure** and **Analysis** on the assembly model
- **Edit** as necessary
- **Follow** the procedures in the Lesson Lecture
- **Save** the completed assembly (and its required components) as a single **Zipped** file
- **Upload** the **zipped** file to your Course Management System as an assignment if available.

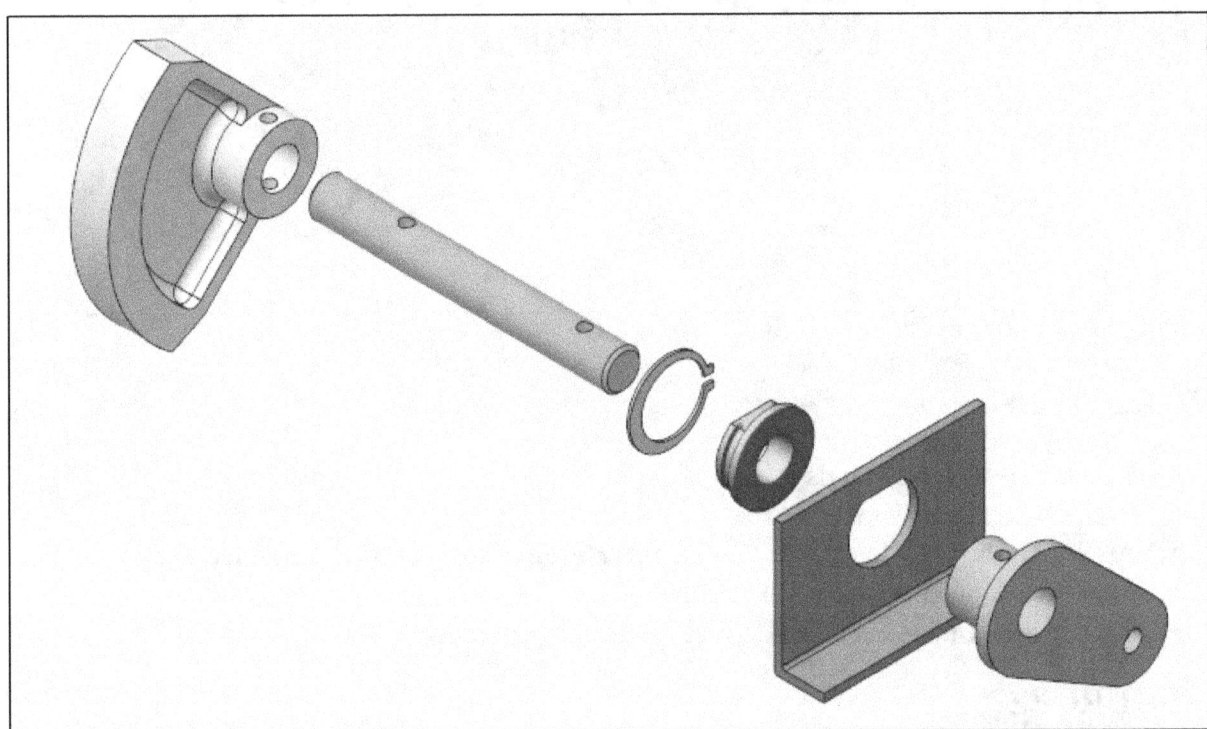

Figure 19.2 Exploded Gear Assembly

Lesson 20 Valve Assembly (Video Lesson)

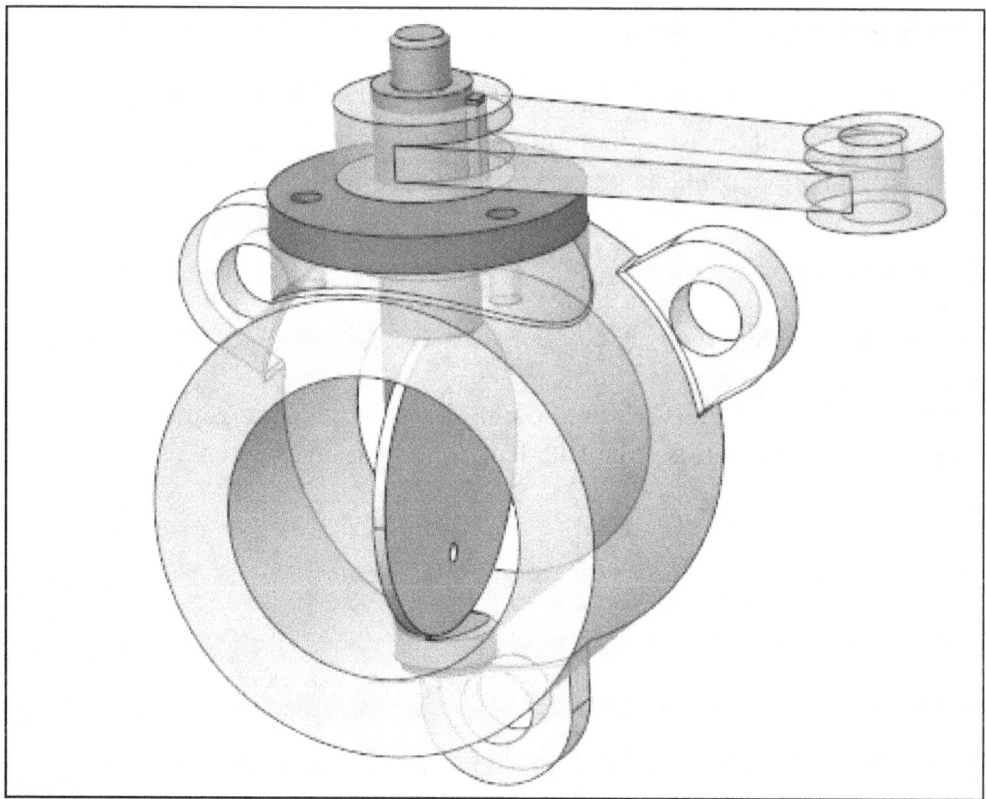

Figure 20.1 Valve Assembly

OBJECTIVES

- Use **Assembly** commands
- Establish and edit **Units**
- Apply **Analysis** to determine **Global Interference** and **Global Clearance**
- **Measure** features to analyze components fit
- Use **Edit Definition** features to resolve design problems
- Use **Top-Down Design** to create the Key component

RESOURCES

- Lesson 20 Lecture at **YouTube Creo Parametric Lecture Videos**

- Lesson 20 3D PDF models embedded in a PDF

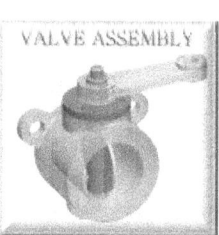

- Choose **commercial** or **academic** Valve Assembly for parts.

811

Steps

Follow the Steps to complete the lesson:

- **Download** the Lesson Lecture onto your hard drive
- **Play** directly from your computer using *Windows Media Player* (**.wmv** file)
- **Create** the assembly using the available parts
- **Check** your Units
- Use **Measure** and **Analysis** on the assembly model
- **Edit** as necessary
- Use **Top-Down Design** to create the **Key** component
- **Follow** the procedures in the Lesson Lecture
- **Save** the completed assembly (and its required components) as a single **Zipped** file
- **Upload** the **zipped** file to your Course Management System as an assignment if available.

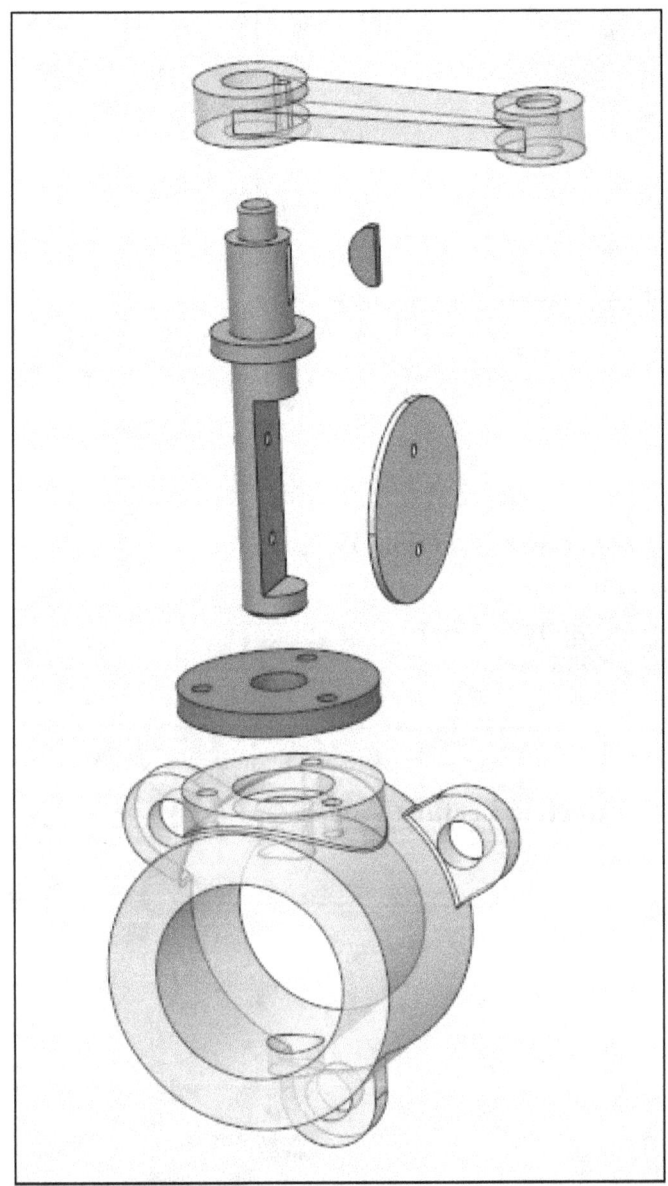

Figure 20.2 Exploded Valve Assembly

Lesson 21 Pulley Assembly (Video Lesson)

Figure 21.1 Pulley Assembly

OBJECTIVES

- Use **Assembly** commands
- Establish and edit **Units**
- Apply **Analysis** to determine **Global Interference** and **Global Clearance**
- **Measure** features to analyze components fit
- Use **Edit Definition** features to resolve design problems
- Use **Top-Down Design** to create the **Pulley** and **Belt** components

RESOURCES

- Lesson 21 Lecture at **YouTube Creo Parametric Lecture Videos**

- Lesson 21 3D PDF models embedded in a PDF

- Choose **commercial** or **academic** Pulley Assembly for parts.

.

Steps

Follow the Steps to complete the lesson:

- **Download** the Lesson Lecture onto your hard drive
- **Play** directly from your computer using *Windows Media Player* (**.wmv** file)
- Use **Top-Down Design** to create the **Pulley** and **Belt** components
- **Edit** as necessary
- **Follow** the procedures in the Lesson Lecture
- **Save** the completed assembly (and its required components) as a single **Zipped** file
- **Upload** the **zipped** file to your Course Management System as an assignment if available.

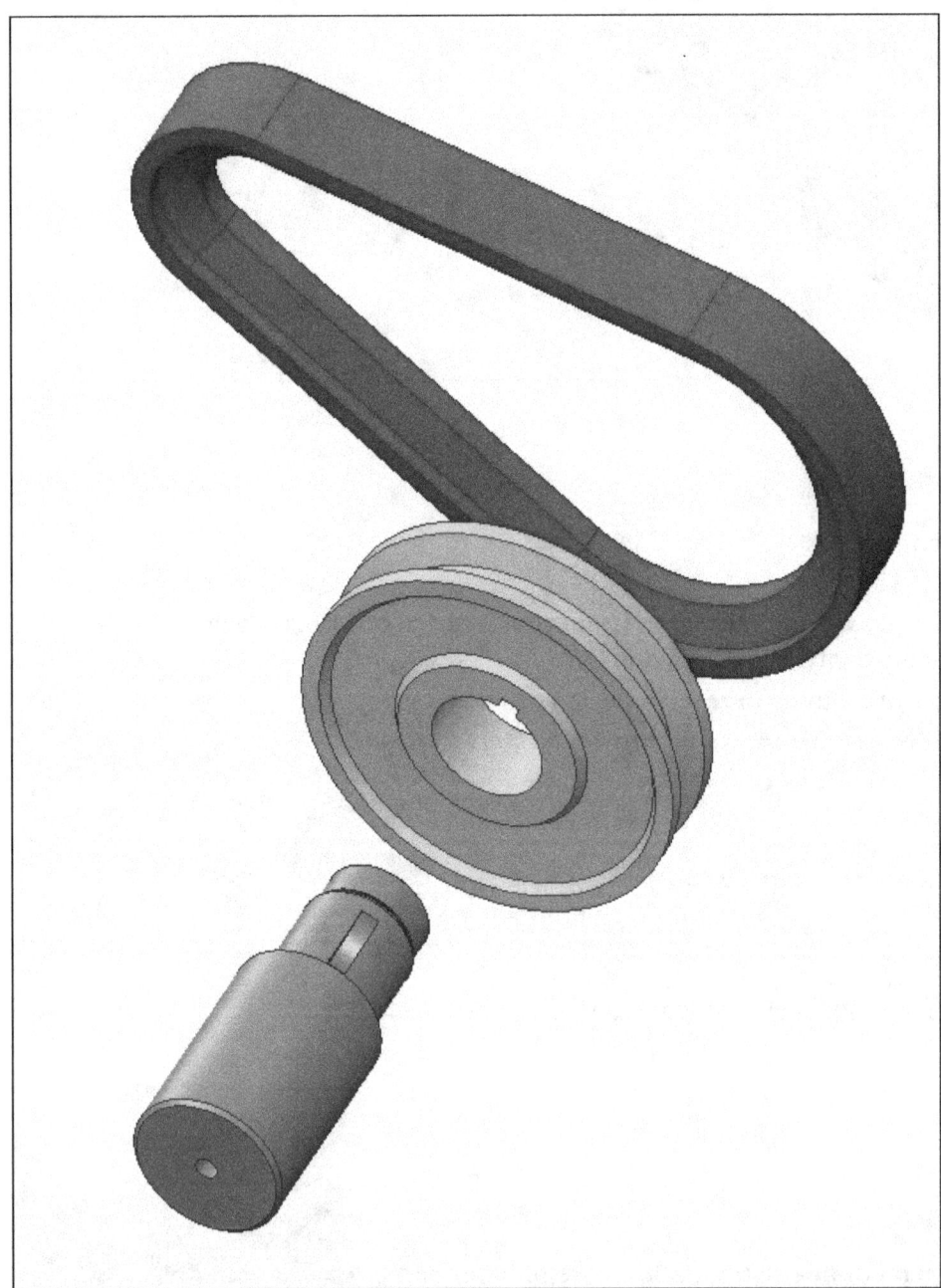

Figure 21.2 Exploded Pulley Assembly

Lesson 22 Coupling Assembly (Video Lesson)

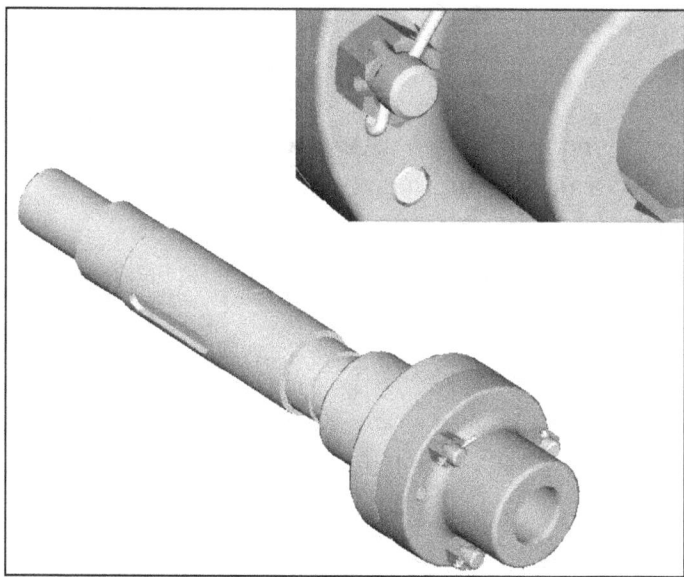

Figure 22.1 Coupling Assembly

OBJECTIVES

- Model the **Shaft**, **Taper Coupling**, and **Straight Coupling** components
- Use **Assembly** commands
- Apply **Analysis** to determine **Global Interference** and **Global Clearance**
- **Measure** features to analyze components fit
- Use **Edit Definition** features to resolve design problems
- Create an **Exploded View**
- Document the components in **Detail Drawings**
- Create an **Assembly Drawing** with **BOM** and **Balloons**

RESOURCES

- Lesson 22 Lecture at **YouTube Creo Parametric Lecture Videos**

  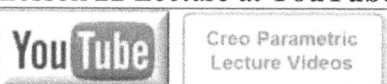

- Lesson 22 3D PDF models embedded in a PDF

  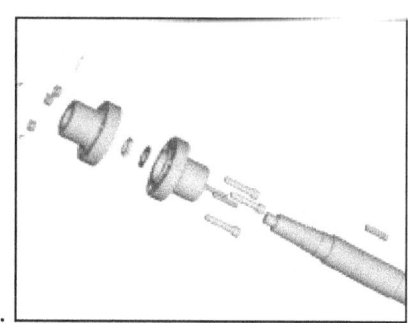

- Choose **commercial** or **academic** for parts.

Steps

Follow the Steps to complete the project:

- **Download** the Lesson Lecture onto your hard drive
- **Play** directly from your computer using *Windows Media Player* (**.wmv** file)
- **Model** the **Shaft**, **Taper Coupling**, and **Straight Coupling** components
- **Create** the assembly using the modeled parts
- **Create** an exploded view
- **Detail** the parts
- **Detail** the assembly
- **Follow** the procedures in the Lesson Lecture
- **Save** the completed assembly (and its components) as a single **Zipped** file
- **Upload** the **zipped** file to your Course Management System as an assignment if available.

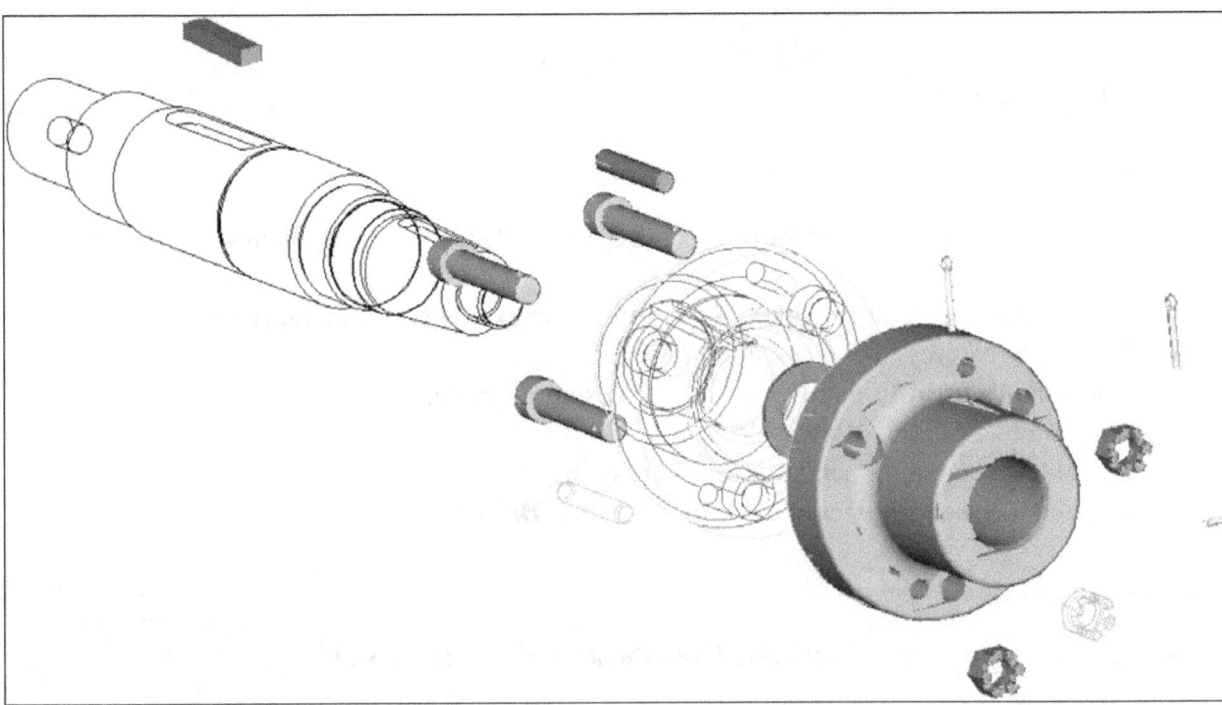

Figure 22.2 Exploded Coupling Assembly

Coupling Shaft

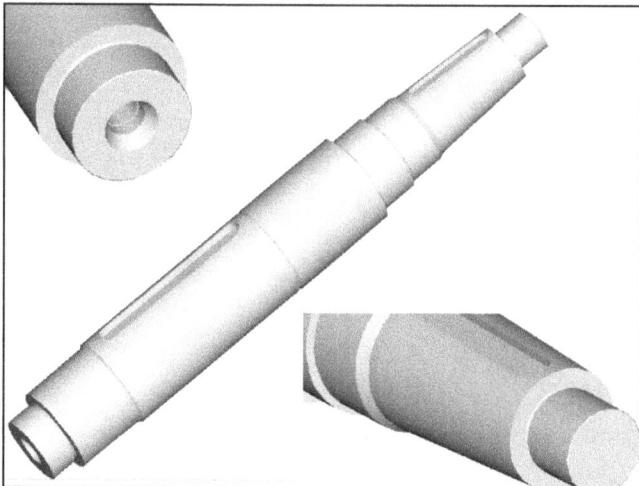

Figure 22.3 Coupling Shaft

Coupling Shaft

Remember to set up the model properties, set datum planes, and add layers to the parts.

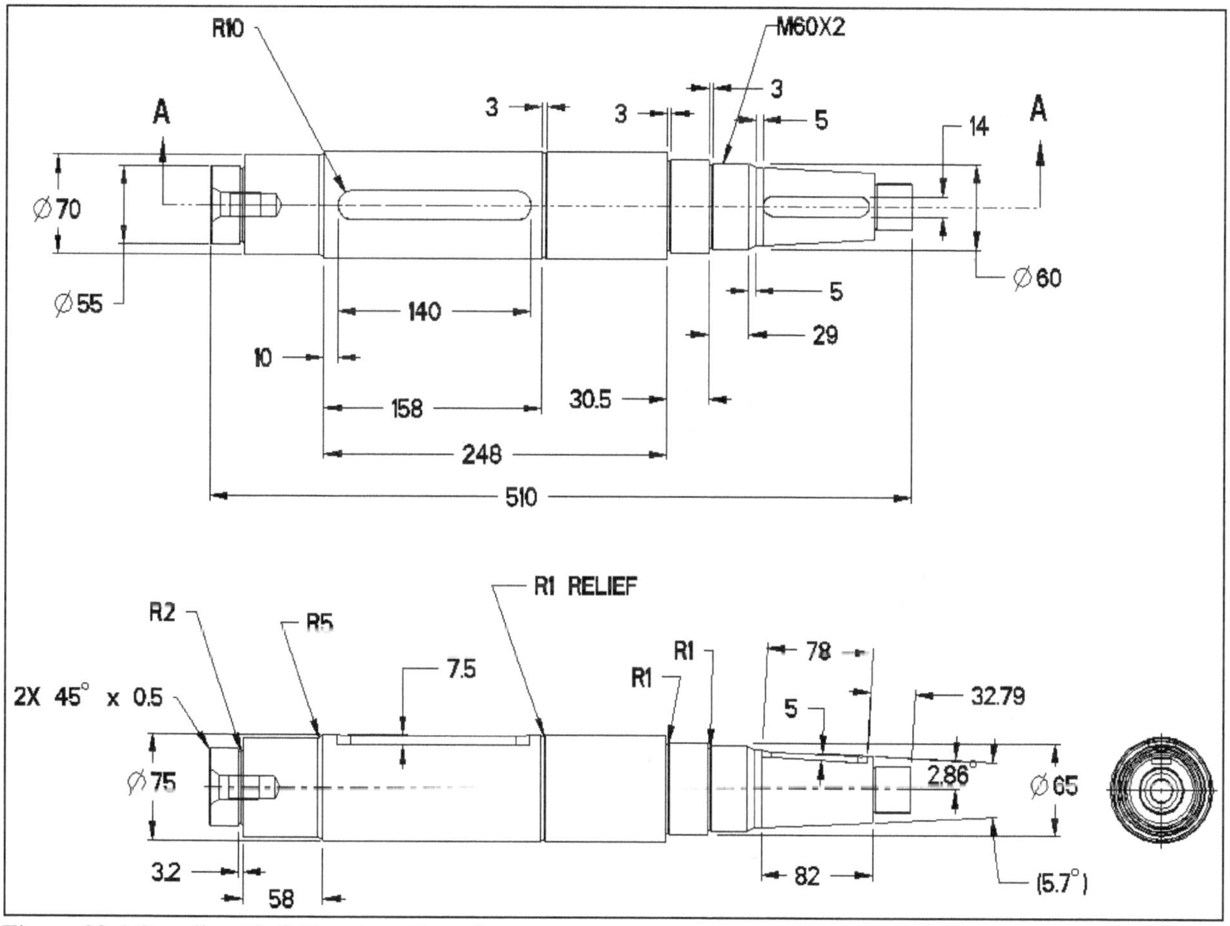

Figure 22.4 Coupling Shaft Drawing, Sheet One

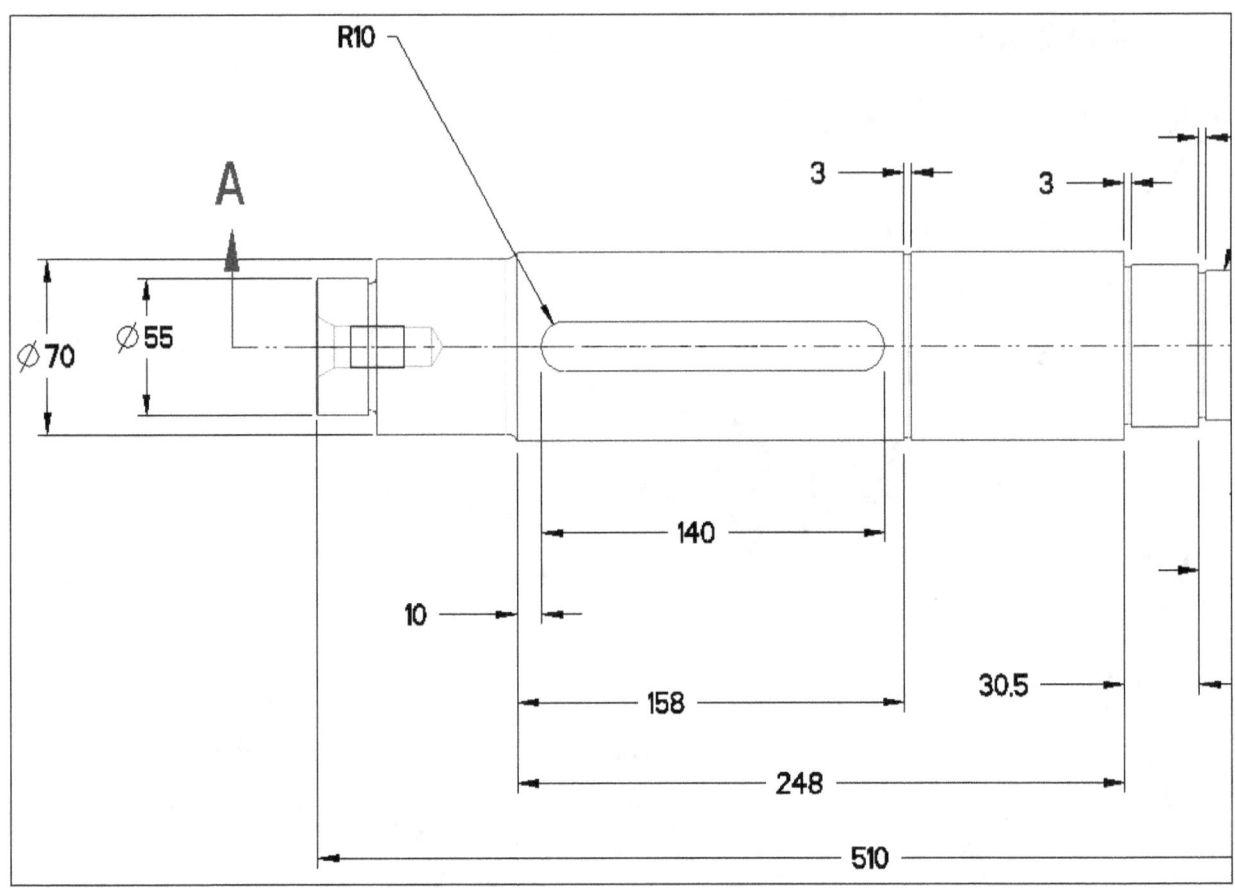

Figure 22.5 Coupling Shaft Drawing, Sheet One, Top View, Left Side

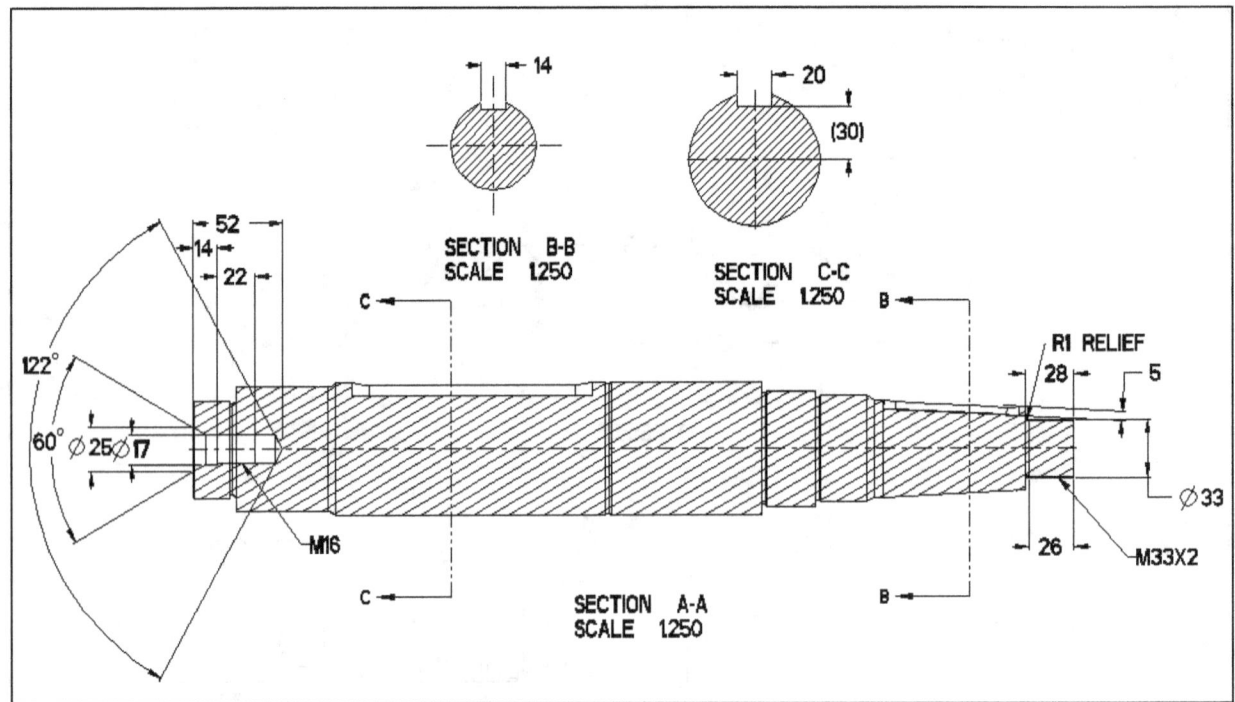

Figure 22.6 Coupling Shaft Drawing, Sheet Two

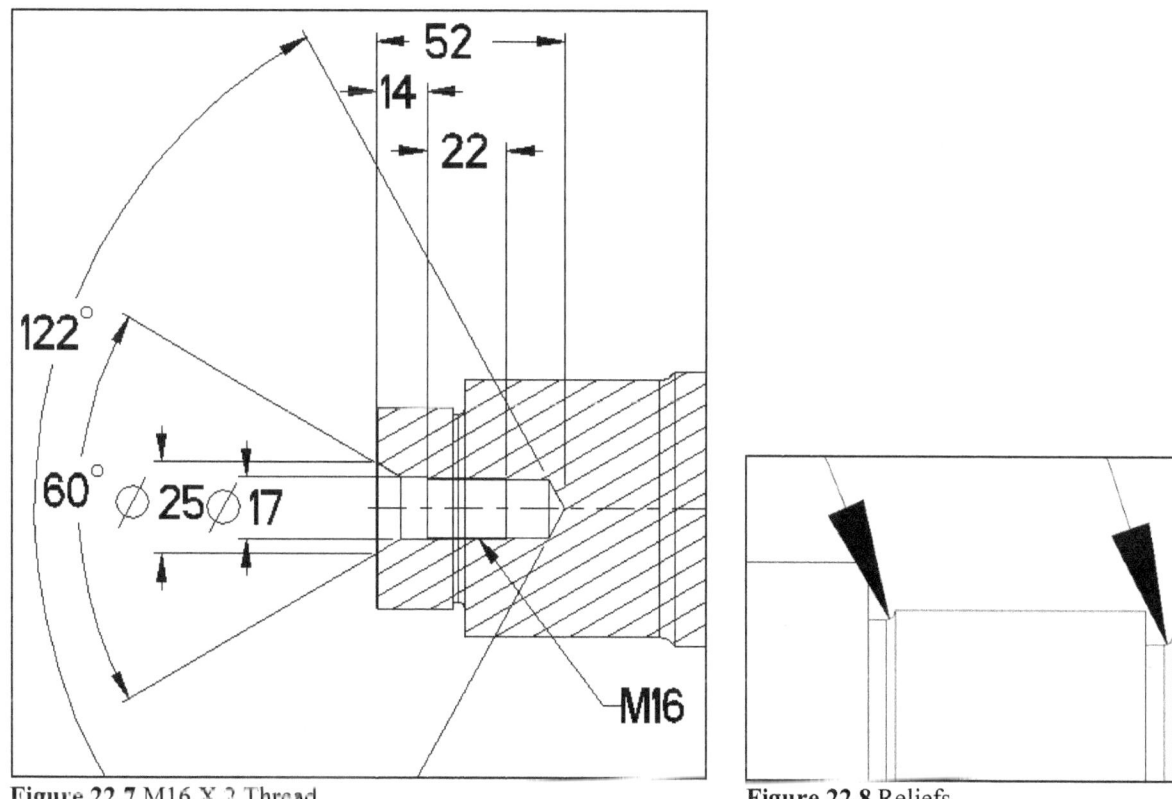

Figure 22.7 M16 X 2 Thread

Figure 22.8 Reliefs

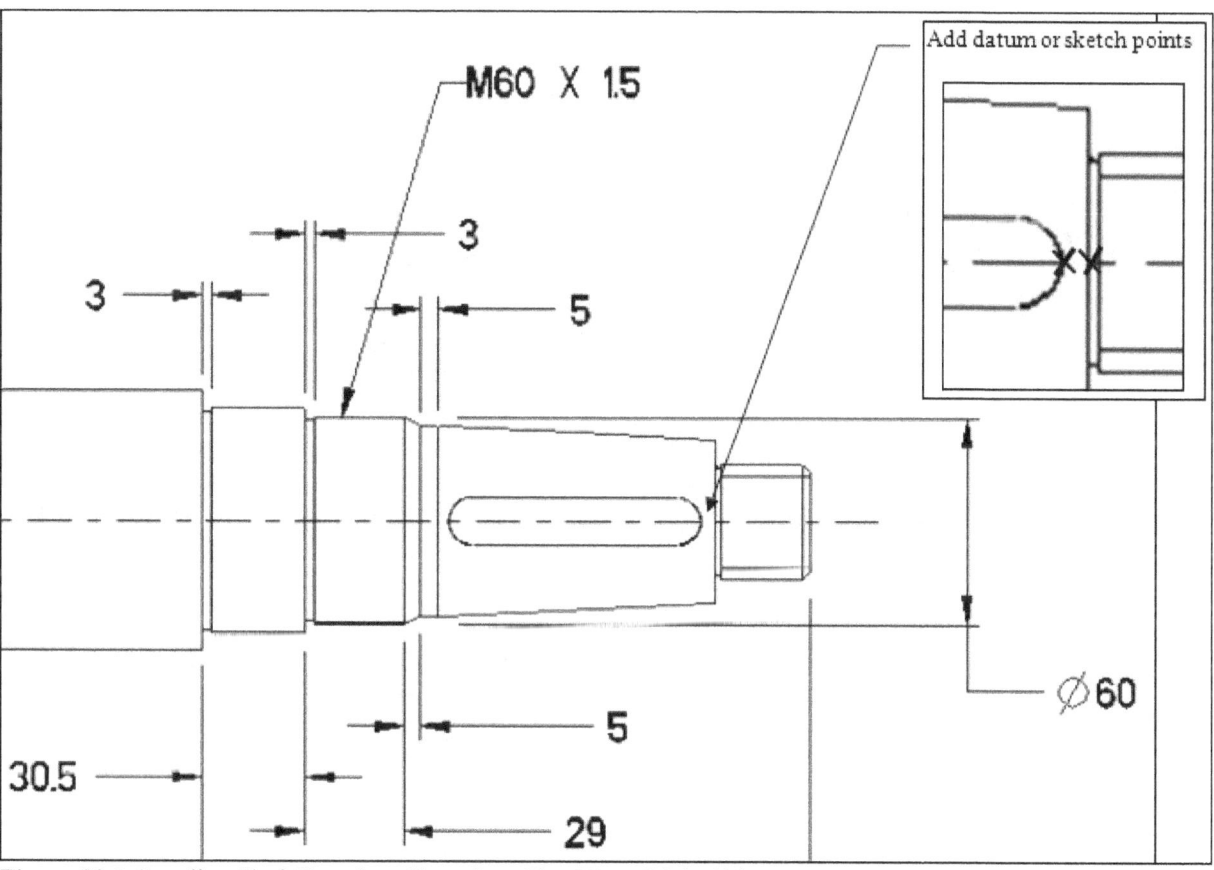

Figure 22.9 Coupling Shaft Drawing, Sheet One, Top View, Right Side

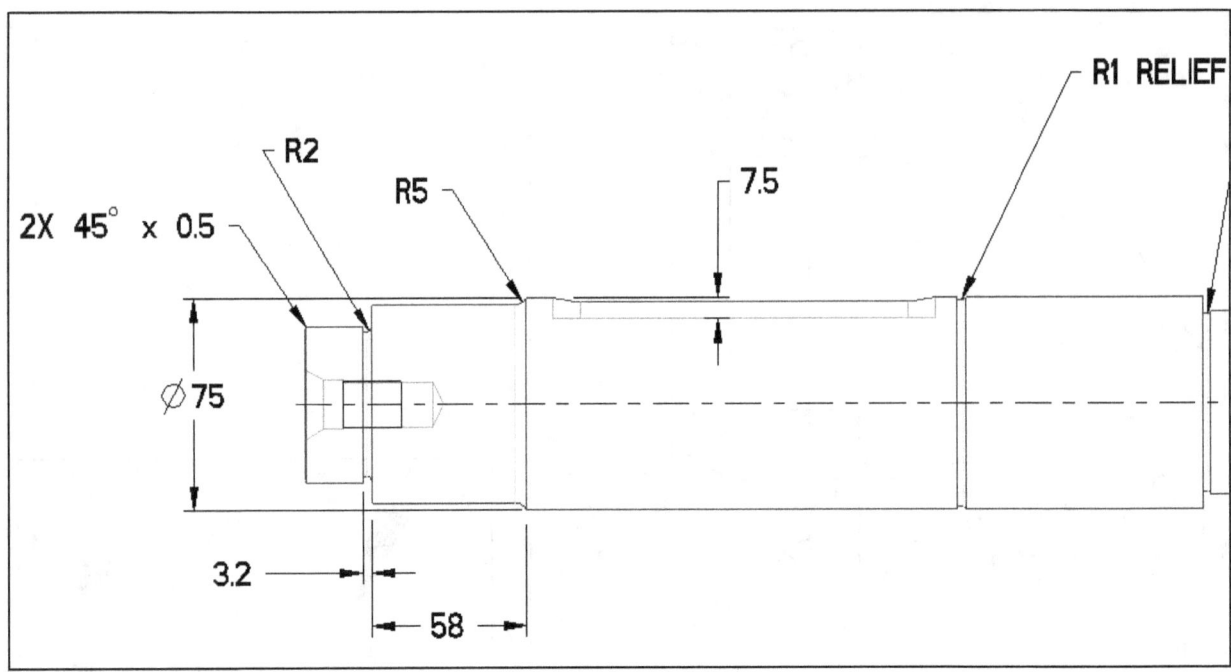

Figure 22.10 Coupling Shaft Drawing, Sheet One, Front View, Left Side

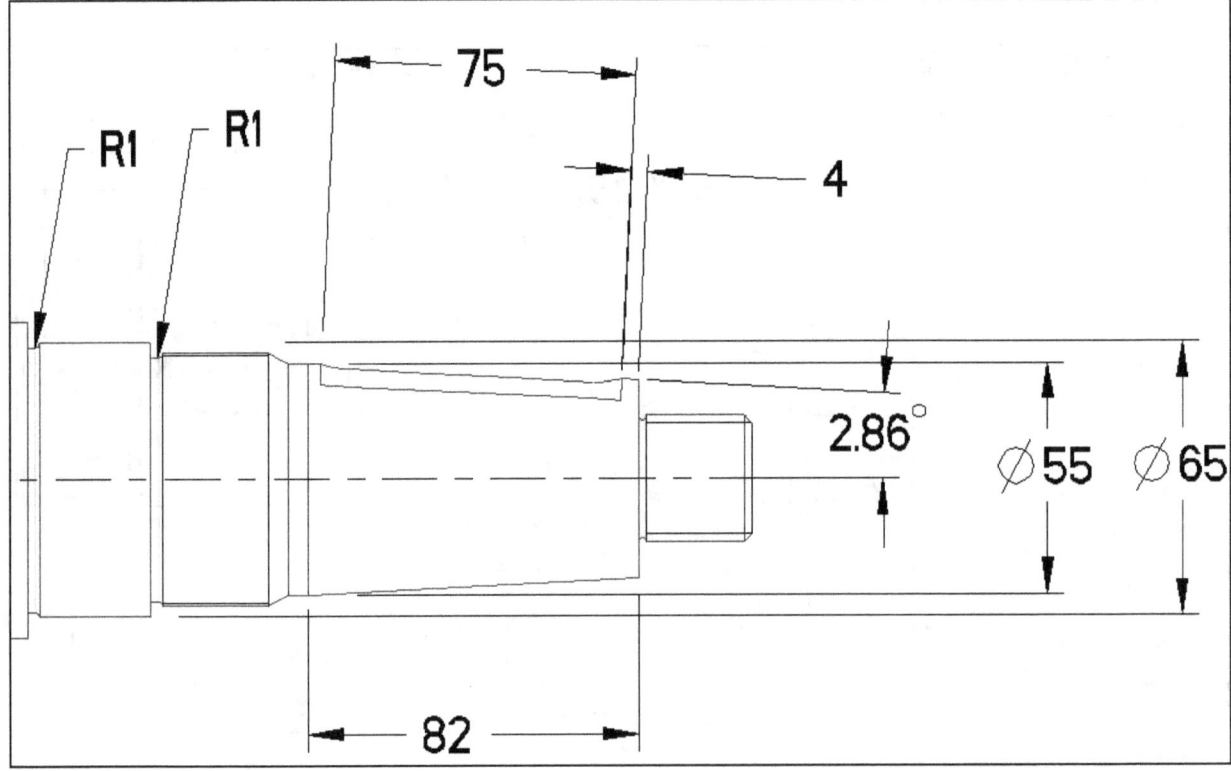

Figure 22.11 Taper

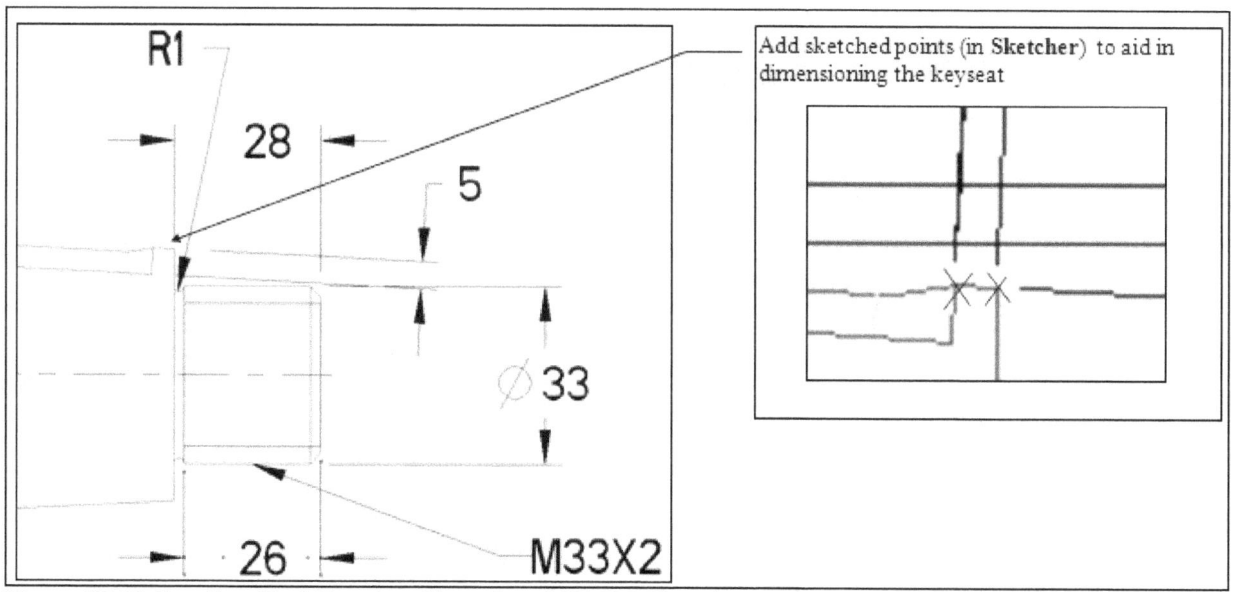

Figure 22.12 Coupling Shaft Drawing, Sheet One, M33 X 2 Threads

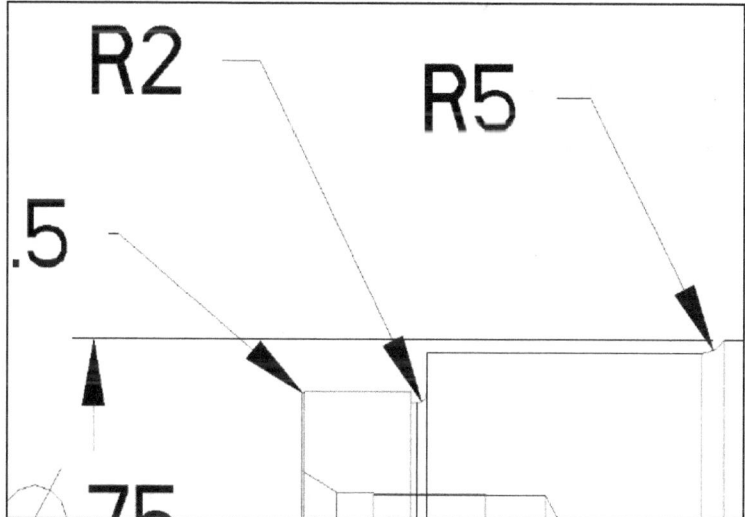

Figure 22.13 Coupling Shaft Drawing, Relief

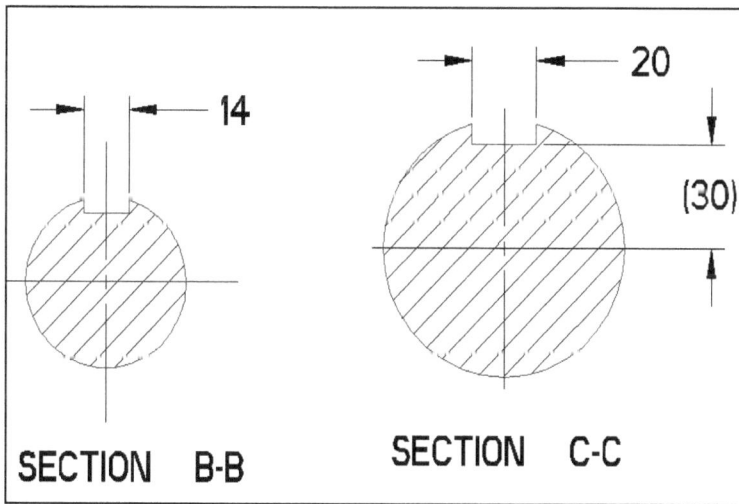

Figure 22.14 SECTION B-B and SECTION C-C

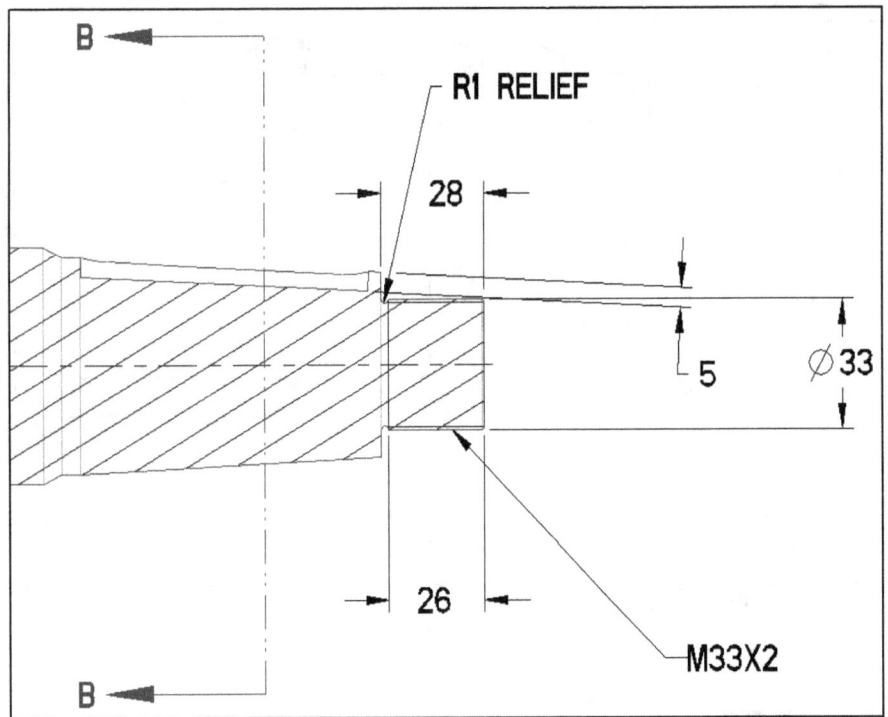

Figure 22.15 Coupling Shaft Drawing, Sheet Two, SECTION A-A Right Side

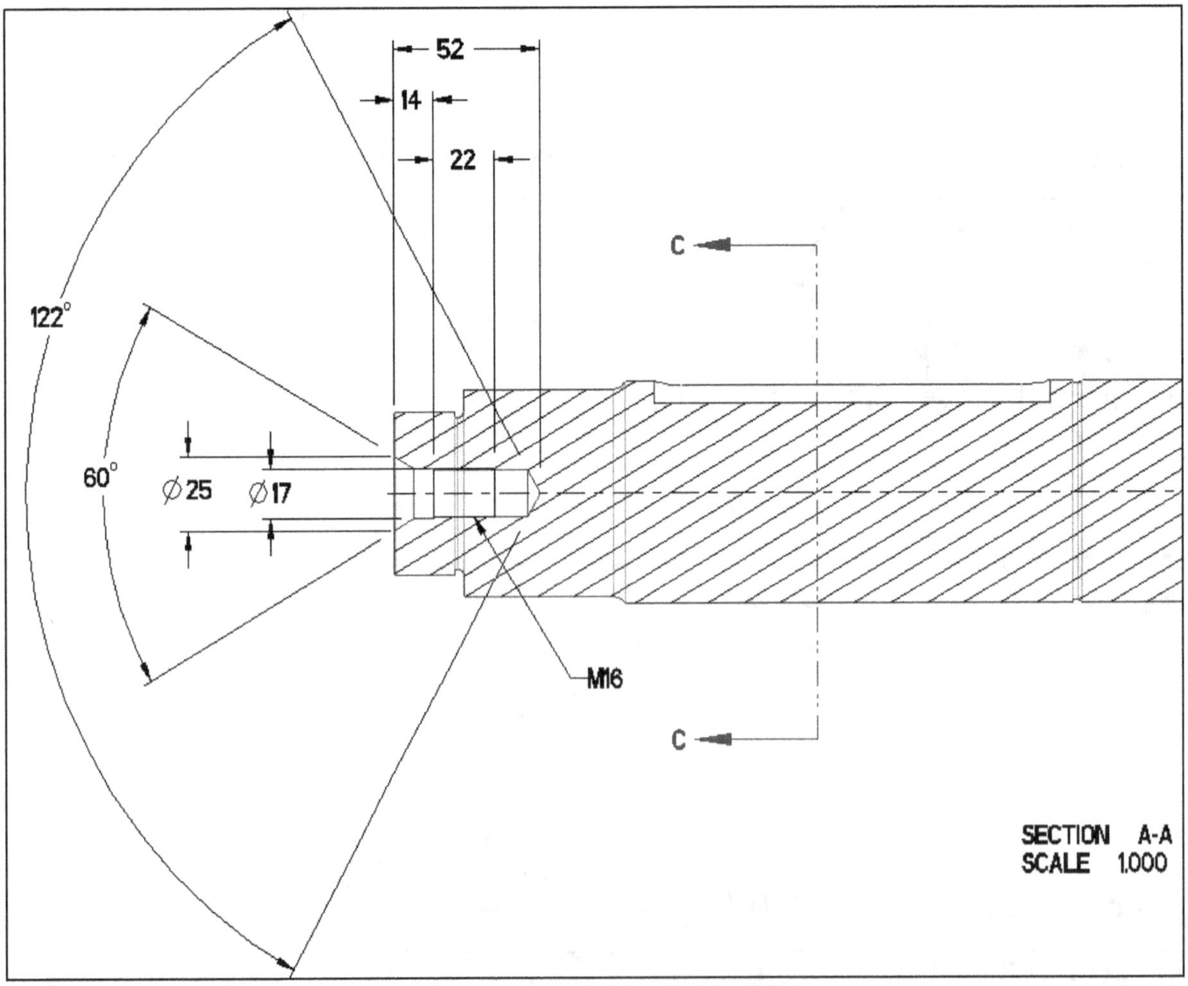

Figure 22.16 Coupling Shaft Drawing, Sheet Two, SECTION A-A Left Side

Taper Coupling

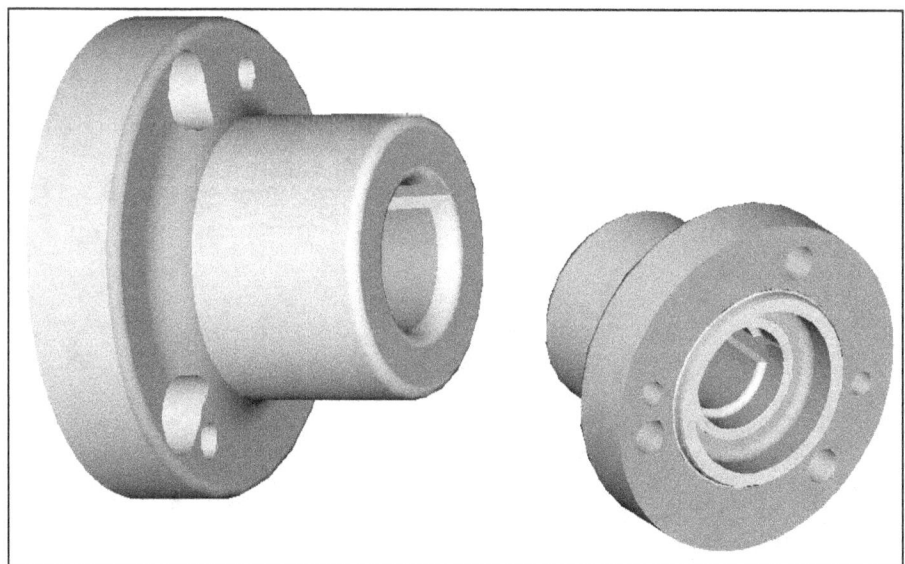

Figure 22.17 Taper Coupling

Taper Coupling

Plan the feature creation and the parent-child relationships for the part. The machined face of the coupling mates with and is fastened to a similar surface when assembled. Plan your geometric tolerancing requirements accordingly. Set the datums to anticipate the mating surfaces.

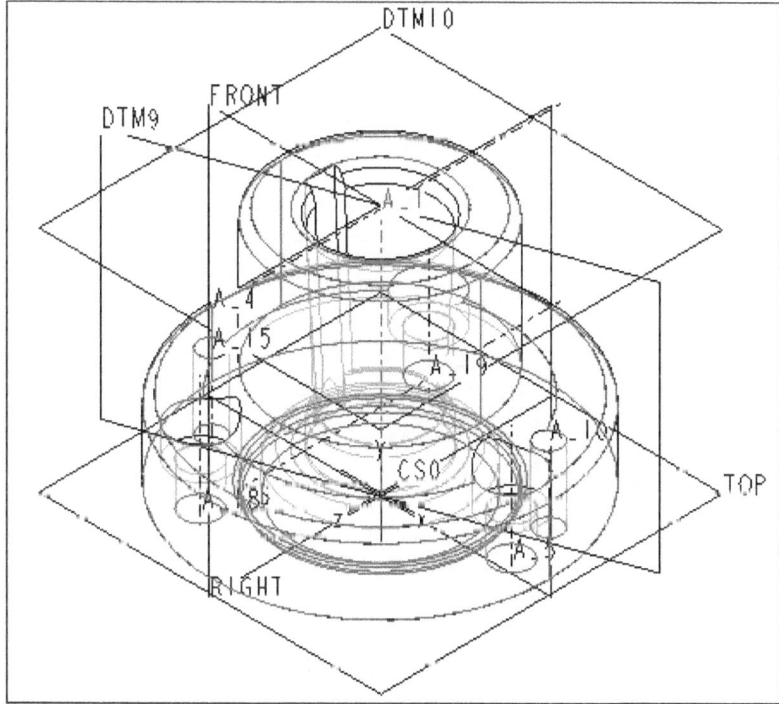

Figure 22.18 Taper Coupling Model with Datum Planes

Figure 22.19 Counterbore

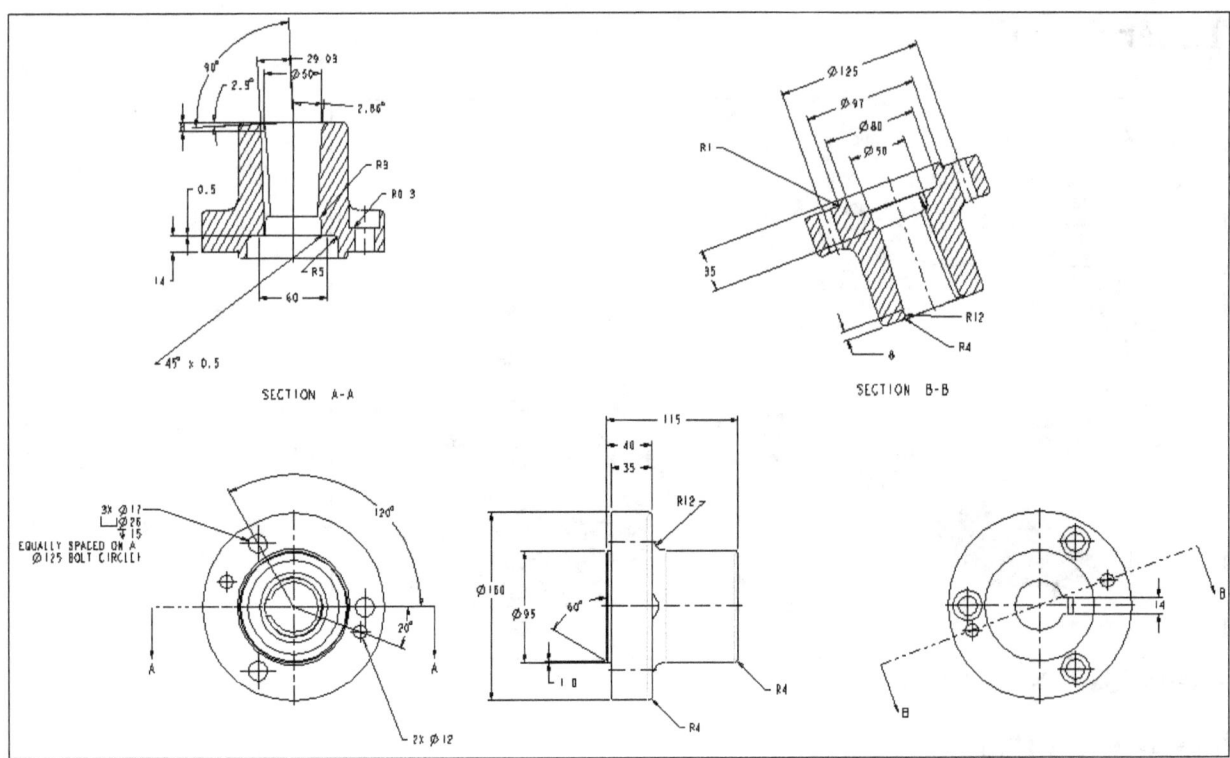

Figure 22.20 Taper Coupling Drawing

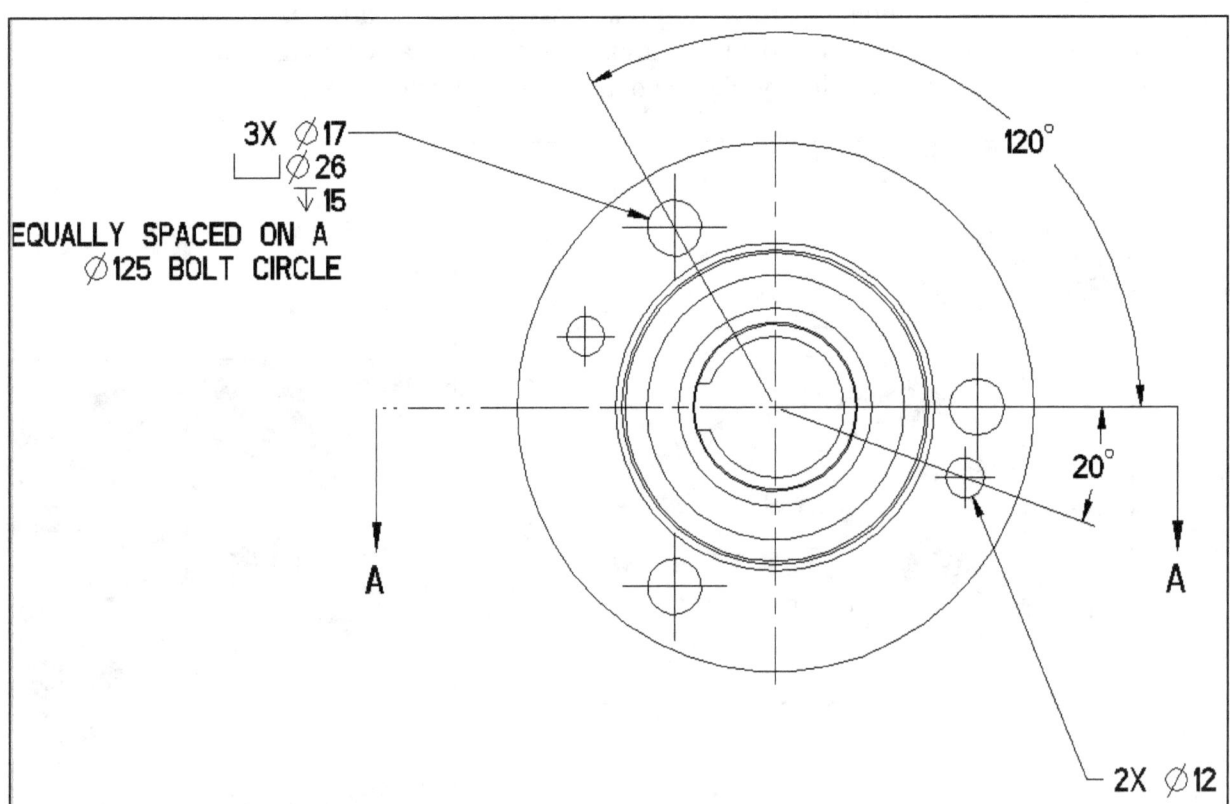

Figure 22.21 Taper Coupling Drawing, Bottom View

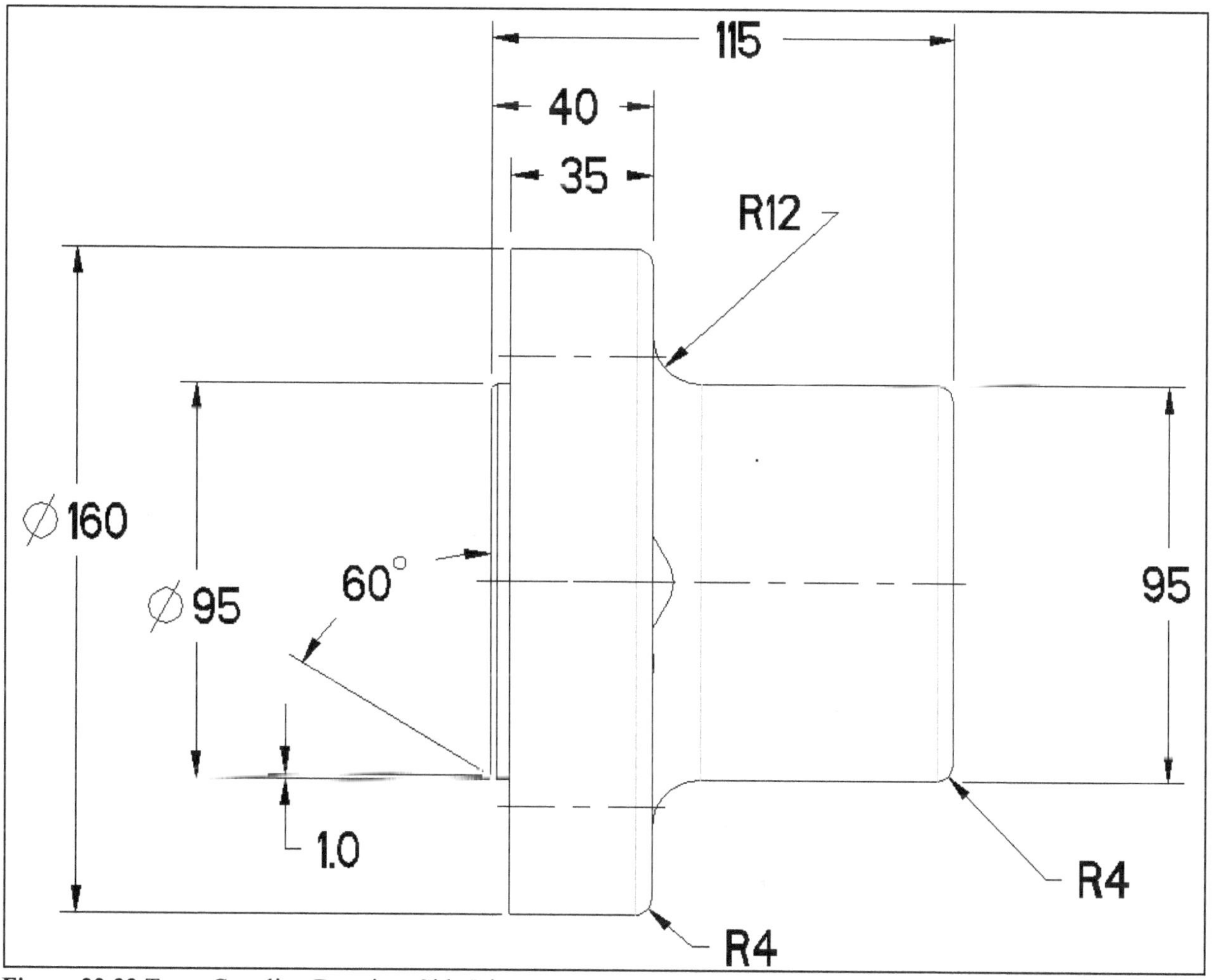

Figure 22.22 Taper Coupling Drawing, Side View

Figure 22.23 Counterbore

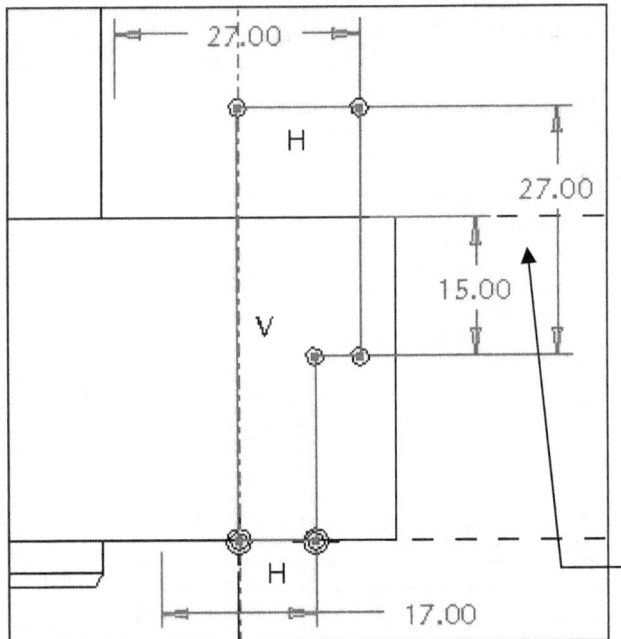

Figure 22.24 Section, Counterbore

Write a relation that will keep this dimension equal to the depth of the counterbore plus the radius of the large round (**R12**).

Dim=15+Radius

d18=d9+d6

Your dim values (**d#'s**) may differ.

Figure 22.25 Taper and Keyseat (14mm Wide)

Figure 22.26 Holes

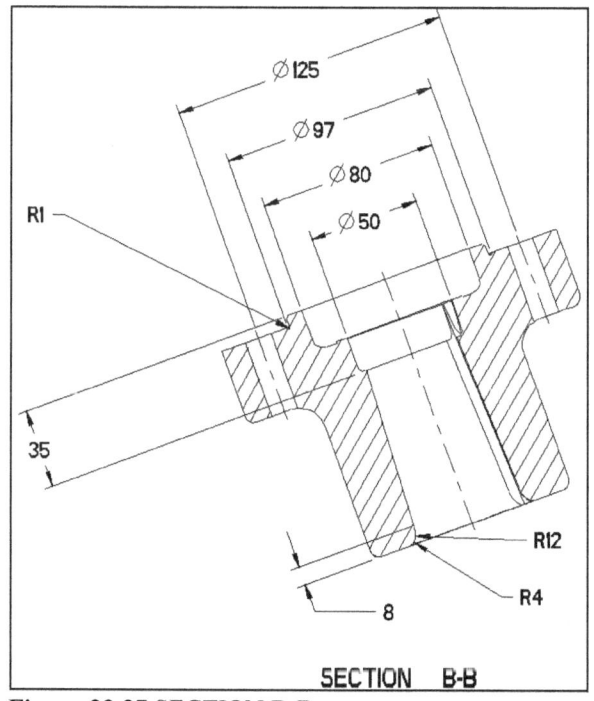

Figure 22.27 SECTION B-B

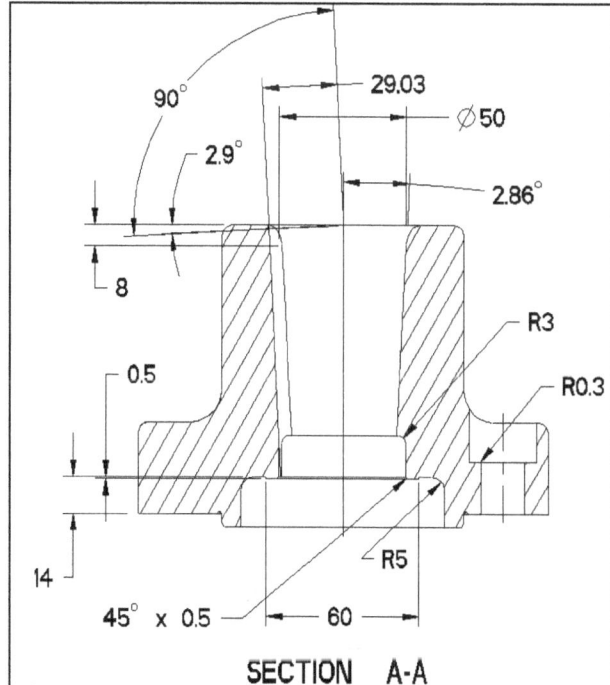

Figure 22.28 SECTION A-A

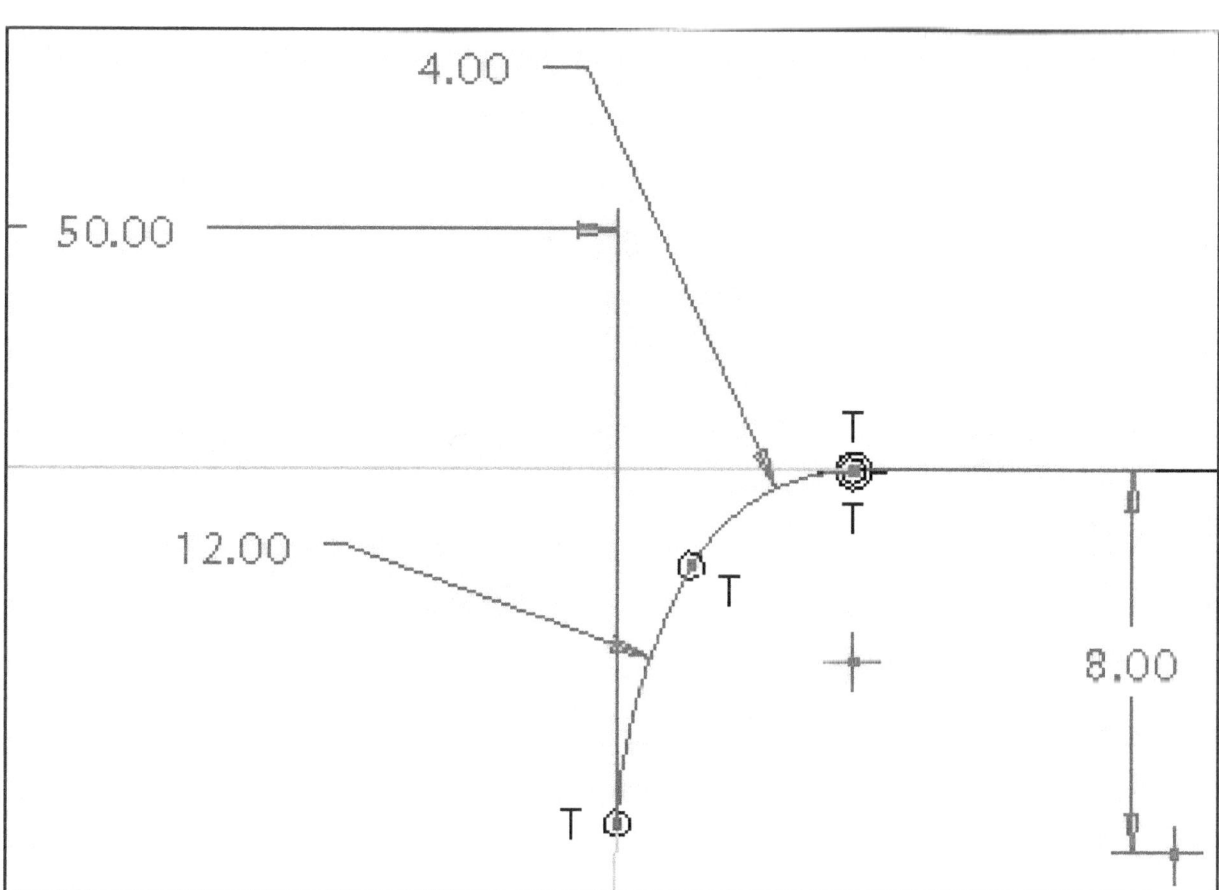

Figure 22.29 Section Sketch Rounds

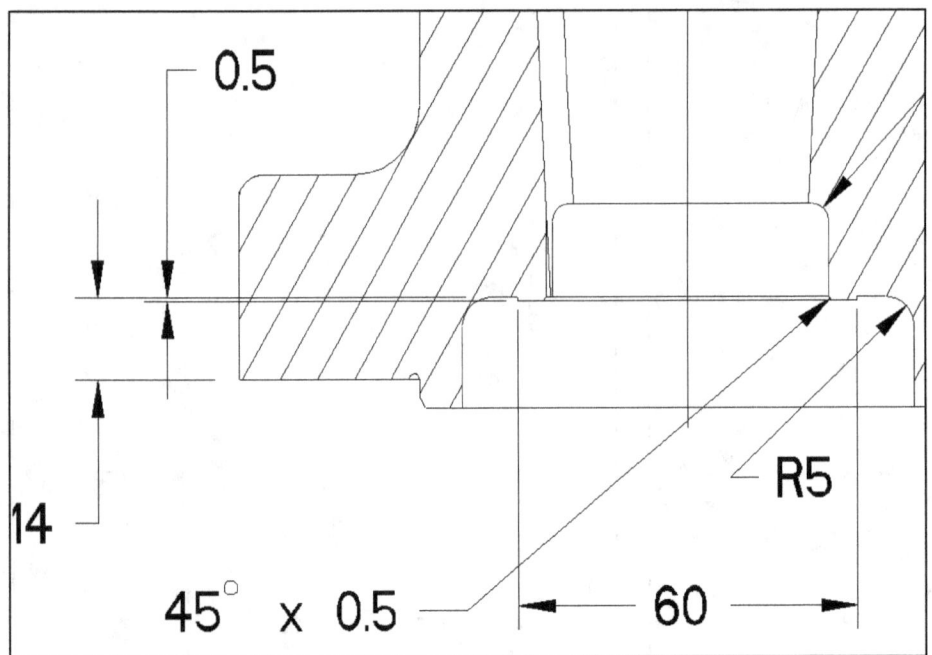

Figure 22.30 Section

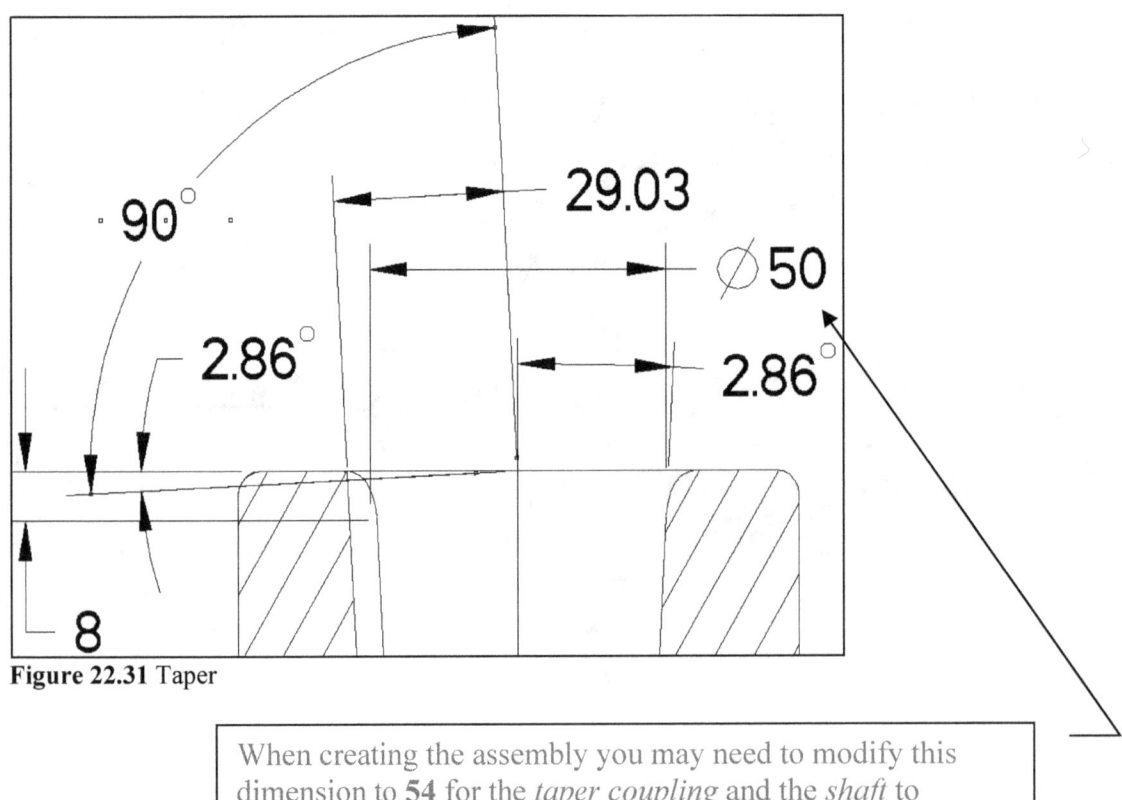

Figure 22.31 Taper

When creating the assembly you may need to modify this dimension to **54** for the *taper coupling* and the *shaft* to correctly seat

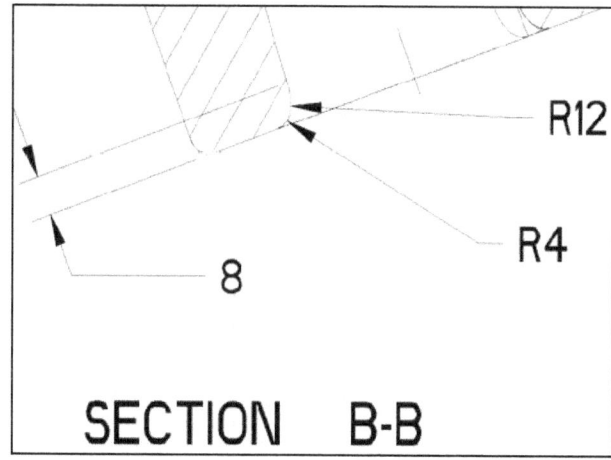

Figure 22.32 Taper Coupling Drawing, Radii

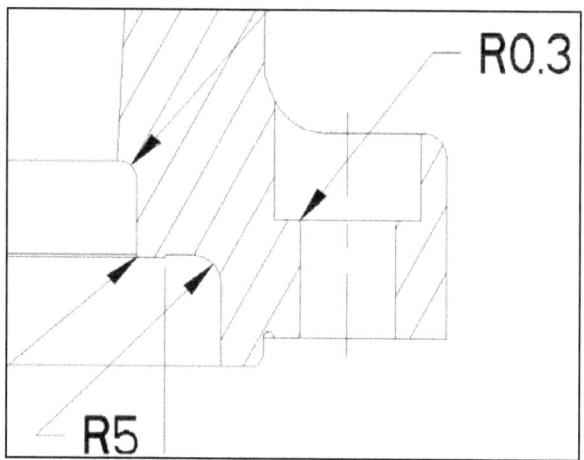

Figure 22.33 Rounds

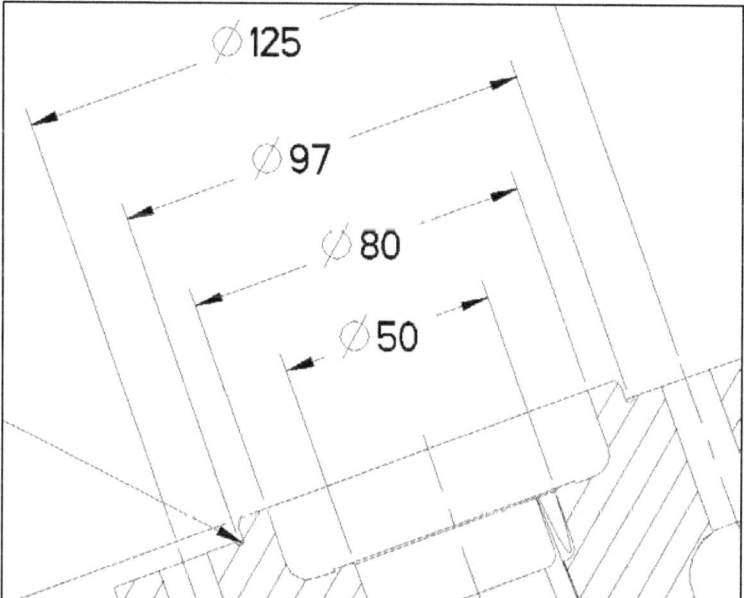

Figure 22.34 SECTION B-B, Mating Diameters

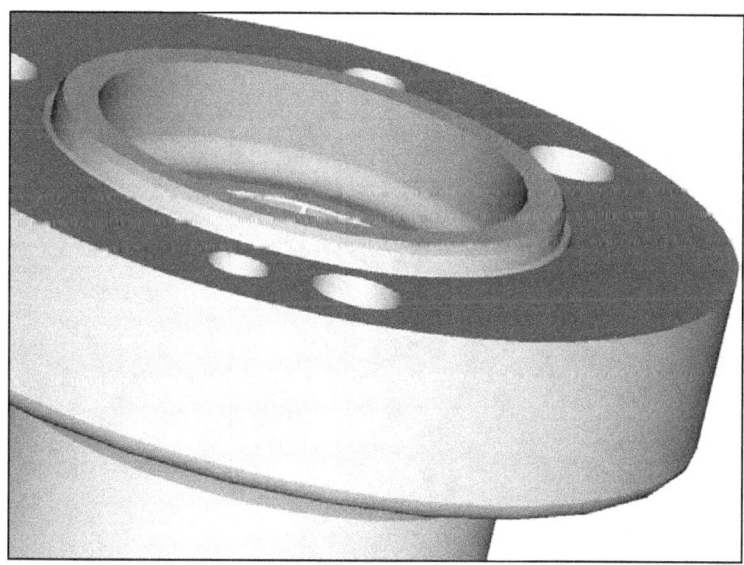

Figure 22.35 Mating Surface

829

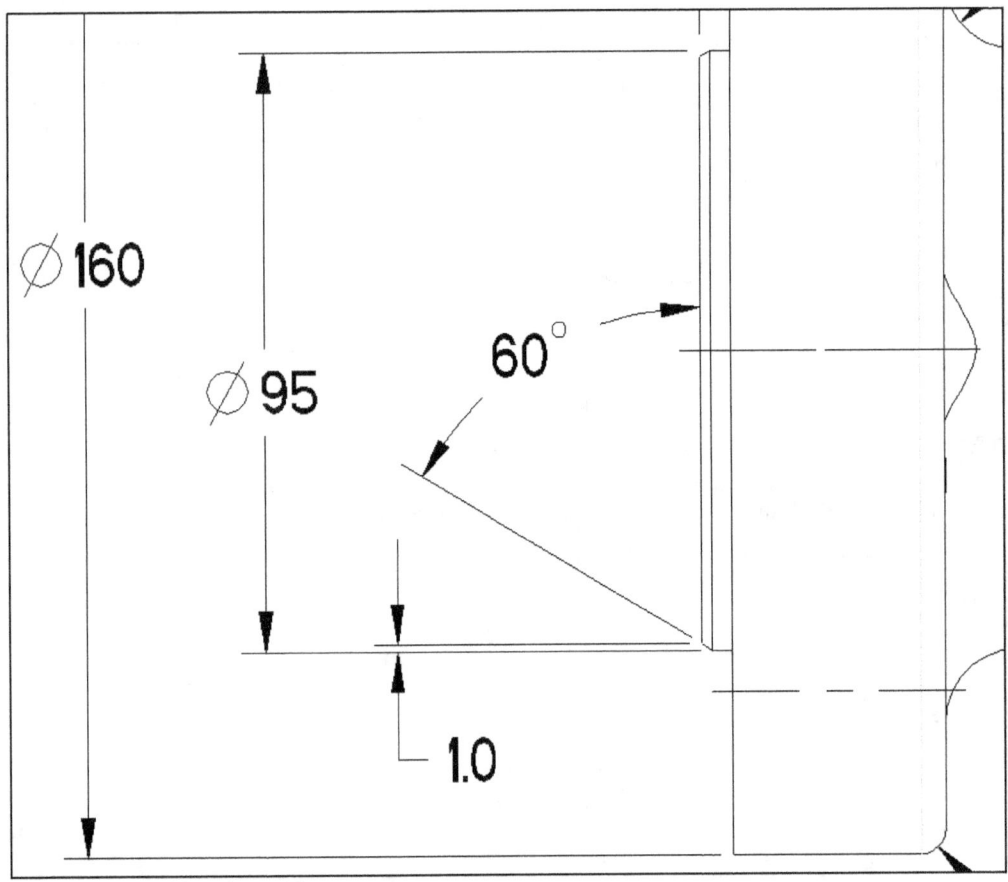

Figure 22.36 Side View, Close-up

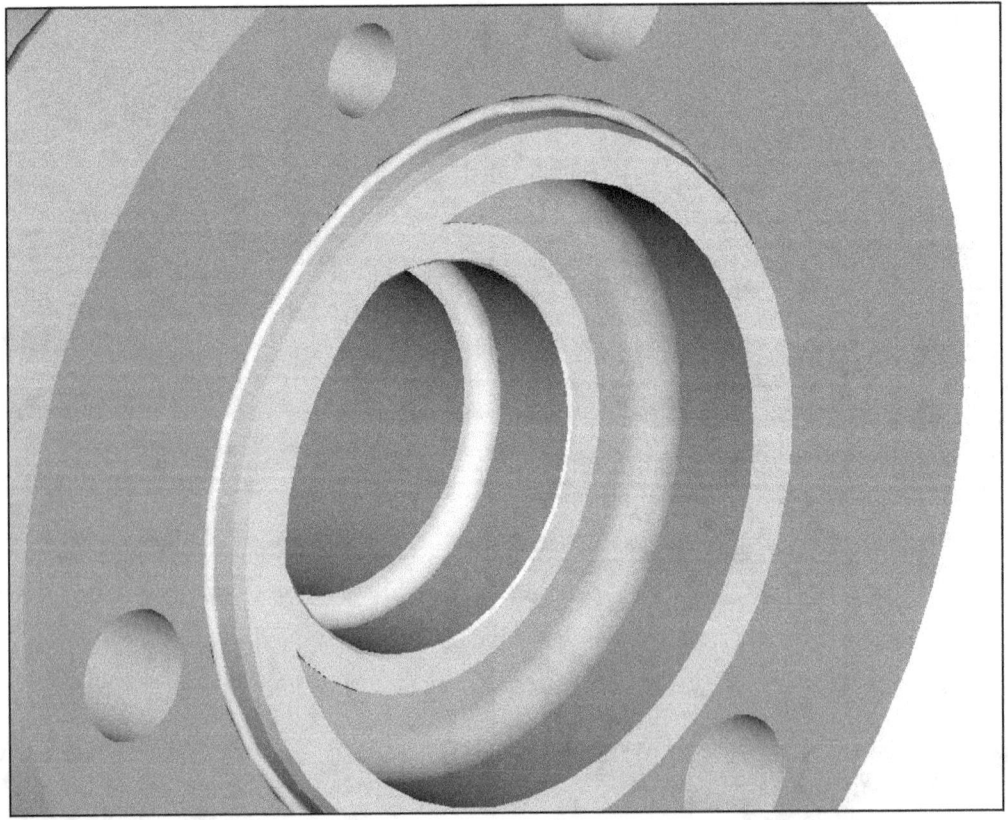

Figure 22.37 Internal View

Straight Coupling

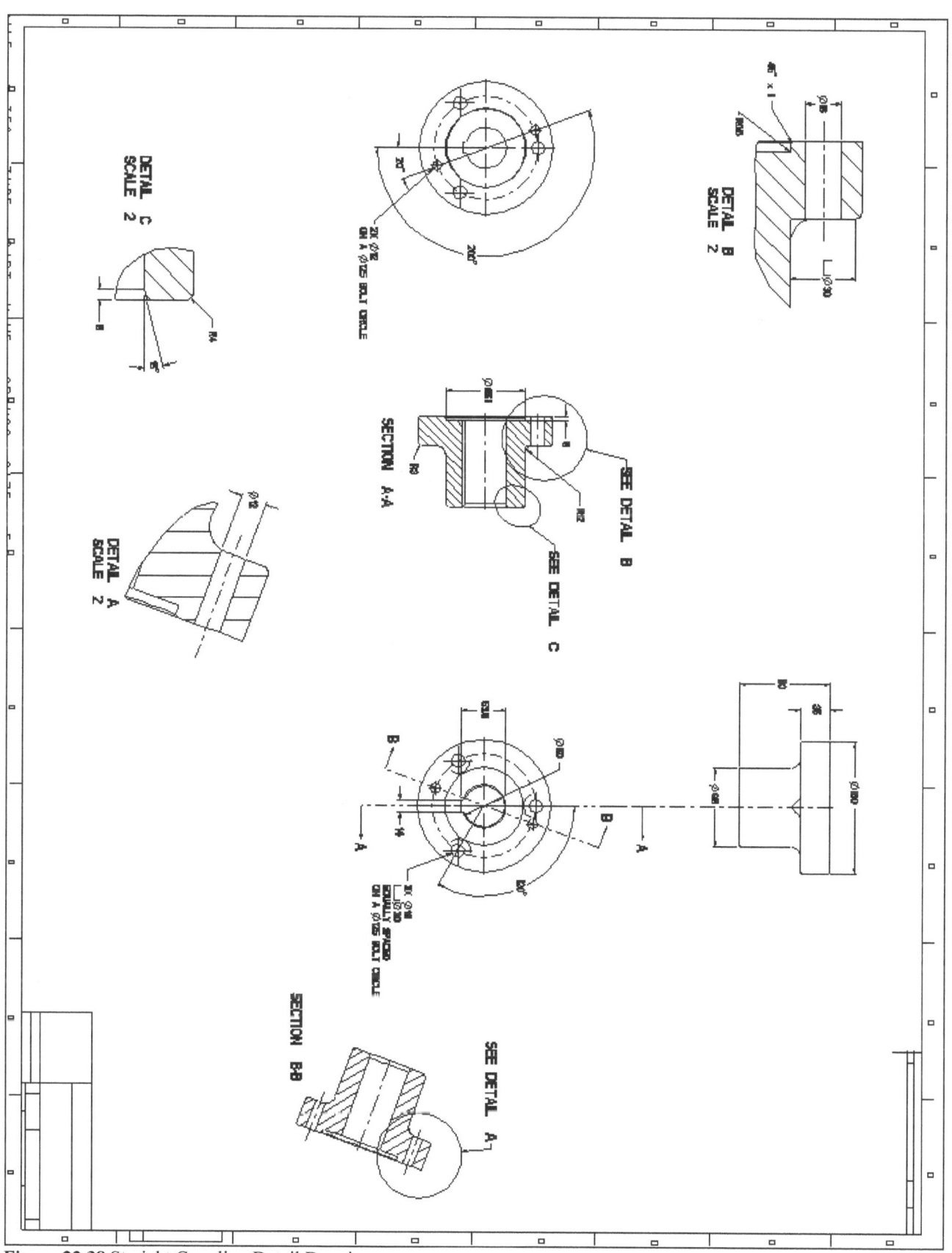

Figure 22.38 Straight Coupling Detail Drawing

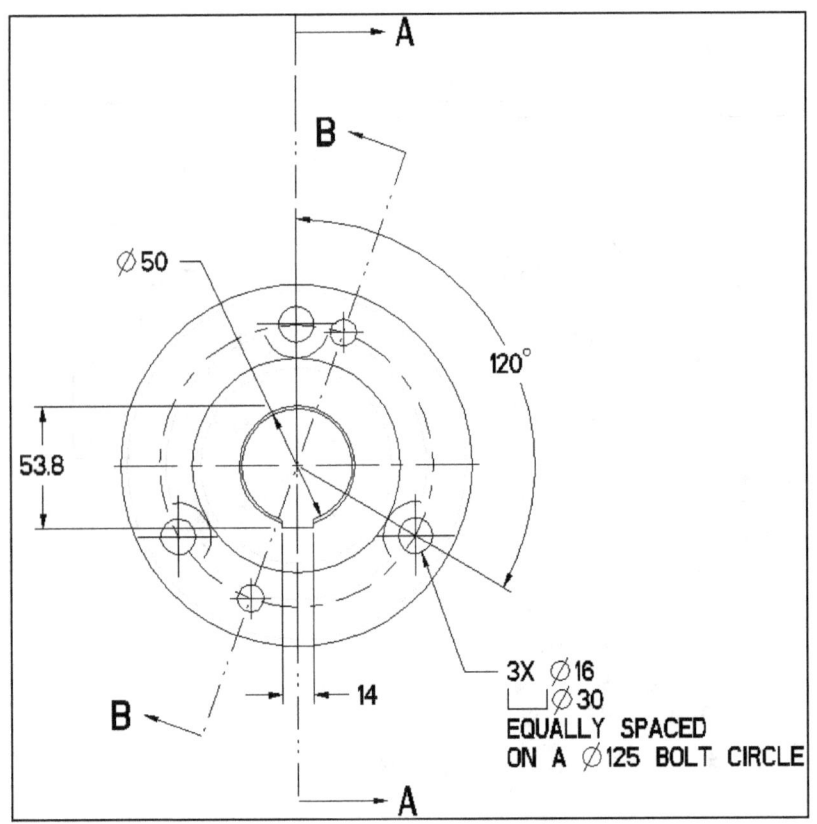

Figure 22.39 Straight Coupling Detail Drawing, Front View

Ø50

53.8

120°

14

3X Ø16
⌴ Ø30
EQUALLY SPACED
ON A Ø125 BOLT CIRCLE

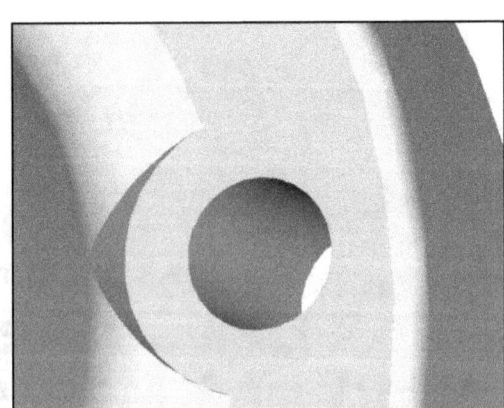

Figure 22.40 Holes

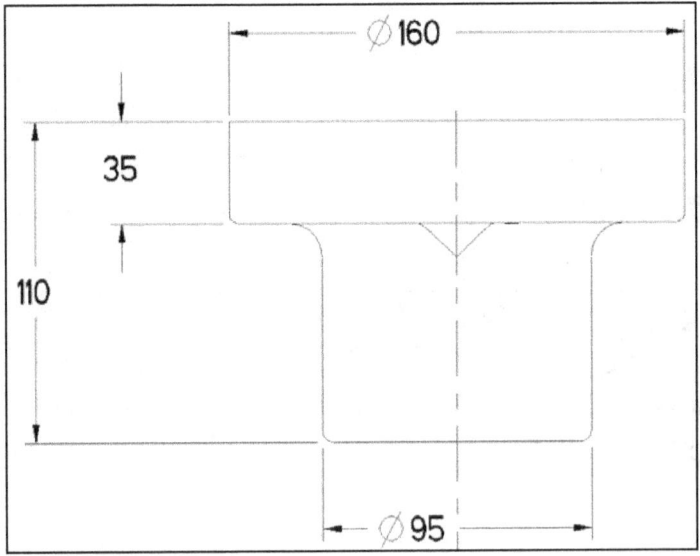

Ø160

35

110

Ø95

Figure 22.41 Straight Coupling, Top View

Figure 22.42 Hole Cutting Round

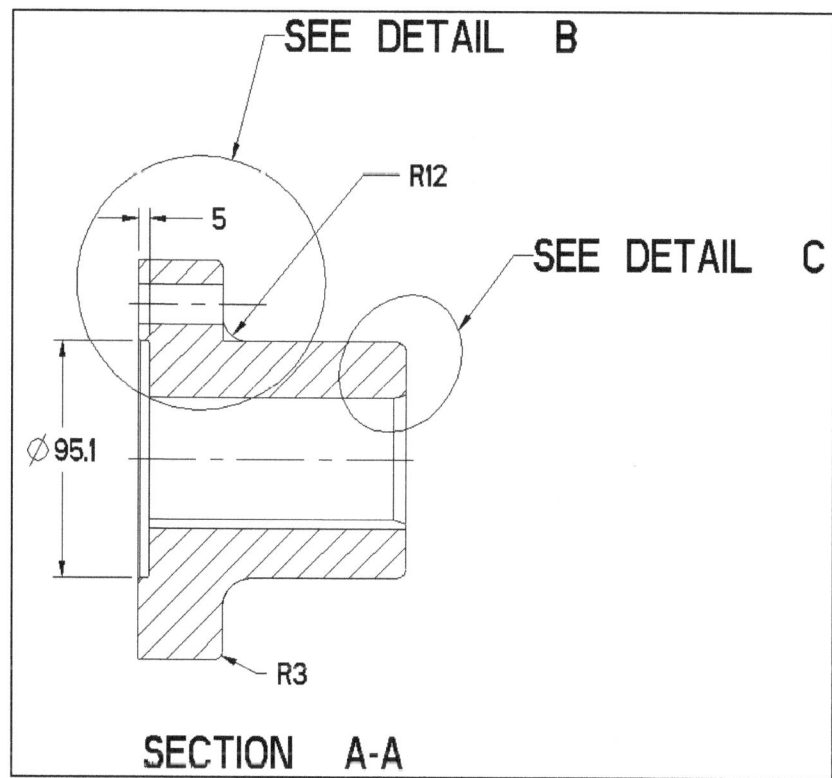

SEE DETAIL B

R12

SEE DETAIL C

5

⌀95.1

R3

SECTION   A-A

Figure 22.43 SECTION A-A

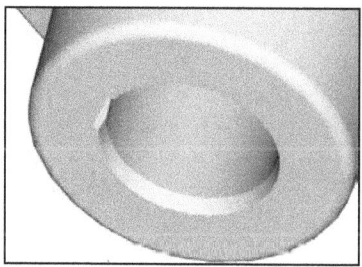

Figure 22.44 Tapered Hole

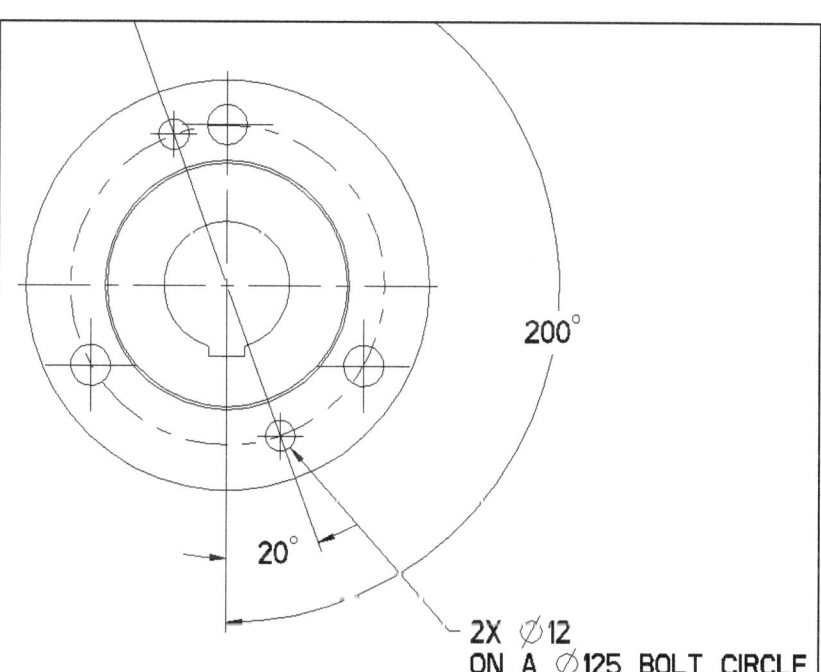

200°

20°

2X ⌀12
ON A ⌀125 BOLT CIRCLE

Figure 22.45 Detail, Back View

Figure 22.46 Holes from Bottom

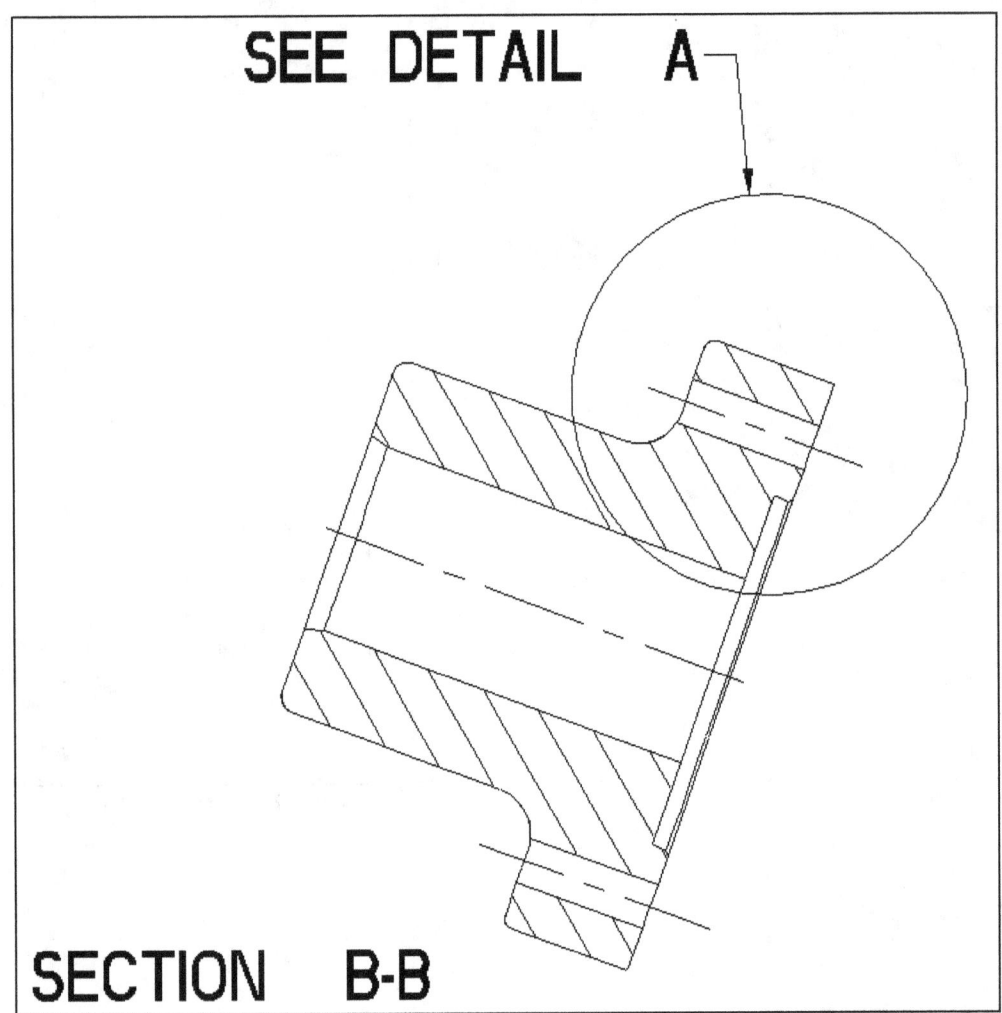

Figure 22.47 SECTION B-B

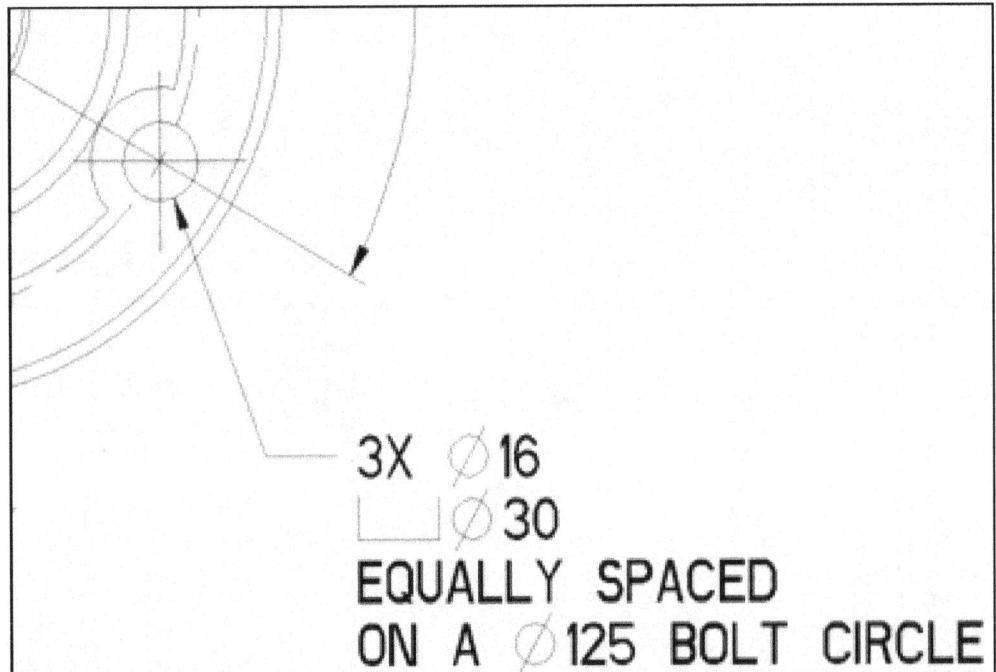

Figure 22.48 Hole Callout

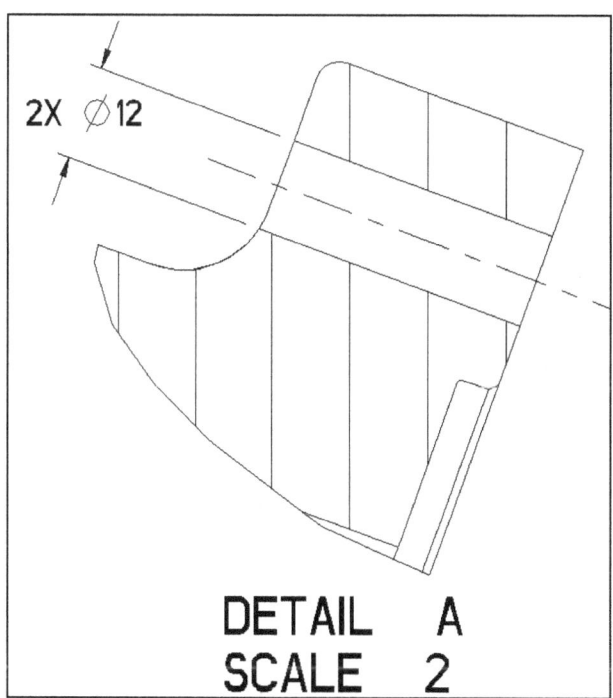

Figure 22.49 DETAIL A

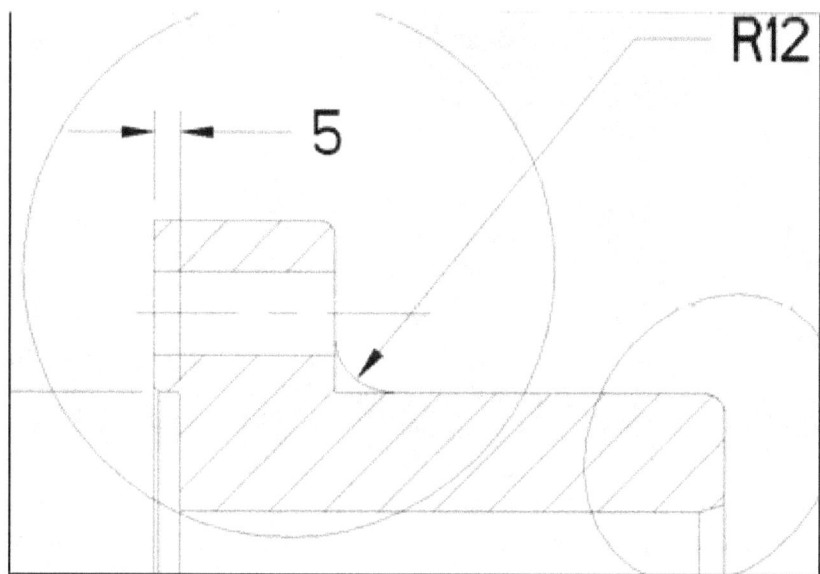

Figure 22.50 Round **R12**

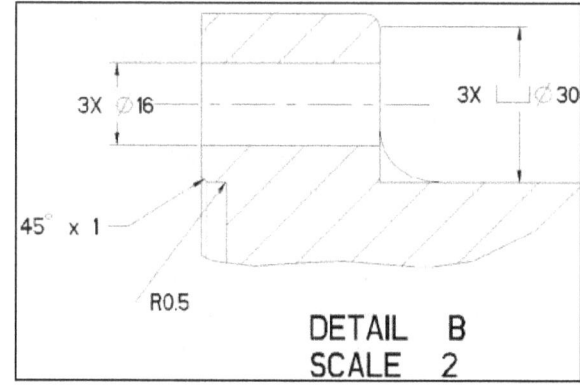

Figure 22.51 DETAIL B

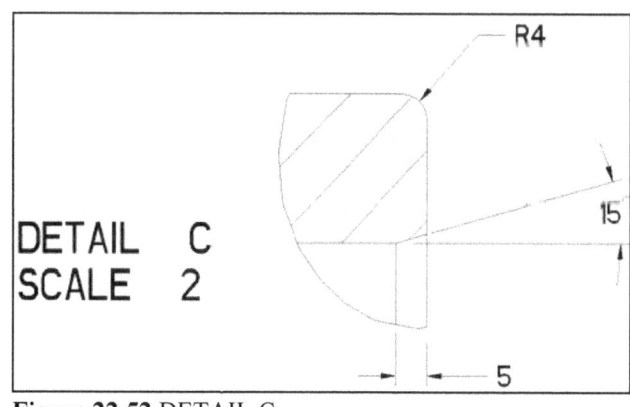

Figure 22.52 DETAIL C

835

Coupling Assembly

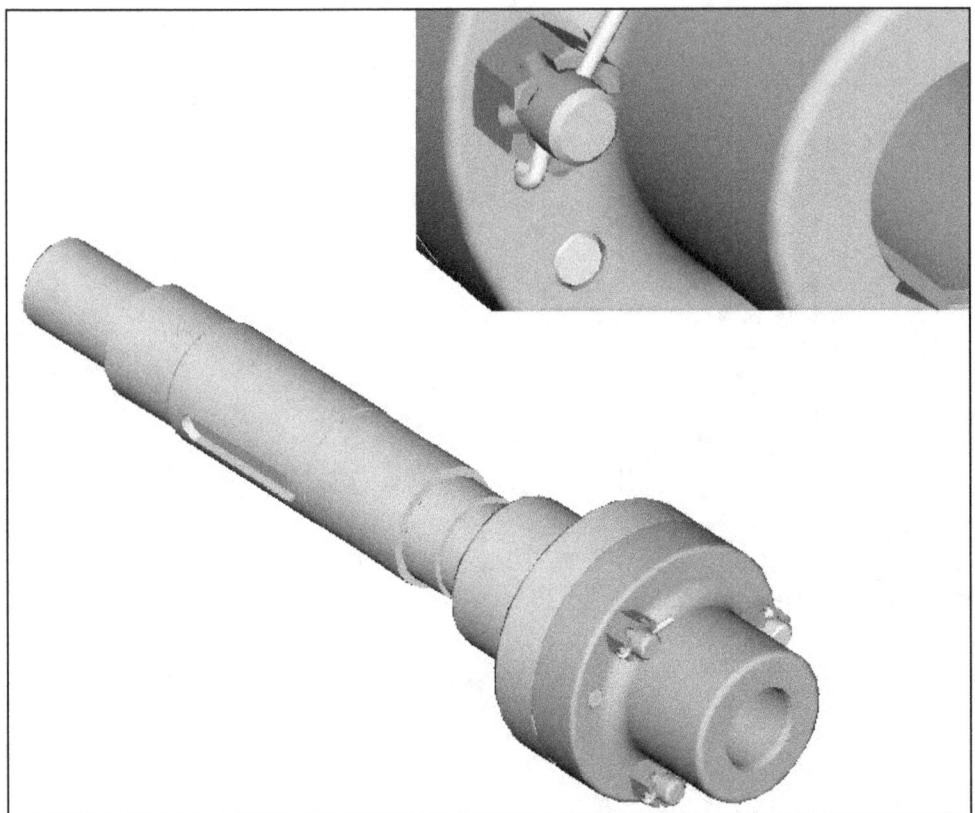

Figure 22.53 Coupling Assembly

Coupling Assembly

The Coupling Assembly requires commands similar to those for the Swing Clamp Assembly. Model the parts and create the assembly. Analyze the assembly and plan the steps required to assemble it. Plan the assembly component sequence and the parent-child relationships for the assembly. After completing the assembly, perform an Analysis using Global Interference. If there is interference between the shaft and the key, modify the key to the correct size.

The Coupling Shaft will be the first component assembled. The Taper Coupling is also used in the assembly. Model the Straight Coupling before you start the assembly. Depending on the library parts available on your system, you may need to model the Key, the Dowel, and the Washer components.

Because not all organizations purchase the libraries, details are provided for all the components required for the assembly, including the standard off-the-shelf parts available in Creo/Pro/E's library. Creo/Pro/LIBRARY commands to access the standard components are provided for those of you who have them loaded on your systems. The instance name is given for every standard component used in the assembly. The Slotted Hex Nut, Socket Head Cap Screw, Hex Jam Nut, and Cotter Pin are all standard parts from the library. The Cotter Pin is modeled in inch units, and the remaining items are metric.

For this lesson, do not assemble the library parts directly from the library. Save each library part in your own directory with a new name, and then use the new part names in the assembly.

Redesign the length of the Coupling Shaft threaded end to accommodate the washer and nut. You decide the new length based on the combined thickness of the two components.

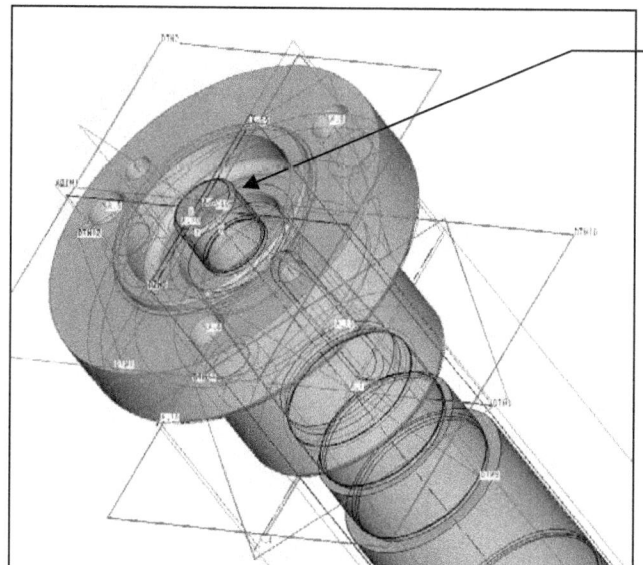

Figure 22.54 Assemble the Taper Coupling

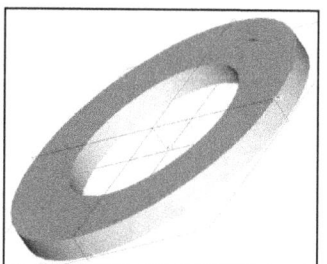

Redesign the length of the Coupling Shaft threaded end to accommodate the washer and nut. *You* decide the new length based on the combined thickness of the two components.

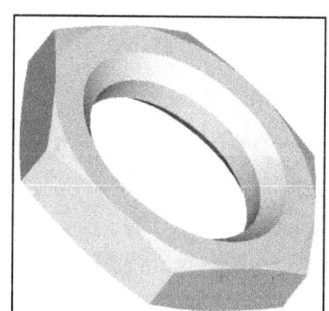

Figure 22.55 Washer

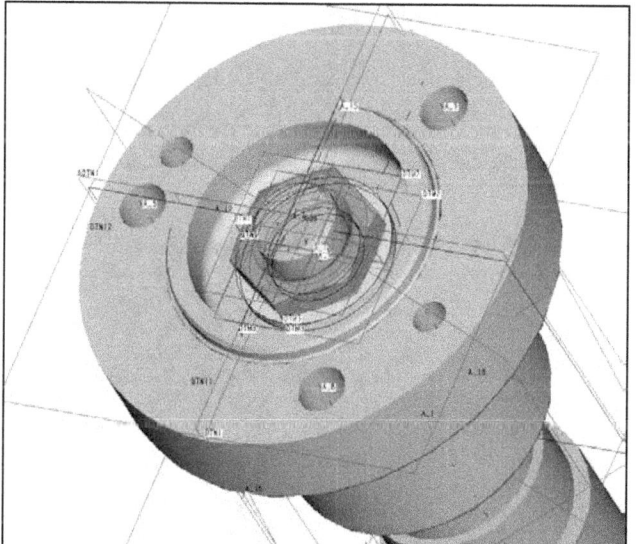

Figure 22.56 Hex Jam Nut and Washer

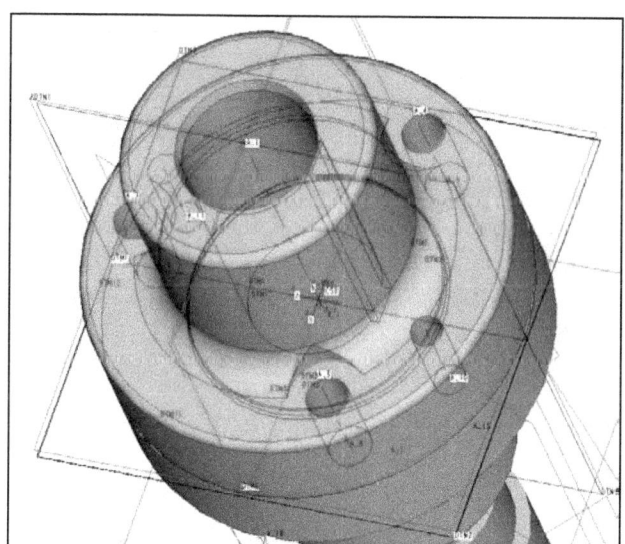

Figure 22.57 Hex Jam Nut

Figure 22.58 Straight Coupling

Figure 22.59 Bearing Surface

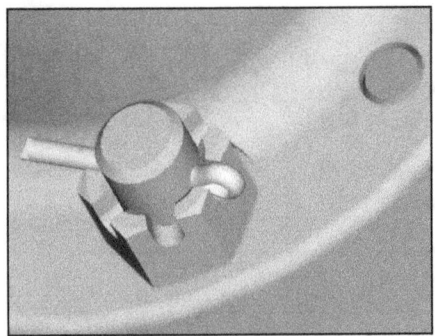 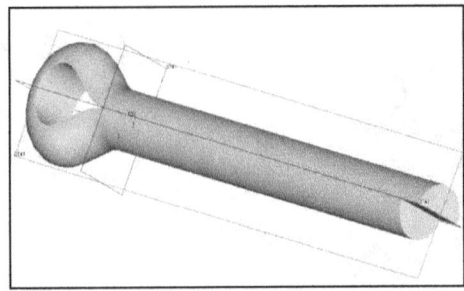 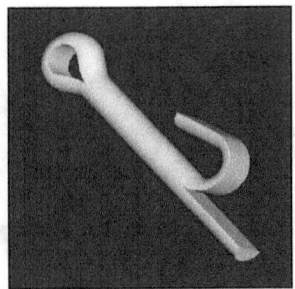

Figure 22.60 Dowel, Slotted Hex Nut, and Cotter Pin. After constraining the cotter pin, try redefining (the trajectory) it to bend one or both of its prongs.

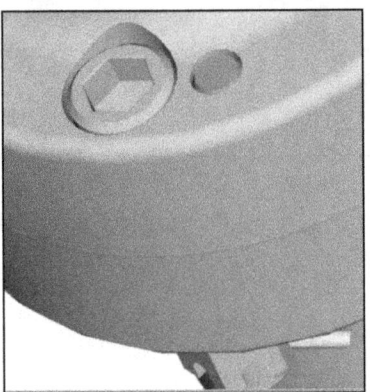 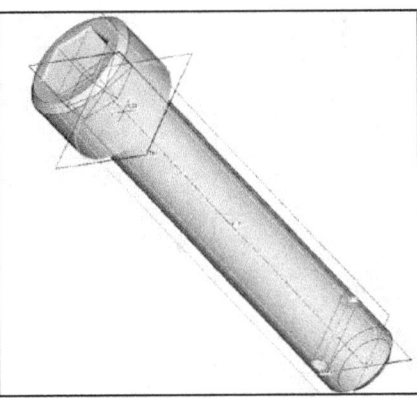 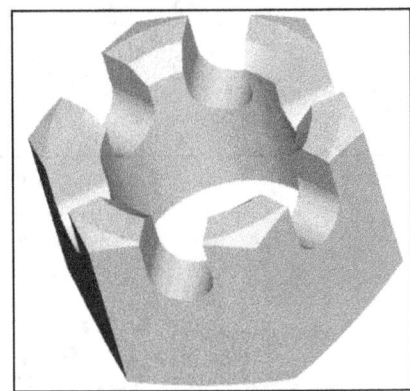

Figure 22.61 Socket Head Cap Screw and Slotted Hex Nut

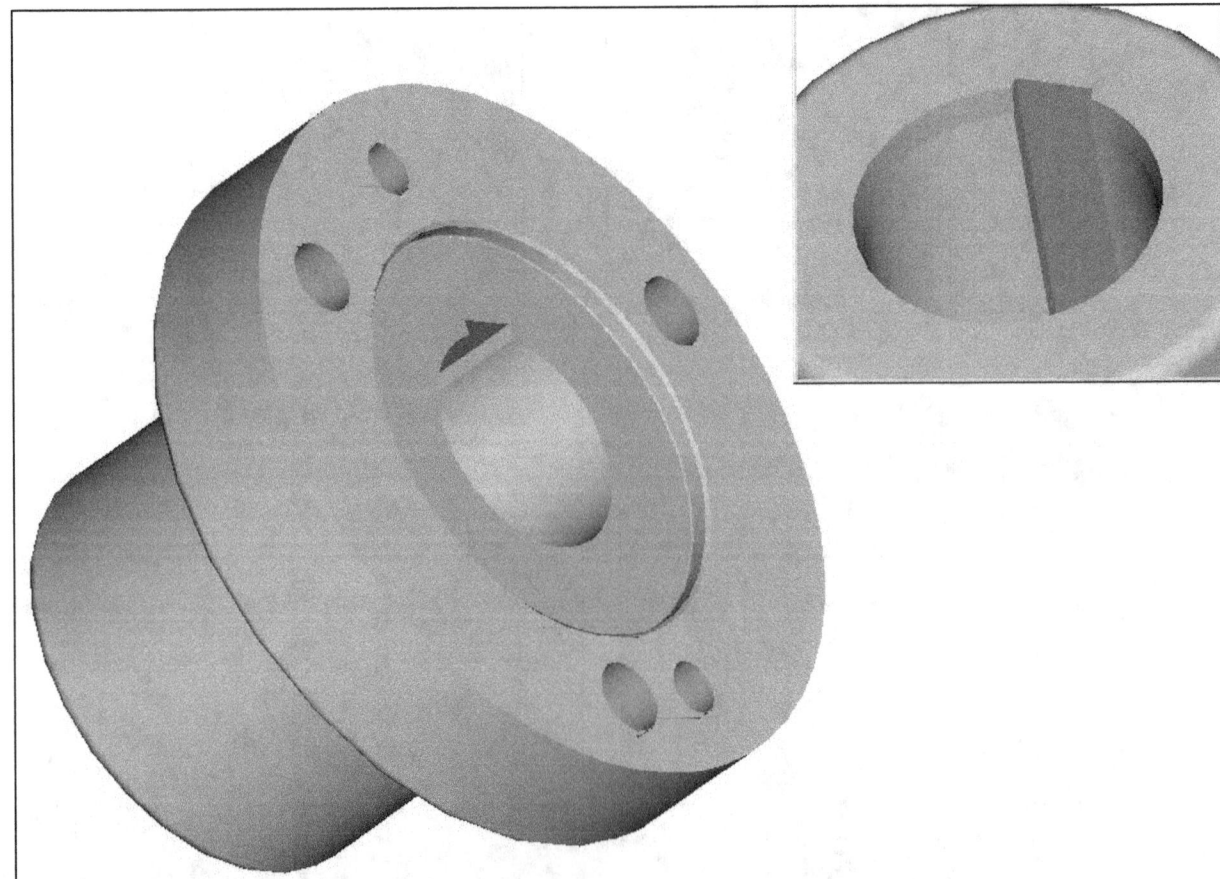

Figure 22.62 Straight Coupling, Part Model

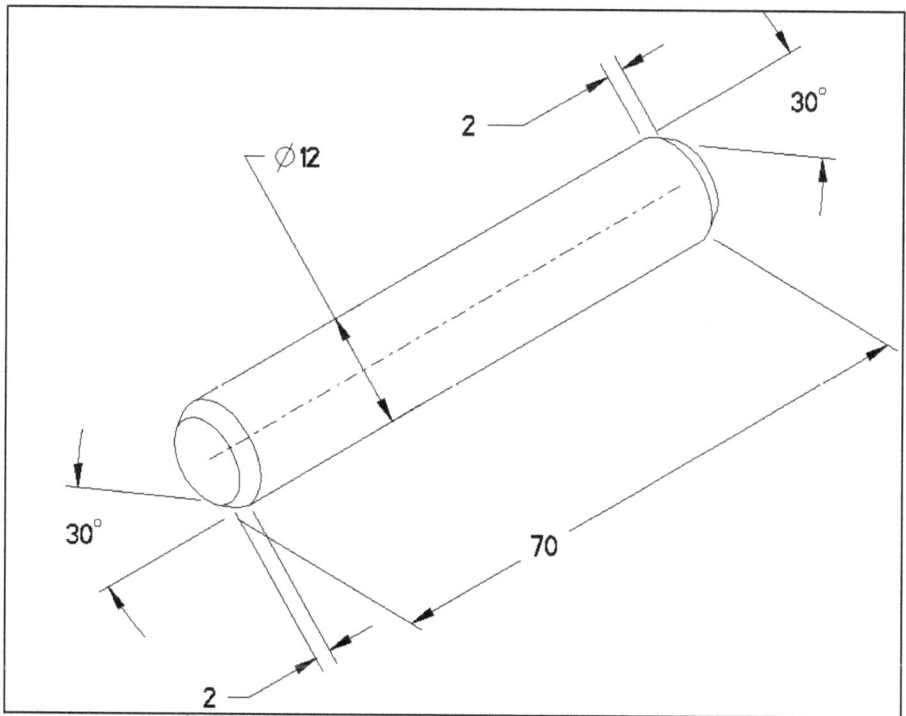

Figure 22.63 Dowel (model this component)

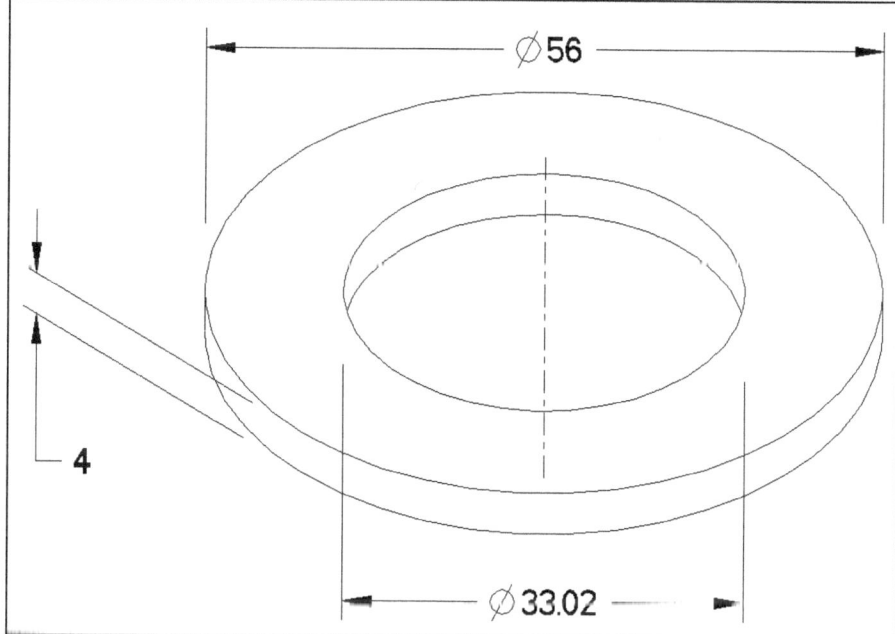

Figure 22.64 Washer (model this component)

Coupling standard parts are also available at www.cad-resources.com >

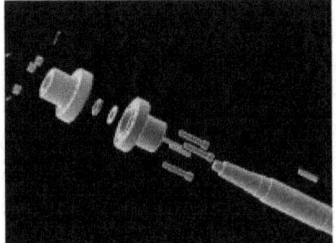

download the zip file and extract into your working directory.

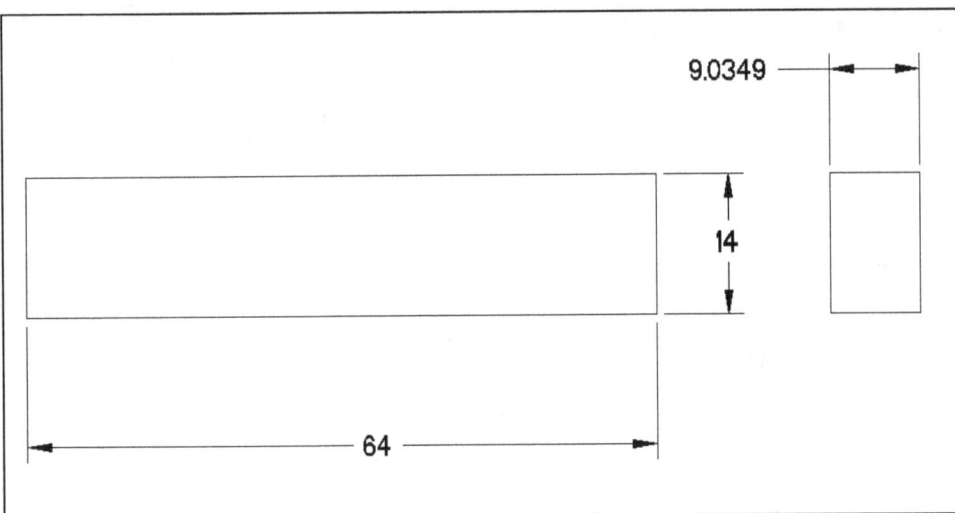

Figure 22.65 Key (model this component)

Figure 22.66 Key

HEX JAM NUT

3DMODELSPACE tab/browser opens **> PTC Creo Standard Parts Library > ANSI METRIC > I ACCEPT > METRIC HEXAGONAL NUTS > HEX JAM NUTS > MHJN10 > lowest level Pro/ENGINEER 2.0 or later > Download CAD Model**

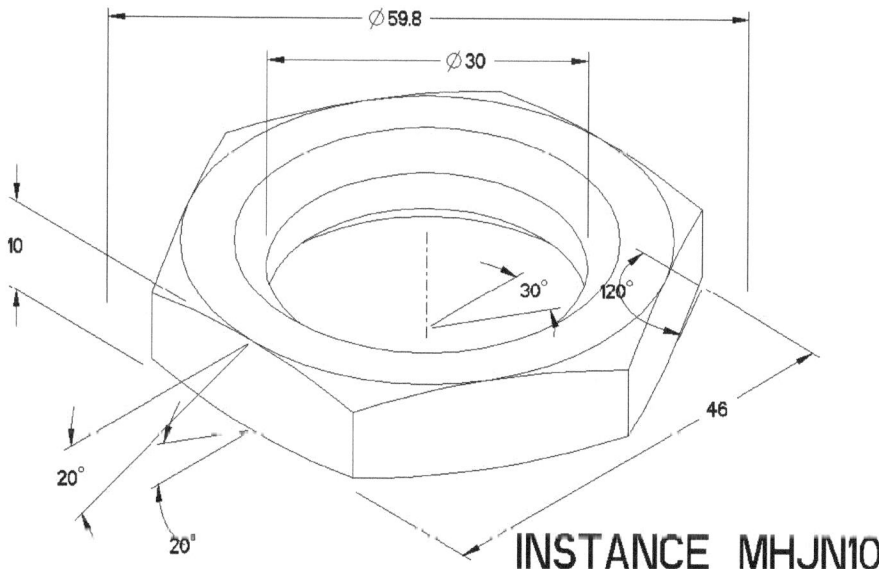

Figure 22.67 Hex Jam Nut

SOCKET HEAD CAP SCREW

3DMODELSPACE tab/browser opens **> PTC Creo Standard Parts Library > ANSI METRIC > I ACCEPT > METRIC SOCKET HEAD SCREWS > SOCKET HEAD CAP SCREWS >** Select the proper diameter and length **> MSCS1210 >** Lowest level **Pro/ENGINEER 2.0 or later > Download CAD Model**

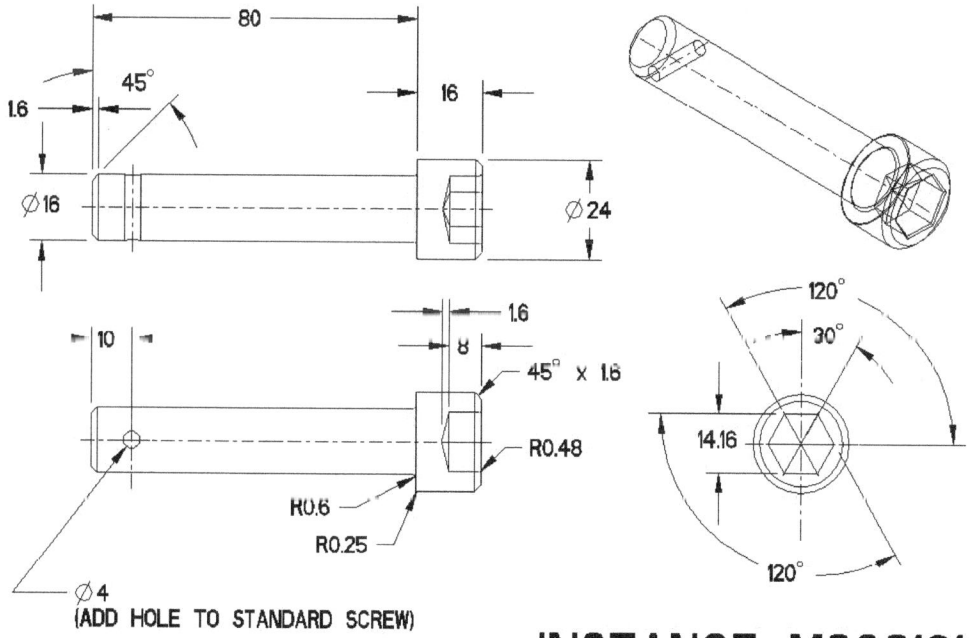

Figure 22.68 Socket Head Cap Screw

SLOTTED HEX NUT

3DMODELSPACE tab/browser opens **> PTC Creo Standard Parts Library > ANSI METRIC > I ACCEPT > > METRIC HEXAGONAL NUTS > SLOTTED HEX NUTS > By Parameter > NOMINAL_DIA_THR_PITCH >** Select the proper size **> M16X2 > INSTANCE = MSHN07**

Figure 22.69 Slotted Hex Nut

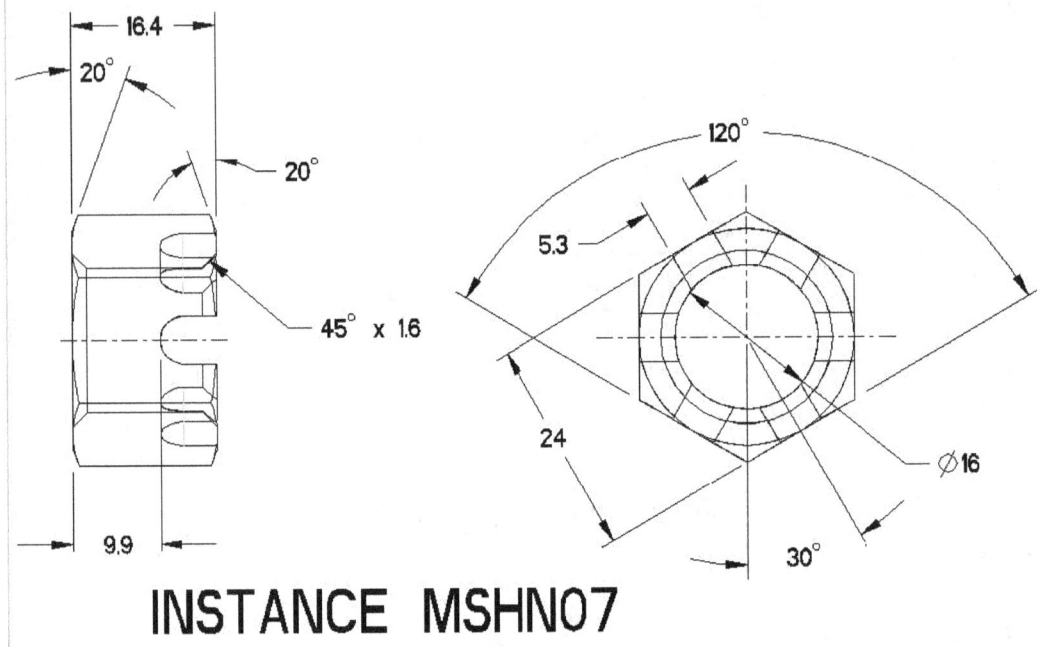

COTTER PIN

File > Open > ANSI ENGLISH > I ACCEPT > Show All Categories > Cotter Pins > Pina.prt > Select the proper size **> PNA09L05**

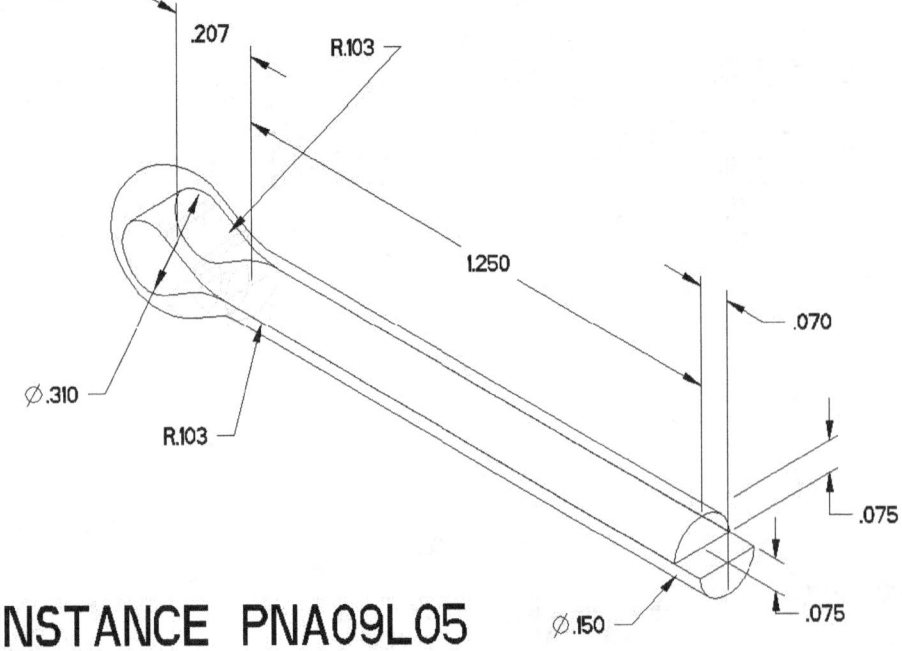

Figure 22.70 Cotter Pin

Coupling Assembly (Exploded)

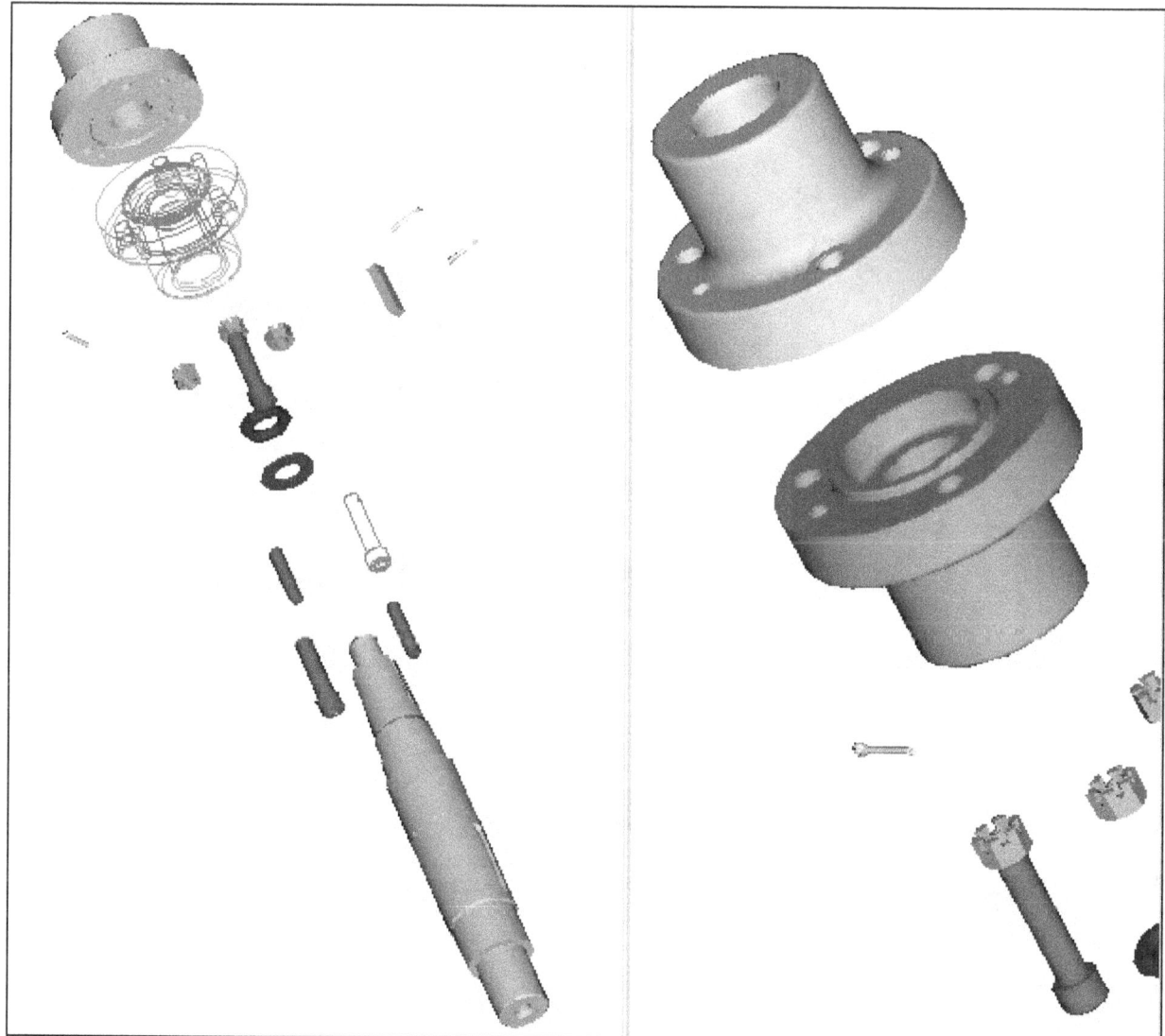

Figure 22.71 Exploded Coupling Assembly

An exploded view needs to be created. Varieties of other views are suggested, including a section of the assembly, a perspective view, and an exploded view with a different component (display) style variation. Each component should have its own color. If you did not color the components during the part creation, bring up each part in Part mode, define, and apply a color. Create three or four View States. You do not need to match the examples shown.

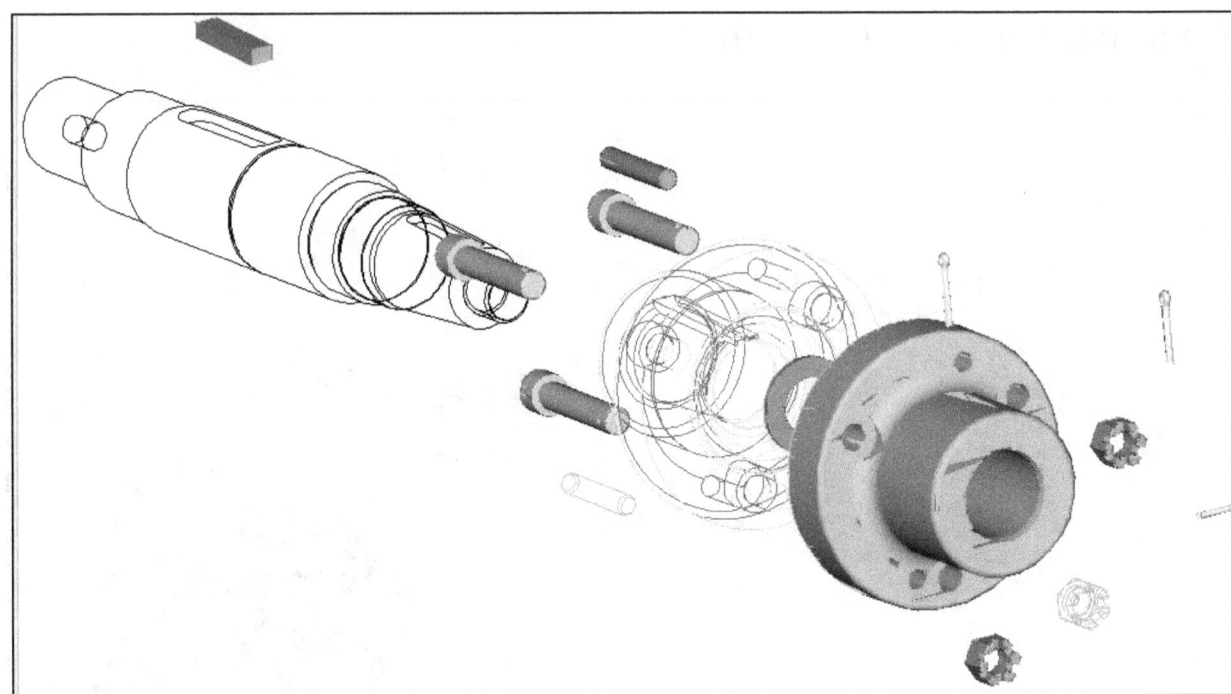

Figure 22.72 Perspective View of Exploded Coupling Assembly with a Style State showing various Component Display Styles

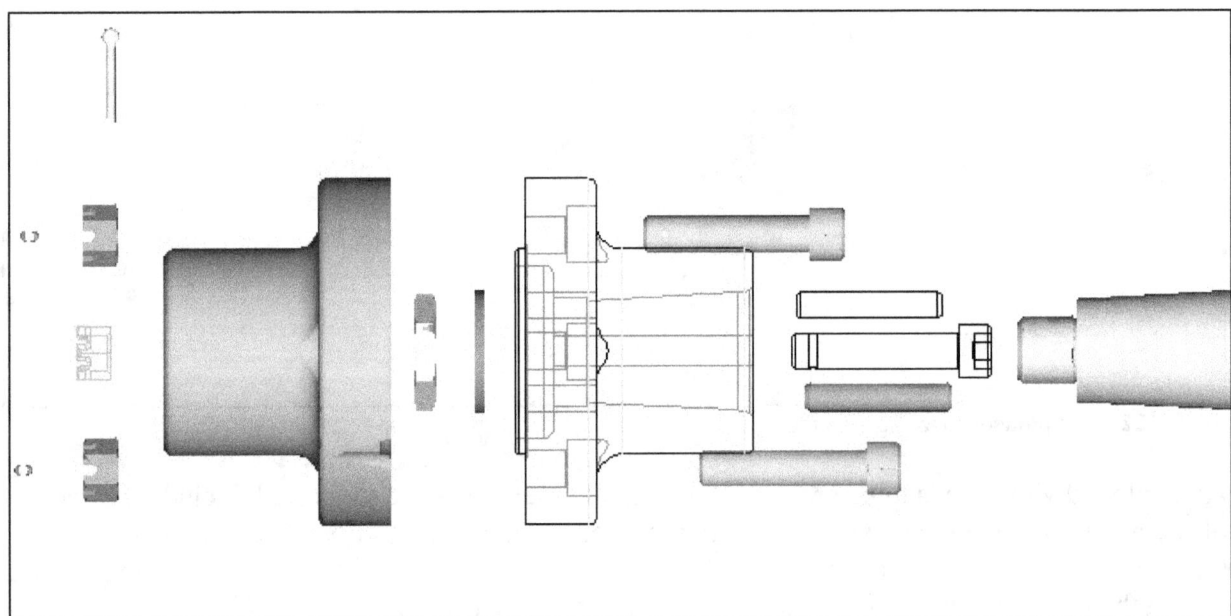

Figure 22.73 Front View

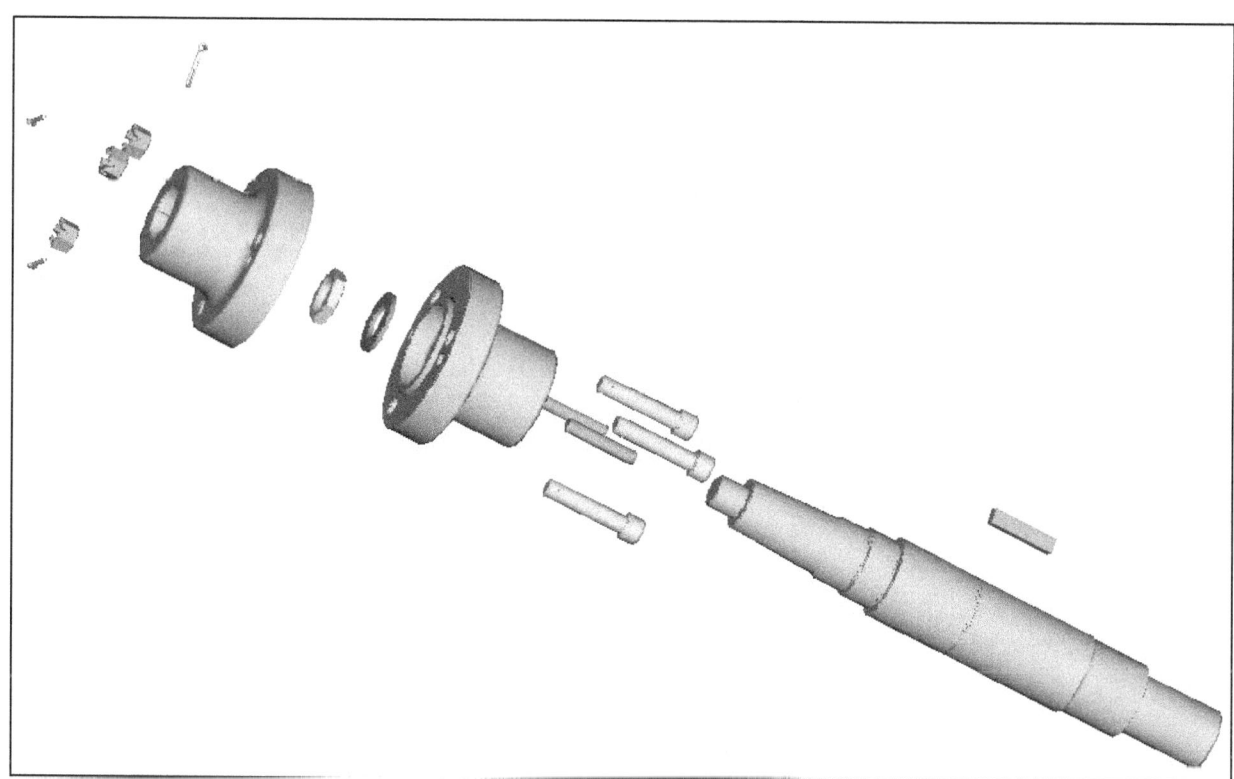

Figure 22.74 Shaded Exploded Coupling Assembly

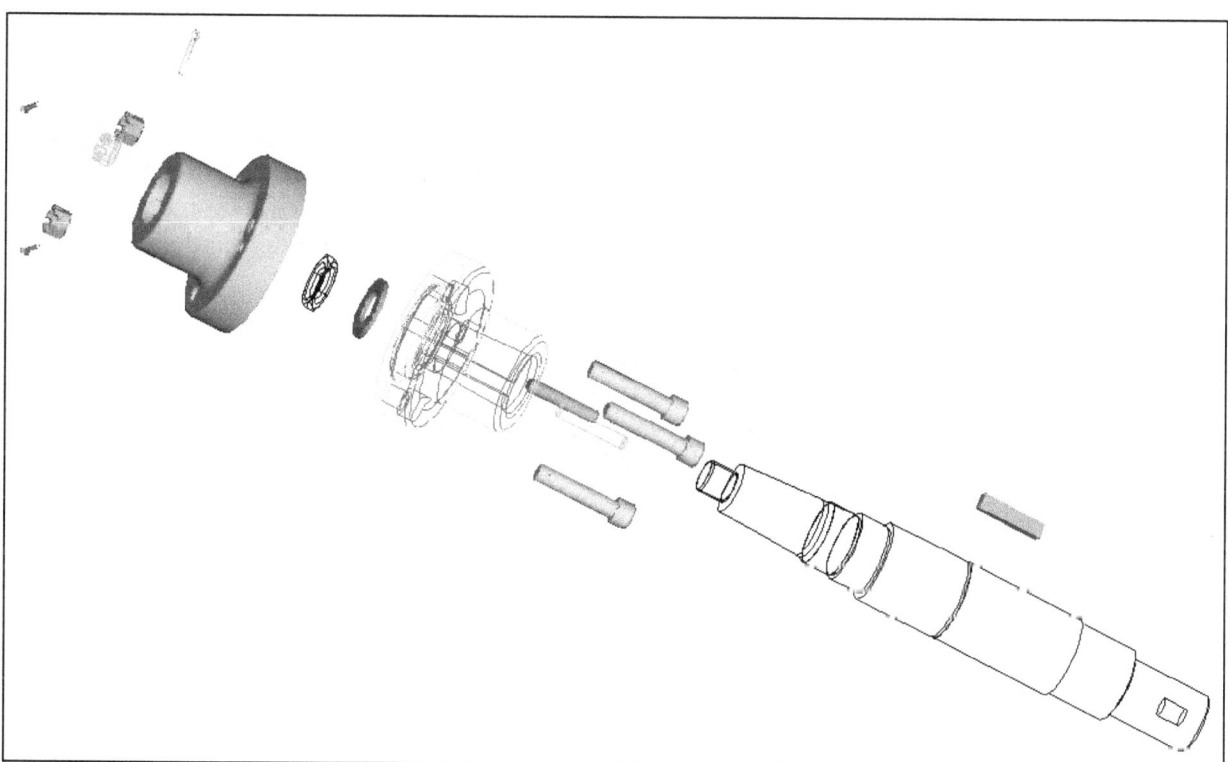

Figure 22.75 Exploded Coupling Assembly with a Different Component Style State

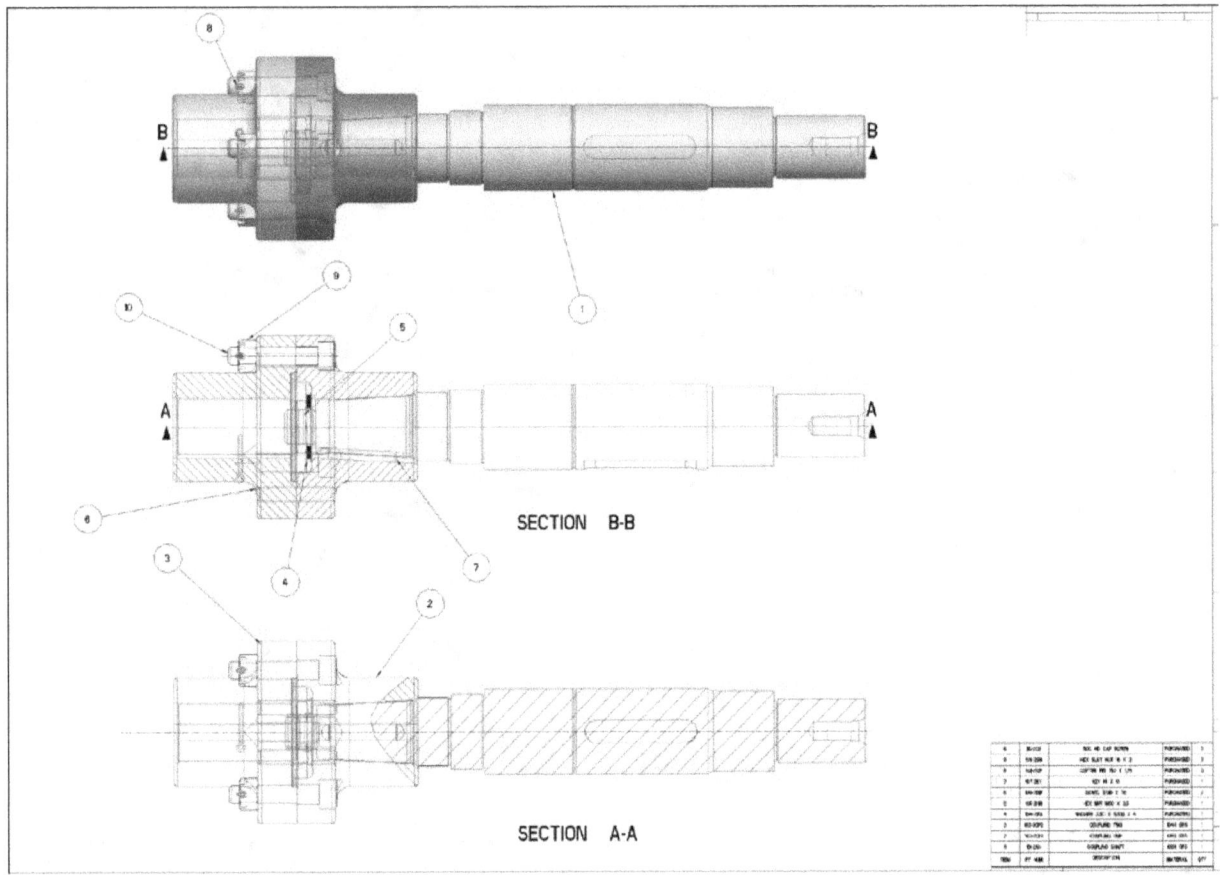

Figure 22.76 Coupling Assembly Drawing

Create a complete documentation package for the Coupling Assembly. The ballooned assembly drawing will have three views and a parts list (BOM). Assign parameters to the parts in the assembly so that they can be displayed on a parts list in the assembly drawing. Some of the items listed here have been created in other lessons. Create or extract existing models and drawings, and plot/print the following:

- *Part Models for all coupling assembly components*
- *Detail Drawings for each nonstandard component, for example, the Coupling Shaft*
- *Assembly Drawing and Parts List (BOM) using standard orthographic ballooned views*
- *Exploded Assembly Drawing of the ballooned assembly*

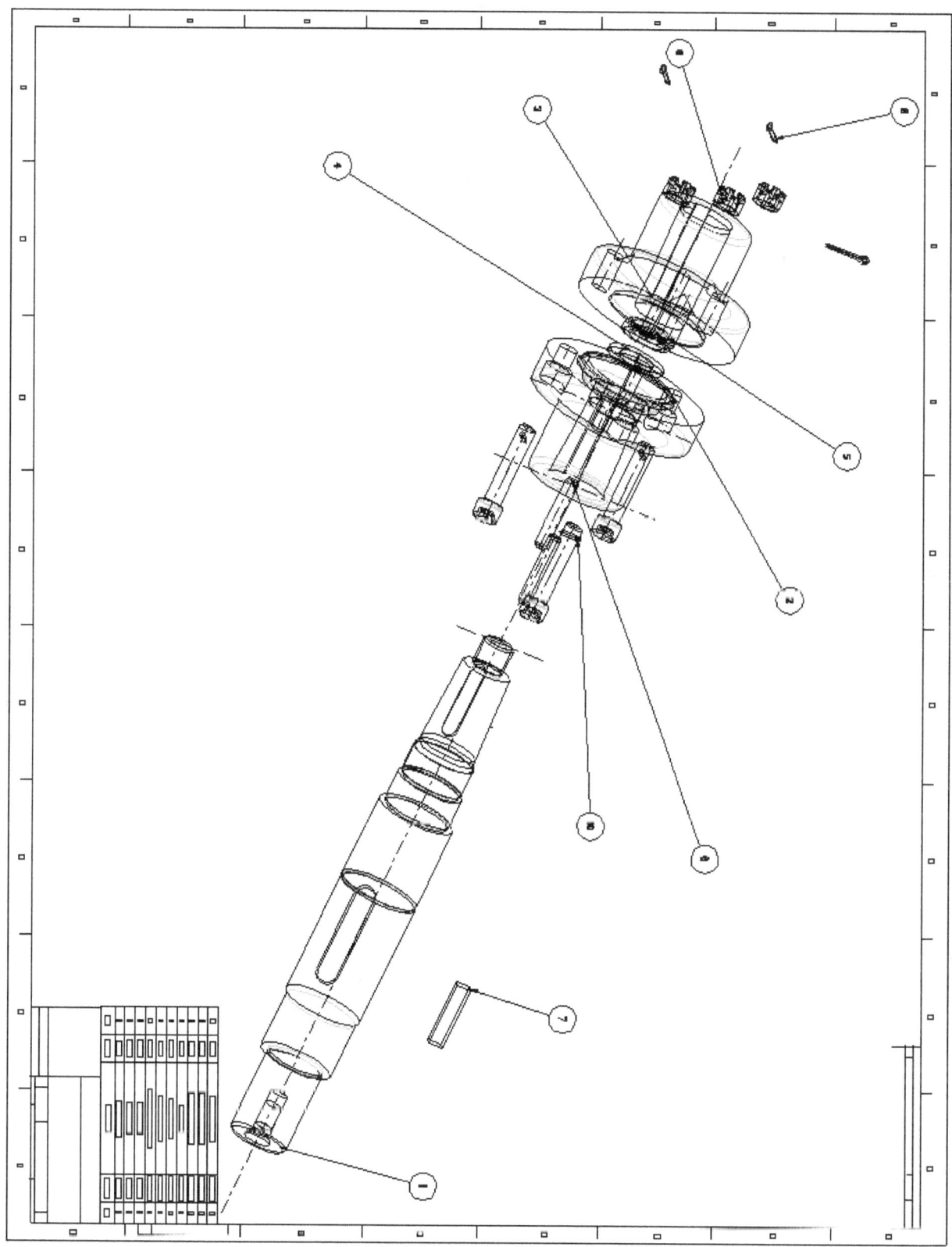

Figure 22.77 Exploded Coupling Assembly Drawing

ITEM	PT NUM	DESCRIPTION	MATERIAL	QTY
10	110-2CS	SOC HD CAP SCREW	PURCHASED	3
9	109-2SN	HEX SLOT NUT 16 X 2	PURCHASED	3
8	108-2CP	COTTER PIN .150 X 1.25	PURCHASED	3
7	107-2KY	KEY 14 X 61	PURCHASED	1
6	106-2DW	DOWEL 120D X 70	PURCHASED	2
5	105-2HN	HEX NUT M30 X 3.5	PURCHASED	1
4	104-2WA	WASHER 33ID X 50OD X 4	PURCHASED	1
3	103-2CP2	COUPLING TWO	1040 CRS	1
2	102-2CP1	COUPLING ONE	1040 CRS	1
1	101-2SH	COUPLING SHAFT	1020 CRS	1
ITEM	PT NUM	DESCRIPTION	MATERIAL	QTY

Figure 22.78 BOM

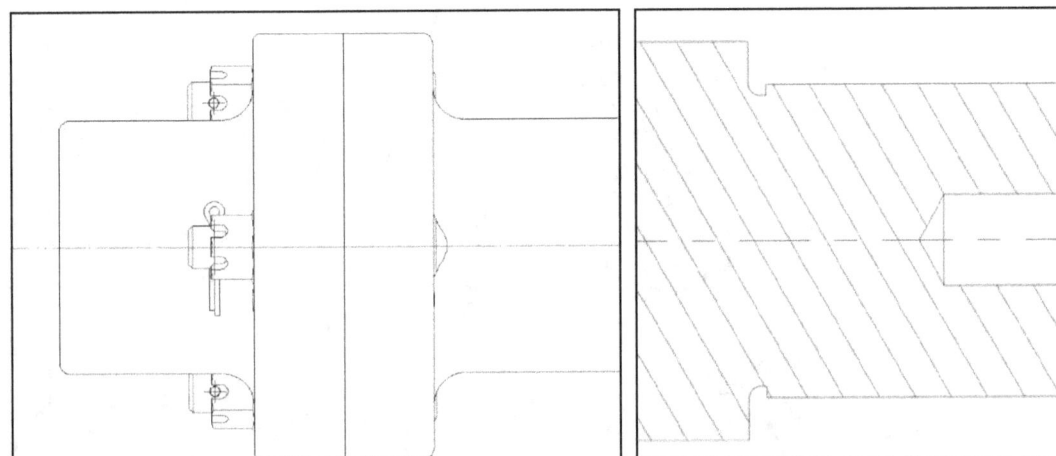

Figure 22.79 Drawing, Slotted Hex Nut   **Figure 22.80** Tapped Hole

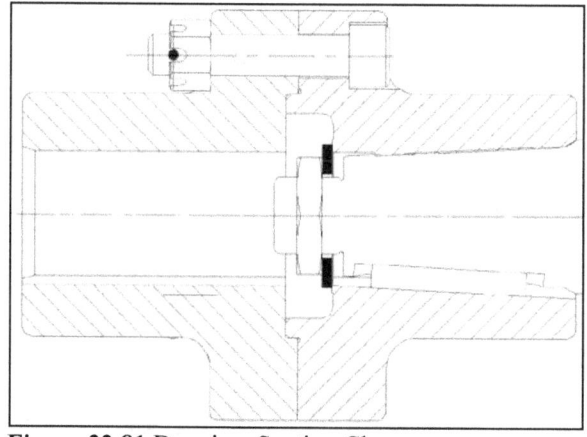

Figure 22.81 Drawing, Section Close-up

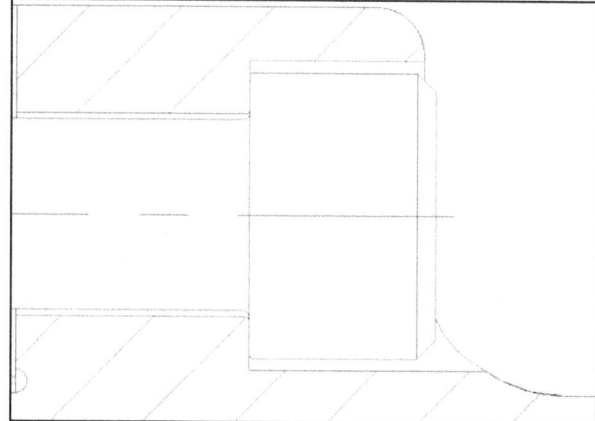

Figure 22.82 SHCS

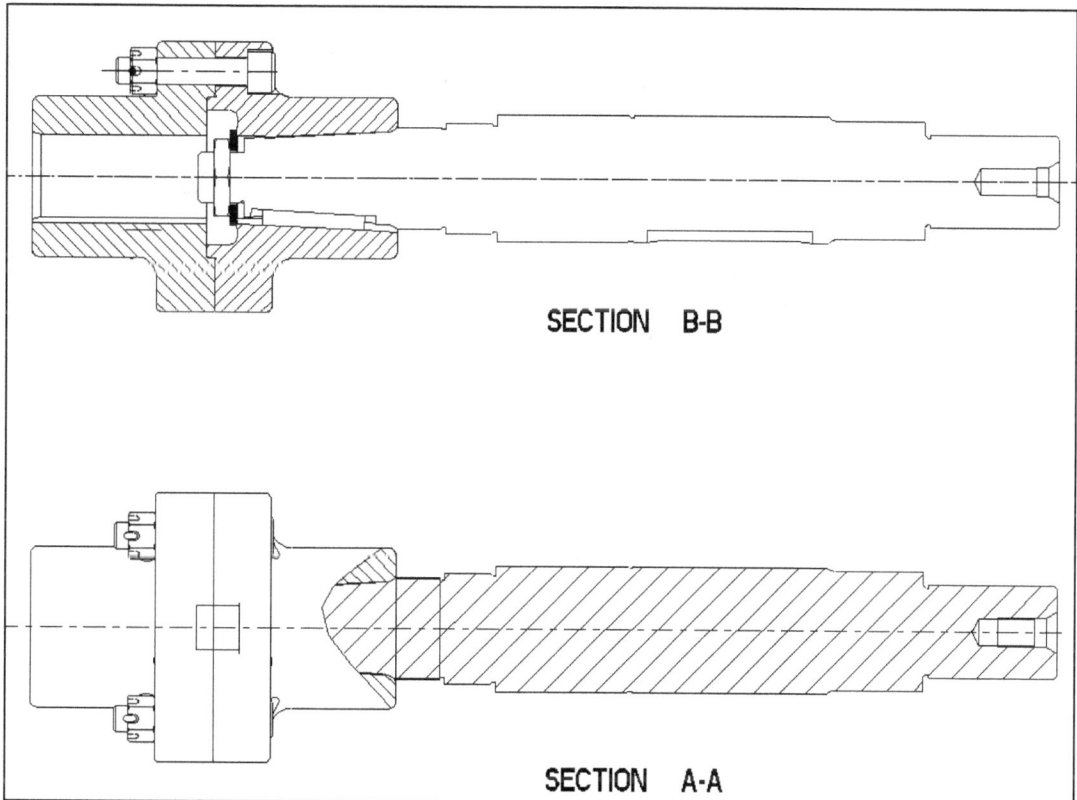

SECTION B-B

SECTION A-A

Figure 22.83 Drawing, Sections A-A and B-B

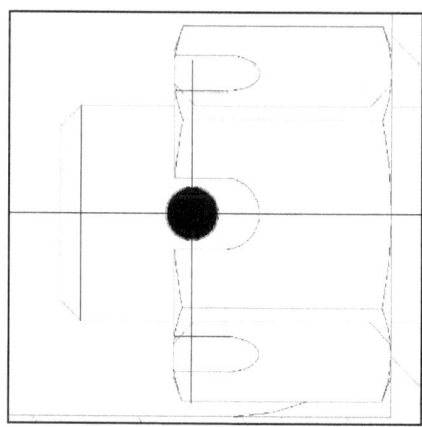

Figure 22.84 Nut

Index

www.ingramcontent.com/pod-product-compliance
Lightning Source LLC
Chambersburg PA
CBHW081717220526
45468CB00008B/1876